D0468140

Remote Sensing for Sustainable Forest Management

Remote Sensing for Sustainable Forest Management

Steven E. Franklin

Library of Congress Cataloging-in-Publication Data

Franklin, Steven E.
Remote sensing for sustainable forest management / Steven E. Franklin.
p. cm.
Includes bibliographical references and index (p.).
ISBN 1-56670-394-8 (alk. paper)
1. Sustainable forestry—Remote sensing. 2. Forest management. I. Title.

SD387.R4 F73 2001
634.9′2′028—dc21 2001029505
CIP

International Standard Book Number 1-56670-394-8
Library of Congress Card Number 2001029505
Printed in the United States of America 1 2 3 4 5 6 7 8 9 0
Printed on acid-free paper

Dedication

for Dawn Marie, Meghan, and Heather

Preface

Remote sensing has been defined as the *detection, recognition, or evaluation* of objects by means of distant sensing or recording devices. In recent decades, remote sensing technology has emerged to support data collection and analysis methods of potential interest and importance in forest management. Historically, *digital* remote sensing developed quickly from the technology of aerial photography and photointerpretation science. In forestry, information extracted visually from aerial photographs is well-understood, well-used, and integrated with field surveys. Information extracted from digital remote sensing data, on the other hand, is rarely used in forest management. It is thought that many remote sensing data and methods are complex, and are not well understood by those who might best use them. The technological infrastructure is not in place to make effective use of the data. The characteristics of much remote sensing data are, perhaps, not well suited to the problems that have preoccupied the forest management community.

But forest management is changing. Today, forest management problems are multiscale and intricately linked to society's need to measure, preserve, and manage for multiple forest values. Population growth and climate change appear likely to create continual pressure on forests, making their preservation, even over relatively short time periods, seem largely in doubt. Human activities threaten the continued physical existence, biodiversity, and functioning of forests. It is probable that no forest on the planet can survive intact without conscious human decision making, and actual on-the-ground treatments and prescriptions that consider ecological processes and functioning. The forest ecosystem is complex and multifaceted; understanding how forest ecosystems work requires new types of data, and data at a range of spatial and temporal scales not often contemplated. Remote sensing information needs to be integrated with other spatial and nonspatial data sets to form the information base upon which sound forest management decisions can be made. The goal is to predict the effects of human activities and natural processes on forests, and to promote forest practices that will ensure the world's forests are sustainable.

A major issue facing those with forest management questions is not simply the collection of data, but rather the interpretation of information extracted from those data. Converting remote sensing data to information is no simple task. Remote sensing measurements have a physical or statistical relationship to the forest conditions of interest which may be uneconomical, impractical, or impossible to measure directly over large areas. The remote sensing technological approach is an applied perspective — applying remote sensing knowledge to satisfy information needs motivated by a strong desire to understand the implications of management while there is still time to learn from prescriptions and to understand forest conditions and processes. A survey of the field of remote sensing in sustainable forest management may help those in direct operational contact with forests to better understand the potential, and the implications, of adopting certain aspects of this new approach. In some ways, the results and methods of remote sensing reviewed here represent the least possible contribution that remote sensing can make, since the improvement of remote sensing — the sensors, data quality, methods of analysis, understanding of geospatial environments — is the subject of an intensive and ongoing worldwide research agenda. This situation is virtually assured to help make remote sensing contributions stronger in the future. It is this assurance that I have sought to identify by highlighting the principal methods and accomplishments in the field, and by outlining future implications and challenges.

I recognize that even successful conversion of remotely sensed data to forestry information products will not be enough; the process of acquiring vast amounts of new information about forests must be seen as part of the wider responsibility in service of the generation of new knowledge about the current state of the forest and the influences of management and natural processes to further the goal of forest sustainability. This book is written for university and college students with some background in forestry, physical geography, ecology, or environmental studies; but one key audience that I hope will see value in this material is the operational forest managers, practitioners, and scientists working with forest management problems. I perceive that remote sensing can be useful in solving problems that arise when forest planning directs forest activities on the ground, but there is rarely time to consider the larger context, the specific tool, the trade-offs in different approaches. Whether remote sensing can help those in positions of responsibility in forest operations and management understand and improve the management of the forest resource is, perhaps, still uncertain. What does seem likely is that remote sensing, at the very least, can help detect and monitor forest conditions, forest changes, and forest growth over large spatial scales and at relevant time steps. Hopefully, with better information comes greater understanding and, in turn, practical improvements. It is hoped that increased confidence will be generated that sustainable forest management is possible, and politically, economically, and socially desirable.

I have tried to provide an international flavor to the book, but as is evident in forest management and probably many other fields, remote sensing has been disproportionately developed and implemented in temperate and boreal forests, and particularly in Europe and North America. It seems likely, though, that the methods that have proven valuable in these forests can work well in many world forests, and references and examples have been sought to try and emphasize this key point. I owe a great debt to the early pioneers of remote sensing — the physicists, engineers, and natural scientists — who sought to discover, document, and summarize the principles of the rapidly emerging remote sensing field; their papers and books are liberally referenced in this book, and should be consulted by those wishing to complete an understanding of the forestry remote sensing application. Recently, new remote sensing books that focus on the social, geographical, and environmental sciences have been added to the mix. Remote sensing has always benefitted — as has forest management — from the inherently multidisciplinary nature of its practitioners, methodologists, experimentalists, and developers. I sincerely hope that the current book with its focus on remote sensing in forestry is viewed in this positive light.

Acknowledgments

This book is a development of my research and teaching in remote sensing applied to forestry problems. From the time I was a forestry undergraduate student at Lakehead University in the mid-1970s, such work has been marked in no small way by an ever-widening collaborative experience among foresters, geographers, ecologists, physicists, and others arriving with an interest in remote sensing from vastly different and sometimes wildly circuitous routes. I consider myself very fortunate to have had the opportunity to work with many such excellent students, faculty, and colleagues; by their efforts and enthusiasm I have been much inspired. I am particularly indebted to Clayton Blodgett, Jeff Dechka, Elizabeth Dickson, Graham Gerylo, Philip Giles, Ron Hall, Medina Hansen, Ray Hunt, Mike Lavigne, Ellsworth LeDrew, Julia Linke, Joan Luther, Alan Maudie, Tom McCaffrey, Greg McDermid, Monika Moskal, Derek Peddle, Richard Waring, Brad Wilson, Mike Wulder, and the helpful staff and students at the organizations and institutions in which I have studied, worked, or taught: Lakehead University, Ontario Ministry of Natural Resources, University of Waterloo, Ontario Centre for Remote Sensing, Geophysical Institute of the University of Bergen, Memorial University of Newfoundland, University of Calgary, and Oregon State University, for some of the ideas and concepts that are mentioned in this book.

I would like to acknowledge an important influence on the direction and nature of my remote sensing research by the late John Hudak, Canadian Forest Service; his enormous enthusiasm and trust in the quality and significance of our forestry remote sensing work were both a challenge and a reward. Thank you, John.

Extensive reviews of the manuscript were received from Dr. Ron Hall (Northern Forestry Centre, Canadian Forest Service), Mr. Stephen Joyce (Department of Forest Resources and Geomatics, Swedish University of Agricultural Sciences), Dr. Peter Murtha (Faculty of Forestry, University of British Columbia), and Dr. Warren Cohen (Forestry Sciences Laboratory, Pacific Northwest Research Station, USDA Forest Service). Portions of the book were reviewed by Dr. Mike Wulder (Pacific Forestry Centre, Canadian Forest Service), and Dr. Ferdinand Bonn (CARTEL, Department of Geography, Université de Sherbrooke). I am very grateful to these individuals for their dedicated efforts to read through the text and provide many suggestions for improvement. I believe their comments and insights have helped create a more comprehensive and worthwhile contribution, but of course I retain sole responsibility for any errors or oversights that remain.

I thank Graham Gerylo and Medina Hansen for their exemplary work on the figures and plates, respectively. To those who agreed to help by providing images and graphics, thank you: Joseph Cihlar, Doug Davison, Ron Hall, Doug King, Monika Moskal, Derek Peddle, Miriam Presutti, Benoît St-Onge, and Mike Wulder. These numerous contributions were instrumental in ensuring an effective set of plates and figures for the book. I am also grateful to Pat Roberson, Randi Gonzalez, and Sheryl Koral of CRC Press for their help in turning a manuscript into this book. The following organizations granted permission to use figures, tables, or short quotations from their publications: American Society for Photogrammetry and Remote Sensing, Canadian Aeronautics and Space Institute, IEEE Intellectual Property Rights Office, Soil Science Society of America, Academic Press, Natural Resources Canada, Elsevier Science, Taylor & Francis, Heron Publishing, Kluwer Academic Publishers, CRC Press, Canadian Institute of Forestry, American Chemical Society, and Island Press.

Part of this book was written while I was supported by a University of Calgary Sabbatical Leave Fellowship at the National Center for Geographic Information and Analysis, University of California — Santa Barbara. This leave was made possible with administrative support by Dr. Stephen Randall (Dean, Faculty of Social Sciences, University of Calgary), Dr. Ronald Bond (Vice-President Academic, University of Calgary), and Dr. Michael Goodchild (Director, NCGIA, University of California — Santa Barbara). I acknowledge gratefully the financial support of my research activities in forestry remote sensing by the Natural Sciences and Engineering Research Council of Canada and the Canadian Forest Service.

Steven E. Franklin
University of Calgary

About the Author

Steven E. Franklin, Ph.D., is a professor engaged in teaching and research in the field of remote sensing at the University of Calgary, Alberta, Canada. He has studied forestry, geography, and environmental studies, and has received his Ph.D. in geography from the Faculty of Environmental Studies, University of Waterloo in 1985.

Dr. Franklin taught classes in remote sensing at Memorial University of Newfoundland (1985–1988) and has been teaching at the University of Calgary since 1988. He has had visiting appointments at Oregon State University College of Forestry (1994) and the University of California Santa Barbara National Center for Geographical Information and Analysis (2000). At the University of Calgary, Dr. Franklin has held the positions of Associate Dean (Research) from 1998 to 1999 and Head of the Geography Department from 1995 to 1998. He has also been Chairman of the Canadian Remote Sensing Society (1995–1997) and is an Associate Fellow of the Canadian Aeronautics and Space Institute.

Dr. Franklin has published more than 70 journal articles on remote sensing and forest management issues in Canada, the United States, and South America. His papers focused on remote sensing applications such as forest defoliation, forest harvesting monitoring, and forest inventory classification.

Table of Contents

1 Introduction

"Satellite imagery and related technology": one of the top ten advances in forestry in the past 100 years (Society of American Foresters, Web site accessed 17 July 2000, http://www.safnet.org/about/topten.htm)

FOREST MANAGEMENT QUESTIONS

Human activities in forests are increasingly organized within plans that have at their core sustainability and the preservation of biodiversity. These plans lie at the heart of sustainable forest management, whose practices are designed to maintain and enhance the long-term health of forest ecosystems, while providing ecological, economic, social, and cultural opportunities for the benefit of present and future generations (Canadian Council of Forest Ministers, 1995). Sustainable forest management supplements a concern with economic values with concerns that species diversity, structure, and the present and future functioning and biological productivity of the ecosystem be maintained or improved (Landsberg and Gower, 1996). The growing acceptance of sustainable forest management has been characterized as a paradigm shift of massive proportions within the forest science and forest management communities that is only now reaching maturation (Franklin, 1997).

The roots of sustainable forest management can be traced back much earlier; for example, the Royal Ordonnance on Forests was enacted in Brunoy on 29 May, 1346 by Philippe of Valois (Birot, 1999: p. 1):

> *The Masters of Forests […] shall survey and visit all forests and all woods which they include, and they shall effect the sales as needed, with a view to continuously maintaining the said forests and woods in good condition.*

People have long been interested in extracting immediate value today from forests, while preserving their characteristics for future generations. These interests have been at various times, as in the present, heatedly debated. During the seventeenth century in England, for example, the fundamental interests of the realm — prosperity, security, and liberty — were invoked to support the positions of *both* conservationists and developers (Schama, 1995: p. 153): "The greenwood was a useful fantasy; the English forest was serious business." In the century just past, the discussion has been no less intense (Leopold, 1949) as greater understanding of the fundamental concepts of conservation, ecology, community, economics, and ethics has emerged.

In a global context in current times, as populations and technological ability to extract resources from what were once considered vast and inexhaustible forests continue to expand, world forests increasingly appear finite, vulnerable, dangerously diminished, perhaps already subject to irreparable damage. Recently, efforts have increased to provide a social accommodation to the issues of sustainable forest management, and it is thought that this social emphasis will have far-reaching implications for the way in which society will view and use the forest resources of the planet. Perhaps the ideas have not changed as much as the actual practice of forestry and understanding of forestry goals. Current human ability to change the global forest environment is unprecedented. Sustainable forest management may not represent a fundamental shift in the way humans view forests inasmuch as it represents a recognition of this global influence.

The change has already meant a difference in human interaction with natural forest systems. There is increased emphasis on scientific management (Oliver et al., 1999) and the recognition of the need to understand better ecosystem functioning (Waring and Running, 1998; Landsberg and Coops, 1999) and patterns (Forman, 1995) over large areas and long periods of time (Kohm and Franklin, 1997). Such understanding is increasingly required regardless of any philosophical stance assumed on the role of forests, management, and continued human use of forests. Sustainable forest management has encouraged continued wide-ranging philosophical discussion within the forest science, applied forestry, and ecological communities (Maser, 1994).

There may be much more to come. Heeding the clarion calls for new directions in forestry has resulted in many instances of direct action (Hansen et al., 1998; Bordelon et al., 2000). In the past few years, the actual practice of forest management in some parts of the world has moved swiftly to accommodate sustainable forestry with uneven-age (single-tree and group selection) and even-age stand management (clearcutting, shelterwood, seed tree), increased rotation times, reduced harvesting amounts, and new patterns of resource utilization and cooperative management. The changes from an older, traditional forest management approach with an emphasis on timber values to sustainable forest management are profound. The process of change is subject to continuing discussion, understanding, clarification, and modification. This is an exciting and intellectually challenging time in which to consider future directions in forests and forestry.

Some believe that the cumulative effect of human needs has resulted in a world forestry in a crisis that can only deepen as populations continue to rise and the resource base declines. To them, the historical and current mismanagement and eradication of whole forests has all but reached the critical point, after which the damage is overwhelming and final (Berlyn and Ashton, 1996; Meyers, 1997). There is abundant evidence that destructive land use practices, widespread pollution, exploitation of species, and perhaps global climate change (Stoms and Estes, 1993), have caused damage to the Earth's ecosystems and consigned many species to oblivion. The current rate of species extinction is estimated at 50 times the base level of the last 400 years, and 100 times the base level of the last half of the twentieth century (Raven and McNeely, 1998). In some areas, the species extinction rate may be 10,000 times the background level (Wilson, 1988). Exactly how much of this may be traced to unsustainable forest practices — such as clearcutting in areas not able to regenerate — is not known. But the message is unmistakable; the human species must desist in knowingly engaging in destructive forest activities that result in continuing, massive, even irreversible damage to the biosphere.

Others view the current problems as symptomatic of a fundamental shift in the balance of human management of resources and economics. During this time of change, the actual direction is not yet clear. Initial indications are that support is moving away from applying best practices over fine spatial scales, as management is reoriented to address concerns over larger areas and longer time periods (Swanson et al., 1997). To help accomplish this necessary transition, an overarching theme is the continued search for fast, consistent, versatile, accurate, and cost-effective information inputs to management problems, but now with the full knowledge of the wide range of scales — local to global — over which forest communities are affected. There is time to adjust, to experiment, to adapt, to understand better the impacts and implications of forest management — there is time, but not much (Boyce and Haney, 1997). Still others believe that human populations will soon stabilize, then begin an orderly decline to sustainable levels. Resources will not become limiting. To them, the forest resource simply must be managed more "scientifically;" for example, by more careful application of the principles of management science (Oliver et al., 1999). The wide range of opinion and the large number of unknowns suggest the difficulties in charting the future of forest management and the fate of the world's remaining forests.

The future of forest management remains unclear. What will be the end result of changing values in society and the societal view of forests? Will increased economic and social needs be met with forest products? Will foresters and other resource management professionals find better ways to manage forests and to meet these needs? Will there be more or less direct (e.g., prescribed treatments, suppression of natural processes) and indirect (e.g., climate change) human intervention in forest growth patterns? Will our understanding of the characteristics of landscape dynamics improve fast enough to allow the incorporation of natural disturbances in planning sustainable forest management? Can human needs and forests coexist sustainably? Can a sustainable forest management approach — can any management approach — succeed in ending the terminal threat of destruction faced by many, if not most, of Earth's remaining intact old-growth forests?

What *is* clear is that increasing amounts of scientific information must be acquired to support the emerging, practical, ongoing goals and objectives in managing forests (Bricker and Ruggiero, 1998; Noss, 1999; Simberloff, 1999). One goal is to adapt forest management continually to accept new objectives. One goal is to learn how to manage forests sustainably so benefits continue and future generations are not compromised. Another goal is to acquire knowledge about the current state of the forest and about how management and natural processes affect future outcomes. These goals require that new information be obtained by:

1. Increasing understanding of forests through field trials, observations on long-term sampling plots, analysis of historical outcomes, growth, succession, and competition observations and models,
2. Transforming and interpreting data from new and existing forest inventories,
3. Developing and accessing data from various purpose-designed national and regional resources, including forest health networks, decision support system networks, and ecological or biogeoclimatic classification systems,
4. Obtaining new data and insights through development and deployment of a suite of new information technologies, including geographical information systems (GIS), computer modeling and spatial databases, and remote sensing of all types and descriptions.

Driving these demands for new information are basic science and management questions, coupled with evolving models of forest economics and forest certification initiatives. However, if generating and accumulating data were the only impediments to sustainable forest management, humanity's problems would be over. A common view in the natural sciences is that there is difficulty handling the data currently available without losing critical key components; in forestry, "We now have more data than we can interpret" (Lachowski et al., 2000: p. 15). There are probably enough data in all the critical areas, and *spatial* data types and volumes are not immediately limiting (Graetz, 1990; Vande Castle, 1998). With increased intensive/extensive monitoring at instrumented research sites, this situation will continue to exist, perhaps developing beyond capabilities to manage the data flow. But data are not information. What may be lacking is a way of understanding these data, of finding the right interpretation of the data, of ensuring the conversion of data to information, and ultimately, the conversion of information to usable knowledge about the current state of the forest and the influences of management and natural processes. It appears that converting data to information is the highest priority for remote sensing to contribute to sustainable forest management.

It has been suggested that compared to previous forest management approaches, all new forest management strategies will require even more record keeping and even wider access to information (Bormann et al., 1994). It is not yet known if the new spatial information technologies — such as GIS, remote sensing, computer modeling, decision support systems, and digital databases — are going to be able to handle all of the new data requirements (Bormann et al., 1994; MacLean and Porter, 1994). Can remote sensing data provide the required information with the greatest accuracy for a given cost? Even if this challenge can be met, information is not understanding; it is not yet known whether increased human understanding of the central issues will result such that management will be improved. There is a critical lack of understanding in several key areas related to human activities on the landscape. For example, it is not clear in what ways, if at all, human-altered spatial structures can mimic natural disturbance regimes, or what consequences human-induced climate change will have on net ecosystem productivity.

The spatial information on patterns of disturbance and productivity is relatively easy to come by; as will be shown in this book, mapping forest insect defoliation, patterns of forest harvesting, and changes in photosynthetic capacity across broad forest ecosystems are operational with current satellite and airborne remote sensing technology and methods. What is not so simple is the understanding of what these patterns mean, and how to implement the more certain power to make decisions that such understanding confers on those responsible for sustainable forest management.

A Technological Approach

Remote sensing has been a valuable source of information over the course of the past few decades in mapping and monitoring forest activities. As the need for increased amounts and quality of information about such activities becomes more apparent, and remote sensing technology continues to improve, it is felt that remote sensing as an information source will be increasingly critical in the future. A powerful line of thought in remote sensing is to consider problems in a technological approach (Curran, 1987). By this, it is meant that remote sensing can sometimes proceed differently than traditional scientific deductive and inductive approaches. These are considered the pure science approaches, and can be contrasted to the scientific technological approach in which the emphasis is shifted to a *methodological* or *applied* perspective. A successful remote sensing application proceeds from the design of methodology. In a remote sensing technological approach, the goal is the application of knowledge — the use of what has been learned — to solve problems.

Forest managers are concerned with the spatial distribution of forest resources within their management area and in the surrounding ecosystems; with the timely acquisition of information on conditions and changes to these resources; with the small and large impacts associated with changing patterns and processes at different scales in time and space; with interpretation of the effect of those changes on unmapped components such as wildlife; and with economic, social, and environmental implications of human activities and impacts on forests. There is a need to have as much relevant information as possible on the conditions of the forest to prescribe treatments, to help formulate policy, and to provide insight and predictions on future forest condition, and health. Typically, there are few choices in how to acquire all the different types of information. The goal of remote sensing, then, is to help satisfy as many of these multidimensional needs for information as possible. This is the application of knowledge: that is, the application of remote sensing knowledge in response to forest management questions.

While remote sensing technology must help in providing information to satisfy the needs that forest managers have, remote sensing must be a cost-effective and easily understandable technology. These are probably two of the most important reasons that aerial photographs are still the most common form of remote sensing in forestry; relative to information content, they are inexpensive and easy to use (Pitt et al., 1997; Caylor, 2000). The field of remote sensing began with fully manual methods of analysis applied to aerial photographs, but has since come to rely on new data and methods. As these data and methods evolve and improve, it appears likely that remote sensing will be increasingly useful in satisfying needs for forest management information.

A *methodological design* is necessary to show how remote sensing data can be used to determine the spatial distribution of forest resources, can detect changes in those resources, and predict changes in other aspects of the forest not captured directly. Remote sensing methodology can be a powerful aspect of a technology to detect changes accurately and to help explain more fully the implications of forest changes and activities. To accomplish this, there must be a series of direct mapping and modeling applications of remote sensing consistent with the needs of forest managers. This is the role of methodological design — to convert data to information in a scientific, understandable, and repeatable way. As in all scientific approaches, the hallmark of good science is the use of the knowledge gained to uncover general laws and to predict future conditions; then, the methodological design becomes part of the established scientific method, the analytical approach for the field (Lunetta, 1999).

A useful way to consider the diversity of the remote sensing data input to sustainable forest management is to examine the kinds of issues and questions around which sustainable forest management revolves. In Table 1.1 some example questions are listed; each sustainable forest management question is paired with an inferential hypothesis which can be suggestive of the ways in which remote sensing can contribute to providing an answer or generating new insight into the question. All questions of forest value are first rooted in an accurate description of the resource, and it is the responsibility of the forest inventory to provide that description (Erdle and Sullivan, 1998). This book presents the technological approach of remote sensing to sustainable forest management questions, but does not attempt to illustrate exhaustively how specific answers are best derived. The infinite variety of such questions and answers prevents that level of detail; this is not a remote sensing cookbook. Rather, the idea is to present to remote sensing data users a review of the achievements of remote sensing and forestry scientists and professionals in addressing key mapping, monitoring, and modeling applications in forests.

It is clear that the information needs of the past and the future differ, and new demands will be placed on the forest inventory and other information resources available in forestry. Here, in a few pages, several decades of progress in forestry remote sensing are rushed through, hopefully with due process, in an attempt to enable the reader to understand the full range of possibilities in remote sensing participation in the forest management questions of the day.

REMOTE SENSING DATA AND METHODS

The general role that remote sensing data might play in forest management, within the relatively narrow range of information sources that are available, is probably not well understood (Cohen et al., 1996b). Generally, remote sensing is understood relative to the more familiar aerial photography (Graham and Read, 1986; Howard, 1991; Avery and Berlin, 1992) and of course, field observations (Avery and Burkhard, 1994). In many situations, such as site or forest stand disease assessment (Innes, 1993) and studies in old-growth or other forests where rare or endangered species or ecosystems may occur, field observations are the only possible way to acquire the needed information (Ferguson, 1996). In other situations, aerial photographs are suggested (Pitt et al., 1997); no other source of information would be appropriate. Currently, it is thought that there are few or no substitutes for *in situ* observation;

TABLE 1.1
Sustainable Forest Management Questions and Corresponding Remote Sensing Hypotheses

Sustainable Forest Management Question Driven by Human Need	Remote Sensing Inferential Hypothesis Driven by Technology
What is the spatial distribution of forest covertypes/classes? Species composition?	Remote sensing observations can be used to differentiate forest covertypes on the basis of forest structure and species composition.
Is there a cost-effective way to map annual changes resulting from harvesting operations and natural disturbances?	Multitemporal remote sensing observations can be used to separate forest management treatments (such as cutovers, thinnings, plantings), new roads, insect damage, windthrow, burned or flooded areas, from surrounding covertypes over time.
How can remote sensing data be compared to existing forest inventory data stored in the GIS?	For some attributes (e.g., stand density) over large areas or within forest stands the information content of remote sensing data is consistent with the accuracy and level of confidence that we now possess in the GIS database. For other attributes (e.g., leaf area index) the remote sensing data are superior.
Can we map more detail within each forest stand, but also see the big picture — the ecosystems in which stands are embedded and areas surrounding my management unit?	Remote sensing observations acquired at multiple scales and resolutions can be used to continuously estimate forest conditions from plots to stands to ecosystems.
Can habitat fragmentation and connectedness be measured and quantified?	Landscape pattern and structure can be detected and quantified using remote sensing observations.
What is the best way to monitor forest production?	Remote sensing observations can be used to obtain precise estimates of driving variables (e.g., LAI, biomass) for use in initiating and verifying functioning ecosystem process models.

and, there are few or no acceptable alternatives to aerial photography. Only under certain rare circumstances is it appropriate or productive to consider remote sensing a *substitute* information source. But is remote sensing a legitimate method of acquiring forest information that cannot be obtained in other ways? There appears to be insufficient awareness of the complementarity of field observations, aerial photography, and specific remote sensing data sources and methods, and the ways these various information sources can work together (Czaplewski, 1999; Oderwald and Wynne, 2000; Bergen et al., 2000).

There are many situations in forest management in which managers and forest scientists are concerned with larger areas and differing time periods; field observations are necessary, but not sufficient; aerial photographs are necessary, but not sufficient. How do field observations, aerial photography, and digital remote sensing fit together? What kind of remote sensing can and should be done in support of sustainable forest management? Information is a management resource. Understanding what can and cannot be remotely sensed with accuracy and efficiency is a key piece of knowledge which those faced with management problems should possess.

Definition and Origins of Remote Sensing

The field of remote sensing has throughout its development been relatively poorly defined (Fisher and Lindenberg, 1989). Any number of reasons for this situation could be cited: the truly multidisciplinary nature of the field; the phenomenal growth of automation in the various "founding" disciplines, such as cartography (Hegyi and Quesnet, 1983); the increasing dominance of GIS in the marketplace (Longley et al., 1999). Has a loose definition of remote sensing led to poor remote sensing science and applications? Some might feel that there is at least one advantage of working in a "poorly defined field" — the feeling among remote sensing practitioners that virtually anything goes. Without a restrictive definition of what is and is not remote sensing, there is great freedom in selecting approaches, methods, even problems to address; accordingly, there are few boundaries to remote sensing established by convention. In remote sensing, one strong focus has always been on methodology; how sensors, computers, and humans can be used together to solve real-world problems (Landgrebe, 1978a,b). This situation has persisted since the early days in remote sensing, perhaps leading to, or at least not preventing, great creativity and breadth in the emerging field.

An original problem in defining remote sensing can be traced to a fundamental philosophical problem, long since resolved, of whether remote sensing was solely the reception of stimuli (data collection) or whether it also included the collective (i.e., analytical) response to such stimuli (Gregory, 1972). But a glance at the titles of early papers in remote sensing journals, or at any of the many remote sensing conferences and symposia throughout the 1960s and 1970s, provides ample evidence that to the pioneers in the field, remote sensing was always much more than simply collection of data — that remote sensing was "the science of *deriving information* about an object from measurements at a distance from the object, i.e., without actually coming into contact with it" (Landgrebe, 1978a: p. 1, italics added). Or this one from Avery (1968: p. 135, italics added): "Remote sensing may be defined as the *detection, recognition, or evaluation* of objects by means of distant sensing or recording devices."

By specifying the key words "deriving information" and "detection, recognition, or evaluation" these definitions suggested that the most important contribution promised by remote sensing was in the conversion of the collected data to information products; the true value and challenge of remote sensing would be realized in the data interpretation and subsequent applications. The developers of applications of remote sensing data understood from the beginning that remote sensing was both technology and methodology. The term "remote sensing" has come to be strongly associated with Earth-observing satellite technology, but more properly has been understood to include all sensing with distant instruments, and that is the meaning that is assumed in this book.

Today, remote sensing is usually defined as comprised of two distinct activities:

1. Data collection by sensors designed to detect electromagnetic energy from positions on ground-based, aerial, and satellite platforms, and
2. The methods of interpreting those data.

Working in those early days of satellites and digital sensors, Avery (1968) intended to cover the emerging field of Earth-orbiting satellite remote sensing separate from aerial photography. Remote sensing is sometimes now considered to encompass aerial photography (Lillesand and Kieffer, 1994) or at least occupy a companion, parallel or perhaps not yet fully integrated position (Avery and Berlin, 1992). Between 1960 and about 1980, at least, there were always two types of remote sensing:

1. Imagery (or visually)-based remote sensing, and
2. Numerically-based remote sensing.

The differences were found in the way the data were acquired, but more significantly, in the way the data were interpreted; in other words, the way information was extracted from the data. To many, this distinction is no longer relevant; the focus has shifted decisively to remote sensing interpretations which best serve the purpose at hand with the available technology (Buiten, 1993). In any given application, a mix of visual and numerical methods is needed.

While the methods and technology of remote sensing have shown tremendous advances, remote sensing is obviously not a completely new idea — having its modern antecedants in the use of cameras and balloons in the nineteenth century (Olson and Weber, 2000). The first use of this technology is not well documented, but there are suggestions that camera and balloon remote sensing technology was used during the Franco-Austrian War of 1859, the American Civil War, and the Siege of Paris in 1870 (Graham and Read, 1986; Landgrebe, 1978a). The first known forestry remote sensing application was recorded in the *Berliner Tageblatt* of September 10, 1887 (Spurr, 1960). The notice concerned the experiments of an unnamed German forester who constructed a forest map from photos acquired from a hot-air balloon. Interestingly, the power of the perspective "from above" was regarded as the principal advantage, and many of the same problems that have since preoccupied aerial mappers and digital image analysts were identified: geometric distortion, spatial coverage, uncertain species identification, within-stand variability, visible indicators of growth and development, and so on. Aerial photography was established as a reconnaissance tool in the first World War (Graham and Read, 1986). The growth of digital remote sensing as a field from these humble but practical beginnings is described in a later section.

By far, the most common remote sensing in forestry, historically and today, is conducted using optical/infrared sensors; and by far the most common of these sensors is the aerial camera (film). However, other sensors in the passive microwave, thermal, and ultraviolet portions of the spectrum are under rapid development and instruments may soon reach operational status. *Li*ght *d*etection *a*nd *r*anging (lidar) instruments, in particular, appear poised to transform forest measurements and remote sensing as an information source (Lefsky et al., 1999a). It is not possible to consider all of these remote sensing devices, but some are introduced in this book and briefly discussed. All of these sensors generate data that are complex and sometimes unique. The forestry user community could no doubt benefit if an entire book were devoted separately to the science, technology, and forestry applications of remote sensing by means of lidar, hyperspectral sensors, thermal sensors, and other instruments. These have not, though, achieved the level of market penetration and user acceptance that aerial photographs, and to a lesser extent, optical/infrared and *ra*dio *d*etection *a*nd *r*anging (radar) sensors have enjoyed. This book is not about aerial photography, instead focusing on the digital remote sensing data and methods generated by multispectral and radar instruments. It is hoped that there may be some common understanding that can emerge from considering the field of remote sensing, as it has been applied in forestry, and focusing on these relatively common sensors and the digital analysis tools that have emerged to extract information from the image data.

Here, those sensors that have been, and will likely continue to be, the main source of digital remote sensing information in support of forest management and practices are discussed. The assumption is that by reviewing forestry remote sensing applications by multispectral and radar sensors, readers will gain an appreciation of some key methodological issues. For example:

1. An appropriate and properly prepared remote sensing database for the task at hand;
2. A fully functional image processing system, perhaps coupled with the ability to write needed computer codes in-house (Sanchez and Canton, 1999); and
3. Access to other sources of digital information, most notably through available geographical information systems.

In this book, the intention is to review remote sensing accomplishments and potential in a way that may make sustainable forest management questions more clear, and their resolution more likely through application of remote sensing technology. What forest practices will ensure our forests are being managed sustainably? Remote sensing provides some of the information that will support management decisions. Several examples drawn from the literature will include case study summaries of work that address some of the forest management questions and remote sensing hypotheses (see Table 1.1). Before examining these issues in detail, an interpretation of two different remote sensing methods used as ways of tackling forestry questions is provided.

The Experimental Method

The experimental method in science is used when the control of variables is possible and desirable (Haring et al., 1992). In many remote sensing applications, the experimental method has been used to better understand the relationship between the forest condition of interest, and the information available about that condition from remotely sensed data. In a remote sensing experiment, the remote sensing observation is the dependent variable, and the independent variables influencing the dependent variable are the forest conditions of interest. If all of these variables can be controlled, then a precise and accurate predictive model can be created by which the independent variable (the forest condition) can be used to predict the dependent variable (the remote sensing observation). Then, the model can be inverted so that the independent variable can be predicted by the remote sensing observations. In forestry applications, a desirable set of independent variables would include forest vegetative, structural, and biophysical conditions such as forest canopy closure, species, density, height, volume, age, roughness, leaf area, and biochemical or nutrient status.

An example of this approach was provided by Ranson and Saatchi (1992) in their study of the microwave backscattering characteristics of balsam fir (*Abies balsamea*). On a movable platform, small balsam fir seedlings were arranged and then observed by a truck-mounted microwave scatterometer at controlled polarizations and incidence angles. By altering the angle of the platform and the spacing of the seedlings, different forest canopy densities were created. With careful biophysical measurements of the seedlings (as the independent variables) and the remote sensing observations (as the dependent variables), strong predictive relationships were developed and used to calibrate a mathematical model of the energy interactions of the microwave beams with the canopies. For example, it was found that the measured leaf area index (LAI — the independent variable) and measured backscattering coefficient (the dependent variable) increased together. Typically, LAI is considered a dynamic forest structure variable, and is estimated by representing all of the upper surfaces of leaves projected downward to a unit of ground area beneath the canopy (Waring and Schlesinger, 1985). The finding that LAI and microwave backscattering were related positively was in accordance with the predictions of the theoretical model relating needle-shaped leaves and microwave energy (Figure 1.1). More leaves created more scattering of the microwave wavelengths, and the deployed microwave sensor was sensitive to this increase in reflected energy.

The control of many confounding variables in an actual forest microwave image acquisition from aerial or space-borne sensors is not possible. In this situation, often it is not known *a priori* which variables will influence remote sensing measurements; for example, the soil background and topography will variably influence the measurements of microwave energy. These influences can overwhelm and confound the influence of the leaves; such influences are not uniquely determined. Typically, the scientists would have little or no ability to control the experiment for topographic or soil differences over large areas. Even the range of leaf area conditions and the type of imagery are typically very difficult to control; often the satellite or airborne sensor that may be available for the mission is not the ideal sensor that one would choose to develop the relationship between leaves and microwave energy. The actual relationship between aerial and satellite remotely sensed microwave backscatter coefficients and leaf area index is typically much less predictable than that obtained using experimental methods (e.g., Ranson and Sun, 1994a,b; Franklin et al., 1994).

A second example of the experimental approach involves the identification of nutrients in foliage using spectral measurements. The remote sensing of foliar

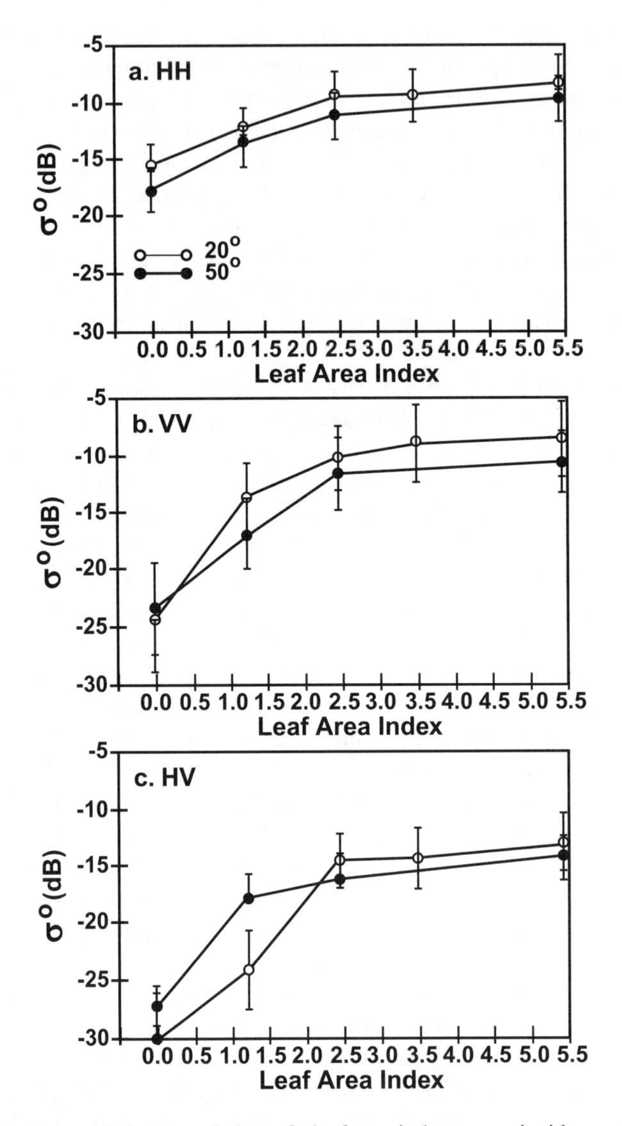

FIGURE 1.1 Relationship between balsam fir leaf area index at two incidence angles and C-band SAR backscatter coefficients measured in a controlled experiment. Higher backscatter is related to higher leaf area index because of increased scattering by the conifer needles. The effect is often more pronounced at lower incidence angle and in cross-polarization data. (From Ranson, K. J., and S. S. Saatchi. 1992. *IEEE Trans. Geosci. Rem. Sensing,* 30: 924–932. With permission.)

nutrients and stress has long been of interest, and new technology has been developed to satisfy the needs of managers for whole leaf, plant, and canopy nutrient estimation (Curran, 1992; Dungan et al., 1996; Johnson and Billow, 1996). Applications might include stress detection (Murtha and Ballard, 1983) and identification of agents of stress before they cause damage or after damage has occurred (Murtha, 1978).

Typically, the first step in these applications is to apply the experimental method in remote sensing to examine the relationship between leaf reflectance properties and pigment concentration (Blackburn, 2000).

An example of this approach is the rapid, nondestructive ground-based foliar N, P, and total chlorophyll estimation technique using measurements obtained from a tripod-mounted spectroradiometer developed by Bracher and Murtha (1994). This approach is based on measurements most useful in nursery and field situations in which light and growth conditions can be controlled. First, treatments of Douglas-fir (*Pseudotsuga menziesii*) seedlings in greenhouses were varied with different solutions of micronutrients and macronutrients. Second, under controlled light conditions leaves were sampled for reflectance, then analyzed for nutrients with various wet chemistry techniques. As pigment concentration of the leaves increased, reflectance measurements in the visible portion of the spectrum decreased in a predictable way. More leaf pigments, such as chlorophyll, meant less red reflectance, thus more red light absorption (more photosynthesis). The relationships to leaf nutrients, such as N and P, were more complex and were indirect; for example, foliar N was not involved directly in the absorption and scattering processes that produced leaf reflectance. Instead, foliar N was highly correlated with leaf chlorophyll concentration. As leaf N increased, leaf reflectance measured in the green portion of the spectrum decreased and the position of the red-edge (the difference between the red and near infrared reflectances) was shifted to longer wavelengths (Figure 1.2). Through the relationships between reflectance, N, and pigment concentrations, usable models (±5% error) of foliar N were developed.

Extending the knowledge gained through experimental methods of remote sensing to create an actual remote sensing application to predict chlorophyll or nutrient status of forests has proven difficult. Even when all factors are controlled, variable relationships have been reported (increased green peak reflectance with increased pigmentation of leaves, for example). Reflectance is influenced by the chemical properties of consitutents and also their geometric arrangement. Field, aerial, and satellite remote sensing measurements of canopy reflectance are subject to a large number of uncontrolled factors that reduce the strength and applicability of the direct relationship between leaf reflectance and chlorophyll. The sensitivity of the remote sensing instruments, as measured by the signal-to-noise ratio, has only recently approached the level required for biochemical estimation (Wessman, 1990; Gholz et al., 1997). Satellite remote sensing instruments are not yet able to acquire data with these characteristics. Airborne (and satellite) remote sensing data are a composite signal from leaves, branches and boles, tree canopies, ground cover and soils, and are influenced by shadowing due to height differences and defoliation, sun-target-sensor geometry, topography, leaf orientation, canopy architecture, and atmospheric turbidity (Bracher and Murtha, 1994; Blackburn, 2000). These and other factors mean that the relationship between leaf chlorophyll and leaf reflectance obtained under controlled conditions cannot be expected to be found in the same form and strength when those conditions of measurement are allowed to vary in an uncontrolled experiment.

What is important to note is that the experimental approach provides superior understanding of the precise form of the relationship, but typically fails in most real

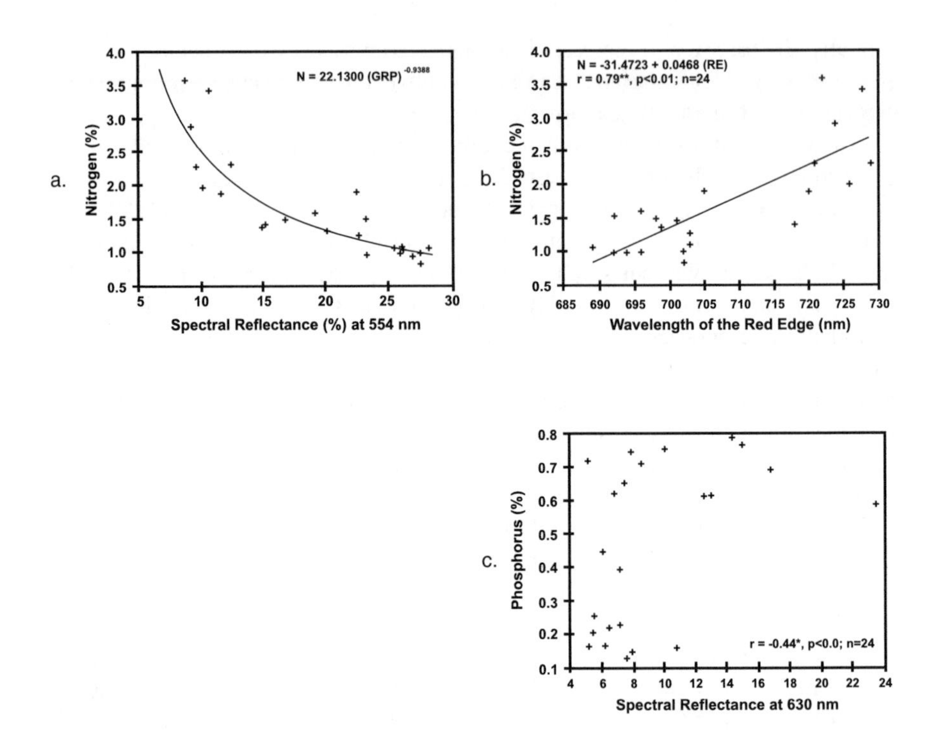

FIGURE 1.2 Relationship between foliar N and percent spectral reflectance for Douglas-fir seedlings in different fertilization treatments measured in a controlled experiment. In (a) a decrease in reflectance at 554 nm (known as the green reflectance peak) is shown with increased foliar N. In (b) the position of the red-edge is shown to occur at longer wavelengths with increased foliar N. Measurements suggested that as foliar N increased, pigment concentration increased, and more absorption of visible light and less reflectance in the green and red portion of the spectrum occurred. The third graph (c) shows a much poorer relationship between foliar P and red light absorption because of the weaker relationship between P and leaf pigment content. (From Bracher, G., and P. Murtha. 1994. *Can. J. Rem. Sensing,* 20: 102–115. With permission.)

applications because there is much less control of confounding variables. What is the point in developing a precise relationship between leaf chlorophyll and reflectance under lab-controlled conditions, if the moment one leaves the lab the relationship cannot be used? The point is that the experimental method reveals the nature and exact form of the relationship which, it is understood, will not be found except under ideal conditions.

The Normative Method

Because of the difficulty, or even the impossibility, of controlling more than a few of the many independent variables during most remote sensing applications, the normative method of scientific procedure (Haring et al., 1992) is often employed. That is, what is sought in remote sensing is an understanding of the normal relationship (for example, between forest characteristics such as leaf area index, canopy chlorophyll or N content, and reflectance or backscatter measured remotely) so that applications of that relationship can be developed to satisfy the practical need for new information. Since not all the independent variables are controlled, the relationships that are found are subject to caveats and constraints that must be carefully documented and described.

An airborne spectrographic image might be acquired to map forest canopy chlorophyll based on the experimental relationship between pigmentation and reflectance discussed in the previous section. If the optical properties of the atmosphere were not observed, and no accounting of the atmosphere was made in the use of the remote sensing measurements, then a description of the effect of the atmosphere on the relationship between remote sensing observation and canopy chlorophyll must accompany the presentation of the relationship. Even if the optical properties of the atmosphere were observed during image acquisition, it is impossible to know all the effects of the atmosphere for every pixel location in the image; assumptions of atmospheric homogeneity are needed to reduce the complexity of the radiative transfer equations through the atmosphere to a manageable level. Alternatively, a number of image processing steps can be taken to simulate control of variables; instead of actual atmospheric measurements and the use of those measurements to correct for the atmospheric contribution to the reflectance, a model or an approximation of the atmosphere conditions can be employed, and an estimate of the atmospheric contribution, perhaps based on the image data themselves, is generated and applied *post hoc*. In this way, the observed relationship can be evaluated against the normal form of the relationship or the form of the relationship that would be expected given complete control of all the variables (in this case, the atmospheric contribution).

An example of this approach was presented by Gong et al. (1995) in estimation of forest LAI at six sites in Oregon using airborne imagery acquired with a variety of bandsets at different altitudes. First, an atmospheric and sensor noise correction was applied with models available in an image processing system. Then, spectral data were compared to field-measured LAI. Imagery with higher spatial resolution had stronger relationships to LAI than did imagery at coarser spatial resolution. The reason suggested was that as the airplane flew at higher altitudes over the sites, the pixels in the imagery covered a larger area. This coarser spatial resolution meant that the sensor measurements contained more non-leaf features, such as background soil, shadows, and different understory characteristics which contributed their own reflectance characteristics, and degraded, subsumed, or hid the direct relationship between leaf area index and reflectance. The differences in spectral, spatial, and radiometric resolution that resulted when the image data were acquired at different altitudes gave rise to very different relationships between spectral reflectance and leaf area index. Typically, a negative relationship between red reflectance and leaf area index was found when the data were averaged over a larger area; but with very small pixel sizes there may be a positive relationship between red reflectance and leaf area, based on pigment concentration differences between individual plant species.

Use of the normative method does encourage the development and use of relationships developed under less than ideal conditions. The approach can also suffer from a lack of generality that could lead to highly spurious local insights.

What is the point in developing a normative relationship between leaf chlorophyll and reflectance on one day in one type of forest, if the moment one leaves those conditions the relationship cannot be used? The point is that, provided the design is sound, estimates of the relationship can be obtained that can be used to solve problems even if the relationship does not generate insight into the "true" relationship. The relationship can be used cautiously, and the methods used to find the relationship can be applied elsewhere.

Many more examples can be found to illustrate these information extraction methods and the scientific procedures in remote sensing. The lessons that have been learned in using the experimental and normative approaches in remote sensing forestry applications are reviewed in later chapters in this book. It seems apparent that a balanced approach, using both experimental and normative methods, has resulted in progress in the application of remote sensing to real-world problems in resource management. It is expected that experimental results will continue to be incrementally tested and refined using the normative approach, and through this fine balance progress in applying new information can be achieved.

CATEGORIES OF APPLICATIONS OF REMOTE SENSING

Applications of remote sensing to sustainable forest management are presented in four categories that include classification of forest covertype, estimation of forest structure, forest change detection, and forest modeling. Each of these categories is further subdivided for discussion and review into individual and separate issues, variables, or model parameters according to the logic shown in Figure 1.3. The consideration of each category proceeds from an understanding of the role that specific forest covertype, forest inventory, change detection, and forest modeling applications may play in meeting information needs for sustainable forest management. It can be anticipated that these information needs will be identified in stand-level and landscape-level planning (Hunter, 1997); generally, to support forest planning there is a need for information related to strategic, tactical, and operational activities in forests, and to monitor the results of these activities. The criteria and indicators of sustainable forest management can be considered a focus for the monitoring issues, and the discussion is tailored to include remote sensing methods, and references to examples of the methods that actually work with known error.

Any categorization of the many remote sensing applications into a small number of categories is constrained by the need to put a limit on the type of application included in the category. The use of remote sensing to estimate some of the more common forest inventory variables — height, age, crown closure — is reviewed before considering other forest information that has potential to be provided by remote sensing technology. Still other variables available by remote sensing, not in high demand presently for forest management, are considered because of their potential role in future management scenarios. An example is the forest leaf area index (LAI); this forest measure is considered by many to be a potentially powerful new forest inventory variable (Buckley et al., 1999) even though LAI is not part of

ORGANIZATIONAL INFRASTRUCTURE

Science and Technology Advancement **Improved Understanding** **Infrastructure Development**

Covertype Mapping

Physiognomic Classification
- Level I & II Classes

Floristic Classification
- Level III Classes

Integrated Classification
- Ecological Classes

Inventory Mapping

Cover, Age, DBH, Height **Biomass** **Volume & Growth**

Change Detection

Harvesting & Silviculture
- Clearcuts
- Partial Cuts

Natural Disturbances
- Defoliation, Damage

Forest Modeling

Ecosystem Process Models **Landscape Structure**

TECHNOLOGICAL INFRASTRUCTURE

Acquisition of Imagery **Image Processing & Calibration** **Ecological Models** **Geographical Information Science**

FIGURE 1.3 Categories of remote sensing applications reviewed in this book organized according to broad categories in forest covertype mapping, inventory variables, change detection, and modeling. These applications are made possible, or at least are more readily accomplished, by the existence and continuing development of a technological and organizational infrastructure.

many forest inventories, is better suited for ecological applications and modeling (Running et al., 1989), and most current applications of LAI could be considered as part of forest change detection and modeling efforts. LAI may never be used directly by managers; rather, the LAI estimate obtained by remote sensing can be used to drive models of forest productivity. These models predict a number of forestry-relevant outputs that would be used in management planning. Here, the distinction between inventory variables and indirect input/output variables is not considered relevant; in future it is expected that forest management demand for ecological applications, models, and comprehensive forest structure information, of which LAI is only one example, will continue to grow until the potential role of such information — alongside the more traditional data in a forest inventory — is better defined and understood.

An early distinction between landcover mapping and biophysical remote sensing (Jensen, 1983) can be seen to persist in the general categories used in this book and in the field in general. A few applications of remote sensing have transcended this distinction, having combined the goals of mapping landcover and estimating biophysical attributes within landcover or newly created landscape units. In fact, increasingly, landscape units of interest are actually based on biophysical variables (Graetz, 1990). While the details of the remote sensing categories used are not critical, what is important is that an opportunity is provided to gain an understanding and appreciation of the essential methods of remote sensing in providing key input and monitoring variables in support of sustainable forest management activities.

Growth of Remote Sensing

The technology of remote sensing originated in the science and technology of aerial photointerpretation (Silva, 1978). Interpretation of aerial photographs had been operational for much of the twentieth century, at least since the first World War, and forestry applications were well established after the second World War (Graham and Read, 1986). In 1968, aerial photointerpretation was thought to have "finally commanded the respect and interest of scientists in many disciplines" (Colwell, 1968: p. 1). Then, this young science was sharply challenged. New imagery and sensing techniques quickly emerged based on various computer technologies, engineering designs, launch and deployment opportunities, and developments in sensor science (Gregory, 1971). The use of the term remote sensing began as a way of describing some of the new imagery and analysis techniques acquired alongside aerial photography during the height of the Cold War. The methods of analysis soon diverged — although, in the view of many, including Buiten (1993) and Beaubien (1994), visual and digital analysis methods have recently reconverged, to the benefit of the overall remote sensing approach. Remote sensing has since developed into a scientific field that can stand on its own, as evidenced by the many journals and scientific societies now devoted to the subject.

In 1985, Curran suggested that rapid technological advancements (e.g., computers) and improved sensor systems (both introduced and envisioned) had propelled remote sensing into a stage of exponential growth. An exponential growth stage can be predicted as the second in the normal progression of a scientific discipline through four distinct growth stages represented by an S-curve model (Figure 1.4). First, a slow steady rise occurs in information and activity as the field begins. This stage is followed by a steeper rise as knowledge accumulates exponentially, followed by a lessening of the curve slope. Finally, a relatively flat or even declining condition exists in which the science is (apparently) exhausted. The field has reached maturity.

At what point on this curve is remote sensing in the year 2001? Vital research activity in remote sensing seems destined to continue for some time; hardly any of the fundamental questions in remote sensing have been completely and exhaustively addressed. Evidently, the exponential growth stage identified in 1985 has continued (Jensen, 2000). Developments are such that remote sensing seems some way from entering the fourth stage, in which "reliable information can, as a matter of routine, be generated for the management of our fragile planet" (Curran, 1985: p. 7). In forestry, digital remote sensing applications have been adopted slowly, if at all (Bergen et al., 2000); few routine applications have been established.

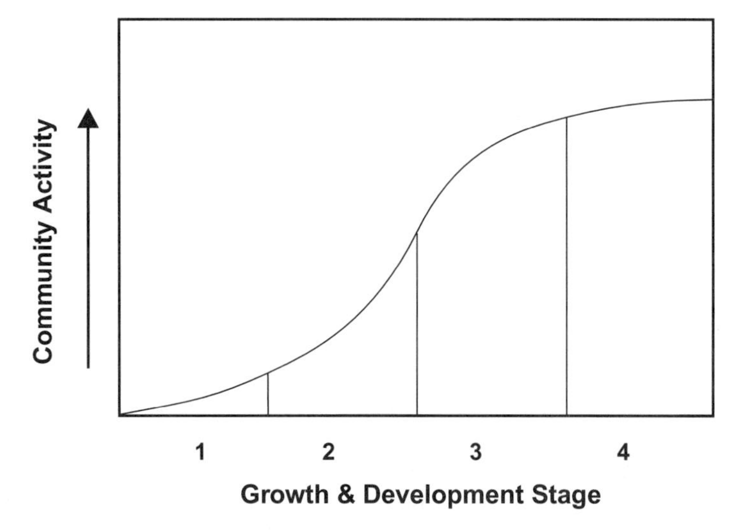

FIGURE 1.4 S-curve of growth in any scientific discipline. Four phases are suggested that involve increasing amounts of community activities, including technological developments and research, that culminate in a mature scientific field. Remote sensing is thought to be in the later stages of stage 2, in which exponential growth continues, or perhaps just entering stage 3, in which growth begins to decline in favor of established protocols and standardization of the methods in the field. (Modified from Jensen, 2000.)

Empirical evidence for the broader interpretation of the growth of remote sensing is found by considering the occurrence of milestones in the development of sensing and space technology (Jensen, 2000), such as the development of radar in the 1940s, thermal imagery in the 1950s, the routine availability of meteorological satellite data in the 1960s, the launch of the ERTS and Landsat satellites in the 1970s, the continuation of medium- and coarse-scale monitoring of the globe in the 1980s, the advent of satellite high spatial detail imagery in the late 1990s (Landgrebe, 1997), the rapid refinement and anticipated widespread application of lidar in the 2000s (Lefsky et al., 1999b). The development of new aerial films and sensors or the launch of a new satellite has often provided an immediate stimulus to those interested in developing the field (Colwell, 1968). "As in the early days of aerial photography, the launch of Landsat 1 in 1972 resulted in a great burst of exuberant effort as researchers and scientists charged ahead trying to develop the field of digital remote sensing" (Congalton and Green, 1999: p. 8). Much attention was given to digital mapping systems and hardcopy or analogue photogrammetric applications such as digital elevation model (DEM) production, exploring the new spectral data, and automating simple image processing tasks. This new interest paralleled continued strong developments in aerial photography, resolution, quality of color, nonvisual imagery, and digital photogrammetric applications.

Other evidence for the growth of remote sensing has been suggested by the increasing attendance at remote sensing symposia and conferences (Lauer et al., 1997). The number of meetings and attendance rose steadily, by an order of magnitude or more, in the 1960s and 1970s. One of the earliest — perhaps the world's first — technical conference on remote sensing was organized in 1962 by the Environmental Research Institute of Michigan (ERIM), then known as Willow Run Labs; there were 15 presentations and 71 participants. Beginning in the 1970s and continuing throughout the 1980s and 1990s, this ERIM-sponsored series of conferences was joined by several other large remote sensing meetings, including the International Geoscience and Remote Sensing Symposia (IGARSS), and the various symposia hosted by the International Society for Photogrammetry and Remote Sensing (ISPRS). National and international meetings focused on remote sensing were regularly sponsored by Canadian Remote Sensing Society (CRSS), the American Society for Photogrammetry and Remote Sensing (ASPRS), the Remote Sensing Society of the United Kingdom (RSS-UK), and many other organizations. Annually, hundreds of papers were presented and thousands of participants registered at these national and international remote sensing meetings worldwide. In addition there are numerous specialized remote sensing meetings — with particular applications or disciplinary foci — sponsored by governments, industry, and scientific societies. A simultaneous trend to more scientific meetings, often controlled by commercial interests, exists in the wider scientific community.

The number of doctoral dissertations with a remote sensing focus provide another good indicator of the explosive growth of the field of remote sensing in recent decades. In 1981, Merideth reported that a review of the Dissertation Abstracts International (DAI) database yielded over 300 doctoral dissertations on aerial photography, remote sensing, and photogrammetry written (primarily) in the U.S. and Canada between 1945 and 1980. During the early years, most theses were directed at developing aerial photography interpretation techniques, mapping applications, photogrammetric and surveying methods (Figure 1.5). A significant portion, perhaps 25%, were devoted to vegetation topics and forestry issues of remote sensing. A quantitative surge occurred in the 1970s, with many theses aimed at Landsat and satellite remote sensing applications, but with continued strong interest in digital mapping, aerial photointerpretation, and analog photogrammetric studies (e.g., aerial triangulation, DEM generation, LSP mensuration, orthophoto, and other photogrammetric applications).

A recent search of the web-based ProQuest Dissertations Abstract (a descendant of the DAI database) using similar (but not identical) keywords to those used by Merideth (1981) revealed that more than 900 doctoral dissertations have been completed on remote sensing and related topics in the period 1981–1999 (Figure 1.5). Even allowing for substantial growth in university graduate program enrolments throughout the 1980s and into the 1990s, research funding trends, new sensors, the computer revolution, and so on, the large increase in dissertation research between the 1945–1980 period surveyed by Merideth (1981) and during the last 20 years is no doubt related to the real growth of remote sensing as a discipline. Interest by the scientific community in remote sensing methods and applications continues to multiply. In recent years, the focus has been increasingly on the development and methodology of digital remote sensing applications in a wide variety of settings, and at virtually every temporal and spatial scale imaginable. For example, Wagner (1998) listed 490 international science and technology agreements, involving 76

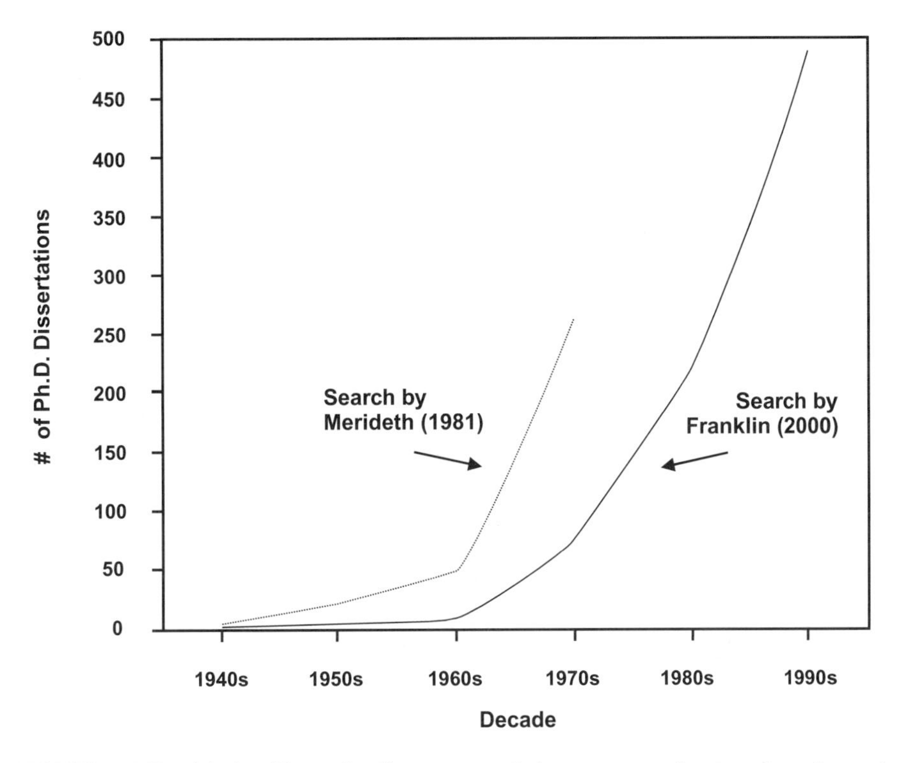

FIGURE 1.5 Empirical evidence for S-curve growth in remote sensing based on doctoral dissertation research in the ProQuest database (a descendent of UMI Dissertations Abstracts International). Activity in remote sensing research would appear to suggest that, on a decadel basis, the field has grown exponentially to the year 2000. More than half the total number of dissertations in remote sensing since the 1940s were completed in the 1990s. Two separate searches are illustrated; while the trend appears consistent, the differences in numbers of dissertations can be attributed to the fact that Merideth (1981) included many photogrammetric terms excluded from the search conducted for this book.

different countries and 32 active satellites, covering the sharing of data and infrastructure in applications as diverse as wildlife tracking (across borders), mapping, water management, natural disasters, monitoring and pollution, and climate change initiatives. Another revolution in remote sensing has begun that seems designed to propel remote sensing into the third and, perhaps, fourth stages of growth as a scientific discipline (Jensen, 2000).

Remote sensing has always been a fascinating technology to produce pretty pictures with a computer, and to many spatial analysts around the world this capability will continue to be important. The public interest has grown enormously; it is reasonably common today to see a satellite image illustrating a magazine or newspaper article, to see remote sensing imagery in films and television, without much particular explanation. People are capable of interpreting such imagery visually, and it seems as natural as looking at a map or photograph. Kirman (2000) found that Canadian public school children in grades 5 and 6 (8 to 12 years old), working in small teams, could find houses, roads, rivers, cars, bridges, and vegetation in planted fields in fine-beam-mode Radarsat imagery with a high degree of accuracy (mean class scores were 76 to 86%) with only a few hours of instruction; earlier work showed that grade 6 (10 to 12 years old) children could work with Landsat digital data (Kirman and Jackson, 1993). It has not always been apparent — nor necessary to understand — that those pictures were rooted firmly in the physics of energy-terrain interactions (Olson and Weber, 2000).

It seems safe to declare, as has Anger (1999), that the remote sensing days of producing only pretty pictures are behind us. A new burst of activity generated by a series of investments in international spaceborne remote sensing (Glackin, 1998) and airborne sensor technology (Anger, 1999) is beginning to accelerate, and the emphasis is squarely on the scientific nature of the data. The quality and resolution of satellite sensor data — already outstanding in the established IRS, SPOT, and Landsat series of satellites — will increase dramatically. Large, complex, expensive civil government and military systems are being replaced or augmented by smaller (Robertson and Cvetkovic, 1991), cheaper, commercially operated systems (Fritz, 1996). Airborne systems are increasingly reliable, cost-effective, and available; worldwide, Anger noted in 1999 that there were more than 30 airborne multispectral and hyperspectral data providers. Less than 10 years previously, only a handful were in business. In the U.S., at least 15 hyperspectral data providers were operational in the year 2000 (Ustin and Trabucco, 2000). In terms of airborne laser (*l*ight *a*mplification by *s*timulated *e*mission of *r*adiation) scanning, there are now more than 40 different data suppliers in the world. These data, when coupled with other sensors (especially digital frame cameras or videographic sensors) are expected, ultimately, to replace photogrammetry as it is currently understood and implemented (Baltsavias, 1999).

These developments, in turn, are leading to an emphasis on remote sensing end-user products. To be driven by client needs requires information products, and this requires new developments in the supporting infrastructures — the graphical user interfaces, image processing software, image compression tools, image browser software, data conversion tools, spectral libraries, standardized output and data products, fast and responsive spatial data architectures, and comprehensive geospatial systems. There are new demands and opportunities for trained applications specialists and image analysts who can understand and effectively use the data and who can view remote sensing from an end-to-end systems perspective — data source to information products. Perhaps equally important, there is a need for those in the user community to understand the technology (Glackin, 1998).

User Adoption of Remote Sensing

In the early days of remote sensing, and continuing even today, many researchers were interested primarily in finding problems to which existing and new remote sensing data, methods, or theories could be applied. The field grew by leaps and bounds as new theories (e.g., Tucker, 1979; Sellers, 1995) and methods (e.g., Landgrebe, 1983) were devised based on the huge volumes of new data and experiences that were generated by each new deployment. Now, remote sensing data and methods have become increasingly complex and varied, and fortunately, increasingly reliable in those areas in which resolvable problems were found. But remote sensing, with the possible exception of meteorological applications (Cracknell, 1998), has not yet succeeded in the practical world — the world of applications.

Interestingly, two contrasting situations appear to have coexisted, perhaps contributing to the creation of no small amount of confusion about the new technology:

1. There appeared to be real differences between problems that were found to be resolvable by remote sensing and the problems that people actually needed to have solved, and
2. The problems that were identified and which have preoccupied the remote sensing community over the last 25 years, and more, have usually turned out to be less simple and less easily resolved than originally thought.

The first is a user-driven perspective; the second is a producer-driven perspective. Generally speaking, progress for users and producers is very noticeable, even dramatic in some areas; much of the material in this book can be seen as an attempt to consider the use of remote sensing from the user's perspective, embodied in the simple questions that forest management has posed: *Can remote sensing provide the necessary information at acceptable cost?* In some ways, the discussion has not been as productive as many would have liked; more along the lines of: *This is what can be done with remote sensing — is it of any value to you?*

On the other hand, the producer-driven perspective has not been without its own set of challenges. Many of the same problems first noticed with the imagery obtained from above have remained. Geometric correction, radiometric correction, atmospheric and topographic influences, automated information extraction, classification, multitemporal image analysis, image mapping, even obtaining a cloud-free image — all have turned out to be much more difficult than originally supposed. This was particularly noticeable in developing applications — enthusiastic scientists observed correlations with a wide range of variables, but turning these into predictive tools for mapping and estimation was (and is) not straightforward. This might be a defining characteristic of remote sensing research … much tougher than it looks!

Much of the difficulty originates in the fundamental measurement unit in digital remote sensing — the picture element, or *pixel*. "In remote sensing, the pixel is much more complicated than one might, at first sight, imagine" (Cracknell, 1998: p. 2045). Twenty years ago, Townshend (1981a: p. 33) defined spatial resolution "as the minimum size of objects on that ground that can be separately distinguished or measured," but then added that "spatial resolution turns out to be a much more complex topic than our initial intuitive definition suggests." Ten years ago, scientists could recall with some amusement the ease with which they thought some of the early problems would be resolved. For example, referring to the problem of modeling radiative transfer using imagery and DEMs, essentially a first-principles issue in remote sensing, Woodham (1989: p. 5) wrote: "You know that problem we've been working on for the last 10 years? Well, it's a lot more difficult than we thought!"

Thus, remote sensing has not yet reached a state of maturity; too much research and development remains to be addressed. Potential remote sensing applications can be found in virtually all natural science disciplines and related engineering disciplines, and increasingly in the social sciences, but real-world operational examples of remote sensing remain relatively rare. In the forestry application, what can be accomplished operationally with the existing stage of development of the field of remote sensing? A minimal list of most-likely, near-future operational forestry remote sensing applications could include (Wynne and Carter, 1997):

1. Forest covertype characterization,
2. Determination of forest stand conditions and forest health,
3. Site characterization, and
4. Fire monitoring.

A few years earlier, in 1991, Rajan presented five uses of remote sensing that were considered operational in Asian tropical forest management:

1. Detection of deforestation,
2. Forest covertype mapping,
3. General assessment of volume and cutting rates,
4. Forest stress, and
5. Fire monitoring.

In India, Raa et al. (1997) also provided a list of five uses of remote sensing; but in their view, all required improved data sets and methods before they could be declared fully operational:

1. Plantation inventory and monitoring,
2. Timber volume estimation,
3. Species identification,
4. Estimation of biomass and productivity, and
5. Biodiversity monitoring.

And finally, in Oregon, Cohen et al. (1996b), focusing not only on the applications but on the fundamental concepts of digital remote sensing, listed three remote sensing applications in forestry that could be considered operational:

1. Mapping forest cover,
2. Measuring and monitoring structure, function, and composition of vegetation,
3. Detecting change in these conditions over time.

The common elements in these lists suggest that remote sensing may play a critical role in forest management in many different forest settings around the world — operational forest covertype mapping, forest structure and change analysis, and forest inventory assessment. A few additional applications appear on the threshold of operational status — notably, landscape structure modeling, defoliation monitoring, and biochemical/biophysical forest inventory (all discussed in this book). It is expected that rather quickly, in the near future in fact, the appropriate role of remote sensing expressed as a complete set of operational remote sensing applications in forestry, will become increasingly apparent (Wynne and Oderwald, 1998).

Current State of the Technological Infrastructure and Applications

Very few remote sensing applications are conducted in isolation, by a single person, without access to prior history or experience. Users of remote sensing operate within an infrastructure, a framework, an environment; with careful planning, the infrastructure can be designed specifically to accomplish successful remote sensing in support of sustainable forest management goals and objectives. The important questions for a forest manager contemplating remote sensing, and for a remote sensing specialist contemplating the forestry applications are (1) can remote sensing data be converted to the information that is needed to understand and manage the forest resources of the planet, and (2) what are the components of a well-designed remote sensing infrastructure to support sustainable forest management? At one end of the spectrum, the end-user would buy raw data from a receiving station and perform all data processing and information extraction in-house. At the other end, a number of intermediate informational products could be generated by data providers or value-added consultants. The end-user would then buy only selected information products. The level of investment in terms of technology, infrastructure, training, and knowledge for the end user is vastly different under these different scenarios. Is there a preferred way to operate remote sensing forestry applications? The issue has not yet been resolved.

One can view the remote sensing infrastructure in terms of both a technological and an organizational infrastructure. An organizational infrastructure is required as the umbrella under which successful remote sensing applications are conducted. An umbrella structure with this wider meaning would perhaps encompass the entire network that is necessary to support a remote sensing facility or laboratory; a remote sensing forestry workstation. Individual components of this workstation or network could include the capital and personnel (and their experience) issues that comprise the institutional response to the increasing importance of remote sensing and GIS in environmental management. In other words: how much support for remote sensing can be expected from the institutions in which the remote sensing work is being conducted? Organizational infrastructure issues, such as spatial data standards and accuracy, provision of training materials and opportunities, and documentation and distribution of test datasets, are important components of the overarching structures under which specific remote sensing activities occur.

Such issues typically appear to lie beyond the immediate reach of most users of the technology — beyond the remote sensing workstation environment. Progress in dealing with these issues will no doubt come from successful industry consortia, universities, and governmental initiatives, supplemented and encouraged by applications specialists and resource managers trying to implement remote sensing in

TABLE 1.2
List of Canadian Remote Sensing Advisory Groups that Have Helped Create an Increasing Emphasis on Applied Remote Sensing and the Development of the Technological Approach to Environmental and Management Issues

1. Agriculture and Land Use
2. Disaster Monitoring
3. Environment
4. Forestry
5. Geoinformatics Technology
6. Geoscience
7. Hydrology
8. Mapping

support of forest management. One example is provided by the list of working groups in the Canadian Advisory Council on Remote Sensing (CACRS), a group comprised of industry, university, and government representatives formed to help advise the Canada Centre for Remote Sensing on the critical issues for funding research (Table 1.2). None of the issues considered by these groups is basic research; all are aimed at practical applications of remote sensing, bringing the technology to bear on real-world problems, applied research issues. The intention is to ensure that the individual remote sensing workstation is supported with progress in these larger issues.

Another specific example in which the larger infrastructure issues are addressed in the U.S. was presented by Estes and Star (1997). The focus in this initiative is on the integration of remote sensing with the enormous resources of data and methods in the realm of geographical information systems. They proceed by recognizing a key fact of life in the existence of both remote sensing and GIS analysts; these two technologies need to talk to each other, find ways of synergy and complementarity, and work more effectively in tandem. Research priorities to improve remote sensing and GIS integration included three broad categories (Table 1.3):

1. Science and Technology Advancement,
2. Improved Understanding, and
3. Infrastructure Development.

The list covers only one component of the work in the National Center for Geographic Information and Analysis (NCGIA), but it is clear that even here there is a preponderance of practical, applied research issues. The focus is now on getting the technology into the hands of the users such that benefits can be immediately derived. It is understood that some of the benefits cannot be derived without improvements in the larger infrastructure. A second list, reinforcing this view and the needs expressed, outlines the research priorities in the University Consortia for GIS (Table 1.3).

The organizational infrastructure is dependent on a link between the research and practice of remote sensing. Numerous attempts have been made and are ongoing

TABLE 1.3
Broad Research Priorities in Remote Sensing GIS Integration

Remote Sensing and GIS Integration (Estes and Star, 1997)

1. Science and Technology Advancement
   - Advanced Feature Extraction
   - Spatial Analysis and Modeling
   - Visualization
   - Lineage or Heritage Tracking
2. Improved Understanding
   - Education and Training
   - Data Format and Structure Conversion
   - Spatial Information Management
   - Error and Accuracy
   - Scale
   - Time
3. Infrastructure Development
   - Standards
   - Spatial Data Catalog
   - Test Datasets

UCGIS (Budge, 1999)

- Spatial Data Acquisition and Integration
- Distributed Computing
- Extensions to Geographic Representation
- Cognition of Geographic Information
- Interoperability of Geographic Information
- Scale
- Spatial Analysis in a GIS Environment
- The Future of Spatial Information Infrastructure
- Uncertainty in Spatial Data and GIS-Based Analysis
- GIS and Security

to bridge the research and practicing communities. In one recent example, Oderwald and Wynne (1998) described a consortium of image suppliers, researchers, and potential users. The intent is to provide those engaged in resource assessment and inventory with information on the many potential sources of spatial data, including aerial photographs, digital elevation models, orthophotos, and satellite images. The goals of the consortium are

- To determine what level of information can be derived given current remote sensing data and methods,
- Develop methods for obtaining more specific information using the improved imagery becoming available,
- Disseminate state of the art methods to resource managers through training sessions, research reports, and publications.

The problem, as they see it, is that very little use of any of these information sources is made; the use of these data and the many types of newer remote sensing data in actual forest management appears minimal. Apparently, resource management continues to rely almost exclusively on field observations, and analogues available through aerial photography. What seemed to some, as remote sensing gathered momentum in the 1970s, 1980s, and 1990s, to be the logical decision to convert to the digital environment and these new imagery, has turned out to be less convincing, less inevitable, and less realistic. Despite growing recognition of the tremendous potential, and an enormous investment in remote sensing and other geospatial information technologies (Bergen et al., 2000), the vast majority of resource assessment and inventory is still performed on the ground (Oderwald and Wynne, 1998).

This situation will change, and in fact, is already changing. There appears to be a new vitality in the remote sensing field that is based on a number of powerful trends: the widespread ability to process imagery on the desktop, the long-term and historical availability of satellite data, increased access to new types of airborne and satellite imagery, and the enormous growth in GIS and spatial data in general. Landsat, for example, has been continuously acquiring imagery since 1972, when most of the people now living on the planet were not yet born. New data are continually coming onstream. Growth in computer technology — memory, storage, speed — is nearly exponential. The engineering marvels that comprise modern-day space technology are nothing short of astonishing. For example, the Landsat 7 ETM+, launched in 1999, produces approximately 3.8 gigabits of data for each scene. The ground component can collect and process 250 Landsat scenes per day, and deliver at least 100 of the scenes to users each day, radiometrically corrected and geometrically located on the Earth to within 250 meters. Is there a demand for such awesome data rates, data volumes, and precision? In some ways, remote sensing continues to be driven by technology; in others, the driver is the increasing demand for better data and methods in a wide range of disciplines.

One outcome of this activity is the construction and maintenance of a technological infrastructure, the foundation upon which all successful remote sensing applications are built. In this book, the components of the technological infrastructure are considered to be comprised of those aspects of the technology required to support the conversion of remote sensing data to remote sensing information by application specialists for input to management, planning, and operational activities. For example, decisions may be needed on the most appropriate way in which to use remote sensing to solve specific forest management problems, such as the mapping and monitoring of forest biodiversity or landscape fragmentation. Therefore, appropriate infrastructure must be in place not only to provide insight and examples to enable remote sensing imagery to be used optimally, but also to suggest how best to acquire imagery with the appropriate characteristics: where the points of integration with GIS technology occur, what should be the role of models and field data, and a host of relevant questions related to radiometric and geometric processing of imagery. It is thought that the era of complete information product provision by external entities is still some way off; more forestry organizations will need to become even more proficient in handling remote sensing data because the benefits of the data source now justify the required investment.

The remote sensing infrastructure must allow the trade-offs in acquiring imagery with various combinations of image resolutions. The effect of scale must be addressed for a given application. A complete remote sensing technological infrastructure will have tools that enable the user to test which are the algorithms that work, when do they break down, when are new algorithms needed, where is the focus for new software development. What are the key components that must be in place to get started in remote sensing? What questions can be answered with the technology available now? What questions can be answered with a little more research? And, what are the big research issues still outstanding?

In short, a remote sensing technological infrastructure is necessary to support users in their search for methods and explanations in how best to:

1. Acquire and prepare a remote sensing data set for a forestry application,
2. Understand the implications of not preparing the data set well,
3. Evaluate the selection and performance of certain recommended processes, and
4. Consider options when the image processing system available does not support the best procedures.

In other words, the infrastructure is required to present clearly the critical decision-points in a remote sensing acquisition and analysis activity.

Acquisition of data, preprocessing systems, image analysis functionality, geographical information systems, and forest ecological models can all be considered key pieces of the hardware and software infrastructure — the immediate environment surrounding and supporting successful forestry remote sensing activities. One of the most important but often neglected pieces of technological infrastructure is the image preprocessing system which must support the correction of imagery for variability not related to the target (atmosphere, view angles, etc.) and must provide tasks that can be used to relate imagery to other georeferenced data.

Figure 1.6 shows graphically the components of these two structures — the technological infrastructure and the organizational infrastructure — which together make up the environment in which remote sensing for sustainable forest management can be optimized. This figure highlights the focus of the material presented in Chapters 3, 4, and 5 of the book. These three chapters will focus on the technological infrastructure necessary in remote sensing. It is not possible to review all available ways of preparing to conduct forestry applications of remote sensing. The idea is to consider the development of various components of the infrastructure that influence individual user's decisions and methods, and to summarize the collective experience of the remote sensing community. In more than 30 years of digital remote sensing applications, it is possible to offer suggestions and guidance into what works, what will be enough to get by, when should the line be drawn, and what investment might be necessary. At the same time, this review can provide insight into the critical elements of organizational infrastructure that overlie and encompass individual efforts in remote sensing applications.

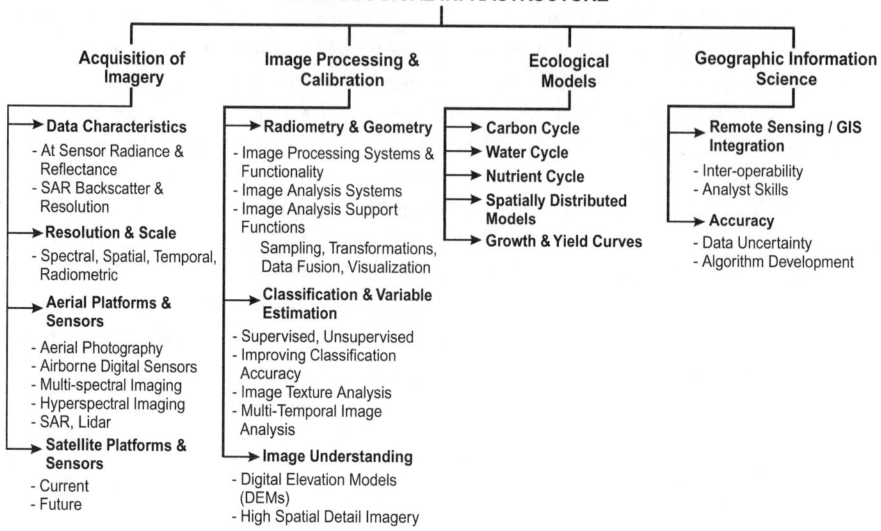

FIGURE 1.6 Components of a remote sensing infrastructure. Four broad categories are shown suggesting that remote sensing applications in forestry require the combined characteristics and continued development, or synergy, in remote sensing image acquisition and analysis, ecological modeling, and geographical information science.

Three Views of Remote Sensing in Forest Management

To many people — not only to remote sensing scientists, the producers of the technology, but to many familiar with the field and its spectacular growth — the potential role of remote sensing as an information resource to support sustainable forest management appears enormous and immediate based largely on two facts:

1. Sustainable forest management requires synoptic and repetitive biophysical and biochemical vegetation data for large geographic areas over long periods of time, and
2. Remote sensing is the only way to acquire such data.

There is no other way! No other technology can possibly provide the data required at reasonable cost, accuracy, and effort. No other technology will promote the integration and synergy of analysis for the wide-ranging data types required to address contemporary forestry questions (Oderwald and Wynne, 2000). Based on these views, and an understanding of the extent to which information about the forest continues to limit a sustainable approach, positive scientific assessments reviewing the beneficial role of remote sensing in forest management, forest science, and ecological applications are not hard to find in the literature. For example (referring to the contribution of aerial and satellite remote sensing to forest management):

1. "Synoptic, timely information, which can be provided only with satellite data, is needed to support local, national, and global decision-makers in the crucial planning efforts designed to preserve the habitability of this planet for generations to come" (Iverson et al., 1989b: p. 140);
2. "... remote sensing and GIS ... help give us a better understanding of forest systems, how they function, and how to manage them with a holistic view" (Cohen et al., 1996b: p. 432);
3. "A strong correlation between stand development and satellite observations would make a valuable contribution to stand development forecasting and thus enable the improvement of forest management strategies" (Ahern et al., 1991: p. 388).
4. "... two time series of satellite images showed the reduction and rapid fragmentation of the giant panda's habitat in China. More than any other factor, it was this perspective provided by satellite imagery that changed the ... manager's views about the main threats to panda survival" (Mackinnon and de Wulf, 1994: p. 130).

The idea, that remote sensing can make a substantial contribution to forest management flows from the unique characteristics that remote sensing data provide — synoptic, repetitive, quantitative, and spatially explicit capabilities. When these features of remote sensing data are considered in light of the full range of remote sensing methods (including satellite, aerial, and field-based methods), there appears to be strong support for the suggestion that remote sensing can satisfy at least some of the projected sustainable forest management information needs.

Immediately, some early and not-so-subtle difficulties are encountered; some expected, others simply nasty surprises. The data are poor in quality, or the methods are just too complex for practical use; the data are not available in the appropriate formats; there are data registration errors; there are clouds and cloud shadows; the time of year is wrong; noise effects; and data volumes are enormous. This is nothing like the way we used to do things. Even the ordering procedure is incomprehensible. Nothing ever works right, there is no hope! Remote sensing (particularly satellite remote sensing) is not a viable information resource in the management of the forest because there are always better and less expensive ways of obtaining the necessary information, particularly when the idea of necessary information is limited and narrowly defined (Holmgren and Thuresson, 1998). There have been a number of scientific assessments noting that results of remote sensing applied to forestry planning and management have not always met expectations (recall that when using the term remote sensing, many refer primarily to satellite remote sensing and do not refer to the contribution of aerial remote sensing, particularly photography):

1. "... application of remote-sensing techniques to routine forest management does not seem feasible in the near future" (Battaglia and Sands, 1998: p. 24);
2. "... the increased resolution of the TM sensor (over the MSS) has not resulted in forest classifications of sufficient detail (i.e., Anderson Level

III [Anderson et al., 1976]) to warrant practical use of this technology by forestland managers" (Wolter et al., 1995: 1129);

3. "It is, perhaps, time to draw the conclusion that current satellite sensors are not in general suitable for forestry planning" (Holmgren and Thuresson, 1998: p. 90).

Is it a problem of data, of methods, of applications, of all three; or of other factors, such as an overreliance on a single source of information, not yet considered? Has there been a misplaced investment by forest managers and scientists (the users of the technology) in the research agenda of the satellite remote sensing aerospace community (the producers of the technology) as some (Holmgren and Thuresson, 1998) have recently insisted?

Often the mythology surrounding a technology like remote sensing acquires a status the field perhaps no longer deserves. In 1989, Aronoff suggested that dispelling some of the more popular (and negative) myths surrounding remote sensing would result in more critical thinking about needs, and less anecdotal justification when deciding on the use of remote sensing. At that point in time, concerns revolved around the spatial resolution and accuracy (thought to be inadequate), cost and complexity (thought to be too expensive and difficult to use), the experimental nature of remote sensing (thought to be unreliable), and data availability (thought to be poor). Such concerns mirrored those documented by Yatabe and Fabbri (1986) to explain the reluctance of many geologists to use remote sensing during the 1980s. More than a decade after Aronoff (1989) presented cogent arguments that effectively demolished some of these more tenacious remote sensing myths, one safe conclusion is some myths die hard! At a minimum, remote sensing technology should be considered, or reconsidered in light of existing capability, synergistic technologies, and evolving needs.

Certainly, satellite imagery cannot be expected to replace aerial photography by providing the same type of forestry information (Roller, 2000). What is the role for remote sensing in providing the required data, when often, what *is* required is not well defined? In a similar context it has been noted that some of the toughest challenges are those that are poorly specified or stubbornly and inherently unanswerable (English and Dale, 1999). Problem definition and reformulation of problems on the fly are all too common and necessary, both in remote sensing (Lunetta, 1999) and in forest management (Erdle and Sullivan, 1998). As well, forest managers may be willing to accept answers which are perhaps less quantitative and less immediate than an investment in remote sensing would provide. Perhaps it *is* better, safer really, to continue using methods that are known to work and which have fixed budgets. Field work goes on, aerial photography are ordered, stands are mapped, the GIS is accessed. The GIS! Now that was an expensive upgrade!

It sometimes appears that there was an assumption that if something could be seen in a remote sensing image, there would shortly be an automated procedure that could translate that vision into products, results, maps, or even fundamentally different changes in approaches to forest resources. But it is rare to convert data to information at a glance; instead, a whole series of learning processes must be initiated, and even then there are no guarantees that the methods, data, or personnel can complete the task.

To many familiar with the challenges of forest management, the potential of remote sensing is not in doubt (Ahern, 1992; Wynne et al., 2000); it is the current role of remote sensing in sustainable forest management that would appear, at best, ill-defined (Oderwald and Wynne, 1998). It would be difficult to invent, develop, and test a technology such as remote sensing and have it applied to a complex problem, such as forest planning and management, in a few decades, without some difficulties encountered along the way. However, it is the context of the issue that must be considered. Earlier failure to meet expectation was often because of inappropriate methods used by inexperienced or new users (naturally, almost everyone was a new user) and applied to the wrong (but available) data set, at the wrong scale, for the given (sometimes poorly defined) application. Upon close inspection, remote sensing data — available globally and at almost any location of the globe, at anything from 1 kilometer to less than 1 meter pixel resolution, and at variable time scales and spatial extents — do appear more promising, but real benefits are intangible, perhaps even out of reach — where to begin?

The appropriate starting position is to understand remote sensing data, and then develop an appreciation of the methods and applications that are possible. Unfortunately, understanding remote sensing data — and the assumptions and limitations in using the data — is not widespread in the user communities (Duggin and Robinove, 1990; Olson and Weber, 2000). As has been noted, remote sensing has expanded enormously, even exponentially in only a few years, and it is not the only field to have done so. It is difficult to keep abreast of even one field in today's explosive technological society, and to many disciplinary scientists, managers, operations, and practicing foresters, remote sensing represents a new field entirely. Acquiring new (remote sensing) skills and keeping them sharp in the face of generally poorly evaluated options (Townshend and Justice, 1981) can be overwhelming. Expectations can increase rapidly in the technological pressure cooker; people are looking for a technological fix, and can sometimes believe they have found one without applying due diligence. For example, in the rush to develop verifiable national biomass estimates with real economic implications for countries, remote sensing is seen as a quick fix, maybe even a last hope.

Compounding the issue is the fact that, historically, it appears the remote sensing community has sometimes not understood the user's problem well enough to offer proper advice (Landgrebe, 1978b). Users and developers of technology have often suffered from a lack of communication, but it is never too late to attempt to bridge these differences. Now might be a reasonable time to drop altogether the distinctions between the two, and focus on the fact that users and producers of a technology such as remote sensing are part of the same team, driven by the same needs, and should work together to understand the perspective and challenges of the complete system.

A related problem appears to be that sustainable forest management is only uncertainly, vaguely, and even negatively understood by many (Yafee, 1999), and its principles are only now beginning to be implemented at the local level (Kohm and Franklin, 1997); perhaps many managers continue operating exclusively at the local scale, and have not yet expanded their information needs to include the ecosystem and landscape scales to which many remote sensing techniques are currently well suited. At the other end of the spectrum are the policy and social problems. Five remote sensing surveys showing alarming reductions of forest extent in the Philippines from 1965 to 1986 were consistently overridden by political concerns. Despite significant investments in remote sensing surveys of the extent and condition of forest cover in the Philippines, no appreciable effect on forest management from remote sensing research and applications can be discerned (Kummer, 1992).

Constructive criticism is a powerful force in preventing error, and criticism from those expected to benefit from remote sensing has been extremely valuable in developing the field. There is a delicate interplay at work here; too often the users have not understood the technology; too often remote sensing has been conducted without reference to the end-user. This latter situation has resulted in an optimistic bias that has caused all kinds of damage in both the forestry and remote sensing communities (Wynne et al., 2000; Olson and Weber, 2000). At least since the inception of Landsat as a potential forest monitoring system, remote sensing has been marketed relentlessly as a forestry tool, even to the point of overselling the technology.

A third view of remote sensing in forest management is that, while there is a lack of experience using remote sensing (and GIS) in an integrated way in the whole process of forest planning and management, the planning and management of the world's forests cannot be done without these technologies (Martinez et al., 1996). This view, which may well represent the majority view of remote sensing scientists and resource management professionals (Urban, 1993; Congalton et al., 1993; Green, 1999), has emerged based on the balance between the increasing need for information by the forest managers and the power of the technological approach to generate that information. In this view of remote sensing, it is understood that in managing the forests the professionals will need access to manual and digital remote sensing from field, air, and spaceborne platforms, and in concert with a whole host of enabling technologies and other data such as Global Positioning Systems (GPS) and GIS (Wessman and Nel, 1993; Michener and Houhoulis, 1996; Cohen et al., 1996b). A common thread in planning and management is monitoring, and it has long been recognized that, in forestry and vegetative landcover at least, remote sensing is an excellent monitoring tool (Tucker, 1979; Dottavio and Williams, 1983; Hayes and Cracknell, 1987; Stoms and Estes, 1993; Trichon et al., 1999; Lunetta and Elvidge, 1999; Stoms and Hargrove, 2000, and many others).

In Sweden, Hansen et al. (1998) outlined the relatively simple procedures which are common in monitoring many managed forest landscapes:

1. Inventory is conducted on all lands at fixed intervals (typically, every 10 years);
2. Inventory methods are heavily oriented toward tree growth but are being adapted to indentify unique ecological areas, threatened or endangered species, and areas that need restoration;
3. A combination of field (nonspatial) and survey (spatial) mapping methods are used to document and track various characteristics of the ecosystem.

The real dilemma here, as elsewhere in the world, is how much investment in measuring and monitoring can forest managers afford to undertake. Where to deploy these scarce monitoring and measuring resources? Where, how much, and how to, measure and monitor — these are critical questions (Fletcher et al., 1998).

Remote sensing is not a panacea for forest management at any scale (Wessman and Nel, 1993). Remote sensing will certainly not provide all the required answers to forest management about monitoring and measuring; nor is remote sensing incapable of providing any of the answers needed to move forward. Instead, remote sensing may have already earned a place as one of the single most important sources of information that those responsible for forest management can access, and remote sensing appears poised to become an even greater asset to those who can understand and use this new tool. A range of monitoring and measuring tools will be required. One of the greatest challenges is to integrate remote sensing with field observations, and to identify the appropriate role for remote sensing as a tool to handle particular forest management problems and opportunities. Only through careful, documented applications of remote sensing methods, data, and results, will the appropriate role of remote sensing as an information resource in forestry become established and recognized by managers.

ORGANIZATION OF THE BOOK

OVERVIEW

A literature review and example applications are presented to illustrate to natural resource managers the information potential of remote sensing at multiple geographic scales, across time, and across disciplines and issues. The focus is on the near term and medium spatial scales because this is where remote sensing has made, and will continue to make, its greatest impact in forest management. Some may argue that remote sensing will ultimately have its greatest impact when global remote sensing provides the inputs needed to balance the Earth's carbon budget, and to provide much needed understanding of how local activities influence global cycles. But this book was designed to convey what is available to the forest manager, operating at the scale of a few hundred or thousand hectares over seasons, years, and perhaps a few decades, to help understand and manage the forest resource. Here, the concern is with the spatial and temporal scales at which forest management appears most likely to change; over one to several forest drainages or watersheds, and over one complete forest rotation (up to 100 years, often much shorter).

Operational or near-operational examples of forest change detection, forest defoliation monitoring, forest classification, and forest growth modeling, are included in the later chapters. It is felt that the wide array of sustainable forest management questions and feedback is demonstrated best through the use of examples, which are reasonably well documented and understood, and are hopefully brought forward to the forest management and remote sensing communities in an accessible format in this book. A synopsis of the current understanding of sustainable forest management, and its subsequent goals and objectives, is presented (Chapter 2). This is no simple task as the range of views and practices currently discussed under sustainable forest management is already enormous, and still growing. However, from the vantage point of first appreciating the information needs, and second understanding the technology, the practical applications of remote sensing in forestry can be identified and methods discussed and updated (Sohlberg and Sokolov, 1986; Eden and Parry, 1986; Howard, 1991). Interspersed throughout the presentation are comments on data sources and methods in remote sensing, but the reader is referred to sources of more complete listings than those presented here (e.g., Jensen, 2000). The review will cover the update of current forest classification and inventory conditions and the analysis of within-forest stand and ecosystem variability by remote sensing through various forms of modeling.

CHAPTER SUMMARIES

Chapter 1: Introduction

This treatment of remote sensing for sustainable forest management begins by reviewing working definitions of remote sensing and sustainable forest management. For different reasons, perhaps by their very natures, these two fields are not well defined. The focus in later chapters is on specific aspects of interest to those in forest management employing a technological approach. By way of introduction to the field and this book some issues are summarized which forest managers must confront in their search for tools and approaches which will, when used appropriately, facilitate continued human and natural world coexistence. It is noted that too often a technological approach is developed in a vacuum from the true potential users; and a gap exists between those who establish and develop the new approaches, and those who need to use a new approach to resolve a new problem or an old, persistent problem.

Chapter 2: Sustainable Forest Management

As more goals and objectives are identified under sustainable forest management and ecosystem-based management approaches, new knowledge is required. The complexity of forest management, like virtually every aspect of human life, has increased, not only due to changing demands on forests but because of rapid technological developments in many areas, including management systems. A new approach is to consider criteria and indicators of sustainable forestry which incorporate social, cultural, environmental, and economic aspects of management. Implicit is an ecosystem perspective; with this comes an expanded monitoring program which considers aspects of ecosystem processes and patterns not previously examined in detail. New monitoring activities, together with the need for detailed decision support systems combined with geographic information systems and models, add to the information burden. Remote sensing methods need to be adapted to the forest science and ecological issues that are emerging through the sustainable forest management approach. Technology does not drive sustainable forest management, but questions about fundamental processes do. A synergistic combination of remote sensing technology, field observations, and human creativity and collaboration is perhaps the only way to answer certain questions at the local to regional scale, and certainly at the global scale.

Chapter 3: Acquisition of Imagery

The massive technological revolution in spatial data — which encompasses GIS, remote sensing, GPS, and computer modeling — has become increasingly formidable, complete with newly developed specialized languages and a context that only the initiated can understand and master. Operational remote sensing in sustainable forest management requires understanding of, and access to, the appropriate remote sensing data for the specific problem at hand. Data acquisition issues relevant in aerial remote sensing missions and in ordering satellite imagery are reviewed.

Chapter 4: Image Calibration and Processing

A fully functional image processing system (calibration, correction, algorithms for analysis, links to GIS and ancillary data, field data input, models, map output, and accuracy assessment algorithms) is a prerequisite for the use of remote sensing images. Even though remote sensing systems and computers are becoming more complex, with steep learning curves and initial investment, a significant level of remote sensing image analysis can be accomplished locally as a result of the continued widespread availability of good data and reasonably inexpensive introductory image processing computer systems. Perhaps, in time even most of the analysis and interpretation of remote sensing imagery and image products can be done by forest managers working in teams given this necessary infrastructure and understanding of remote sensing capability. Sections are included on image analysis, training a classifier — unsupervised, supervised, modified, decision rules — selection, application, testing, accuracy assessment, and applying digital remote sensing classifications and models. The growing trend to integrated systems of multiple spatial data sources is introduced.

Chapter 5: Forest Modeling and GIS

Remote sensing can be considered part of the emerging world of geographical information science (GIScience). In many ways, certain key components of a good remote sensing training background are in common with those working in GIScience; the issues typically relate to understanding the limits of the data, the methods, and the science which support remote sensing and other geospatial data. There exists considerable overlap between a remote sensing image processing system and GIS components, and most commercial image processing systems have comprehensive links to GIS; these extend all the way back to the first systems in which remote sensing data were archived. Remote sensing data are increasingly used to update a GIS, and are analyzed with reference to existing GIS data. The most important problems have to do with spatial data infrastructure, spatial data error and uncertainty, and image analysis functionality. Nowhere is this more apparent than in the increasing use and sophistication of ecosystem process models. These models are poised to create new opportunities in forest management. This situation has developed partially because of the emerging synergy between GIS, remote sensing, and several different types of modeling.

Chapter 6: Forest Classification

Classification is one of the most important ways in which image information is extracted and presented for use in forest management. Here, a brief discussion of general purpose landcover classification and forest classification philosophies and class schemes is presented, leading to a review of remote sensing classifications aimed at mapping landcover, structural, and successional forest classes. Different levels of classification hierarchies are described, with the lowest level (highest detail) corresponding to the classification typically used in forest inventory: derived by considering forest stand species composition, density, crown closure, height, and age.

Chapter 7: Forest Structure Estimation

The second major information extraction technique in remote sensing has come to be known as continuous variable estimation. Typically, models are constructed based on field and image data at known locations; these models can be classificatory or, more likely, designed to estimate a continuously varying forest attribute such as crown closure or stand age. The literature is full of seemingly confident and workable remote sensing models; but their use in actual forest management situations remains marginal, partly because of the complexity of the approach and partly because of the generally weak relationships that have been reported in real-world situations. Some options for improving the relationships for certain forest types are suggested based on empirical vegetation indices and other image transforms, spectral unmixing, and indices related to stand structure.

Chapter 8: Forest Change Detection

This chapter deals with the third major set of information products in remote sensing: forest change detection. Here, image classification and image differencing procedures are described with applications in clearcut mapping, partial harvesting, and regeneration surveys based on different sensor packages. Natural disturbances can be analyzed using remote sensing imagery, such as forest damage, defoliation, and fire. The chapter concludes by looking at the change in spatial structure that can be monitored using remote sensing information and landscape metrics. Forest fragmentation and input to wildlife and biodiversity assessments are increasingly in demand as forest management considers larger and larger areas and includes new forest values in management.

Chapter 9: Conclusion

This chapter concludes the book with a review of the technological approach. Issues related to understanding pixels and the multiscale, multiresolution, multitemporal remote sensing concept are introduced with a consideration of remote sensing research issues. The applied nature of the remote sensing research agenda is highlighted.

2 Sustainable Forest Management

> Whatever relationship [people] choose to adopt with [the] environment in the years that lie ahead, science and technology as typified by the development of remote sensing will increasingly be called upon to meet the crisis of choice so clearly embodied in the rate at which we exploit our resources, develop new industrial strategies and seek to protect the quality of life.
>
> — D. S. Macdonald, 1972

DEFINITION OF SUSTAINABLE FOREST MANAGEMENT

Forest management is necessary because of human needs to balance:

1. The flow of values from the forest, and
2. An unimpaired ability to continue providing those values.

In its present form, forest management adopts a position identical to that of any management activity designed to accommodate a large, open system; forest management is comprised of conscious human actions that lead to a goal. Broadly speaking, the goal of forest management has almost always been stated as the continued flow of benefits from forests to satisfy present and future human needs. In some areas, management is needed to ensure the continued existence and future productivity of the forest in any form. In others, forest management is an ancient practice. Thus, in many forests the results of some of the earliest forest management practices have been known for years; in others, they are only now becoming available to be assessed. In the goal, at least, there seems to exist a remarkable degree of consensus.

The most recent innovations in forest management conform to a sustainable forest management approach. A key facet of this approach is the use of new forestry practices that satisfy the expressed desire or goal that forest management succeed in maintaining forest ecosystems in a sustainable condition; that is, that human activities in the forest do not negatively affect the ability of the forest to continue in virtually the same way as before. Obviously, such a goal is highly idealized; for example, the effect of natural climate change, if it could be discriminated from human-induced climate change, cannot yet be predicted with much confidence. The best information has been obtained from paleoenvironmental records (Shugart, 1998). How can the impact of human activities on the forest be predicted? Equally obviously, a great range of scientific opinion can be accommodated within the sustainable forest management approach; for example, the terms themselves are vague and open to interpretation. What is a forest ecosystem? What condition was it originally found in? How can forest condition be measured? What differences between original condition and present or future condition can be accepted under the sustainable ideal?

Sustainable forest management has been defined by the Food and Agriculture Organization (1994a) as a multidisciplinary task, requiring collaboration between government agencies, nongovernmental agencies (NGOs), and, above all, people, especially rural people. It is of concern at local, regional, national, and global scales. The activity of management is presented in terms of the essential management processes. Sustainable forest management therefore involves:

1. Planning the production of wood for commercial purposes, as well as meeting local needs for fuelwood, poles, fodder, and other purposes.
2. The protection or setting aside of areas to be managed as plant or wildlife reserves, or for recreational or environmental purposes.
3. Ensuring that the conversion of forest lands to agriculture and other uses is done in a properly planned and controlled way.
4. Ensuring the regeneration of wastelands and degraded forests, the integration of trees into the farming landscape, and the promotion of agroforestry.

While there may be as many definitions and descriptions of sustainable forest management as there are forests and managers responsible for them, most definitions of sustainable forest management are based on two commonsense, easily understood principles:

1. Sustainable forest management must be based on understanding and management of ecosystem processes and patterns over long time frames and large spatial scales (Boyce and Haney, 1997) and;
2. Sustainable forest management must be based on goals that are social, as well as ecological (Noss, 1999).

These principles are not controversial, rather like clean air and water. Everyone can agree that more understanding of ecosystem process and patterns can lead to better management. From which direction will this understanding emerge? Typically, what is meant by increased understanding is knowledge based on a scientific approach. Some would argue that even more rigorous application of the scientific methods that have helped create the problems that exist in forestry is wrongheaded (Suzuki, 1989). Is science likely to provide only a fragmented view, rather than the holistic view that is needed? Economic, social, and cultural biases are often more important than the use of actual scientific methods in determining whether scientific results and knowledge are used correctly; Behan (1997: p. 414) suggested that "forestry is as much a political enterprise as it is scientific." The interpretation of scientific results in the face of an always-present degree of uncertainty, and the actions suggested by science, are rarely the sole domain of the scientists, but rather are subject to the distorting prism of the human political process. Clearly, Western civilization is the most advanced scientific society in history, and the effectiveness of the linear, positivist, reductionist, specialized scientific method in dealing with complex systems (including forests) is globally recognized. Perhaps what is needed now is greater reliance on the scientific method, not less; more traditional scientific experimentation, not less (Simberloff, 1999); more emphasis on the relations within the system, not less. Welcome developments would be less reliance on anecdotal beliefs, less subjectivity in interpretation, and greater adherence to a rigorous implementation of scientific findings.

Almost everyone can agree that social objectives must be addressed, that people *are* part of the ecosystem (Weyerhaeuser, 1998). This does not mean that immediate human profit or even enlightened economic self-interest can outweigh every other concern, or that nature is simply a "vast supermarket set up by God for the benefit of the human race" (Manguel, 1998: p. 7). At least one clear step has been taken by society away from "such arrogant nonsense" (Manguel, 1998) — away, that is, from exploitation to responsibility, to a form of ecological conscience (Leopold, 1949) in what many have viewed as the ongoing political, spiritual, and economic battle to save the planet. Because people are part of the system does not mean that all continued and even increased human activity is only natural, and is not potentially dangerous. With current and increasing levels of population and human activity, large forests cannot be unmanaged; only conscious decision making by humans will provide for sustainable forest management. For example, the exclusion of humans activities *is* possible. The management prominence of areas in which human activity is excluded can be reduced (Simberloff, 1999). Clearly, the right decision making by humans is required to ensure the sustainable use of resources. How are the right decisions made?

What appears to be the main point of contention is not the philosophy, but the practical directions that flow from the two principles of sustainable forest management; for example, how best to balance human use and preservation, to maintain biodiversity, and to achieve economic benefits are at the heart of the desired goal of sustainable forest management. How best to proceed with obtaining economic benefits in the light of uncertainty, even ignorance, of the true consequences of our actions. How best to consider the needs of current and future generations. The definition of sustainable forest management is less important than what has come to be understood by managers and the public as a sustainable forest management plan (Phillips and Randolph, 1998). These plans are where the answers to how best questions can be found. Will the proposed management procedures:

- Aim to maintain viable populations of native species *in situ*?
- Acknowledge ecological patterns and diversity in terms of the processes and constraints generating them?

- Sustain ecosystem diversity, health, and productivity at different geographic and time scales?
- Be based on a broad, integrative, interdisciplinary approach?
- Include public involvement in planning and decision making?
- Include results of recent scientific research and technology?
- Be adaptive management techniques (including monitoring and evaluation)?
- Include educational programs?
- Involve setting priorities based on societal demands within the constraints of ecosystem patterns and processes?

Even if the proposed management plan is carefully devised with these aims in mind, there can be problems in implementation and in evaluation. Forest management, like many complex environmental management activities (e.g., consider fisheries management or urban planning), remains an imperfect science with a limited history (Hobbs, 1998). Perfecting management through science will continue to be a source of frustration; there will be mistakes, uncertainties, and unpalatable trade-offs. Science is necessary, but not sufficient (Kohm and Franklin, 1997) — broadly speaking, the endeavor is nothing less than humans attempting to understand the planet well enough to coexist sustainably with their environment.

FORESTRY IN CRISIS

There is ample evidence for a global failure by society to practice sustainable forest management, regardless of the management paradigm that is invoked to support human activities in forests (Berlyn and Ashton, 1996; Landsberg and Gower, 1996; Boyce and Haney, 1997; Rousseau, 1998). By some accounts, almost half of Earth's original forest cover is gone, much of it removed within the past 30 years, with only one fifth remaining in large tracts of relatively undisturbed forest — what the World Resources Institute calls *frontier forest* (Bryant, 1997). Forests continue to be degraded, damaged, eliminated, and converted to nonforest use. Why is this? Perhaps the goal of sustainable forest management can seem futile in the face of the long list of problems facing the world and its forests. Arguably, the list of problems is headed by population growth. Population is certainly not the only issue, although it could be argued that population increases underlay virtually every major human crisis or concern, including climate change; also poverty, hunger, debt, overdevelopment, underdevelopment, and political instability — the list of human travails is virtually endless. Over time, many of these challenges can be seen as intricately linked to the central environmental/population problem — can humanity coexist sustainably with the environment?

Rationally, achieving sustainable forest management under a constantly growing human population should be considered impossible (South, 1999). By 2100, the most optimistic scenarios for a stable world population range from 8.5 to 14 billion people, with worst case estimates of more than 100 billion people (United Nations, 1992; Raven and McNeely, 1998). If human populations continue to increase, sustainable forest management is likely to be neither possible nor important. For large numbers of people and the resources that sustain them under a changing climate and continued overpopulation pressure, sustainable forest management would be, perhaps, the least of their concerns. Instead, they would be concerned with finding food, clean water, fuel, and shelter. In the near term, human responses to regional and local population crises can result in species extinction, soil erosion and degradation, desertification, deforestation, loss of biodiversity; the effects can be immediate and devastating. The complete destruction of the world's forests might seem to be a minor problem compared to the starvation and death of millions of the world's poorest people. The link between a healthy forest environment and successful human lives has rarely been made explicit.

Perhaps the most important aspect of the search for sustainable forest management practices in light of possible world population trends is the growing recognition of the scale of the problem. There is a pressing need for solutions, locally, regionally, and globally. Over the next 20 years, the global wood demand is expected to increase by an average of 84 million cubic meters annually; almost doubling current levels of wood consumption (Kimmins, 1997). Where will these resources be obtained? Can they — under any stretch of the imagination — be provided without a continued or even increased rate of degradation of the world's remaining forest resources? In one view, the increased wood demand by growing populations can only be satisfied by increased management for single-use — a massive and immediate investment in forest plantations (Sutton, 1999). Presently, perhaps 10% of global wood demand is satisfied in this way (Kimmins, 1997). Would such an approach be sustainable?

A fundamental concern is the rate at which forestland continues to be converted to other uses (Waring and Running, 1998). Obviously, the conversion of forest land to other uses is driven by human needs, as is human dependence on fossil fuels. More humans, more needs — need for land, need for resources, need for continued development. Both land conversion and fossil fuel use are driving climate change, thought to be the main factor in altering fundamental ecosystem processes (Waring and Running, 1998). It is possible that climate change may create a new source of uncertainty in forest management; our current understanding of ecosystem processes may be undermined. Much of our current understanding of the environment, designed to allow accurate predictions of future states, is based on the assumption of continued growth under reasonably constant climate conditions. The most significant human impact on climate results from the emissions of gases (particularly carbon dioxide, nitrous oxide, methane, chlorofluorocarbons, and ozone) into the atmosphere, but there are numerous other impacts the significance of which are largely unknown (e.g., urban heat, high-altitude aircraft condensation trails). Together, these impacts appear to be responsible for a global temperature increase of about 0.6°C over the past century. A further rise of 1 to 4.5°C is expected by the 2030s unless human impacts are greatly reduced immediately (Canadian Institute of Forestry, 2000). The greatest warming is expected at high latitudes in both hemispheres in the winter months.

Current and past climatic changes have occurred at various rates, with species responding individually across different regional settings (Schoonmaker, 1998). For example, the impact of climate change has been examined on a national scale by many countries. In Canada, potential (positive and negative) impacts of climate change on trees and forests include (Canadian Institute of Forestry, 2000):

1. A northward-migrating tree line (estimates range up to 100 km for every Celsius degree of warming);
2. A northward movement of the zone of maximum growth of a given tree species;
3. Enhanced growth of some species and forest types (CO_2 fertilization); reduced growth of others (particularly those with southern limits);
4. The introduction of new species, varieties, and forms which may evolve as a result of the climate changes, species migrations, and the exposure to new habitats;
5. Altered disturbance regimes and human activities, such as harvesting, silvicultural, and planting operations;
6. Changes in the physical characteristics (e.g., snow cover), growth, and composition of forests and associated ground vegetation;
7. Changes in wildlife populations;
8. Changes in Canada's network of ecological reserves and parks (which may have to be reevaluated because of changing distribution of ecosystems).

Climate change is presently best understood at the continental scale, but there are already suggestions that changes can and will be profound at local and regional scales. For example, within this larger climate change scenario in Canada, Thompson et al. (1998) and Parker et al. (2000) suggested that in the province of Ontario there are expectations of profound impacts on forest ecosystems because of increased temperature, altered disturbance regimes, and widely varying local anthropogenic factors, such as increased fire suppression and harvesting. In Canada's westernmost province of British Columbia, Hebda (1997: p. 13-1) suggested that profound impacts under a warming trend could be expected including "up-slope migration of tree lines and ecosystem boundaries, disappearance of forest ecosystems in regions of already warm and dry climates, northward migration of forest types in the interior, replacement of biogeoclimatic zones by zones with no modern analogues, and increased fire frequency." These findings confirm, and provide regional details, of global trends which have been reported in international climate change planning scenarios covering many different regions of the globe (Houghton et al., 1990; Singh and Wheaton, 1991). Few management decisions have yet been made; for example, in preparing for the effects of climate change on forest biodiversity in British Columbia, the critical needs are for data and understanding (Hebda, 1998).

The search for an approach to management of the forests of the world that is sustainable must continue as a top priority for the forest scientists and managers, because there is little doubt that sustainable forest management is a fundamental requirement if human use of natural resources is to continue at anything near the current rate of consumption. Two recent political signposts have been erected that the world's forest community cannot ignore:

1. **The United Nations Conference on Environment and Development, the 1992 Earth Summit in Rio de Janeiro, Brazil** — this meeting of global leaders and environmental organizations focused intense world scrutiny on a wide array of international environmental issues, including

forest conservation and management. In addition, specific conventions were signed concerning world climate change and the conservation of biodiversity. These agreements have led to an emphasis on monitoring criteria and indicators of sustainable forest management (FAO, 1994a), and on forest certification (Coulombe and Brown, 1999). Although the Forest Principles document was voluntary, rather than the binding Forest Conventions document that was originally sought, the net effect of the Rio Earth Summit was that the whole process of forest management was opened up with the consequent healthy questioning of conventional wisdom and practices.

2. **The Kyoto Protocol on Climate Change (1997)** — created a new, legally binding treaty for industrialized nations to meet the voluntary emissions targets set at Rio de Janeiro. This meeting led to an emphasis on national reporting of carbon budgets as the focus of national contributions to global climate change and has introduced the possibility of international trading of carbon credits (Pfaff et al., 2000). Governments set 2012 as the deadline for reductions of six greenhouse gases. The Clean Development Mechanism (CDM) offered an opportunity to reduce greenhouse gas emissions and forest loss.

Each of these political agreements has given the world's forest scientists and managers much to contend with, not the least of which has been a workable monitoring system to provide key information on local, regional, and global performance in managing forests and forest ecosystems.

ECOSYSTEM MANAGEMENT

Managing forests with ecosystems and landscapes as the basic management units represents a major shift in thinking and practice in some parts of the world, leading many to believe this is the required stimuli to develop a sustainable forest management approach. Ecosystem management has emerged in scientific, political, and economic arenas as perhaps one of the most important changes in history in human natural resource management. In 1991, a Society of American Foresters Task Force endorsed ecosystem management; their support was firmly based on views championed by conservationists and foresters decades ago, but perhaps long neglected in actual forest practices. This vision of natural resource management integrates human needs for forest products and services with needs for long-term conservation of environmental quality and ecosystem health.

Ecosystem management is a process-oriented approach to resources management, meaning the emphasis on management is to understand and maintain the essential processes that create and sustain ecosystem conditions. Unfortunately, understanding of ecosystem processes is neither complete nor simple, and so it has been difficult to identify just what is, or should be, ecosystem management. The wide range of definitions of ecosystem management "has caused confusion and even threatens its future as a management paradigm" (National Research Council, 1998: p. 208). Ecosystem management appears to be a very basic concept that, as so often happens with basic concepts, appeals to common sense yet defies simple rational definition. The advocates of ecosystem management are heterogeneous and their approaches a complex mix (Cooke, 1999). Such factors, while contributing to a delightful and stimulating intellectual challenge, have helped create a paradigm of ecosystem management that "is not founded on specific scientific tests, and prescriptions are vague" (Simberloff, 1999: p. 102). To many, ecosystem management is a never-ending process that will depend significantly on our ability to always learn, change, and improve our management (Boyce and Haney, 1997). The central premise of ecosystem management is sustainability. How is it possible to determine sustainability? Over time, various activities will be judged sustainable because they can be done sustainably. This presupposes a lot of trial and error; because of relatively long rotations and the still-evolving modeling tools, there may not be much time left to view the results (and create realistic simulations) then make appropriate adjustments.

Ecosystem management is managing over longtime scales, over multijurisdictional spatial scales, and for a wide range of values. It is holistic in its view of natural and human resources (Franklin, 1997); it is site specific in that it deals with the local conditions, but always in the context of larger patterns. Ecosystem management transcends boundaries, since much of the forest is partitioned or segmented, and there must be an assessment of this larger whole, rather than an isolated view of the particular conditions in stands, sites, ecosystems, or landscapes. Ecosystem management, perhaps most importantly, attempts to integrate societal constraints while contributing to an increase in scientific knowledge (Maser, 1994). Another way of viewing the ecosystem management paradigm is to consider the kinds of activities and information needs that managers face under ecosystem management plans and guidelines. What kinds of problems are forest managers typically called on to solve in their everyday management function? How does a forest manager balance recreation and other nonconsumptive values and the increasing demand for timber products?

The differences in management paradigms over the past century seem more related to implementation than philosophy or design. Virtually every forest management approach either states explicitly or implies that forest management is designed to sustain production and avoid environmental deterioration. Management may be based on the annual allowable cut (Morgan, 1991), which is defined as the average volume of wood that may be harvested annually under sustained yield management (Expert Panel on Forest Management in Alberta, 1990). Under a sustained yield management paradigm, the challenge for the forest manager is to determine the appropriate amount, distribution, and location of timber to cut within a defined area (e.g., lease), by considering harvesting, regrowth, and natural disturbances. Typically, sustained yield decision making is based on a calculation that a given unit of land is managed to provide a specified amount of resources, usually expressed as a volume of timber, over a specified amount of time (the rotation age) and over a specified area. Globally, there are clear limits to sustained yield management. Obviously, it is difficult to maintain production on a sustained yield basis if permanent damage is caused by forestry practices; no sustained yield forest management plan would support complete removal of the forest resource. Unfortunately, that is exactly what seems to have occurred in many forests (Berlyn and Ashton, 1996; Bryant, 1997).

In some areas of the world, the amount of available timber far exceeds the ability to harvest; sustained yield continues to be the highest goal of forest management. In such areas, sustained yield is a component of a sustainable forest management strategy. In other areas, increasing tension developed between timber and other forest values. The sustained yield forest management strategy practice was modified, including new values such as maintenance of biodiversity, preservation of wildlife, habitat, and human recreational enjoyment of forests. As many forest areas were converted to other land uses, the smaller forestland base must provide increased yields; in many countries the amount of land in the forestland base is considerably reduced from historic levels. Yet increased yields have been provided sustainably through several rotations in such areas.

The multiple-use strategy was designed to provide the largest sum of social, economic, and spiritual benefits. This management plan was one in which sustained yield was measured not solely on the basis of timber products, but included other valued attributes of the forest. The idea of land capability was introduced formally into the planning process. While measuring land capability is difficult, especially when considering capability values other than timber (e.g., wildlife habitat suitability or habitat effectiveness), the intention was to create management plans in which forestland was allocated to a variety of purposes to meet different demands simultaneously. Priorities might be needed to sort out the competing demands. Priority-setting exercises gave rise to the idea of primary and secondary uses of forest areas. Under multiple-use the main problem for the forest manager was to manage these priorities; in essence, to determine which was to be the primary forest use, how it could best be implemented, and, where desirable, how it could be modified to accommodate secondary and incidental uses.

The forestland base is probably inadequate to meet all future demands in light of increasing populations and economic needs. This alone appears to eliminate sustained yield and multiple-use management planning as viable forest management strategies for large areas of the world. Can these strategies ensure that forest biodiversity is not compromised? Will anyone believe such predictions under these management plans? What is needed now is far more complex than could be considered under these management paradigms (Larson et al., 1997); no less than a way of managing forests such that their essential processes, their biological functioning in the local to global scheme, is preserved for all time. In the forestry community today, there is widespread agreement that ecosystem management is on the right track (Behan, 1997). There is also a growing sense of urgency in implemention "… our future existence on this planet depends on it" (Boyce and Haney, 1997: p. 12).

Forest Stands and Ecosystems

Traditionally, management activities are applied to discrete parcels of forest. The forest stand has come to represent the fundamental management unit under the sustained-yield and multiple-use management strategies. Managing forest stands first required their definition and delineation on the landscape; one principle underlying this delineation is that stands are homogeneous or acceptably heterogeneous for the purpose of management treatments. Spies (1997: p. 12) put it this way:

> The definitions and spatial boundaries of stands and ecosystems are typically determined for specific purposes of management and science [respectively]. A stand typically has been defined as a unit of trees that is relatively homogeneous in age, structure, composition and physical environment. The characteristics used to delineate stands often refer to the tree layer since this traditionally has been the focus of forest management and is relatively easily mapped using aerial photographs. Soil and topographic features also frequently are used to delineate stands, especially if they have a strong effect on stand productivity or harvesting operations. Specific stand definitions, sizes, and shapes will vary depending on management intensity and objectives and the spatial heterogeneity of the vegetation, soil and topography.

A basic assumption was that a forest could be partitioned into sensible units for management. This idea had global applicability. For example, in France the general approach was to structure the existing forested surfaces, or the areas susceptible to be forested, into homogeneous units called sites, where a site is a piece of land of variable surface area homogeneous in its physical and biological conditions (mesoclimate, topography, soil, floristic composition, and vegetation structure) (Becker, 1999). A forest site justifies that, for a given species, a specific silvicultural method may be applied, which can be expected to result in a productivity bound within known limits. Many forest studies — not simply remote sensing studies of forests, but studies of forest management, forest growth, forest disturbance, and forest change — begin with statements such as these:

1. "A forest type is an area of forest which exhibits a general similarity in tree species composition and character. Maps of native forest that detail the distribution of forest types have traditionally been made using aerial photographs supported by ground surveys." (Skidmore, 1989: p. 1449).
2. "Planning should be based on natural forest compartments defined and delineated by applying criteria such as soils, topography, forest composition, regeneration capacity, usable timber volume, and existing local uses." (Kuusipalo et al., 1997: p. 115).

Organizing the landscape into homogeneous units — or acceptably heterogeneous units for the purpose of management — requires an understanding of the forest structure and the role of the resulting strata on the effects of treatments and prescriptions which might be applied to achieve management objectives. Sometimes, units of land become homogenous because common treatments are applied within their boundaries. But a comprehensive documentation of forest classification and strata (or attribute) mapping logic does not exist, and the actual effect of the stand delineation on the effectiveness of management has rarely (if ever) been examined systematically. As Kimmins (1997) noted in reviewing forest classification systems, the classification of forests is purposive, and the purpose is often as varied as the products that are generated to help achieve it. In fact, it seems increasingly obvious that the rules of forest mapping as practiced over the past few decades are not particularly logical at all, but are strongly dependent on the skill of the analyst, the local nature of the forest condition, and the cultural tradition in the particular jurisdiction responsible for fulfilling demands for forest information.

The aerial photointerpretation method, at the heart of the delineation of forest stands, is labor intensive and subjective, and may result in inconsistencies in the assignment of forest type boundaries and names between different aerial photointerpreters, and over time with individual interpreters (Skidmore, 1989). The use of stands identified in this way has been questioned on the grounds that they are rarely in fact homogeneous, and they do not always have stable and recognizable boundaries (Holmgren and Thuresson, 1997). There is a growing recognition of the arbitrariness and difficulty of working with forest stands understood and applied on the landscape in this traditional way.

It is clear that, after organizing the forested landscape in this manner, it is possible to develop an efficient economic model of forest resource value (Erdle, 1998). For example, Weintreb and Cholaky (1991) developed strategic and tactical models for decision making in forest planning on the assumption that zones are divided into management units, and then stands, which are considered homogeneous. The stands are required for accounting purposes, but likely also for operational management prescriptions. One of the key features is that stands, for the purpose of management, can be used to organize the forest into a spatial hierarchy (Oliver et al., 1999). Despite the potential for "value-conflict," is the forest stand spatial hierarchy likely to be replaced anytime soon with a different system? Or perhaps the appropriate question is, Is it likely that in future operational management, the variability of properties of interest within stands will not exceed that between stands?

If timber volume were to continue to be the overriding principle underlying forest planning, with constraints imposed by other values, then perhaps forest stands delineated in this traditional way will continue to be a suitable way, or even the only suitable way, of organizing the landscape for management. By focusing on the regulation of forestry (by which is meant forest treatments) in a sustained yield and multiple-use forest management, all other values can be seen as simple constraints. Is there any need to better understand ecosystems under this system of management? Under this scenario, there are few or no problems that cannot be resolved with existing management treatments, existing ways of organizing the forest into discrete parcels or stands, and existing levels of understanding and information. But perhaps the constraints will continue to increase in complexity, ultimately overwhelming any and all forms of management in their demand for additional knowledge and scientific information upon which to base decisions.

Ecosystem management, on the other hand, considers multiple forest values over a full range of spatial scales over time. Forest stands do not seem to have as central a role to play under this management paradigm; instead, spatial structures which correspond to physical features or intrinsic characteristics of processes occur at a wide range of nested scales (Bellehumeur and Legendre, 1998). Ecosystems, from stream reaches to regional biomes, are the most likely operable management units (Kohm and Franklin, 1997). Typically, forest stands are component parts of forest ecosystems; forest stands and forest ecosystems are not synonymous and are certainly not equivalent concepts. Forest ecosystems may or may not be comprised of forest stands (Shugart, 1998), delineated in the traditional way based on the concepts of forest structure (Spies, 1997). Several new difficulties arise:

1. What is an ecosystem? Adopting a simple relationship between vegetation and its total environmental system, the ecosystem, and then confusing the two (de Laubenfels, 1975; Graetz, 1990), has much less applicability in a management system that attempts to explain ecological functions rather than assume them. In effect, the context defines the limits of the definition of an ecosystem for forestry applications — the forest ecosystem is an abstract concept or a natural unit of certain areal constraints (Shugart, 1998).
2. What is the spatial hierarchy (O'Neill et al., 1986) suitable for managing forest ecosystems? The traditional ecological organization of levels in a hierarchy (organism, population, ecosystem, landscape, etc.) appears less and less useful, almost irrelevant, as the role of ecological scale is clarified and integrated into a spatial hierarchy for operational purposes (Simmons et al., 1992; Peterson and Parker, 1998).

Managing forest ecosystems might be one of the more difficult endeavors that humans have attempted, not only because of the range of scales over which the issues remain pertinent (local to global), but also because of the continued operation in a data rich but information poor environment. Ecosystems are probably far more complex than any other system that humans have tried to manage or understand, including financial systems upon which humans spend extraordinary amounts of time and money annually (Woodley and Forbes, 1997). A common belief is that forest ecosystems will never be understood completely; obviously then, management cannot wait for complete and total knowledge of the effects (Larson et al., 1997). From the field and remote sensing perspective, it is not yet known with certainty or great confidence what to measure to provide the answers needed for "urgent and long-term management questions" (Noss, 1999: p. 136). This theme will occur repeatedly as the literature and prospects for remote sensing are examined.

Achieving Ecologically Sustainable Forest Management

Forest management is the process of "designing and implementing a set of actions which is deemed likely to result in a set of forest conditions which is deemed likely to provide the desired values in the desired amount over time" (Erdle and Sullivan, 1998: p. 83). The long-term evolving nature of forest management is, perhaps, not widely appreciated (Fedkiw and Cayford, 1999), but the process can be simplified by considering four basic elements (Figure 2.1; Smith and Raison, 1998):

1. The definition of forest values,
2. The description of the forest (the inventory),
3. The identification of treatment alternatives, and
4. The description of the biological response to treatment.

In the previous section items one and two were briefly discussed, but the entire process by which forests are managed is more fully elaborated here. Typically, forest values are captured in a set of goals or guidelines — typically called the Codes of

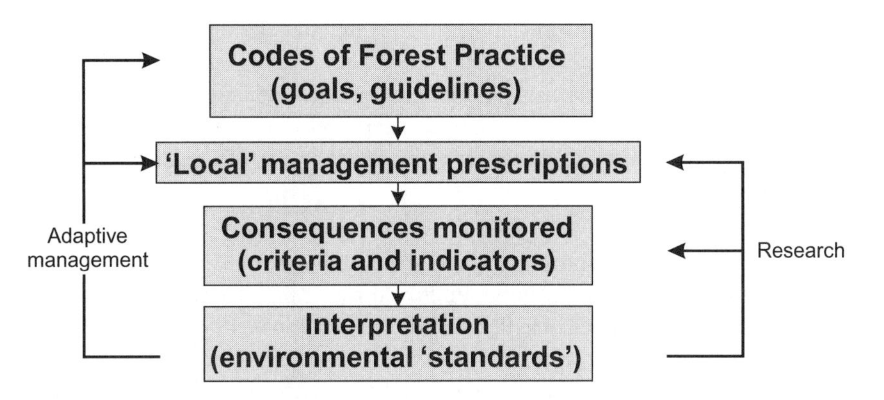

FIGURE 2.1 The basic elements of forest management include linkages between values, treatments, monitoring criteria and indicators, continued research, and adaptive management. Adherence to careful consideration of each of the components and steps in the process are necessary in achieving ecologically sustainable forest management. (From Smith, C. T., and R. J. Raison. 1998. *The Contribution of Soil Science to the Development and Implementation of Criteria and Indicators of Sustainable Forest Management,* pages 121–135, Soil Science Society of America, Madison, WI. With permission.)

Forest Practice — which could be construed as approved ways of achieving environmental care given certain economic realities (Smith and Raison, 1998). When considering the goals and objectives of sustainable forestry, the comprehensive Codes of Forest Practice would include statements on timber values, wildlife habitat, aesthetics, biodiversity, water regimes, ecological health, and recreation (Erdle and Sullivan, 1998). Such statements are required before any attempt is made to relate these to the information needs in a broad way, for example, through strategic planning (Weintreb and Cholaky, 1991). The goals, and the way the goals are achieved, should not be confused.

The Codes of Forest Practice are implemented on the ground via local management prescriptions. A broad set of treatment alternatives will create greater flexibility to influence stand composition, structure, and forest pattern and, by extension, flexibility to influence resulting forest values (Erdle and Sullivan, 1998). The results of local management prescriptions implemented to produce desired forest conditions, and hence values, must be rigorously monitored. But while it is probably impossible to keep track of all aspects of forest conditions and their relationships to forest values, new efforts have been made to allow a more complete monitoring program to be designed. In the next section, measuring progress and change through the use of criteria and indicators is discussed (Anonymous, 1995; Noss, 1999). Interpretation of change in forest conditions must be validated scientifically and subject to testing (environmental standards). Tying these four components together are adaptive management and research activities, discussed more fully in the following sections.

Examining the flow of decisions and output products that result from forest management has led some to consider that the best way to determine sustainability is through certification (Fletcher et al., 1998; Vogt et al., 1999). Motivation and interest in forest certification have included (Coulombe and Brown, 1999):

1. Improving auditing and assessments of the performance of forest management,
2. Strengthening credibility and public acceptance of forestry,
3. Improving overall business and forest practices, and
4. Exploring market incentives through development of demand for forest and wood products.

A significant aspect of forest certification efforts has been that most are voluntary, nonregulatory approaches to promote improved forest practices and forest management systems (Fletcher et al., 1998; Coulombe and Brown, 1999). There is general and wide agreement in the forest community that appropriate standards' mechanisms be used in the development of a certification protocol (Lapointe, 1998). For example, under the auspices of the Canadian Standards Association (CSA), a member of the International Organization for Standardization (ISO), a sustainable forest management certification program has been developed with participation by four groups:

- Producers — industry, woodlot owners, and cooperatives;
- Professional and scientific — academic, research, and professional groups;
- Public interest — consumer and environmental nongovernmental organizations, and aboriginals;
- Regulatory — federal, state, and provincial groups.

The approach has been to develop standards that apply to the environmental management system, the performance on the ground, or both. The management system and performance standards that emerged in Canada were based on six broad criteria and many specific indicators published by the Canadian Council of Forest Ministers (1997) (described in the following section). Ways of measuring specific indicators were audited and tested before acceptance as the National Standards of Canada by the Standards Council of Canada, the organization charged with all aspects of standards development and implementation in the country. This included the accreditation of registrars (those empowered to certify), auditors, and associated training programs. Issues that can arise during acceptance of the standards are related to the perceived level of commitment by the participating organizations, the degree of public participation in developing the standards and implementing the certification program, all management system elements, and a built-in continual improvement mechanism.

For example, in forest planning, explicit forecasts and assessment of outcomes relative to those forecasts are required; the auditor must include on-the-ground examination of the forest, and the result of forestry activities in relation to planned objectives and environmental impacts. This can considerably increase the costs and complexity of the certification process. Application by an organization for an audit leading to management system or performance certification — perhaps leading to product certification and labeling — is voluntary (Lippke and Bishop, 1999). As of May 2000, there were more than 16 million hectares of Canadian forest land certified under one of three such third-party auditing systems (Natural Resources Canada, 2000).

One of the key activities required to support forest certification is effective monitoring and evaluation. Monitoring systems need to be sensitive to anthropogenic and environmental changes; that is, there must be a way to determine cause of change. For example, ecosystem responses to climate changes can be grouped by their impact on biodiversity and productivity. Designing and implementing a monitoring system that can provide insight into all these possible changes, to separate and attribute cause and effect, and to do so with enough warning to allow mitigation (e.g., invoking tradable emissions credits) to be used, is an immense task.

CRITERIA AND INDICATORS OF SUSTAINABLE FOREST MANAGEMENT

A key feature of sustainable forest management is a monitoring program to ensure performance and management goals are met. One possible approach is based on purpose-designed experiments; for example, limited management objectives, such as high regeneration success in plantations, could be examined by experimenting with species and planting techniques in a traditional experimental design and survey method. In complex systems, monitoring across a broad range of activities and experimental results can be efficiently narrowed down to the measurement of indicators within broad categories, or criteria. This criteria and indicator (C&I) approach has been vigorously pursued by many national and international entities in recent years with the result that discussion of C&I monitoring of sustainable forest management has reached global significance (Wallace and Campbell, 1998; Noss, 1999).

According to work reported by Smith et al. (1999: p. 4), environmental indicators of sustainable forest management should have the following attributes:

- Easy to measure
- Cost effective
- Accommodate changing conditions
- Scientifically sound and based on functional ecological relationships
- Forest ecosystem specific, yet able to be scaled up (e.g., using spatial statistical techniques and GIS)
- Integrative of ecosystem functional relationships (e.g., many indicators chosen to represent selected key ecosystem processes or fewer key indicators integrate across the entire ecosystem)
- Related to management goals or values

Taken together, the criteria and indicators are intended to provide a common understanding and scientific definition of sustainable forest management (Food and Agriculture Organization, 1998). However, despite widespread agreement on the utility and need for this approach, many indicators within each of the general criteria cannot be reported nationally, regionally, or even locally; one significant problem is that the necessary data often do not exist, and the ecological processes may not be known with enough certainty to decide which data are required. Another problem is the "massive commitment to collecting, processing, storing, retrieving, analyzing, and documenting huge quantities of data [...] needed to evaluate whether or not management of forest resource is sustainable" (Whyte, 1996: p. 204). Few indicators have been adequately tested or validated (Noss, 1999).

This issue relates to the use of the national criteria and indicators in the development of local criteria and indicators in a wide variety of ecological settings and policy frameworks. It is difficult to create indicators that can operate effectively in a wide variety of forests. This has given rise to the proliferation of local-level indicators. If used for certification purposes, there may be difficulty in achieving consensus (Coulombe and Brown, 1999). Such local-level indicator lists may run well into the hundreds for the next few years. For example, by 1999 most of the 12 Canadian Model Forests had initiated discussions aimed at narrowing the broad national criteria and indicators to suit local and regional conditions (Anonymous, 1999). While many of the selected indicators in one Model Forest would be applicable elsewhere, differences soon emerged that would need to be reconciled if local lists were used to "roll-up" to the national level. The overriding concern in developing such local-level indicators appears to be the need for monitoring on-the-ground changes; typically, in local settings, performance evaluation is where the action must be (Erdle and Sullivan, 1998).

In Canada, the national criteria and indicators of sustainable forest management are broad areas identified by common agreement among forest stakeholders. Then, specific elements and indicators are developed and reported. Elements are divided into different indicators that represented measurable forest and economic variables (Table 2.1). The six criteria represent agreed-upon social (and cultural), environmental, and economic principles:

1. Conservation of biological diversity,
2. Maintenance and enhancement of ecosystem condition and productivity,
3. Soil and water resources conservation,
4. Forest ecosystem contributions to global ecological cycles,
5. Multiple benefits of forestry to society,
6. Accepting society's responsibility for sustainable development.

The 6 criteria, 22 elements, and 83 indicators comprise a system of reporting that can be used to highlight trends or changes in the status of forests, and forest management, over time (Canadian Council of Forest Ministers, 1997).

A second example of this approach is the International Food and Agriculture Organization (1994a) list of criteria and indicators (Table 2.2). This list was compiled from five separate sources (the International Tropical Timber Organization, the Tarapoto Process, the Center for International Forestry Research, the African Timber Organization, and the Central American Commission for Environment and Development). The differences in the two tables of criteria and indicators — which stem from the different types of forests they are meant to address — are less important than the agreement on the approach.

TABLE 2.1
A Canadian Approach to Criteria and Indicators of Sustainable Forest Management in Boreal and Temperate Forests

Criterion 1: Conservation of Biological Diversity

Element 1.1 Ecosystem Diversity

Indicator 1.1.1	Percentage and extent, in area, of forest types relative to historical condition and to total forest area
Indicator 1.1.2	Percentage and extent of area by forest type and age class
Indicator 1.1.3	Area, percentage, and representativeness of forest types in protected areas
Indicator 1.1.4	Level of fragmentation and connectedness of forest ecosystem components

Element 1.2 Species Diversity

Indicator 1.2.1	Number of known forest-dependent species classified as extinct, threatened, endangered, rare, or vulnerable relative to total number of forest-dependent species
Indicator 1.2.2	Population levels and changes over time of selected species and species guilds
Indicator 1.2.3	Number of known forest-dependent species that occupy only a small portion of their former range

Element 1.3 Genetic Diversity

Indicator 1.3.1	Implementation of an *in situ/ex situ* genetic conservation strategy for commercial and endangered forest vegetation species

Criterion 2: Maintenance and Enhancement of Forest Ecosystem Conditions and Productivity

Element 2.1 Incidence of Disturbance and Stress

Indicator 2.1.1	Area and severity of insect attack
Indicator 2.1.2	Area and severity of disease infestation
Indicator 2.1.3	Area and severity of fire damage
Indicator 2.1.4	Rates of pollutant deposition
Indicator 2.1.5	Ozone concentrations in forested regions
Indicator 2.1.6	Crown transparency in percentage by class
Indicator 2.1.7	Area and severity of occurrence of exotic species detrimental to forest condition
Indicator 2.1.8	Climate change as measured by temperature sums

Element 2.2 Ecosystem Resilience

Indicator 2.2.1	Percentage and extent of area by forest type and age class
Indicator 2.2.2	Percentage of successfully naturally regenerated and artificially regenerated

Element 2.3 Extant Biomass

Indicator 2.3.1	Mean annual increment by forest type and age class
Indicator 2.3.2	Frequency of occurrence within selected indicator species (vegetation, mammals, birds, fish)

Criterion 3: Conservation of Soil and Water Resources

Element 3.1 Physical Environmental Factors

Indicator 3.1.1	Percentage of harvested area having significant soil compaction, displacement, erosion, puddling, loss of organic matter, etc.
Indicator 3.1.2	Area of forest converted to nonforestland use, for example, urbanization

TABLE 2.1 *(Continued)*
A Canadian Approach to Criteria and Indicators of Sustainable Forest Management in Boreal and Temperate Forests

Indicator 3.1.3	Water quality as measured by water chemistry, turbidity, etc.
Indicator 3.1.4	Trends and timing of events in stream flows from forest catchments
Indicator 3.1.5	Changes in distribution and abundance of aquatic fauna
Element 3.2 Policy and Protection Forest Factors	
Indicator 3.2.1	Percentage of forest managed primarily for soil and water protection
Indicator 3.2.2	Percentage of forest area having road construction and stream crossing guidelines in place
Indicator 3.2.3	Area, percentage, and representativeness of forest types in protected areas
Criterion 4: Forest Ecosystem Contributions to Global Ecological Cycles	
Element 4.1 Contributions to the Global Carbon Budget	
Indicator 4.1.1	Tree biomass volumes
Indicator 4.1.2	Vegetation (non-tree) biomass estimates
Indicator 4.1.3	Percentage of canopy cover
Indicator 4.1.4	Percentage of biomass volume by general forest type
Indicator 4.1.5	Soil carbon pools
Indicator 4.1.6	Soil carbon pool decay rates
Indicator 4.1.7	Area of forest depletion
Indicator 4.1.8	Forest wood product life cycles
Indicator 4.1.9	Forest sector CO_2 emissions
Element 4.2 Forestland Conversion	
Indicator 4.2.1	Area of forest permanently converted to non-forestland use
Indicator 4.2.2	Semipermanent or temporary loss or gain of forest ecosystems (for example, grasslands, agriculture)
Element 4.3 Forest Sector Carbon Dioxide Conservation	
Indicator 4.3.1	Fossil fuel emissions
Indicator 4.3.2	Fossil carbon products emissions
Indicator 4.3.3	Percentage of forest sector energy usage from renewable sources relative to total sector energy requirements
Element 4.4 Forest Sector Policy Factors	
Indicator 4.4.1	Recycling rate of forest wood products manufactured and used in Canada
Indicator 4.4.2	Participation in the climate change conventions
Indicator 4.4.3	Economic incentives for bioenergy use
Indicator 4.4.4	Existence of forest inventories
Indicator 4.4.5	Existence of laws and regulations on forestland management
Element 4.5 Contributions to Hydrological Cycles	
Indicator 4.5.1	Surface area of water within forested areas
Criterion 5: Multiple Benefits of Forests to Society	
Element 5.1 Productive Capacity	
Indicator 5.1.1	Annual removal of forest products relative to the volume of removals determined to be sustainable
Indicator 5.1.2	Distribution of, and changes in, the land base available for timber production

TABLE 2.1 *(Continued)*
A Canadian Approach to Criteria and Indicators of Sustainable Forest Management in Boreal and Temperate Forests

Indicator 5.1.3	Animal population trends for selected species of economic importance
Indicator 5.1.4	Management and development expenditures
Indicator 5.1.5	Availability of habitat for selected wildlife species of economic importance
Element 5.2 Competitiveness of Resource Industries (Timber/Nontimber-Related)	
Indicator 5.2.1	Net profitability
Indicator 5.2.2	Trends in global market share
Indicator 5.2.3	Trends in research and development expenditures in forest products and processing technologies
Element 5.3 Contribution to the National Economy (Timber/Nontimber Sectors)	
Indicator 5.3.1	Contribution to gross domestic product of timber and nontimber sectors of the forest economy
Indicator 5.3.2	Total employment in all forest-related sectors
Indicator 5.3.3	Utilization of forests for nonmarket goods and services, including forestland use for subsistence purposes
Indicator 5.3.4	Economic value of nonmarket goods and services
Element 5.4 Nontimber Values (Including Option Values)	
Indicator 5.4.1	Availability and use of recreational opportunities
Indicator 5.4.2	Total expenditures by individuals on activities related to nontimber use
Indicator 5.4.3	Membership and expenditures in forest recreation-oriented organizations and clubs
Indicator 5.4.4	Area and percentage of protected forest by degree of protection
Criterion 6: Accepting Society's Responsibility for Sustainable Development	
Element 6.1 Aboriginal and Treaty Rights	
Indicator 6.1.1	Extent to which forest planning and management processes consider and meet legal obligations with respect to duly established aboriginal and treaty rights
Element 6.2 Participation by Aboriginal Communities in Sustainable Forest Management	
Indicator 6.2.1	Extent of aboriginal participation in forest-based economic opportunities
Indicator 6.2.2	Extent to which forest management planning takes into account the protection of unique or significant aboriginal social, cultural, or spiritual sites
Indicator 6.2.3	Number of aboriginal communities with a significant forestry component in the economic base and the diversity of forest use at the community level
Indicator 6.2.4	Area of forestland available for subsistence purposes
Indicator 6.2.5	Area of Indian reserve forestlands under integrated management plans
Element 6.3 Sustainability of Forest Communities	
Indicator 6.3.1	Number of communities with a significant forestry component in the economic base
Indicator 6.3.2	Index of the diversity of the local industrial base
Indicator 6.3.3	Diversity of forest use at the community level
Indicator 6.3.4	Number of communities with stewardship or comanagement responsibilities
Element 6.4 Fair and Effective Decision Making	
Indicator 6.4.1	Degree of public participation in the design of decision-making processes

TABLE 2.1 *(Continued)*
A Canadian Approach to Criteria and Indicators of Sustainable Forest Management in Boreal and Temperate Forests

Indicator 6.4.2	Degree of public participation in the decision-making processes
Indicator 6.4.3	Degree of public participation in implementation of decisions and monitoring of progress toward sustainable forest management
Element 6.5 Informed Decision Making	
Indicator 6.5.1	Percentage of area covered by multi-attribute resource inventories
Indicator 6.5.2	Investments in forest-based research, development, and information
Indicator 6.5.3	Total effective expenditure on public forestry education
Indicator 6.5.4	Percentage of forest area under completed management plans/programs/guidelines which have included public participation
Indicator 6.5.5	Expenditure on international forestry
Indicator 6.5.6	Mutual learning mechanisms and processes

Source: From Canadian Council of Forest Ministers, 1997. *Criteria and Indicators of Sustainable Forest Management,* Canadian Forest Service, Natural Resources Canada. With permission.

What is the role of remote sensing in monitoring criteria and indicators of sustainable forest management? This question has only recently been addressed as the credibility and usefulness of the C&I approach becomes better known (Hall, 1999). Referring to the Canadian Council of Forest Ministers C&I in Table 2.1, Goodenough et al. (1998) suggested that 22 indicators (of the 83) could be addressed partially or wholly by remote sensing technology (Table 2.3). In their assessment, the emphasis was clearly on indicators that could be readily obtained by satellite remote sensing, using the current suite of Earth-observing satellites (see Chapter 3), and methods of processing the available imagery (see Chapter 4).

What follows is a brief review of the criteria with suggestions for specific remote sensing applications. These may lead to the development of monitoring protocols for each of the indicators for which remote sensing can contribute information. These applications provide the focus for the discussions in later chapters of this book.

Conservation of Biological Diversity

Biodiversity is the variability among living organisms and the ecological complexes of which they are a part (Canadian Council of Forest Ministers, 1997). To some, biodiversity has come to mean the whole expression of life on Earth (Lugo, 1998). Consequently, there is no single measure of biodiversity, or even agreement as to how biodiversity should be measured (Silbaugh and Betters, 1997). Elements of biodiversity occur at multiple scales of biological organization including genetic, population, ecosystem/community (Boyle, 1991), and regional landscape (Noss, 1990). From the remote sensing perspective, there may be a role for remote sensing technology in managing for forest biodiversity at the population, ecosystem, and regional landscape scales.

The precise form that the potential remote sensing contributions may take in these tasks is complex because of the lack of understanding of biological diversity

TABLE 2.2
Example of Criteria and Indicators of Sustainable Forest Management in Tropical Forests

Criterion 1: Extent of Forest Resources and Global Carbon Cycles

Area of Forest Cover
Wood Growing Stock
Successional Stage
Age Structure
Rate of Conversion of Forest to Other Use

Criterion 2: Forest Ecosystem Health and Vitality

Deposition of Air Pollutants
Damage by Wind Erosion
Incidence of Defoliators
Reproductive Health
Insect/Disease Damage
Fire and Storm Damage
Wild Animal Damage
Competition from Introduction of Plants
Nutrient Balance and Acidity
Trends in Crop Yields

Criterion 3: Biological Diversity in Forest Ecosystems

Distribution of Forest Ecosystems
Extent of Protected Areas
Forest Fragmentation
Area Cleared Annually of Endemic Species
Area and Percentage of Forestlands with Fundamental Ecological Changes
Forest Fire Control and Prevention Measures
Number of Forest-Dependent Species
Number of Forest-Dependent Species at Risk
Reliance of Natural Regeneration
Measures *in situ* Conservation of Species at Risk
Number of Forest-Dependent Species with Reduced Range

Criterion 4: Productive Functions of Forests

Percentage of Forests/Other Wooded Lands Managed According to Management Plans
Growing Stock
Wood Production
Production of Non-Wood Forest Products
Annual Balance Between Growth and Removal of Wood Products
Level of Diversification of Sustainable Forest Production
Degree of Utilization of Environmentally Friendly Technologies

Criterion 5: Protective Functions of Forests

Soil Conditions
Water Conditions
Management for Soil Protection
Watershed Management

TABLE 2.2 *(Continued)*
Example of Criteria and Indicators of Sustainable Forest Management in Tropical Forests

Areas Managed for Scenic and Amenity Purposes
Infrastructure Density by FMU Category
Criterion 6: Socioeconomic Functions and Conditions
Value of Wood Products
Value of Non-Wood Products
Value from Primary and Secondary Industries
Value from Biomass Energy
Economic Profitability of SFM
Efficiency and Competitiveness of Forest Products Production
Degree of Private and Non-Private Involvement in SFM
Local Community Information and Reference Mechanisms for SFM
Employment Generation/Conditions
Forest-Dependent Communities
Impact on the Economic Use of Forests on the Availability of Forests for Local People
Quality of Life of Local Populations
Average Per Capita Income in Different Forest Sector Activities
Gender-Focused Participation Rate in SFM
Criterion 7: Political, Legal, and Institutional Framework
Legal Framework that Ensures Participation by Local Government and Private Landowners
Technical and Regulatory Standards of Management Plans
Cadastral Updating of the FMU
Percentage of Investment on Forest Management for Forest Research
Rate of Investment on the FMU Level Activities: Regeneration, Protection, Etc.
Technical, Human, and Financial Resources

Source: From *Readings in Sustainable Forest Management,* Forestry Paper 122, Food and Agriculture Organization, 1994. With permission.

issues at almost any scale, but particularly at scales above the stand level. Noss (1999: p. 135) suggested that "managers and policy makers need to be cognizant of the biological significance of the forests they manage in a broad context; otherwise they may inadvertently compromise global biodiversity by managing their forests inappropriately." In essence, this encourages foresters to view local management activities in a regional context. This larger biodiversity issue constrains some forest management activities at the strategic level; for example, Waring and Running (1998) have noted that decreased harvesting amounts in the U.S. Pacific Northwest has generally meant increased harvesting amounts elsewhere in the world.

It is reasonable to assume that remote sensing will be increasingly used in providing baseline and temporal monitoring data for various forest area indicators, such as (Goodenough et al., 1998: Table 2.3):

- Percent and extent, in area, of forest types relative to historical condition and to total forest area.

TABLE 2.3
A Suggested List of CCFM Criteria and Indicator Products that Can Be Partially or Substantially Obtained by Remote Sensing Data and Methods in Support of Sustainable Forest Management Initiatives in Canada Based on a New Program Called Earth Observation for Sustainable Development (EOSD)

1.1.1	**Percent and extent, in area, of forest types relative to historical condition and to total forest area.**
1.1.2	**Percent and extent of area by forest type and age class.**
1.1.3	**Area, percent, and representativeness of forest types in protected areas.**
1.1.4	**Level of fragmentation and connectedness of forest ecosystem components.**
2.1.1	**Area and severity of insect attack.**
2.1.2	**Area and severity of disease infestation.**
2.1.3	**Area and severity of fire damage.**
2.2.1	**Percent and extent of area by forest type and age class.**
2.2.2	*Percent area successfully naturally regenerated and artificially regenerated.*
2.3.1	*Mean annual increment by forest type and age class.*
3.1.2	**Area of forest converted to nonforestland use, e.g., urbanization.**
3.2.1	**Percent of forest managed primarily for soil and water protection.**
3.2.3	**Area, percent, and representativeness of forest types in protected areas.**
4.1.1	*Tree biomass volumes.*
4.1.3	**Percent canopy cover.**
4.1.4	**Percent biomass volume by general forest type.**
4.1.7	**Area of forest depletion.**
4.2.1	**Area of forest permanently converted to nonforestland use, e.g., urbanization.**
4.2.2	**Semipermanent or temporary loss or gain of forest ecosystems, e.g., grasslands, agriculture.**
4.4.2	*Participation in the climate change conventions.*
4.5.1	**Surface area of water within forested areas.**
5.1.1	*Annual removal of forest products relative to the volume of removals determined to be sustainable.*
5.1.2	*Distribution of, and changes in, the land base available for timber production.*
5.1.5	*Availability of habitat for selected wildlife species of economic importance.*
5.4.4	*Area and percent of protected forest by degree of protection.*

Note: Table entries in **boldface** can substantially be met by remote sensing, whereas those in *italics* can only be met partially by remote sensing. It is assumed that remote sensing is combined with geographic information provided from other sources. Remote sensing can not directly measure age. However, broad classes, such as mature and immature forest stands, can be identified by current remote sensing data and methods.

Source: Goodenough, D. G., A. S. Bhogal, R. Fournier et al., 1998. *Proc. 20th Can. Symp. Rem. Sensing,* Canadian Aeronautics and Space Institute, Ottawa.

- Percent and extent of area by forest type and age class.
- Area, percent, and representativeness of forest types in protected areas.

If the concern is with biodiversity at the scale of the operational forest management unit — the forest stand and the forest ecosystem — then the pertinent questions might include:

What is the biodiversity of the current and historical landscape?
What is the likely future biodiversity potential of the landscape under different management regimes?
Specifically, what is the effect of natural disturbance on biodiversity?
Can human-induced disturbance be organized such that biodiversity is enhanced or preserved?
How can biodiversity objectives be included in forest management planning?

At the landscape level, measures of biodiversity may be constructed from landscape metrics and understanding of patch dynamics. The link between current ecological understanding of landscape structure (Haines-Young and Chopping, 1996) comprised of patch and matrix dynamics expressed in terms of ecosystem patterns and processes and elements or measures of biodiversity, is incomplete though improving (Simberloff, 1999). Natural and human disturbances, ecological succession, and recovery from previous disturbances are all forces that modify ecosystem pattern or patches within the landscape. Forman (1995) described five disturbance processes that change a landscape, influence habitat loss, and that can occur simultaneously. The types of alterations to the landscape structure produced by these processes are distinctive, providing target patterns to detect and monitor over time by remote sensing. For example:

1. Perforation — the creation of holes in the patch or landscape;
2. Dissection — cutting a landscape area or matrix into equally wide linear features, such as roads and pipelines;
3. Fragmentation — breaking and separating the matrix into smaller, non-contiguous segments or patches;
4. Shrinkage — the sizes of patches decrease;
5. Attrition — patch disappearance.

Can these patterns be remotely sensed? If so, can these patterns be related directly to descriptions of biodiversity? Some general ideas have emerged which have helped focus the issue of stand- and ecosystem-level forest management for biodiversity objectives. For example, species richness is what the public has come to understand as biodiversity (Simberloff, 1999); species abundance and distribution is relatively simple to measure, but the variability of those measures over relatively small spaces and short time frames is not well understood. Remote sensing data can be used as an explanatory framework in which point measures of species richness can be embedded (Griffiths et al., 2000).

Improvements in understanding of factors controlling biodiversity may be possible if remote sensing can provide improved spatial and temporal data on species richness (Stoms and Estes, 1993). It should be possible to measure and track features such as the patch size and diversity, the distance and connectivity between like patches, and the edge/area ratios of patches within the landscape in order to understand the biological diversity of a given landscape (Noss, 1990). Edges have been recognized as an important structural attribute of the landscape; for example, the varied thrush (*Ixoreus naevius*) is a forest interior species that avoids forest edges

(Hansen et al., 1991). If a large percentage of this bird's forest habitat were fragmented, its ability to survive would markedly diminish. On the other hand, higher biodiversity will be found in landscapes of diverse patches; spatial heterogeneity can be positively correlated with species richness of an area (Turner, 1989; Urban et al., 1987; Wickham et al., 1997). Thus, biodiversity will likely be positively influenced by the edge/area ratios of patches within the landscape, but this influence may be detrimental to the survival of some individual species.

This discussion flows naturally from the consideration of patches and the mosaic of forest in which they are embedded (O'Neill et al., 1988; McGarigal and McComb, 1995; Forman, 1995). Forest landscapes have been described as hierarchical (Baskent and Jordan, 1995) and as vegetative oceans with islands (patches) of habitat. As forests become fragmented by disturbance, patches become smaller and more distant from one another (Harris, 1984). Species richness would be a direct function of the "forest island" area and colonization rate in a direct analogy with island biogeography theory (MacArthur and Wilson, 1967). At equilibrium, the local extinction rate will be inversely related to area; that is, higher rates of extinction will occur in smaller areas, rates of immigration to islands will decline with distance from the "mainland colony," and therefore, larger islands closer to the mainland will have greater species richness than smaller more distant ones. These patterns may nest at smaller and smaller scales (larger and larger areas).

Species richness in a landscape may be a function of the number of niches since every species has a set of environmental requirements for life and reproduction, and increased number of niches may be correlated with increased number of species (Wickham et al., 1997). Dispersal between populations on a regular basis promotes gene flow and helps to decrease the probability of population extinction and fluctuations in population size. Landscapes having corridors that act as conduits and connected patches that act as stepping stones for moving objects are likely to have greater biodiversity. The composition, structure, and quality of the corridors affect the connectivity, that is, the degree to which organisms can move through the landscape matrix (Anderson and Danielson, 1997). To maintain balance, the number of migrating individuals should not exceed that of those emigrating. Potential breeding habitat must exist and reproduction must successfully replace loss due to mortality, otherwise local extinction (extirpation) will occur. The sizes and shapes of habitat patches are important because area-related edge effects may influence reproduction.

Only two of these landscape concepts have been sufficiently understood to lead to the creation of measurable indicators of sustainable forest management: fragmentation and connectivity. The monitoring of forest fragmentation is complex since there is no single measure of fragmentation that is insensitive to scale and patch input (Brown et al., 2000); instead, fragmentation is an interpretation of several individual metrics (see Chapter 5). What is needed is a spatially explicit data set that links patch, mosaic, and ecosystem functioning; spatial heterogeneity can be measured by considering different types of patch diversity within a landscape at various scales, using vegetation composition and structure (Jorgensen and Nohr, 1996), geomorphology (Burnett et al., 1998; Nicols et al., 1998), ecoclimatic stability (Fjeldsa et al., 1997), levels of photosynthetic activity (Walker et al., 1992), or a thermal/energy balance regime (Bass et al., 1998). In landscapes with factors affecting productivity and reproduction, positively greater biodiversity will occur (Wickham et al., 1997). Areas of high net primary productivity (NPP) have more resources to partition among competing species, and can thus support a greater number of species and larger populations than areas with low net primary productivity (Walker et al., 1992). However, the relationship between species richness and productivity is not completely straightforward, depending on the level of productivity, the particular species, and the balance between available soil nutrients and light (Rosenzweig and Abramsky, 1993; Tilman and Pacala, 1993).

Maintenance and Enhancement of Forest Ecosystem Condition and Productivity

Forest condition, forest health, and forest productivity are vague terms that can mean very different things to different people. To ensure sustainability, it will be necessary to characterize forest ecosystem condition with respect to condition (or structure) (Jupp and Walker, 1997) and health (Sampson and Adams, 1994). Such an assessment of a forest implies an understanding of the forest management objectives for that forest. Biological elements that strongly influence forest ecosystem condition, health, and productivity include levels of disturbance and stress, ecosystem resilience (Ludwig et al., 1997), and extant biomass (Canadian Council of Forest Ministers, 1997). Of interest are indicators that measure energy transfers, nutrient cycling, recovery potential, and species productivity. Key indicators that pertain to disturbance and productivity measures and are potentially amenable to remote sensing methods include (Goodenough et al., 1998: Table 2.3):

- Area and severity of insect attack.
- Area and severity of fire damage.
- Percent and extent of area by forest type and age class.
- Area and severity of disease infestation.
- Percent area successfully naturally regenerated and artificially regenerated.
- Mean annual increment by forest type and age class.

Characterizing forest ecosystem condition requires field data, remote sensing data, and models based on current ecophysiological understanding. Forests are not static, but instead are in a constant state of flux. The number of dimensions over which fluxes take place is large, and not yet fully known for a wide range of forest conditions and ecosystems. For example, an integrating measure of net performance (e.g., accumulation) and trophic status is *biomass*, often taken to represent the mass of living organisms inherent in an ecosystem. In the Canadian Council of Forest Ministers' (1997) C&I system, net biomass production expressed as the mean annual increment (MAI) is considered a measure of forest ecosystem condition.

The relationship between age (or ecosystem development stage) and productivity is complex (Figure 2.2). In forests, this curve is usually simplified using empirical models developed at permanent sample plots (PSPs), which express growth and yield as a function of age. This approach relates mean annual increment to site

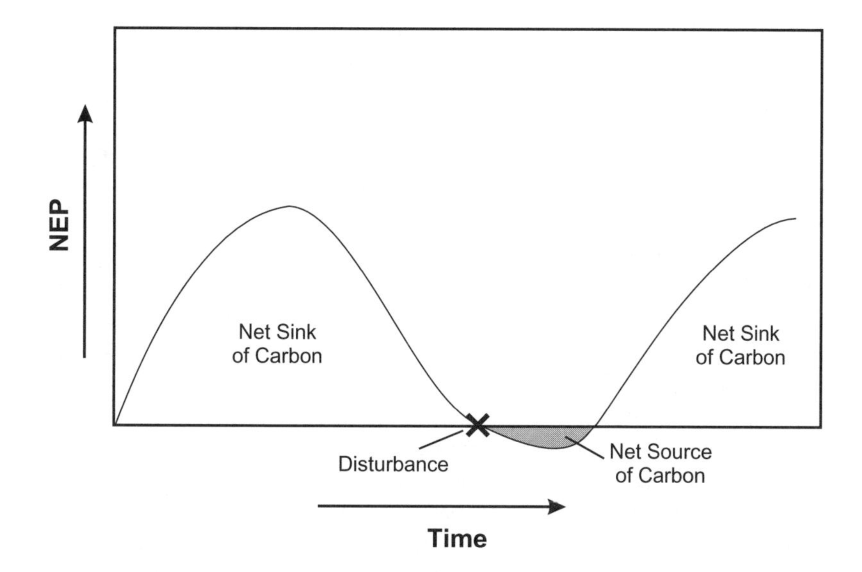

FIGURE 2.2 The relationship between age and net ecosystem productivity can be represented by an idealized curve that provides target patterns for remote sensing and modeling. As the forest matures, carbon is stored and the amount of carbon sequestered can be estimated by a combination of remote sensing and process modeling based on climate and initial conditions. Disturbance often creates a net source of carbon that can be remotely sensed (by observing forest changes) or modeled. Over time the forest recovers, begins regeneration, and the processes in continued growth (carbon, nutrient, water cycles) can be monitored by remote sensing and modeled.

conditions to provide a coarse estimate of productivity. Net primary production (NPP) is the increase of foliage, branches, stems, roots, and plant reproductive organs, and is dependent on the pattern of carbon allocation. Annual NPP is calculated from gross photosynthetic production (GPP) by accounting for autotrophic respiration (Waring and Running, 1998) which can, under certain conditions, be assumed to be a constant fraction (Prince, 1991; Hunt, 1994; Waring et al., 1998). Autotrophic respiration is the process by which plants expend photosynthetic products for the synthesis and maintenance of living cells.

Early successional stages of ecosystems and disturbed ecosystems are energetically inefficient as measured by longwave radiation (Schneider and Kay, 1994; Bass et al., 1998). Stressed forest ecosystems will also differ from healthy ecosystems in energy interactions. For example, because of increased latent heat transfer, a transpiring canopy appears relatively cool compared to a water-stressed canopy (Coughlan and Dugan, 1997). As forests grow they can increase in structural complexity and productivity; overmature stands may become more susceptible to disturbance (Luther et al., 1997) and may show a decline in productivity, species richness, and biodiversity. Disturbance and stress influence the chemical and structural conditions in forest ecosystems, and influence ecosystem functioning. After disturbance, a forest ecosystem will typically show an initial increase in net primary productivity, peak at a site-dependent age, and then decline over time. The initial site quality and cause and severity of disturbance is important; in many forest ecosystems different disturbances will create different successional pathways (Jakubauskas 1996a,b) and in this way influence the form of the NPP:age relationship.

Ecosystem resilience reflects the persistence of ecosystems and their capacity to absorb changes while maintaining productivity and biodiversity (Canadian Council of Forest Ministers, 1997). Biotic (e.g., insects and disease) and abiotic (e.g., fire, pollution, wind, ice) disturbance and stress affect forest ecosystem resilience, and hence, general health, productivity, and succession. Many forests require periodic disturbance to maintain stability (Ludwig et al., 1997); ecosystems with greater regenerative capacity and a balanced distribution of forest types and age classes are considered to be more resilient, and therefore more sustainable (Canadian Council of Forest Ministers, 1997).

In Oregon, five forest stand types ranging from open areas where new trees were being established to old-growth structures, were chosen as management targets to represent forest development that historically resulted from patterns of disturbance (Bordelon et al., 2000). This structure-based management approach coupled with a landscape design system allowed managers to consider changes in wildlife habitat, such as interior habitat area and patch placement, that resulted from management treatments such as stand thinning, shelterwood, group selection cuts, and patch cutting. The forest management system was considered sustainable because the pattern and arrangement of forest conditions — the structure of the landscape — was incorporated into the overall management objectives for each specific area.

Conservation of Soil and Water Resources

Forestry practices must be designed and implemented to conserve soil and water resources, for these characteristics of ecosystems are fundamental to sustainability. Maintenance of appropriate levels of soil oxygen, nutrients, moisture, and organic matter is tied to productivity and water quality. Anthropogenic and environmental influences on the quantity and quality of water may be a function of annual and seasonal variations in precipitation and temperature, forest disturbance, land use, and site-specific factors such as the type and use of harvesting equipment, road location, and pollution discharge. One obvious indicator of the conservation of soil and water resources is the amount of forest land converted to nonforestland use (particularly urbanization, including roads and various industrial uses). Many forests contain areas that are managed primarily for soil and water protection (e.g., major urban watershed supply areas). Remote sensing may be an ideal tool to use in mapping such areas and contrasting their condition with areas under different management treatments; one possible key is to link an understanding of effective strategies to achieve water quality objectives with information on the spatial distribution of changes in land use (Scott and Udouj, 1999).

Monitoring soil properties is not simple. Remote sensing may have only a minor role in the measurement of forest soil quality indicators. Soil is inherently variable and no single measurement is likely to accurately reflect the productivity or quality of a site (Payn et al., 1999). One requirement is to distinguish between direct and indirect (i.e., inferred from changes in forest growth) measures of soil quality (Smith and Raison, 1998). The links between the soil resource and ecosystem health and productivity are multiple and not completely understood. It appears likely that only locally calibrated indicators will be sufficient to provide quantitative inferences about soil health and productivity; for example, the way in which "soil organic matter affects ecosystem processes (e.g., nutrient supply, infiltration, storage of water) will vary in different forests" (Smith and Raison, 1998: p. 129). The relationship between soil organic matter and soil fertility has been shown to differ in native forests and plantations with differing treatments and management. Soil disturbances (e.g., compaction, erosion) are highly variable, and can provide negative feedbacks to hydrology, aquatic resources, and forest productivity (Richardson et al., 1999).

Vegetation is a major factor in the hydrologic cycle, as shown in Figure 2.3. The individual processes which combine to create the forest hydrologic cycle are rea-

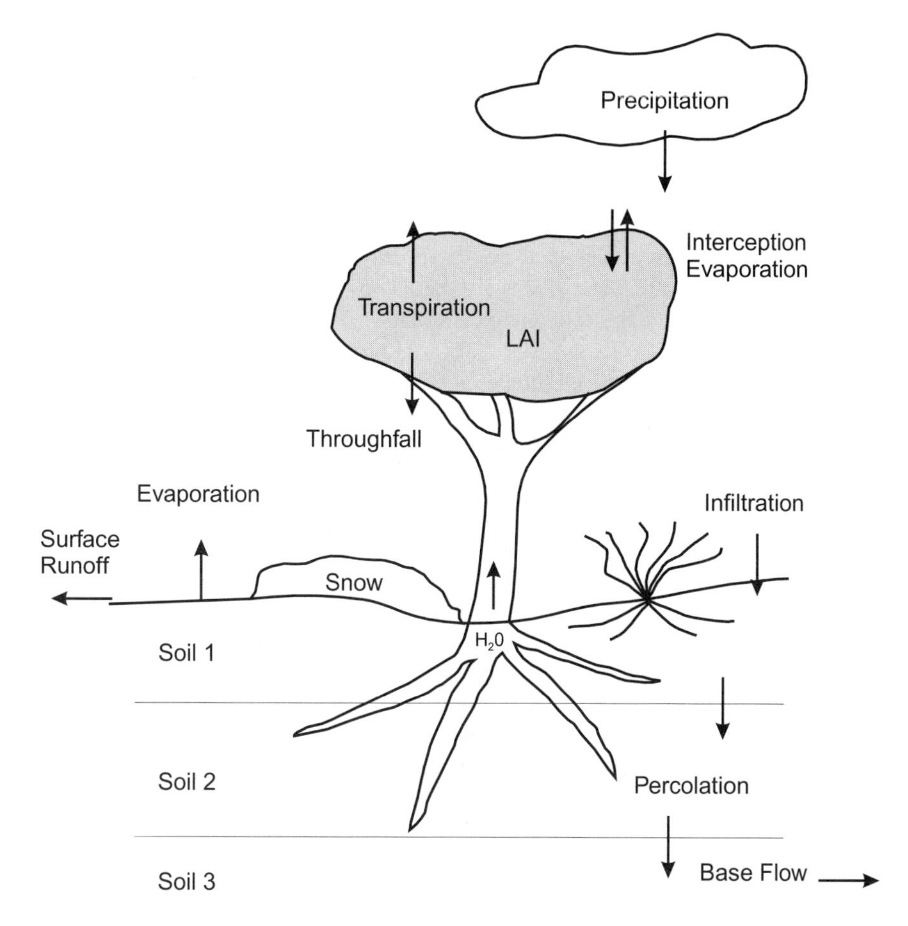

FIGURE 2.3 The forest hydrologic cycle represented as a vegetation water balance. Shown are the main processes of interception, infiltration, runoff, evaporation, and transpiration. Issues in remote sensing relate to the ability of sensors to detect and quantify the exchanges. (From Waring, R. H., and S. W. Running. 1998. *Forest Ecosystems: Analysis at Multiple Scales,* 2nd ed., Academic Press, New York. With permission.)

sonably well known (Waring and Running, 1998). Water, in the form of precipitation, is intercepted and evaporated from the surface of vegetation and the litter layer, and infiltrates the soil. Extraction by roots, surface runoff, percolation through to the water table — these processes are affected by the density and depth of root channels and organic residue incorporated into the soil (Waring and Running, 1998). Water taken up by plants may be stored temporarily within the stem, branches, and foliage, but most is transpired from the leaves to the atmosphere through leaf stomata (Waring and Running, 1998). As soils dry out, resistance to water flow into tree roots increases, and stomatal conductance is reduced (Peterson and Waring, 1994). Heat and water vapor transport processes, such as transpiration, are strongly influenced by LAI and by vegetation roughness. One of the key indicators related to the hydrologic cycle that is likely to be obtained by remote sensing is the quality and amount of foliage.

Forest Ecosystem Contributions to Global Ecological Cycles

Forests play a major role in the global carbon cycle (Canadian Council of Forest Ministers, 1997) by virtue of the fact that they occupy one third of the land surface, but account for two thirds of the net annual photosynthesis (Berlyn and Ashton, 1996). As forests grow and die, large fluxes of carbon occur between forests and the atmosphere through processes of photosynthesis, respiration, and decomposition (Figure 2.4). Carbon and water fluxes are regulated by climate, but are significantly impacted by natural and anthropogenic disturbance. Whether a forest and associated forest soils are a sink or a source of carbon is not a simple question. The carbon fluxes are confounded by local climate, age, standing biomass, species and successional dynamics, nutrient status (site quality), disturbance regime, and management.

Photosynthesis is the process by which plants assimilate atmospheric CO_2 into reduced sugars (Waring and Running, 1998). Respiration releases CO_2 back into the atmosphere. Ryan et al. (1997) estimated that autotrophic respiration can be more than 50% of the carbon fixed in photosynthesis in forest ecosystems, and may regulate productivity and carbon storage because respiration increases with temperature. Carbon products are used in net primary production (NPP): the growth of foliage, branches, stems, roots, and plant reproductive organs. As living cells die and accumulate on the forest floor, a substrate is formed that supports animals and microbes which, through their heterotrophic metabolism, release CO_2 back into the atmosphere (Waring and Running, 1998).

The carbon and water cycles are intricately linked with the cycling of minerals through forest ecosystems. Waring and Running (1998: p. 101) described the broad pattern:

> *Precipitation washes minerals from the atmosphere and deposits them on leaves and other surfaces. Water carries dissolved minerals into the soil where they are taken up by roots and transported in the transpiration stream. Water also carries minerals out of the system through erosion and by leaching. Plants respire carbon obtained through photosynthesis to convert minerals from elemental to biochemical forms and to recycle nutrients from older to newer tissues. Heterotrophic and symbiotic organisms rely on carbon supplied from roots and that extracted from detritus to acquire their energy*

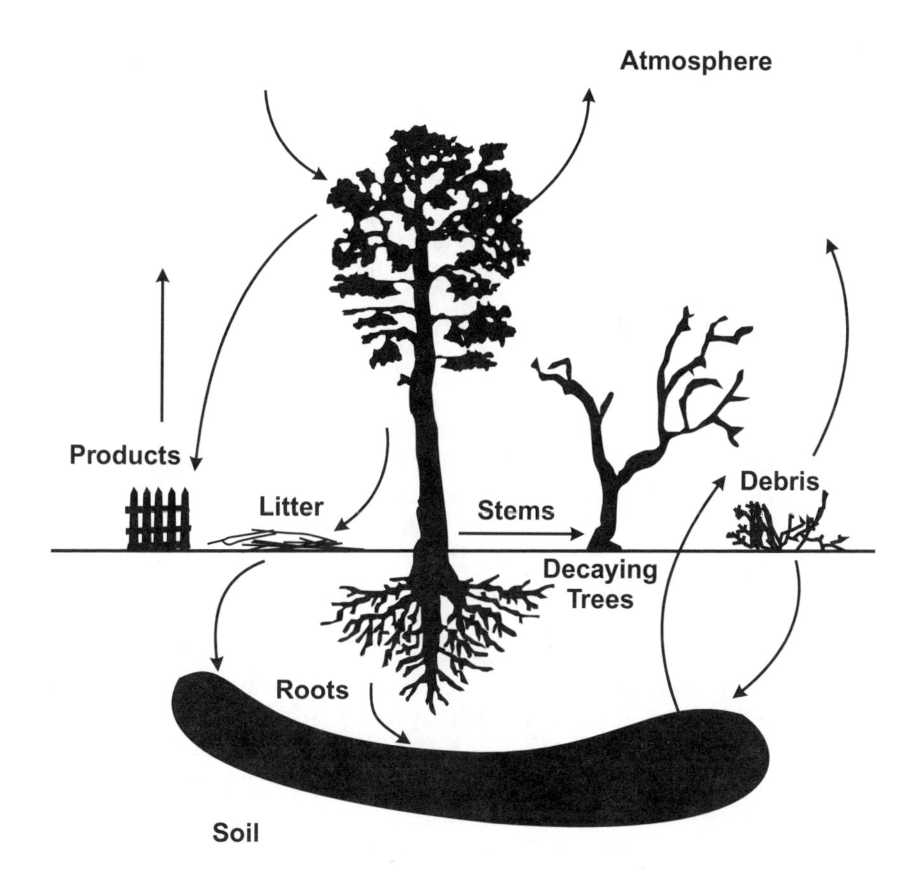

FIGURE 2.4 The forest carbon cycle is comprised of the carbon fluxes caused by photosynthesis and respiration. Issues in remote sensing relate to the ability of sensors to detect and quantify the exchanges. (From *Criteria and Indicators of Sustainable Forest Management*, Canadian Council of Forest Ministers, 1997 Policy, Canadian Forest Service, Natural Resources Canada, Ottawa. With permission.)

> *supply and nutrients. Low molecular weight acids produced as metabolic products enhance the release of additional minerals from soil and rock. Other products of microbial decomposition contribute to the accumulation of soil humus.*

Symbiotic, asymbiotic, and biological N-fixation, for example, are the major sources of N in many types of ecosystems, while weathering of rocks produces most of the other essential minerals (Waring and Running, 1998).

Several ways of determining the forest ecosystem contribution to global ecological cycles exist; none is yet precise enough to influence local management. From a remote sensing perspective, there are many possible contributions that can be made in understanding and monitoring forest ecosystem contributions to global ecological cycles at a wide range of scales, from the forest stand to the biome level. Of the 20 indicators in the Canadian Council of Forest Ministers (1997) C&I listing under this criterion, remote sensing was thought by Goodenough et al. (1998: Table 2.3) to be a contributing information source to seven indicators (some of which were mentioned in the previous sections):

- Percent canopy cover.
- Percent biomass volume by general forest type.
- Area of forest depletion.
- Area of forest permanently converted to nonforestland use, e.g., urbanization.
- Semipermanent or temporary loss or gain of forest ecosystems, e.g., grasslands, agriculture.
- Surface area of water within forested areas.
- Tree biomass volumes.

An obvious and complementary approach to direct measurement of indicators relies on general mechanistic ecosystem models of forests that simulate the fluxes of carbon, water, and nitrogen. Such models have been tested in a wide range of conditions and have been available for use with remote sensing data for some time (Running and Coughlan, 1988; Running and Gower, 1991). "These models are based on a series of interlinked hypotheses that specify the way that water, carbon and nitrogen move through a forest ecosystem and the mechanisms that regulate these fluxes" (Peterson and Waring, 1994: p. 212). The models are essential in developing carbon budget estimates at a variety of scales. Improvements in data inputs and process models may be an important element of the reporting required under the Kyoto Protocol; credible estimates of carbon in forest ecosystems remains a high priority before a viable carbon market and the provisions of the Clean Development Mechanism can be invoked (Pfaff et al., 2000).

A model of forest productivity designed to estimate global carbon budgets is expected to interface to global climate models, providing the feedback between the biosphere and the atmosphere that is critical in closing the cycle. A key variable in the process models is leaf area index (LAI), quite difficult to obtain in the field (Buckley et al., 1999), and nearly impossible to obtain for large areas with the precision required by the models. LAI may be estimated by climate modeling, but for large areas of the world there is a large and unpredictable difference between the LAI predicted by climate, and the actual LAI of the vegetation. Management activities may be one source of the differences. This situation prompted Running et al. (1989) to suggest that of all possible forest or vegetation variables obtainable from remote sensing, LAI may be the variable of greatest importance. This is because of the value of LAI in parameterizing models of hydrological, carbon, and nutrient cycles.

Multiple Benefits of Forestry to Society

Understanding the flow of economic and other benefits to society from the finite but renewable forest resource is a critical component of sustainable forest management (Canadian Council of Forest Ministers, 1997). Of particular interest are issues related to interplay between economic and recreational values of forests, the social and political processes by which such values are incorporated into planning procedures, and the comparison between different management approaches which might be designed to optimize the return of certain values. Many of these measures will be qualitative, by virtue of the near-impossibility to measure them quantitatively.

It is not expected that remote sensing will have a major role to play in monitoring such indicators. For example, of the 16 indicators in this criteria, only 4 were considered to be even partially obtainable by remote sensing methods (Goodenough et al., 1998: Table 2.3). These were restricted to general indicators of changes in the land base (such as the amount of land available for timber production), and economic indicators, such as the annual removal of forest products relative to the volume of removals determined to be sustainable.

The amount of forestland held in various states of protection can also be an indicator of the value that protected forests have in the current economic climate; but remote sensing can only document the physical state of forest conditions, not political status. The issue of habitat and wildlife species sustainability also arises in the context of this criterion, but it is unclear if the historical distributions of habitat should be used, or if there are reasons to consider only the current status. Certainly, remote sensing methods can be used to help understand and document habitat. Historical methods of analysis (aerial photographs, predictive models) have been developed but, of course, are severely restricted in that available time frames tend to be only the most recent.

Accepting Society's Responsibility for Sustainable Development

Like the previous criterion, this aspect of sustainable forest management is aimed at the societal response and collective responsibility to manage forest resources sustainably. Why forests are destroyed or conserved by societies that are dependent on them is a social question (Cooke, 1999). Remote sensing can contribute to understanding this larger question, but none of the suggested Canadian Council of Forest Ministers (1997) 16 indicators under this criterion can be obtained by current remote sensing technology (Goodenough et al., 1998: Table 2.3).

There is a need to be more certain of the effects of different management treatments, a need to understand disturbance regimes and the requirement under every management plan for predictability, and the need to include other stakeholders in decisions about the landbase in the wholesale move to sustainable forestry practices (Rice, 1996). In the former Soviet Union, centralized planning is thought to have led to unsustainable practices (World Bank, 1997) as site-specific decisions were controlled at the regional or national level; a sharp discontinuity occurred between the scales of personal perception and management decision. Management is best coordinated across a spatially nested organization (Oliver et al., 1999), the core of which is the individual operations on a specific stand, considered as part of a landscape, contained within a region (subforests and forests), and a component part of a national forest policy or strategy. This approach assumes that forests can be considered as comprised of units in a spatial hierarchy, the smallest of which is the forest stand or ecosystem. If decisions at this lowest level of the hierarchy propagate through the system and are used cumulatively at each succeeding level, there is a greater likelihood that overall coordination will be achieved in sustainable practices. For example, a decision to reduce harvesting in one area to sustainable levels will not result in a decision to increase harvesting in another area to unsustainable levels, simply to make up the difference.

Role of Research and Adaptive Management

The role of research in forest management may be changing in response to the emerging and evolving need and desire for sustainable management (Vogt et al., 1996). In the past, environmental research and monitoring networks were separate activities, each typically with greater resources, individual emphasis, and excitement associated with the research endeavor (Noss, 1999; Bricker and Ruggiero, 1998). Now, it appears that a new relationship between managers and scientists is emerging based on the broad applied information needs in sustainable management (Bormann et al., 1994; Dale 1998). It is suggested that management and research will gradually become more interrelated. Each can benefit from the other (Hobbs, 1998). Management will benefit — with better information at the appropriate scale; research will benefit — with greater resources and sharpening of research hypotheses.

A key feature in sustainable forest management is adaptive management — the feedback and adjustment parts of the management paradigm. Essentially, when adaptive management is employed in managing a forest area, all management activities are viewed as experiments or are considered part of an ecological test of existing theories and models. This requires a minimum level of scientific understanding by managers, which must continue to increase under adaptive management approaches (Bormann et al., 1994). Such understanding must be acquired and communicated within a changing social and policy context for research and the practice of scientists. Adaptive management also creates new demands on scientists to clarify their hypotheses and to collect appropriate data (Likens, 1998). In many ways, it is no longer acceptable to create broad research programs with vague principles as goals, designed to collect enormous amounts of data and with few tangible benefits. The research agenda of funding agencies is increasingly tied to focused research initiatives with identifiable benefits and products. Some scientists resist this trend; others embrace the challenge.

Perhaps the greatest challenge lies not with ensuring increased scientific understanding, but with recognizing and ensuring social diversity and acceptance in the larger community (Keen, 1997). Scientists increasingly must engage the communities in which they work; resisting the impulse to avoid review in the wider public venues.

The greatest technical difficulty in reviewing many of these criteria and indicators is defining the significant degree of change for each indicator (Smith and Raison, 1998). The goal is to be in a position of measuring indicators at several points in time, and using the change in various indicators to assess the sustainability of a given forestry operation. Since it is known that various indicators will require local calibration to determine their relationships with important ecosystem processes (e.g., forest growth, level of suspended sediment in water), the levels of change (thresholds) associated with a forest practice will change in different forests. How much change is a significant change? A technical challenge relates to these thresholds, since for many of the indicators proposed (e.g., fragmentation, erosion, soil organic matter) it is presently not known what specific quantitative values represent the threshold between sustainable and unsustainable levels. Again, adaptive management seems likely to provide a workable management strategy (Gregersen et al., 1998): define indicators, measure indicators, conduct research on indicator thresholds in a variety of settings, and change practices as unsustainable activities become apparent (Boyce and Haney, 1997).

INFORMATION NEEDS OF FOREST MANAGERS

What data are needed to apply and assess adaptive and sustainable forest management? This is a very large question to which only partial answers exist — simply, it is not yet known for certain, with confidence, what data are required. However, many key data are well understood in terms of their role in developing and monitoring sustainable forestry practices. The question can be narrowed and simplified; still, there are no easy answers. For example, for those managing forests under climate change scenarios, a pertinent question would be, What data are needed to modify the management approach and forest practices if the climate changes? It is probable that for many forests the current relationship between climate and disturbance, climate and spatial distribution of forest covertype, or above-ground biomass, is not well understood. What about the influence of climate on forest productivity (Hunt and Running, 1992)? Evapotranspiration? Runoff?

To begin answering such questions, one can look at the C&I lists, suitably modified for the local environment, and begin to construct a list of the data — and the subsequent information obtainable from those data — that would be required in order to understand if current or future practices are sustainable in managing a forest. An example has been provided by Landsberg and Gower (1996) in their discussion of the role of remote sensing, GIS, and ecological process models in sustainable forest management (Table 2.4). In essence, based on physiological principles, they list management-level information in the form of layers of data that managers will require to handle concerns with production forestry, water yield, wildlife protection, or the impact of management practices such as clearcutting or burning. Their list is made possible by research aimed at increasing understanding of ecosystem functioning, and the use of this understanding in management (Waring and Schlesinger, 1985). Together with the progress in identifying social and economic criteria and indicators of sustainable forest management, it is now possible to begin to identify the kinds of information needs that exist now, or will exist in the near future, in order to satisfy the approach to sustainable forest management.

Forest managers and operational foresters need information on at least three separate dimensions or scales in order to execute properly and within a sustainable context any operational forest activities (Weintreb and Cholaky, 1991):

1. Operational information needs are specific to local areas; operational information is required to answer such questions as where to harvest, where to implement silvicultural treatments, and where to find certain forest structures such as old-growth stands. At this level, for example, foresters need to know species distributions within forest covertypes,

TABLE 2.4
Information Layers Necessary in Sustainable Forest Management and the Most Likely Source of Direct Information from Four Different Yet Synergistic Resources: GIS, Remote Sensing, Modeling, and Field Data

	Data Source			
Data Layer	**GIS**	**Remote Sensing**	**Model**	**Direct Observation**
Direct Information				
Topography	*	*	*	
Soil Nutrient Status	*		*	*
Soil Water-Holding Capacity	*		*	*
Forest Type		*		
Stand Density		*		*
Average Tree Height		*		*
Standing Biomass		*		*
Leaf Area Index		*	*	
Solar and Net Radiation				*
Temperature				*
Air Humidity				*
Precipitation				*
Derived Information				
Absorbed Photosynthetically Active Radiation				
Water Use				
Water Balance				
Water Constraint on Growth				
Temperature Constraint on Growth				
Nutritional Constraint on Growth				
Biomass Increment				
Partitioning Coefficients				
Mass of Tree Components				
Stem Volume				

Source: From Landsberg, J. and S. T. Gower, *Applications of Physiological Ecology to Forest Production,* Academic Press, San Diego, 1996. With permission.

forest stands, and forest ecosystems. It is at this level that the highest detail and the greatest uncertainty often exists, but perhaps the consequences of that uncertainty are not considered too great since the areas involved are small — the homogeneous or acceptably heterogeneous forest stands and ecosystems of which they are a part. These operational information needs are embedded in a larger context.

2. At some higher level, tactical information needs represent a lower level of detail but cover a broader area; for example, the management information necessary to plan or determine optimal road location, or to consider biodiversity management in which the landscape structure plays a part.

3. These information needs are embedded in an even larger framework of strategic information needs. This level might include requirements for broad, general information, such as an understanding of forest covertypes and ecoregion distribution over large areas. To make a strategic decision, such as the allocation of a certain area of land to timber production and another area to wildlife habitat, a relatively broad level of information is required initially.

Forest managers recognize the inherent uncertainty of the present and future (Franklin, 1997); but they still have a job to do, a day-to-day function, and a career perspective that requires a constant stream of information at the operational, tactical (or planning), and even the strategic level. Their many subsequent activities are based largely on that information and the world view that informs and shapes their lives. But the complexity of forest management — even in traditional areas — has increased due to changing demands on the forests and the technological developments in many areas, including management systems (Bachelord and Griffith, 1994). This has increased the knowledge requirements for forest resource management professionals. For example, there is an increasing need to understand and use computer simulations and geographic information systems; such systems are not simple to understand and use, requiring education and training. For certification as a professional forester in the U.S., the Society of American Foresters (SAF) requires at least one course in the general area of Surveying and Mapping (includes photogrammetry, remote sensing, land surveying, mapping and area determination, and Geographic Information System applications) (Sader and Vermillion, 2000).

Increased exposure to these topics is likely to be valuable, but obviously new areas of teaching should not be implemented at the expense of valuable, older or traditional forestry education. Of course, forestry is not unique in struggling with the common problem of the limited curriculum vs. the real needs of the professionals in the field (Bachelord and Griffith, 1994).

In traditional areas, information requirements encompass traditional forestry knowledge areas, such as silviculture and harvest planning, coupled with increased demands for comprehensive forest ecological understanding. In the middle of the last century, Leopold (1949) remarked that increased ecological understanding of land could be found in a wide range of disciplines — geography, agronomy, history, botany, and even economics … not necessarily in prepackaged education, but more likely a result or product of social evolution. Clearly, there is an urgent need to understand the development and dynamics of patterns in ecological phenomena: the role of disturbance and the characteristic spatial and temporal scales of vegetation (Urban et al., 1987). Such issues cannot be addressed adequately without an understanding of the likely sources of data that will be used. Franklin and Woodcock (1997: p. 146) put it this way: "What are the spatio-temporal scaling properties of geographic phenomena (vegetation) that affect the integration of remote sensing with biospatial data in [I]GIS?" Can this question even be understood without significant training beyond the traditional knowledge areas of forestry and ecology?

There is a requirement to maintain detailed tree and stand knowledge within the context of a landscape-level perspective, including knowledge of the consequences of some relatively poorly understood concepts, such as spatial heterogeneity (Weishampel et al., 1997), forest fragmentation (Saunders et al., 1991), and connectivity (Forman, 1995). Today's forest managers require a detailed knowledge of ecological land classification, and the multiple relationships between vegetation communities and forest stands as represented in the forest inventory. Ecological land classification provides information on the structure of ecosystems and the relations between landforms, soils, and climate, but these relations must be interpreted, sifted, sorted, and used to explain or understand forest stand conditions and processes. The forest inventory itself is a key requirement or information need that must be understood; its strengths and weaknesses, its limits, and its potential and future development. The relationship between inventory variables — easily measured and readily available — and physiological variables, such as LAI and stand growth, must be explicit; there is a need for greater precision and understanding in these relationships and their use. Multidisciplinary, large-scale, multiownership approaches have increased, drawing on traditional forestry (such as silviculture) and genetics, physiology, microclimate, and ecology (Buckley et al., 1999; Larsen et al., 1997). There is a need for forest managers to understand historical distributions and natural heritage systems (Delorme, 1998).

Few expect that individual forest managers will be fully versant in all the knowledge areas pertaining to traditional forestry, physical, and biological understanding of forest ecosystems, technological approaches, and human/environment relations. Rather, teams of professionals are called upon to contribute expertise and insight. Thus, social information needs such as knowledge of team dynamics and group approaches to problem solving, could be added to the list of growing knowledge requirements of forest managers (Table 2.5).

Many of these knowledge requirements will coincide exactly with the information needs required to allow existing, sometimes unsustainable, management practices to continue. In other words, often the information needed to make sustainable forest management decisions already exists. Perhaps this information is not used, is poorly understood or ignored, or is overruled by other considerations. Often, the management problems may be stated independently of the information needs that are required to enable foresters to offer sustainable management solutions. For example, many forests cannot be managed without access to information on areas surrounding the actively managed forest — the multijurisdictional regional setting. Yet such information is often not available, and when available, is rarely compatible with the information available inside the forest management area. This highlights one specific need for forest managers — the need for information on their forest region in the context of larger area patterns. At the same time, there is ever-increasing pressure to extend management decisions and to make such decisions more soundly and with projections over space and time (Waring and Running, 1998).

Technological developments have created an entirely new class of information needs for forest managers; the need to understand new methods, applications, and potential new sources of information. Managers must possess the ability to think critically about these technological developments, to interpret relative worth, and to actually accomplish or supervise tasks with the new tools. This means creating an awareness of the infrastructure and applications environments in which these tools

TABLE 2.5
Summary of the Information Needs of Forest Managers

Traditional Forestry

- Forest Structure
- Disturbances
- Silviculture
- Harvest Scheduling

Increased Physical and Biological Understanding of Forest Ecosystems

- Large Area Landscape Structure and Dynamics
- Physiology, Genetics, Biodiversity
- Multiple Spatial-Temporal Scales

Technological Approaches

- Remote Sensing
- GIS
- Spatial Statistics
- GPS
- DSS

Human/Environment Interaction

- Anthropogenic Factors
- Noneconomic Benefits

Social Interaction

- Public Participation
- Interdisciplinary Team Environments

operate — understanding the standards and accuracy issues, the complexity and utility of GIS, the role of sampling forests and the use of geostatistics, and the role and function of ecosystem process models. While some researchers appear to have taken a narrower view of the value and potential of individual components in this panoply of technology (Holmgren and Thuresson, 1998), many users and researchers of the technological approach have understood for some time that the integration of different technological developments, such as the use of GIS, ecosystem models, decision support systems, and remote sensing, together represent the true value of the evolving technological infrastructure that has emerged in support of sustainable forest management. This integration is part of the larger need to unify planning, assessment, monitoring, and research (Noss, 1999). If sustainable forest management is to succeed, there is a need to understand the methods and potential of the technological developments in remote sensing.

Some Views on the Way Forward

It is apparent that forest management in the future will be increasingly tied to forest practices, measured or modeled on the ground, and in several different ways. For example, there will be increased accountability of forest management practices to a wider cross section of society, and there will likely be additional independent monitoring of plans and adjustments as new ideas and feedback become more available in open public venues. The intensity and the diversity of scrutiny will increase. It seems likely that new skills and new technologies will play an increasingly important role in sustainable forest management. Two separate projects can be used to illustrate the new direction of forest management and the requirements for information that managers will need to master in order to make progress toward sustainablility.

The first project, in Oregon, makes use of a newly developed forest model called FORECAST (Seely et al., 1999). This approach is based on a decision support system framework comprised of a user interface, an ecosystem simulation model, and a rule-base (or knowledge base). Each unique regional forest type requires its own calibration data set containing historical bioassay data in four categories: tree data, plant data, bryophyte data, and soil data. These data define the historical growth characteristics and rates of various soil processes and are used by the model to simulate ecosystem development within a range of silvicultural systems, management activities, and disturbance events. Output allows the user to consider temporal patterns of change in standard growth-and-yield parameters, ecosystem state variables, and process rates. A unique feature of the model is the ability to provide summary statistics of forest resource values, including economics, employment, energy consumption, carbon, and nutrient budgets. This allows direct comparison of the impact of alternative stand-level management treatments.

This approach shows much promise and is clearly an extension of current management practices, but with heavy emphasis on better computer models, a richer field data set, and a comprehensive knowledge base. The approach can use remote sensing data in several ways: first, to map spatial distributions of forest types; second, in obtaining strata in some of the bioassay categories; and third, in monitoring model outputs (such as area of disturbance or silvicultural changes). In emphasizing new tools such as computer models, several scientists have provided a note of caution against relying too heavily on models and — an all-too-common problem — marginal data. Walters (1998) honestly reported a common feeling among those trying to integrate mechanistic biophysical understanding into a framework possible of making predictions over multiple time and space scales. The "frightening and humbling experience" reported when involved in modeling the hugely complex reality of forest ecosystems might be a sobering antidote to indiscriminate and wrongful use of the emerging technologies.

A second example of the way forward was provided by Kuusipalo et al. (1997) in a lowland Indonesian dipterocarp forest management area. Sustainability indices were calculated by considering weights for the importance of environmental, social, and economic elements of sustainability. These indices try to balance the complexity of the C&I approach with the reality of having available only a few real measures. The data were very qualitative. Participants from the industry and community were asked to define priorities on each of three elements relative to different management strategies including plantation establishment, multiple use, and forest reclamation, among others. The three elements used were based on environmental, social, and economic conditions. The environmental elements included the effects of different management alternatives on (a) atmospheric carbon balance, (b) soil erosion and nutrient leaching, and (c) overall species richness. The social elements were expressed as ordinal rankings of religious, cultural, and traditional values. The economic elements were divided into two decision attributes related to short-term profitability (10 years), and a rotation-related valuation (45 years).

A hierarchical decision analysis method was used to find global weights based on participant input; together these priorities and weights were combined mathematically to reveal the best management strategy to achieve sustainability. The forest reclamation strategy emerged as a good compromise between the three elements of sustainability for this region and current situation; forest reclamation contained the best balance between short- and long-term economic output, and was strongly influenced by the environmental elements of sustainability. Kuusipalo et al. (1997: p. 112) suggested in their case study that the method represents one possible starting point for a sustainable forest management planning approach in the region, but they pointed out that "the final choice is always a task of human decision-makers."

ROLE OF REMOTE SENSING

Remote sensing can be designed to support sustainable forest management in the presentation and reporting on the criteria and indicators of sustainable forestry, and in the modeling and projections at a variety of scales based on a common understanding of biophysical and ecological principles (Berry and Ripple, 1996). Remote sensing, together with GIS and computer simulation models, appear poised to make significant contributions to the way in which the remaining forests of the world are managed. Time and experience will be necessary before the appropriate role of remote sensing and GIS technology is understood in sustainable forest management.

A lack of experience in using GIS and remote sensing in an integrated way is natural (Martinez et al., 1996); these are emerging technologies, only very recently considered as real tools for use by those concerned with the whole process of forest planning, operations, and management. GIS and remote sensing should be viewed in the same way that other emerging technologies in forest management are viewed; their adoption and continued development will be similar to the adoption and development of other emerging technologies in forestry (Mater, 1998) such as technologies for low-impact harvesting, new drying, short-piece and scrap utilization, scanning, wood hardening, efficient logger programming, and fiber and solid wood waste conversion. Why should remote sensing technology be any different? If anything, these other technologies have a more focused purpose, and are more readily shown in economic terms to be viable technological improvements over existing procedures. Few would consider the potential impact on forestry of new scanning or wood hardening technology on a par with the potential impact of remote sensing. What are the potential impacts of remote sensing? This is a large question to which only partial answers are available.

The first step toward sustainable development of the forest resources of any country or region is to clarify the status of the resource base; regionally appropriate use of GIS and remote sensing mapping, monitoring, and modeling, and direct field measurements are requirements for effective forest planning and management (World Bank, 1997). Beyond this inventory requirement, there are many operational and research uses of remote sensing in forestry applications, but the role of remote sensing cannot yet be definitively stated. This is simply a result of the extremely wide range of potential applications, and the still-rapidly changing technology. Remote sensing and related (e.g., GPS, GIS) technology continues to develop at a rapid pace, and the applications continue to be tested and developed, particularly in the research community.

In the meantime, new satellites and airborne sensors have been built, new algorithms to extract information from these images have been added to commercial systems, and a continuing research agenda has been pursued worldwide at multiple scales and resolutions. These efforts continue to provide new insights and understanding into the possible role of remote sensing in forestry. But internal developments are not the only source of new knowledge in remote sensing. Externally, things are also moving fast. The field of landscape ecology has shown tremendous growth (Forman, 1995) and a tremendous need for spatially explicit data of the type remote sensing seems ideally suited to provide (Urban, 1993). Forest management to maintain or enhance biodiversity has emerged as a global theme (Simberloff, 1999), requiring new structural data and context for understanding species-richness observations. The whole issue of sustainable forest management certification (Vogt et al., 1999) through monitoring biophysical and social C&I has given remote sensing new demands for information to consider and attempt to meet.

Now, the relevant question for forestry appears to be not whether remote sensing can contribute to forest resource management, but rather, What is the best way for remote sensing to contribute to forest resource management? Routine applications are still rare, but research opportunities have been plentiful and continue to grow in sophistication and planning. General assessments of the role of remote sensing in estimation of criteria and indicators (e.g., Goodenough et al., 1998) are rapidly being followed by specific indicator assessments. For example:

1. Scarth and Phinn (2000) evaluated the use of geometrical-optical (GO) modeling in estimating Australian eucalyptus forest cover indicators (Montreal Process indicators "extent of area by forest type and by age class or successional stage," and "extent of forest type in protected areas defined by age class or successional stage"), and
2. Franklin et al. (2000b) used change detection analysis to determine New Brunswick forest clearcut and partial harvesting patterns, and subsequently reported on the the CCFM indicator (4.1.7) "area of forest depletion."

As will be shown in this book, there are many specific studies in remote sensing of forest condition, productivity, health, and change that will lead to quantitative monitoring of value to forest management. The most significant features for forest management offered by remote sensing continue to be the quantitative form of data, and the repetitive and synoptic coverage provided by the technology (Hunter, 1997). With multitemporal data acquisitions — and the large historical Landsat database — it is possible to virtually time-travel to review landscapes a (human) generation old. With data acquired in a consistent format over time, a number of previously difficult applications have been made more feasible.

Two Hard Examples

Much of this book is concerned with presenting a view of changing forest management juxtaposed with the data, methods, and potential of remote sensing. There is a great challenge in interpreting remote sensing data, in developing remote sensing methods, in finding ways to address the need for remote sensing information as part of the monitoring of criteria and indicators of sustainable forest management. There is no doubt concerning the contribution of aerial photography (Avery, 1968). But a reasonable question to ask, after considering the experiences of others and before considering investment in remote sensing is, Have digital remotely sensed information products been used in forest management planning?

Two specific examples are provided here that suggest the future directions that sustainable forest management and remote sensing are helping to create. These examples contain the core characteristics of successful remote sensing applications in forestry (Congalton et al., 1993):

1. There must be a need for up-to-date and timely information on the forest resource over an area too large or otherwise difficult to survey on the ground;
2. There must be the ability to use innovative (i.e., beyond the conventional) image analysis techniques;
3. The best available remote sensing data with a fully loaded GIS should be provided or acquired with the necessary data;
4. The use of models — that can be run iteratively (and that are adaptable), and that are sensitive to the appropriate changing model variables and parameters;
5. The willingness and ability of the project team to adopt the new technology and shift the analytical paradigm to consider the new data and methods in light of the information products, not in light of conventional procedures and products.

An automated land-use allocation project in Ohio has been reported as the prototype for integrating technology and public input in decision making in National Forests (Zeff and Merry, 1993). The model required a remote sensing forest covertype classification, a digital elevation model, and a GIS soils data layer. The objective was to create scenarios of timber harvest sites and haul road locations. The process was based on five steps:

1. Clustering Landsat TM spectral response patterns;
2. Merging classes to create six forest covertypes;
3. Converting image data to GIS format and grid resolution;
4. Building a model to allocate haul roads from harvest sites (incorporating physical site factors, professional forester expertise, and public input);
5. Determining the relative cost of implementing plans.

The end result of the modeling experiment was to provide three or four options for building roads, with associated costs and environmental sensitivity for haul road suitability. The modeling framework was robust and flexible, allowing multiple iterations of the models in which feedback and modifications were provided. This was not an academic modeling exercise; actual road locations and additional potential impacts such as erosion were identified in the planning. Perhaps the most critical component in the study was the provision by remote sensing of a current map showing six different forest classes as the base upon which the harvest and road layers could be designed.

A project in Oregon, described by Congalton et al. (1993), incorporated the use of remotely sensed forest covertypes and a GIS database to build maps and management plans of old-growth forests. The objective was to understand old-growth forest fragmentation conditions, landscape structure and diversity, spatial distribution of forest cover, and the potential response to management decisions. Planning required access to a database that included slope, aspect, elevation, hydrology, location of inventory and research plots, crown closure (four classes), size class/stand structures (30 classes), species (20 classes), current vegetation-type polygons, suitable northern spotted owl (*Strix occidentalis*) habitat, suitable lands for timber production, flight line maps, habitat conservation areas, forest boundaries, potential Pacific yew (*Taxus brevifolia*) habitat, historical distribution of vegetation, and old growth. Color aerial photographs and orthophotography were available. The satellite data were used in an image classification strategy to generate four classification information products: (1) crown closure, (2) species, (3) structure, and (4) size classes. This process was presented as one way to mimic aerial photointerpretation; by delineating homogeneous areas created by overlaying these four remote sensing classifications, smooth polygonal structures were imposed on the landscape. The final product was not dissimilar to the forest stand maps created and used by aerial photointerpreters.

The ability to create the necessary information products required by management in their decision making for Oregon old-growth forests was attributed to four factors:

1. Increased computer power and subsequent reduction in iterative time during the classification process;
2. Fully integrated GIS, remote sensing, and statistical software that could be readily used to understand the relationships between spectral variations and vegetation patterns on the ground;
3. Experienced personnel — the foresters and geographers who could not only understand what caused a variation in image spectral values, but also understood what caused a variation in vegetation on the ground;
4. The quality and information content of the SPOT and Landsat TM imagery — in their view, the data were "so good that it will be several years before image processing methodology is capable of making full use of it" (Congalton et al., 1993: p. 534).

What foresters require to accomplish their task of sustainable forest management begins with good information of the forest resource with a known error structure. Many types of data are needed that can be used to construct key pieces of information required for management in a timely and cost-effective format. The primary role of forest managers is to integrate a broad range of relevant information so that they can produce and implement practical management plans within an approved policy to best meet the needs of the community at the local, regional, and global scales. Often it is simply a new perspective "from above" that provides the key to understanding the role of remote sensing in forestry.

The central hypothesis of this book is that in order to ensure the successful application of sustainable forest management to the world's forests, a combination of science, technology, and human creativity must be applied to remote sensing, GIS, models, and field data. Remote sensing is one piece of the puzzle, which can be seen in (at least) three different dimensions:

1. As an original data collection tool,
2. As a suite of techniques for spatial analysis and modeling, and
3. As an experimental and normative scientific method in and of itself.

This latter role is perhaps the one with the most potential; remote sensing has "the potential to alter our models, our methods of analysis, and, in essence, influence if not change our paradigms" (Wessman and Nel, 1993: p. 174). In forestry, at least, the role of remote sensing can be conceived as, first and foremost, a way of posing questions that can be answered, i.e., remote sensing as a normative scientific method for all to use and develop; this methodological perspective is provided because remote sensing can be used to provide a spatially coherent answer to many of the sustainable forest management questions posed to fulfill information needs (i.e., remote sensing as a tool and a technique). Finally, remote sensing can be considered an ideal consistent source of data at spatial and temporal resolutions useful for resource monitoring and management — a critical, required input to a wide range of models and mapping projects (remote sensing as an input variable).

A logical first step in the application of remote sensing to sustainable forest management is "to assess the biophysical characteristics of the site that are correlated with the attribute for which information is required, and to apply appropriate image analysis methods" (Gemmell, 1995: p. 303). In other words, acquire the right type of data and apply the appropriate information extraction techniques to those data. But what data are the right type, and what techniques are appropriate — and who decides? Few sure-fire prescriptions yet exist, but much common understanding has emerged that can help users assess whether the information provided is suitable for the decision-making process. From this assessment, it should be possible to begin to define the optimum remote sensing methodologies to answer specific questions about the forest at a range of spatial and temporal scales. A good starting point is to consider the acquisition of imagery, followed by the image analysis options, the integration with GIS and modeling approaches, and the available experience and examples.

3 Acquisition of Imagery

We can look forward to the translation of these capabilities of space vehicles and associated remote sensors into a variety of applications programs.

— E. M. Risley, 1967

FIELD, AERIAL, AND SATELLITE IMAGERY

Digital remote sensing images of forests can be acquired from field-based, airborne, and satellite platforms. Imagery from each platform can provide a data set with which to support forest analysis and modeling, and those data sets may be complementary. For example, field-based remote sensing observations might be comprised of a variety of plot or site photographs or images (Chen et al., 1991) and nonimaging spectroscopy measurements (Miller et al., 1976) which, together with airborne or satellite data, can be used to extend the detailed analysis of a small site to larger and larger study areas. Many types of ground platforms (e.g., handheld, tripod, ladder, mast, tower, tramway or cable car, boom, cherry picker) have been used in remote sensing of forest canopy spectral reflectance (Blackburn and Milton, 1997). The variety of free-flying airborne platforms that have been employed in collecting remote sensing observations is nothing short of astonishing: at various times, airships (Inoue et al., 2000), balloons, paragliders, remotely piloted aircraft, ultralight aircraft, and all manner of fixed-wing light aircraft have all been used with varying degrees of success in remote sensing. While not all of these have operational potential, it is a virtual certainty that in supporting sustainable forest management activities in a forest region, a variety of imagery and data from field-based, airborne, and satellite platforms will be required.

Photographic systems have been designed for plot or site hemispherical photography to characterize canopy conditions (Figure 3.1). A hemispherical photograph is a permanent record and a valuable source of canopy gap position, size, density, and distribution information (Frazer et al., 1997). Measurements on the photograph can lead to estimates of selected attributes of canopy structure, such as canopy openness and leaf area index, and have a role as a data source at specific sites initially or repeatedly measured for the purposes of forest modeling. One model, designed to extrapolate fine-scale and short-term interactions among individual trees to large-scale and long-term dynamics of oak-northern hardwood forest communities in northeastern North America, is based on the provision of key data obtained by

FIGURE 3.1 Field-based remote sensing by hemispherical photography. This image is a closed-canopy black spruce stand in northern Saskatchewan taken from below the base of the live crown, looking up. Data extracted from this image include standard crown closure measurements and estimates of leaf area index. (Example provided by Dr. D. R. Peddle, University of Lethbridge. With permission.)

hemispherical (fish-eye) photography to estimate light limitations (Pacala et al., 1996). The model calculates the light available to individual trees based on the characteristics of the individual's neighborhood.

Field spectroscopy (Figure 3.2) can be used in remote sensing in at least three ways (Milton et al., 1995). First, field spectroscopy can be used to provide data to develop and test models of spectral reflectance. For example, field spectroscopic measurements may be helpful in selecting the appropriate bands to be sensed by a subsequent airborne remote sensing mission. Second, a field spectroscopy design may be used to collect calibration data for an airborne or satellite image acquisition (Wulder et al., 1996). And finally, field spectroscopic measurement may be useful as a remote sensing tool in its own right. Examples of this latter application are common in agricultural crop and forest greenhouse studies designed to relate disease, pigments (Blackburn, 2000), or nutrient status to spectral characteristics of leaves (Bracher and Murtha, 1994). Because field-deployed sensors do not cover large areas

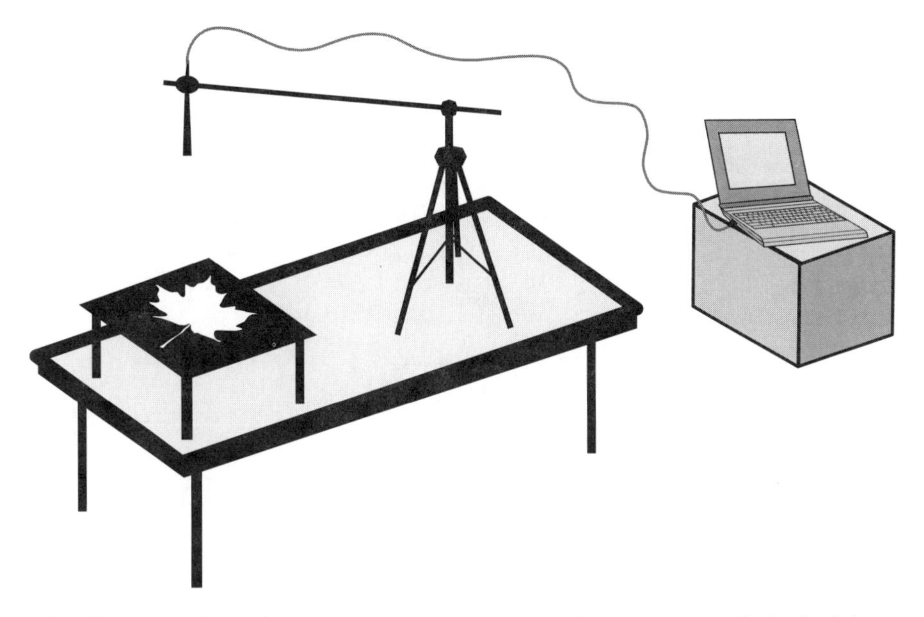

FIGURE 3.2 Field-based remote sensing by spectroscopy. Instrument setup in the field shows the experimental design used to collect spectrographic measurements of vegetation *in situ*. These data are nonimaging remote sensing measurements, and can be used to calibrate other remote sensing data (airborne or satellite imagery).

in the same way that imaging sensors do, sampling must be considered in order to determine the appropriate way to collect the data over surfaces of interest (Webster et al., 1989). The problem is a familiar one: How to determine the appropriate number and locations of measurements to capture the information on forest variability?

The principles of field spectroscopy have been extended through new instrument designs to the emerging remote sensing applications by imaging spectrometers (Curran, 1994) and spectrographic imagers (Anger, 1999). These sensors and applications are considered in more detail in subsequent sections of this book.

One use of airborne systems is to acquire data to validate satellite observations (Biging et al., 1995). Airborne sensors typically offer greatly enhanced spatial and spectral resolution over their satellite counterparts, coupled with the ability to more closely control experimental design during image acquisition. For example, airborne sensors can operate under clouds, in certain types of adverse weather conditions, at a wide range of altitudes including low-and-slow survey flights (McCreight et al., 1994) and high-altitude reconnaissance flights (Moore and Polzin, 1990). In addition, airborne sensors usually exceed satellite systems capabilities in terms of their combined spatial resolution/spectral resolution/signal-to-noise ratio performance (Anger, 1999). Basically, airborne data are of higher quality. Longer exposure times are available to airborne systems. More bands, and optimal bands, can be selected for measurement. Reflectance targets can be deployed with simultaneous measurements of downwelling irradiance at aircraft level which, in theory, creates the possibility of obtaining calibrated, atmospherically corrected surface reflectance data.

TABLE 3.1
Checklist of Flight-Day Tasks for Airborne Mission Execution

Pre-Flight
Location and geometry of flight lines
Azimuth
Length
Survey GCPs and/or Markers
Spatial Resolution
Elevation (across-track pixel size)
Aircraft velocity (along-track pixel size)
Spectral Resolution
Selection of bandwidths
Number of bands
Number of look directions (if applicable)
Location of looks (if applicable)
Bandwidth of scene recovery channel (if applicable)
During Flight
Collection of atmospheric data
Collection of PIFs
Incident light sensors
Geometric positioning data
GPS basestation (differential)
Post Flight
Radiometric processing of image data
Conversion to spectral radiance units
Spectral reflectance determination
Processing of PIFs
Processing of incident light sensor data
Geometric processing of image data
Attitude bundle adjustment
Vertical gyroscope or INS
Differential correction of airborne GPS to basestation

Source: Modified from Wulder et al., 1996.

Flight planning and field-based remote sensing data collection are not infinitely variable, depending on many factors such as local topography and platform capability, but airborne sensors are not limited by orbital characteristics (Wulder et al., 1996). A checklist of the flight-day tasks involved, perhaps following a reconnaissance visit and the detailed flight planning, would include provision for geometric and radiometric ancillary data (e.g., GPS base station, field spectroradiometer for calibration) (Table 3.1). On the other hand, numerous remote sensing service providers exist, able to work from a list of objectives or needs to generate the necessary parameters for the acquisition of the data.

Satellite image providers have developed standard protocols to handle orders. For users, the essential issues relevant to ordering imagery or executing a remote sensing mission are

1. Understand the data characteristics and output formats (e.g., analogue vs. digital products, storage media, and space requirements);
2. Specify the level of processing the imagery will receive before delivery (e.g., radiometric calibration and georeferencing);
3. Specify the environmental conditions (e.g., maximum tolerable cloud and cloud shadow coverage);
4. Consider compatibility with existing imagery and other relevant data.

This final point is an important but perhaps often overlooked issue; data continuity with prior remote sensing data and expected future imagery should be considered part of the investment in remote sensing data acquisition.

The cost of remote sensing is often difficult to determine beyond the acquisition costs, which are usually fixed at a per line or per square kilometer amount. That cost might be more or less directly proportional to the cost of the instrument. Generally, sensor quality is more important than initial sensor cost, particularly in applications where the final cost of the information product is critical (Anger, 1999). This is because much of the cost of remote sensing is embedded in the analysis of the imagery to produce information products. The higher quality (and higher cost) sensor may deliver the information at a lower product cost if those data are more readily converted to the needed information products by requiring less processing. The issue here is a correct matching of the appropriate sensor package and the needs of the user, and a recognition of the trade-off between measurement capability and cost discussed by most system developers (Benkelman et al., 1992; King, 1995). If hyperspectral imagery were required for a forest area it would be very costly to fly an airborne videographic sensor package, since the entire mission cost would be spent on a sensor that cannot deliver the necessary product. But can a satellite hyperspectral sensor acquire the data less expensively than an airborne system? The answer would depend on the ability of the satellite system to generate data of the quality required for the final product.

Criteria for evaluating the cost-effectiveness of information have been suggested as a delicate balance between the characteristics of the information (e.g., unique or new, more accurate, comparable information but different format, and so on) and the cost of producing those characteristics (Bergen et al., 2000). In one early study, Clough (1972) divided 75 mapping or monitoring applications into whether satellites could provide:

1. The same information as currently being used (usually from a combination of field and airborne collection systems),
2. Better information than currently being used, or
3. New kinds of information.

TABLE 3.2
Typical Costs for Different Types of Remote Sensing Imagery per Square Kilometer

Sensor	Coverage (km²)	Acquisition Cost ($)	Analysis Cost Range ($)
NOAA AVHRR	9,000,000	0.0001	0.00005
Landsat TM	26,000	0.02	0.001
SPOT HRV	3.6000	0.75	0.25–0.5
Color IR Photography (new)		5–6.00	2.5–3.0
Aircraft digital imagery		5–10.00	2.5–5.0

Source: Modified from Lunetta, 1999.

Benefit/cost ratios for satellite remote sensing programs ranged from 1.0 to more than 20.0 depending primarily on the quality of the data and the type of application considered. If the application was heavily dependent on field data, but remote sensing observations could replace or augment those data, then the cost savings were large. This principle is still in effect and requires that field data be seriously examined; are they always necessary? Can remote sensing data be used instead (this is rare), as a partial replacement (more likely), or as a way of augmenting other data (very likely)? Are remote sensing data unique such that their very use can suggest new applications not previously possible? Is it valuable to envision different phases or sampling intervals — first, satellite data; second, partial coverage by aerial sensors; finally, field sampling?

Early discussions of the cost of launching and delivering satellite data compared to airborne data often resulted in first, one platform, then, the other platform proving to be more cost-effective; the most pertinent comparison considers these remote sensing data with aerial photography in areas of the world not well covered by historic air photo databases (e.g., Thompson et al., 1994). But rather than focus on image acquisition costs, a more realistic idea of the true cost of remote sensing is to consider typical per hectare costs for different types of remote sensing imagery, with estimated image analysis costs to generate equivalent products (Table 3.2). In this admittedly simplistic rendering of the broad costs there is much flexibility to deploy different sensors to arrive at the same information product. Satellite sensors are obviously much cheaper in data acquisition and analysis, but can they be used to generate the information product that is required? If not, the cost savings (over airborne data) are completely fictitious. The cost of aerial photography and airborne digital data diverge when analysis costs are considered, but these two data sources offer the same information content.

DATA CHARACTERISTICS

A basic understanding of the characteristics of remote sensing data is necessary to consider the relevance of the multispectral or hyperspectral view of the forest. Such

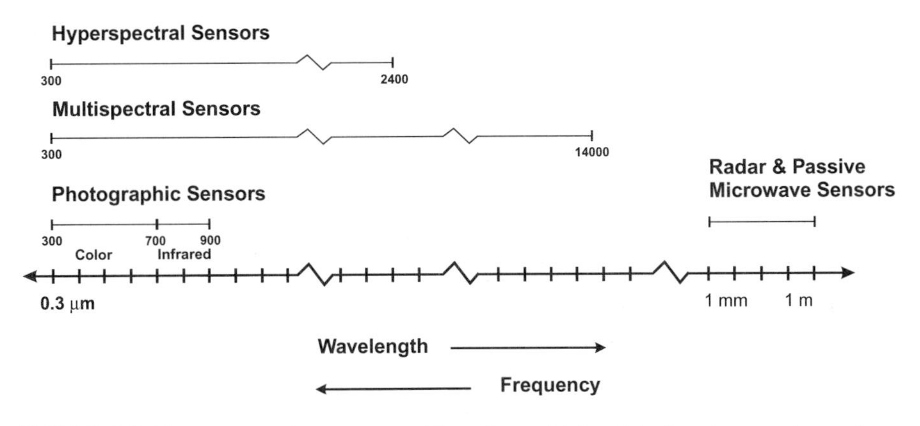

FIGURE 3.3 Electromagnetic spectrum with regions of interest in forestry remote sensing. Although many sensors operate in different regions of the spectrum and provide data useful in forestry applications, the main regions of interest are the optical/infrared and microwave portions of the spectrum.

understanding is required to judge when the remote sensing perspective from above is the most appropriate view to select in a given problem context. In earlier chapters, some sense of the various data characteristics was provided, but now it is appropriate to become more specific. The comments are restricted to the two main portions of the electromagnetic spectrum (Figure 3.3) currently used in remote sensing forestry applications: (1) optical/infrared, and (2) active microwave. Of these two, optical/infrared imagery are presently the most common, and this will likely continue to be so in the future.

Other remote sensing image data acquired using other sensors or in different regions of the electromagnetic spectrum have specific characteristics that must be considered prior to their use in forestry applications. For example, lidar data are not yet operational in any region of the world yet their potential is enormous — the promise of accurate and reliable tree and canopy height information. Imagery acquired in the thermal, UV, and passive microwave regions are typically used in specialized applications rather than as a general-purpose information source in forestry. In some applications, these other types of data are absolutely necessary — for example, thermal imagery can be used in reconstruction of surface temperature patterns which in some forests can be related to vegetation water stress and biodiversity (Bass et al., 1998). In other applications it is useful to be aware of the characteristics of these imagery as substitutes or ancillary information for the main optical and microwave imagery.

Optical Image Formation Process

In an ideal world, a remote sensing image would be formed directly from the reflectance provided by a target, and received by a perfectly designed sensor. The only limiting factor would be the wavelength sensitivity of the sensor. Of course, reality means that remote sensing imagery is acquired in a process that is much more complex. Major complications arise from the quality of the sensor and the recording medium, and in the process of acquiring the actual spectral measurement. An image, formed by observations of differing amounts of energy from reflecting surfaces, is affected by the original characteristics of these reflecting surfaces (such as leaves, bark, soil) and a whole host of other factors, such as the atmosphere and the adjacent surfaces involved in the image formation process. The principles of optical reflectance interaction with forests have been summarized by Guyot et al. (1989) and have received more detailed treatments in textbooks by Curran (1985), Jensen (1996, 2000), and Lillesand and Kiefer (1994), among others (e.g., Avery and Berlin, 1992; Richards and Jia, 1999).

The most important aspect of the image formation process is to understand how it is possible to create imagery in which it is not clear what element of the process — the spectral characteristics of the target, the illumination geometry, or the atmosphere — has caused the particular appearance of the image. Ideally, the process should be completely and singularly invertible; that is, based on the appearance of the image it should be possible to reconstruct the cause of that appearance and, as noted, in the ideal world the sole cause of image appearance would be the influence of the target. Unfortunately, the appearance of targets in imagery is affected by the fact that remote sensing measurements are typically acquired at specific angles of incidence (e.g., the solar and sensor positions). Surfaces reflect incoming energy in a pattern referred to as the bidirectional reflectance distribution function (BRDF): this effect is best considered as the difference in reflectance visible as the position of the viewer changes with respect to the source of light. Forests, in particular, are strongly directional in their reflectance; it is not just the geometry of the sensor and the source of illumination that are important, but the target as well. The BRDF effect is seen across the image as the target position changes within the field of view of the sensor. Therefore, knowledge of the position of the sensor, the target, and the originating energy source may be critical in using the collected measurements.

In Chapter 4, this factor and others which affect the use of remotely sensed observations are discussed; but the discussion is limited to considering the image processing tools that are available to deal with the uncertainties in measurements that result. This is not a discussion of the physics involved in remote sensing, which can be obtained elsewhere (e.g., Gerstl and Simmer, 1986; Gerstl, 1990). Rather, issues are considered that can be dealt with by applications specialists and remote sensing data product users. The only requirement is access to generally widely available image processing tools. For example, radiometric processing of imagery can range from little or no consideration of atmospheric effects to a fully functional radiative transfer model of the atmosphere which considers atmospheric constituents, angular effects, and optical paths. Much progress has been made in the development of an automatic and user-friendly procedure to correct specific sensor data — particularly Landsat TM — for atmospheric absorption, scattering, and adjacency effects (e.g., Ouaidrari and Vermote, 1999). On the other hand, Hall et al. (1991b) provide a good example of an alternative image processing approach to atmospheric radiative transfer codes and sensor calibration when reliable atmospheric optical depth data or calibration coefficients are not available — which, unfortunately, is often the case. It is this level of image processing that is of interest to those using remote sensing imagery, since it relies on approximations and simplifications of the more complex tools which are sometimes not readily available to all users of remote sensing data.

Roughly speaking, the factors affecting remote sensing spectral response include (in general order of importance):

1. The spectral properties (reflectance, absorption, transmittance) of the target (Guyot et al., 1989);
2. The illumination geometry, including topographic effects (Kimes and Kirchner, 1981);
3. The atmosphere (O'Neill et al., 1995);
4. The radiometric properties of the sensor (e.g., signal-to-noise ratio);
5. The geometrical properties of the target (e.g., leaf inclination).

The spectral response curve of green leaf vegetation and idealized biochemical compound reflectance curves are presented in Figure 3.4. These curves illustrate the portions of the spectrum in which absorption and reflectance dominate for different compounds. For a green leaf, there is typically a small green peak reflectance (at approximately 550 nm), and a small red well of absorption by chlorophylls (at approximately 650 nm). The rapid rise in reflectance in the near-infrared (before 1000 nm) is known as the red-edge (Horler et al., 1983), and there are several water absorption bands at longer wavelengths. These curves are idealized representations of the measurements; here, the concern is with gaining an appreciation of the sum effect that these factors and the different forest components such as bark, leaves, and soil can have on the expression of these spectral measurements contained in remote sensing imagery. Understanding this basic pattern of reflectance and absorption can help with the interpretation of remote sensing imagery in forestry applications.

At-Sensor Radiance and Reflectance

Remotely sensed data are typically presented to the user in the form of digital numbers (DN). These digital counts are consistent internally within the image and between different bands (or wavelengths), and therefore can be used in many image analysis tasks without further processing (Robinove, 1982; Franklin and Giles, 1995). However, to facilitate comparison between the same or different sensors at different times, and the comparison between satellite, airborne and field-based sensors, conversion to physical units (standardized) is required. At-sensor radiance factors may be calculated from the digital numbers with the use of appropriate sensor calibration coefficients (Teillet, 1986). These are published for civilian satellites following in-flight procedures using absolute calibration tests over terrestrial targets such as White Sands, New Mexico (the Landsat platforms) and La Crau, France (the SPOT satellite platforms). The coefficients are stored in the image data header files and are updated regularly by the various satellite operations groups. The at-sensor radiance equation may take the following form:

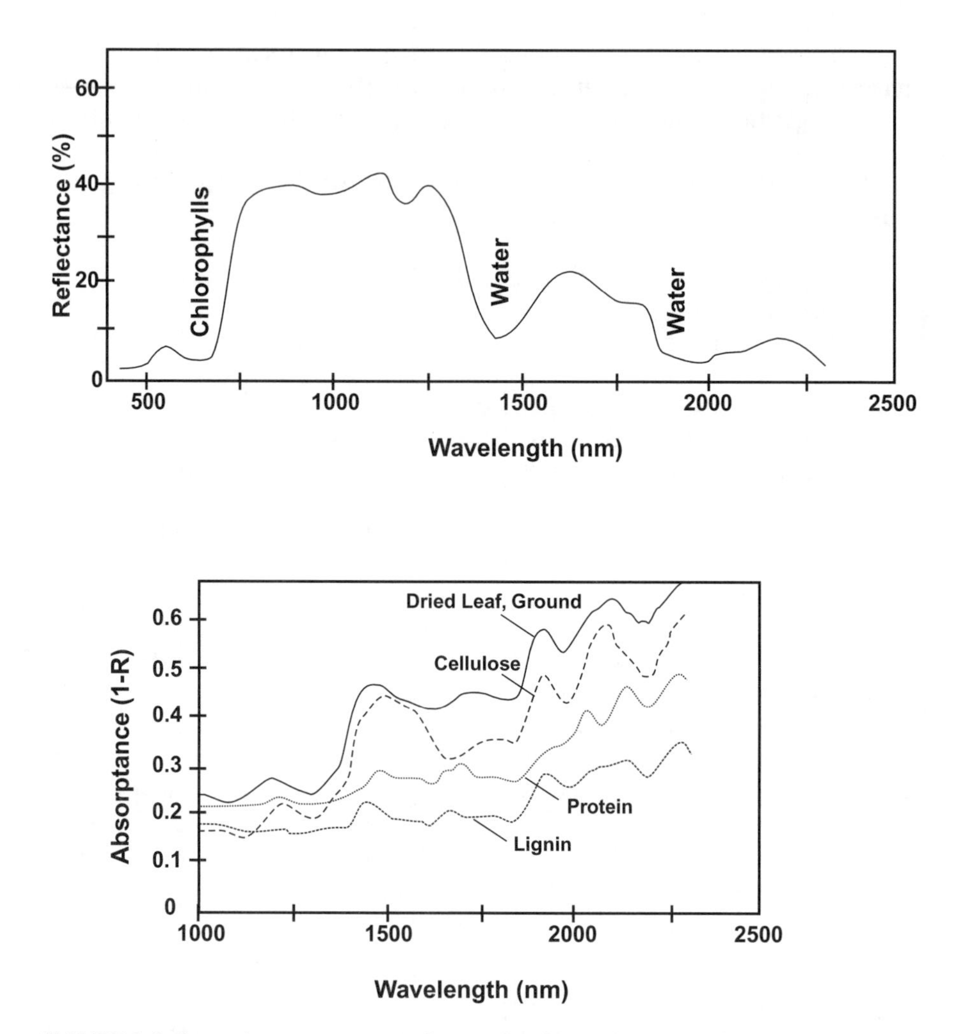

FIGURE 3.4 Spectral response curves of vegetation illustrating the portions of the spectrum in which absorption and reflectance dominate. In (a) the total hemispherical spectral reflectance of conifer needles (whole, fresh, and stacked five deep before data acquisition) is shown. Note the small green peak reflectance (at approximately 550 nm), the absorption by chlorophylls in the red region of the spectrum, the rapid rise in reflectance in the near-infrared (before 1000 nm), and the water absorption bands at longer wavelengths. In (b) a comparison is shown of the absorptance of oven-dried, ground deciduous leaves measured in a laboratory spectrophotometer compared to the absorptance characteristics of three biochemical compounds (lignin, protein, cellulose). The same features are visible in these curves, which differ primarily in the amount of absorption and reflectance. The original curves have been shifted up and down slightly to improve clarity. (From Peterson, D. L., J. D. Aber, P. A. Matson, et al. 1988. *Remote Sensing of Environment*, Vol. 24, pages 85–108, Elsevier, New York. With permission.)

$$L_s = a_0 + a_1 \, DN \tag{3.1}$$

where: L_s is the at-sensor radiance (W m^{-2} µm^{-1} sr^{-1}),
DN is the raw digital number, and
a_0 to a_1 are the absolute calibration coefficients for the particular satellite or airborne sensor system under consideration.

A common approach to computing at-sensor radiances has been to use a normalization equation with the maximum and minimum DN recorded in the scene. Then, a_0 to a_1 would be equivalent to a simple gain and offset, based on a scaled measure of the range of DN in the image plus a spectral reference. Spectral calibration targets are designed and deployed more easily during airborne remote sensing missions than during satellite overpasses. Similarly, radiometric calibration can be accomplished more easily in sensors that are returned periodically to the laboratory.

The measurement that is most useful in physical applications in forestry is reflectance, which is a property of the target alone. At-sensor reflectances can be calculated (after Qi et al., 1993):

$$\rho = \frac{\Pi d^2 \, L_s}{E_0 \cos \theta_z} \tag{3.2}$$

where: ρ is the apparent reflectance,
d is the normalized Earth/Sun distance,
L_s is the at-sensor radiance (W m^{-2} µm^{-1} sr^{-1}),
E_0 is the irradiance (W m^{-2} µm^{-1}), and
θ_z is the solar zenith angle.

For an airborne sensor, E_0 is estimated or recorded coincidently with image acquisition by an incident light sensor measuring incoming solar irradiance; for a satellite sensor, E_0 is the exoatmosphere irradiance. The apparent reflectance does not consider atmospheric, topographic, and view angle effects (described in more detail in Chapter 4).

SAR Image Formation Process

The principles of microwave interaction with forests have been summarized by Henderson and Lewis (1998). SAR sensors are active remote sensing devices; energy at known wavelengths is both generated and recorded by the instrument. The recorded energy is generally referred to as backscatter. Relationships between microwave backscattering coefficients and forest conditions have been reported as a function of the scattering properties of forests experimentally (Ranson et al., 1994), and empirically (Wu, 1990; Durden et al., 1991; Kasischke et al., 1994; Waring et al., 1995b). Here the interest is in gaining an appreciation of the principal mechanisms involved in radar beam interactions with a forested landscape (Figure 3.5); this will include volume scattering (from leaves and branches), direct scattering

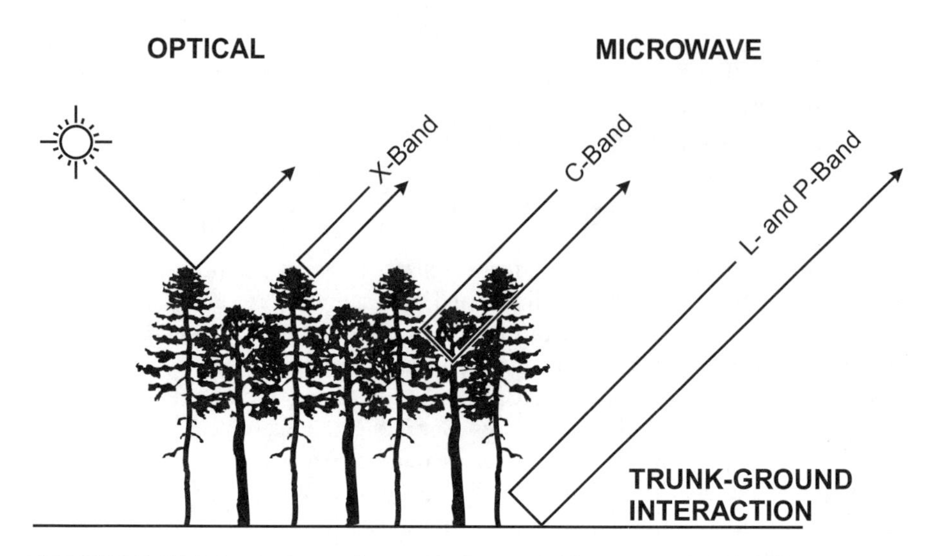

FIGURE 3.5 SAR image interactions with forests. Different wavelengths of microwave energy have different penetrating ability in forest canopies; X-band data (short wavelengths) are dominated by tree leaf interactions in much the same way that optical wavelength data are influenced by closed canopies. C-band data (slightly longer wavelengths) are dominated by twig and small branch interactions; L- and P-band data (much longer wavelengths) are dominated by the trunk-ground interactions. Many other effects, including those caused by topography and incidence angles, can dominate or influence the SAR image data of forests. (From JPL Publ. 86-29. 1986. Jet Propulsion Laboratory, Pasadena, CA. With permission.)

(from the ground and the stem/ground double-bounce), and other radiative transfers within the scene.

The wavelengths used in microwave sensing are typically long enough that they pass unimpeded through most atmospheric constituents, and of course, since the source of illumination is provided, these sensors can operate independent of the Earth's rotation.

SAR Backscatter

The most common mode of operation for active microwaves is the synthetic aperture radar (SAR) in which the forward motion of the platform is used to artificially generate a long antenae for reception of the microwave beam. This long antenae effectively increases the spatial detail of the subsequent image products. Microwave energy has a wavelength range of approximately a centimeter to several meters; radar system wavelengths are designated with letters on the basis of the military code (Table 3.3). In all of these systems, the radar equation is used to estimate the strength of the returning signal following emittance of a pulse:

$$P_s = \frac{P_t G_t}{4\pi R^2} \tag{3.3}$$

where: P_s is the power density of the scatterer,
P_t is the power at the transmitter,
G_t is the antenna gain, and
R is the distance from the antenna.

The power reflected by the scatterer in the direction of the receiving antenna (S) is equal to P_s times the radar cross section, which will differ by cover type, wavelength, polarization, and surface geometry.

Typically, image analysts are presented with a two-dimensional array of pixel intensities recorded as 8-bit or 16-bit digital counts, which are proportional to the backscattered amplitude (square-root of power), plus a range-dependent noise level (Ahern et al., 1993). Backscattering coefficients for typical forest components (leaves, bark, soil) are presented in Table 3.4.

SAR image data contain shadowing (layover) effects and specular and Lambertian surfaces; in areas of significant relief topography can dominate satellite SAR image data to the point where they may be useless in forestry applications (Domik et al., 1988; Rauste, 1990). In one study, Foody (1986) found that the topographic effect could reach 50% of airborne SAR tonal variations of a vegetated study site. Simple empirical corrections for this effect have been reported with mixed results (Teillet et al., 1985; Hinse et al., 1988; Bayer et al., 1991; van Zyl, 1993; Franklin et al., 1995a). Further discussion is presented in Chapter 4.

RESOLUTION AND SCALE

Resolution is a quality of any remote sensing image and can be referred to as the ability of the sensor system to acquire image data with specific characteristics. There are four main categories of resolving power applicable to remote sensing systems (Jensen, 1996). Each is discussed in the sections below, followed by a brief presentation of the implications of these resolutions and image scale.

TABLE 3.3
Radar Wavelength Military Code Designations

Code	Wavelength Range (cm)	Imaging Wavelengths (cm)[a]
X-band	2.4–3.8	3.0*, 3.2
C-band	3.8–7.5	5.3**, 6.0
L-band	15.0–30.0	23.5***, 24.0, 25.0
P-band	30.0-100.0	68.0

Note: NASA/JPL AirSAR is a multifrequency system.

[a] Commercial examples: *Intera Star-1 airborne mapping system, **ERS-1 Active Microwave Imagery (AMI) and Radarsat, ***JERS-1 SAR sensor.

TABLE 3.4
Typical Backscatter Coefficients (in dB) for Different Features of Interest in Remote Sensing

Feature	Range	Wavelength	Polarization
Wet grass and loblolly pine stands[a]	–5.0 to –9.0	C-band	VV
Dry grass and loblolly pine stands[a]	–7.5 to –11.5	C-band	VV
Pine and hemlock forests[b]	–3.0 to –12.0	P-band	VV and HH

[a] ERS-1 SAR observations (Lang et al., 1994).
[b] Backscatter model results (Wang et al., 1994).

Source: Adapted from Lang et al. (1994) and Wang et al. (1994).

Spectral Resolution

Spectral resolution is the number and dimension of specific wavelength intervals in the electromagnetic spectrum to which a sensor is sensitive. Particular intervals are optimal for uncovering certain biophysical information; for example, in the visible portion of the spectrum, observations in the red region of the spectrum can be related to the chlorophyll content of the target (leaves). Broadband multispectral sensors are designed to detect radiance across a 50- or 100-nm interval, usually not overlapping in a few different areas of the optical/infrared portions of the electromagnetic spectrum. Hyperspectral sensors are designed to detect many very narrow intervals, perhaps 2 to 4 nm wide. A hyperspectral sensor may record specific absorption features caused by different pigments, such as the chlorophyll a absorption interval.

Spatial Resolution

Spatial resolution is the projection of the detector element through the sensor optics within the sensor instantaneous field of view (IFOV). This is a measure of the smallest separation between objects that can be distinguished by the sensor. A remote sensing system at higher spatial resolution can detect smaller objects. The spatial detail in an optical/infrared image is a function of the IFOV of the sensor, but also the sampling of the signal, which determines the actual pixel dimension in the resulting imagery. Historically, spatial resolution from polar-orbiting terrestrial satellites has been on the order of 20 to 1000 m or more; recent advances (and military declassification) in sensor technology, as well as the lower orbits selected for many of the new platforms, mean that satellite sensor spatial resolution can approach 1 m or less.

Radar image resolution in ground range (R_{gr}) is determined by the physical length of the radar pulse (t) emitted and the depression angle of the antenna (θ):

$$R_{gr} = t/(2 \cos \theta) \quad (3.4)$$

The sensor depression angle (θ) is a constant which differs for each of the available side-looking SAR systems, and may also differ for a single sensor with mission

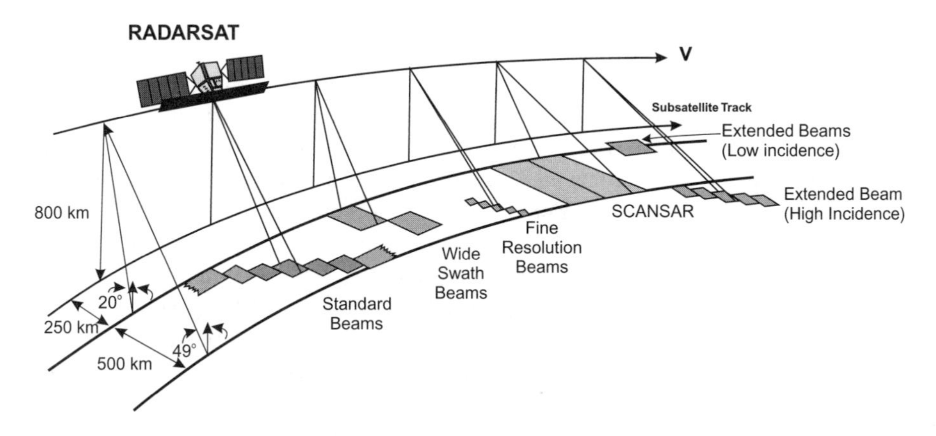

FIGURE 3.6 Radarsat, Canada's first remote sensing satellite, has been operational since 1995. The system provides multiple spatial resolutions in C-band like-polarized format. The beam modes and acquisition parameters were designed initially to provide all-weather imagery of sea ice and ocean phenomena and have been used successfully in some forest applications. These data are in high demand in areas with high cloud cover conditions and are often used in concert with optical/infrared data. (From Luscombe, A. P., I. Ferguson, N. Shepperd, et al. 1993. *Can. J. Rem. Sensing*, 19: 298–310. With permission.)

design. For example, the Radarsat sensor package can be programmed during image acquisition to permit a wide range of incidence angles on the ground (Luscombe et al., 1993) (Figure 3.6). The ERS-1 satellite, launched in 1991, was programmed to alter the Active Microwave Imager (AMI) SAR sensor depression angle after the first year of operation.

Azimuth or along-track resolution (R_{ar}) is limited by antenna length (D_a) at any given wavelength (λ) and slant range to the target (R_s):

$$R_{ar} = (0.7\ \lambda\ R_s)/D_a \quad (3.5)$$

This resolution is improved in SAR systems by a Doppler shift which permits the collection of (nominally) square-area size pixels in the range and azimuth directions.

Temporal Resolution

The image frequency of a particular area recorded by the sensor is referred to as the temporal resolution. The recorded frequency of an image determines the type of environmental change detected by the sensor and the rates of change that can be estimated. Many new satellites (in the past 10 years) and most future satellites could increase revisit capabilities with programmable sensors that can look to one or the other side of the nominal flight path. This has given rise to consideration of angular spectral response patterns (Diner et al., 1999); multiple observations can mean that different angular characteristics of the reflectance distribution pattern are captured. This could be considered a different kind of resolution altogether.

Radiometric Resolution

The sensitivity of the detector to differences in the signal strength of energy in specific wavelengths from the target is a measure of radiometric resolution. Greater radiometric resolution allows smaller differences in radiation signals to be discriminated. The detector signal has an analogue gain applied before quantization with an analog-to-digital converter. The quantization determines the number of bits of data received for each pixel; and determines the number of levels that can be represented in the imagery, but is not the radiometric resolution directly.

This resolution is analogous to film speed in the analogue photographic systems; the same light conditions will seem brighter and create more contrast when captured on faster film because the film is more sensitive to the radiant flux. Color changes that seem obvious in aerial photographs are sometimes not readily apparent in some digital imagery because of the relatively large differences in radiometric resolution between film and digital sensors; color aerial photography, for example, can theoretically provide many times the radiometric resolution of satellite sensors. A reflectance change of a small percent can cause a dramatic change in color visible to the eye and recorded on color film (say, from green needles to red immediately following insect defoliation of conifers). However, those reflectance differences in the green and red portion of the spectrum recorded by satellite sensors hundreds of kilometers above the target would be minimal.

Relating Resolution and Scale

There are certain trade-offs in considering the resolving power of remote sensing systems from aerial or satellite platforms. For example, an increase in the number of bands is often accompanied by a decrease in the spatial detail (spatial resolution). To acquire more or narrower bands, the sensor must view an area on the ground for a longer period of time, and therefore, the size of the area viewed increases from a constant altitude. If the radiometric resolution is increased (so that smaller differences in radiance can be detected), the spatial detail, the number of bands, the narrowness of the bands, or all three, must be reduced. In addition, the size of the viewed area (pixel size) will influence the relationship between image objects and reflectance. In other words, the amount of energy available for sensing is fixed within the integration time of a detector. The trade-off in sensor design is between spectral resolution (how much the energy is divided into spectral bands), spatial resolution (how large an area is used to collect energy), and the signal-to-noise ratio. Divide the energy into too many bands over too small an area, and the signal within each band is weak compared to the (fixed) system noise. In sensor design, it is the SNR that should be maximized, rather than any one of spectral or spatial resolution.

Typically, satellite data are medium to low spatial resolution data; using the terminology suggested by Strahler et al. (1986), these data are low- or L-resolution. The objects are smaller than the pixel size, and therefore, the reflectance measured for that pixel location is the sum of the objects contributing radiance. Robinove (1979) used this idea with coarse resolution Landsat MSS data (80 m pixels) to generate maps of landscape units covering large areas that were comprised of all features contributing reflectance — vegetation, soils, and topography. In some satellite sensor studies, this generalizing characteristic of relatively coarse spatial resolution satellite data can be considered an advantage, at least up to a certain point, after which the data are too general for the intended use (Salvador and Pons, 1998b). The lower spatial resolution provided more stable and representative measurements over large areas of high spatial heterogeneity (Woodcock and Strahler, 1987); the point is spatial heterogeneity governs the analytical approach given a constant pixel size (Chen, 1999). In manual interpretation of Landsat imagery, for example, Story et al. (1976) suggested that suitability of the imagery is a function of the detail in which it portrays the subject (in their case, Australian land systems). But too much detail can distract the interpreter with unnecessary information that is not significant for the scale of the study (or the purpose of the mapping exercise). Detail in imagery can be a mixed blessing, perhaps even more so when imagery is to be processed digitally.

Airborne data are often high-spatial-resolution data; these data are high- or H-resolution (Strahler et al., 1986). Typically, the objects are larger than the pixel size, and therefore the reflectance measured for a given pixel location is likely to be related directly to the characteristics of the object. In airborne remote sensing, trade-offs in flight altitude, speed of the plane, and data rates for both scanning and recording result in constraints on the range of spatial detail that can be acquired. Some satellite systems provide a similar though more limited range of options in spatial and spectral resolution; users must match the appropriate data acquisition parameters to the application at hand, often by selecting imagery from different satellites or a combination of satellites and aerial sensors for multiple mapping purposes on the same area of land. For example, if the objective was to map leaf area index within forest stands it would be possible, though perhaps not optimal, to acquire and process very high spatial resolution airborne imagery with individual trees visible. The approach is to build the LAI estimate for a stand or given parcel of land from individual tree estimates. A completely different yet complementary strategy would be to acquire satellite imagery at a coarser spatial resolution and attempt to estimate LAI for larger parcels of the stand, then aggregate (classify) or segment the image (Franklin et al., 1997a).

Although the methods would almost certainly become more complex, using aerial and satellite data — or more generally, H-resolution and L-resolution data — in combination may provide results which are more accurate than relying on only a single image source. Four different image spatial resolutions are illustrated in Chapter 3, Color Figure 1* using data acquired from the high (space) altitude NOAA Advanced Very High Resolution Radiometer (AVHRR), Landsat Thematic Mapper (TM) satellite, medium altitude Compact Airborne Spectrographic Imager (CASI), and low altitude Multispectral Video (MSV) airborne system. At the level of the satellite image, broad patterns in vegetation communities and abiotic/biotic/cultural features are clearly visible. Less clear are the variations within these groupings. In forested areas, for example, differences in dominant species and in productivity can be discerned through careful analysis of the relationships between cover and geomorphology. As an illustration, alluvial fans in this area tend to be good sites for

---

* Color figures follow page 176.

deciduous cover, appearing a brighter pink in the false color image. As the spatial resolution increases, the information content increases, but the area covered decreases. At the highest spatial detail (25 cm spatial resolution with the digital video system) individual trees are seen as discrete objects with clear separation from surrounding features; but only a tiny fraction of the area covered in the coarser resolution imagery can be reasonably mapped with this level of detail. This multiple resolution approach can yield a powerful data set that can be scaled from ground data to one image or aerial extent to the next.

Scale is a pervasive concept in any environmental monitoring, modeling, or measurement effort (Goodchild and Proctor, 1997; Peterson and Parker, 1998) and has a direct spatial implication in remote sensing. Scale is related to spatial resolution but is not an equivalent concept. Where resolution refers to the spatial detail in the imagery that might be used for detection, mapping, or study, scale refers to the resolution and area over which a pattern or process can be detected, mapped, or studied.

Scale implies measurement characteristics, typically referred to as grain or, sometimes confusingly, resolution. In essence, scale consists of grain (resolution) and extent (area covered) and these two aspects of scale must be considered whenever scale is of interest. By geographic convention:

1. Small-scale refers to large area coverage in which only a small amount of detail is shown (for example, maps with a representative fraction of 1:1,000,000);
2. Large-scale refers to small area coverage in which a large amount of detail is shown (for example, maps with a representative fraction of 1:1000).

One way in which to relate scale (as a mathematical expression) and image detail or resolution is to categorize levels of image spatial resolution which can be described based on the scale at which environmental phenomena can be optimally identified or estimated:

- *Low spatial resolution imagery* — Optimal applications are in the study of phenomena that can vary over hundreds or thousands of meters (small scale) and could be supported with GOES, NOAA AVHRR, EOS MODIS, SPOT VEGETATION, HRV, and Landsat data. Examples of the use of this type of imagery include mapping objectives at the small-scale: forest cover by broad community type (coniferous, deciduous, mixed wood); abiotic/biotic characteristics; Level I physiographic and climatic classifications (Anderson et al., 1976; Chapter 6).
- *Medium spatial resolution imagery* — Optimal applications are in the study of phenomena that can vary over tens of meters (medium scale) and could be supported with imagery from Landsat, SPOT, IRS, and Shuttle platforms, and by aerial sensors. Examples of the use of this type of imagery might include mapping objectives at the medium scale: patch level characteristics and dynamics; tree species; crown diameters; tree density; the number of stems; stand-level LAI; Level II forest covertype and vegetation type classifications (Anderson et al., 1976; Chapter 6).

- *High spatial resolution imagery* — Optimal applications are in the study of phenomena that can vary over scales of centimeters to meters (large-scale), are currently supported by aerial remote sensing platforms, IKONOS, and very specific applications of coarse resolution satellite imagery (e.g., coarse pixel resolution unmixing studies). Examples of the use of this type of imagery might include mapping objectives such as individual trees and other discrete ground objects (understory assemblages); forest structure; forest cover (crown diameters, closure); LAI; understory composition or rare species detection. In future, it is expected that additional satellite sensors will be deployed with high spatial resolution; these data will approach the level of detail now available routinely from aerial photography.

Of course, the data source is only one of several variables that must be considered in any monitoring application at any particular scale (Wulder, 1998b). For instance, although multispectral satellite imagery is now available at less than 1 m spatial resolution, the spectral resolution (number, width, and position of the bands) and the radiometric resolution (dynamic range) of the data may not be appropriate for some applications. This is particularly the case when satellite data with few bands are compared to the corresponding aerial sensor capability (hyperspectral). Earlier experience in the late 1980s showed that the first SPOT satellite High Resolution Visible (HRV) sensors had a low dynamic range compared to data from three nearly spectrally equivalent Landsat TM sensor bands. The coarser spatial resolution Landsat TM sensors were preferred in forest defoliation studies because of the higher dynamic range and the presence of two additional bands in the shortwave infrared portion of the spectrum (Joria et al., 1991; Franklin and Raske, 1994).

In addition, the optimal available method for the identification of phenomena at a particular scale may not be the most accurate method possible. The best choice of methods and remote sensing data is a function of the accuracy (positional and thematic) and scale (resolution and geographic extent of the project), together with issues associated with personnel (e.g., technical training), and cost (data acquisition, data storage requirements, and time for analysis).

AERIAL PLATFORMS AND SENSORS

A quick glance at the remote sensing literature provides numerous examples of the use of a wide variety of aerial platforms (balloons, helicopters, fixed-wing airplanes, drones), and sensors ranging from analogue cameras to charge-coupled devices (CCDs) applied to forestry problems. An early review by Schweitzer (1982) identified multispectral scanners, airborne laser sensors (now commonly referred to as lidar), and aerial photography as promising environmental monitoring tools. In the 1990s, airborne digital videography, digital frame camera systems (King, 1992; Neale and Crowther, 1994), imaging spectrometry (Vane and Goetz, 1993; Curran, 1994; Curran and Kupiec, 1995), and airborne synthetic aperture radar (Thompson and MacDonald, 1995; Dobson, 2000) were added to the list.

Airborne radar has continued to improve since the early development of environmental and mapping radar in the 1960s, partly because of the need to prepare users for operational satellite radar imagery from platforms such as Canada's Radarsat (Luscombe et al., 1993), the Shuttle Imaging Radar (SIR) missions, and the imagery produced from the European Radar Satellite (ERS-1) and Japanese Environmental Remote Sensing (JERS-1) systems. These improvements clearly show that the development of a market for one remote sensing product (airborne radar) invariably introduces opportunities for other types of remote sensing data and products (satellite radar).

Continued work in specialized forestry applications using ultraviolet and thermal imagery has been reported, but the main emphasis has been on aerial photography, digital multispectral sensors, lidar, and radar remote sensing. A key feature of this wide range of remote sensing data collection technology is to remember that these are tools designed for different jobs within the purview of the forest manager (as illustrated by the scaling of imagery in Chapter 3, Color Figure 1; each has a place, partly based on historical uses, but also based on continual improvements and potential developments, that will satisfy the increasing demands for data and information products. Reviewed in the following sections are the main operating issues of sensors designed to collect data in these portions of the spectrum; developments, principal methods of analysis, and potential.

Aerial Photography

Aerial photographs have been used extensively in forest management since the 1940s (Spurr, 1960; Lachowski et al., 2000). Acquisition of aerial photographs for use in forestry must first be planned with reference to altitude (desired photo scale), vantage point (e.g., vertical, oblique), camera type (e.g., metric, panoramic, small format), filtration, and film emulsion type (Jensen, 2000). The purpose of the photography is important; higher quality and greater control of photography is needed for forest inventory, as compared to forest updates, for example (Gillis and Leckie, 1996). Progress in photographic science has provided continual improvements in aerial camera technology, film speed, contrast, resolution, processing, and exposure latitude (Fent et al., 1995; Hall and Fent, 1996). Conventional black and white, color, and color infrared metric aerial photography are acquired routinely over extensive forest regions at a range of scales in support of forest management operations and planning.

Early forestry applications of aerial photography were restricted to locational surveys and logistical support for field crews, however, users quickly became aware of the tremendous power and flexibility afforded the analyst by the aerial perspective, and came to rely on the near-permanent data record contained in an aerial photograph (Heath, 1956).

> *Their value lies not in a cut and dried technique or even in easily accessible results but in saving the forester's time day after day in many minor ways and in permitting ... better-informed decisions without delay* (Spurr, 1960: p. 348).

A complete list of management uses for aerial photography has yet to be compiled; Spurr (1960) listed their use in creating basic forest maps, cadastral surveys, forest inventory and record keeping, insect and disease surveys, silvicultural surveys, forest administration (e.g., timber sales), road location, fire protection, forest recreation, and range and wildlife management. The interpretation of stereoscopic aerial photographs for forestry is currently a skill highly valued by industry and governments in the resource sector (Avery and Berlin, 1992). Aerial photointerpretation is taught at the university level in virtually every forestry, environmental, and geography department in the world. Interpretation of aerial photographs is considered an important component in forestry education as part of the training in remote sensing and GIS (Sader and Vermillion, 2000).

The interpretation of aerial photographs relies overwhelmingly on the general ability of the trained human analyst to identify features and areas of interest (Figure 3.7). To help develop these abilities, many agencies provide interpretation and certification programs. Lueder (1959) outlined the process of photointerpretation by

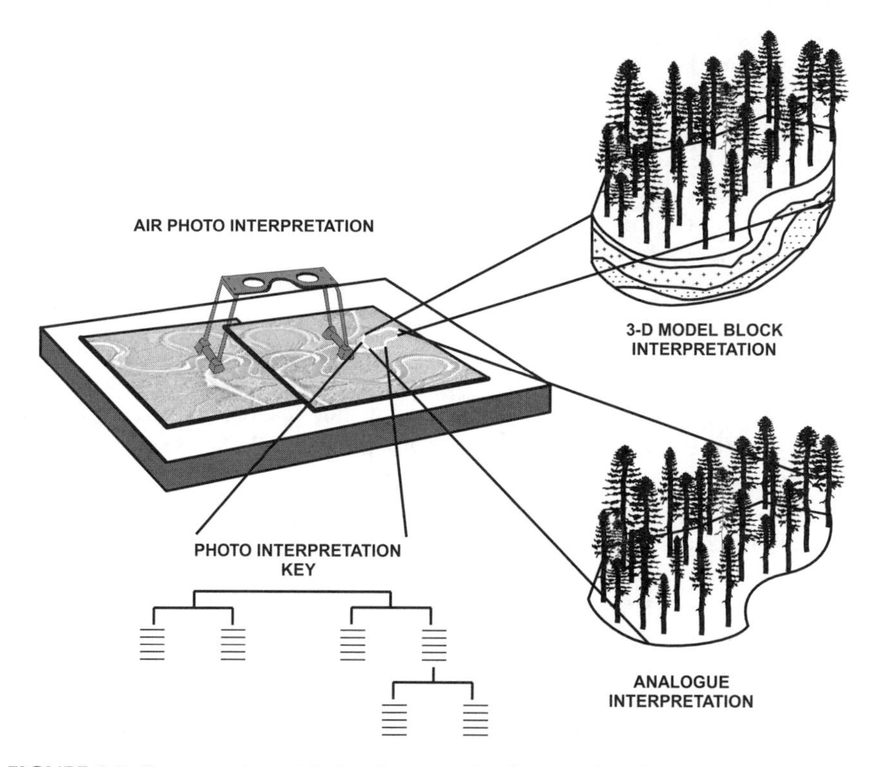

FIGURE 3.7 Stereoscopic aerial photointerpretation by use of models, analogues, and photokeys has a long and valuable tradition in forestry. Generally, the approach is to outline photomorphic areas using different photo-elements (tone, texture, patterns, shapes, and so on) and then identify the tree species, crown closure conditions, density, and other forest conditions of interest (e.g., soils) by examining the photography in more detail and with field surveys. This approach is strongly dependent on the skill of the analyst and the availability of appropriate (e.g., seasonal) high-quality photography. As remote sensing continues to mature it is thought that digital methods will increasingly find application in providing information traditionally accommodated through the use of these aerial photointerpretation methods.

trained and skilled interpreters. Photointerpretation relies on the deductive and inductive evaluation of aerial photo patterns. The photomorphic approach is the basis of most land use, land cover, and forest inventory mapping; the analyst identifies objects or areas by outlining distinctive tone, texture, pattern, size, shadows, sites, shapes, or associations (Lillesand and Kiefer, 1994). Typically, a hierarchical approach is used to organize the interpretation; general covertypes are separated; familiar objects are identified first as the interpreter moves from the known to the unknown features and from the general to the specific (Spurr, 1960). Subsequently, those areas are subdivided into smaller units, and labeled according to the level of detail desired or attainable given the image resolution (Ahearn, 1988). The final product in forestry photomorphic interpretation would be a forest stand identified and labeled (usually) through a comprehensive system of classification based on species composition, density or stocking, canopy height, and age classes (Gillis and Leckie, 1993). (Different classification schemes and approaches are discussed in Chapter 6.)

On typical photography acquired for forest mapping, perhaps five to ten forest types can be recognized consistently, with three to five height classes, up to ten density classes, and between five and ten sites (Spurr, 1960). More or less field work would be used to generate the description for the stand. Interpreters use selection or dichotomous keys (e.g., Avery, 1978; Hudson, 1991), or perhaps a checklist-based interpretation key (Avery, 1968; Kreig, 1970). Use of one type of key or another might depend on the existing state of knowledge for forests in the area, as well as the heterogeneity of the landscape.

Standard mapping photography in actively managed forests is usually augmented with supplemental aerial photography (Zsilinszky, 1970), high-altitude (Moore and Polzin, 1990), and large-scale (or small-format) aerial photography (Spencer and Hall, 1988) for specialized purposes such as forest inventory (Aldred and Lowe, 1978; Hall et al., 1989b), pest damage and defoliation mapping (Hall et al., 1983), regeneration (Hall and Aldred, 1992), and cutblock surveys. The use of these small and medium formats in technical forestry applications is expected to continue to generate favorable reviews (Graham and Read, 1986; Gillis and Leckie, 1996). Rowe et al. (1999) suggested that the most common formats for photography in resource management (other than standard metric formats) are the 35 to 70 mm small-format camera systems. This technology can be operated by virtually anyone without significant training. In their application, logging road length was obtained by scanning the small-format photographs into a computer system and manually interpreting roads with a CAD package. Small-format cameras and photography are very low-cost, relative to most other systems.

Aerial camera technology has seen significant technological improvements with respect to improved lens resolution, forward motion compensation, computer-based exposure control, integration with GPS receivers, and gyro-stabilized camera mounts (Mussio and Light, 1995; Hall and Fent, 1996; Light, 1996). When combined with improvements to aerial film and processing technologies, the photo quality can now be more easily controlled, and this will have a significant influence on the resultant accuracy of the information based upon which forest management decisions are made (Fent et al., 1995). Aerial photo quality is particularly important as digital capture is increasingly being undertaken for the production of orthophotos, and producing images to be used as a backdrop for on-screen image interpretation and feature delineation. Workstations have now been developed for mono or stereo interpretation that greatly improves the efficiency by which the digital capture of photointerpretation can be made (Graham et al., 1997; International Systemap Corp., 1997).

In a recent review of remote sensing for vegetation management practices on large (>10 ha) clearcuts, Pitt et al. (1997) suggested that among currently available sensors, aerial photographs continue to offer the most suitable combination of characteristics. Aerial photography provides high spatial resolution, stereo coverage, a range of image scales, a variety of film, lens, and camera options, capability for geometric correction, versatility, and moderate cost. The authors predicted future wider demands for remote sensing in forest vegetation management, and emphasized a series of activities to prepare for what they termed the imminent digital era. One initial strategy has been to attempt forestry work with digitized aerial photographs (Meyer et al., 1996; Holopainen and Wang, 1998; Bolduc et al., 1999), digitized satellite photographs (King et al., 1999), and orthophotography products (Duhaime et al., 1997). Earlier, Leckie et al. (1995) suggested that the use of digital high-resolution (<1 m) multispectral imagery (from a variety of sensors) as an alternative to aerial photography for forest inventory mapping "is a possible revolutionary innovation." New data from high spatial resolution satellite sensors (Glackin, 1998) and new aerial digital sensors (Caylor, 2000) are now competing directly in the mapping and monitoring markets with aerial photographs; this competition will quickly grow more fierce as greater confidence and experience in the new data accumulates. Several such satellite systems, the IKONOS platform among them, are currently, or are poised in future to generate, photo-quality imagery from low-Earth-orbiting platforms (Glackin, 1998).

Users of high spatial detail satellite imagery, and some of the new types of airborne digital imagery, have quickly experienced a major stumbling block: the necessary image processing tools to use such digital imagery are not yet fully developed or even available (see Chapter 4). This has prompted various attempts to develop a transitional product based largely on human interpretation skills, but with some aspects of a digital approach. Perhaps one of the best examples is the work of Madden et al. (1999) in the development of a photointerpretation key for the Florida Everglades. Using manual interpretation of vegetation polygons based on a selection key of color infrared photography, each Everglade vegetation type was keyed using standard photomorphic tools (color, tone, texture, pattern, height, shape, and context). Representative sections of the air photos (tiles) or photo-chips were digitized at high spatial resolution for specific vegetation classes. This digital photointerpretation key proved highly useful in training new interpreters and in decreasing the learning curve that typically exists in any new vegetation mapping and classification project.

The key to successful digital use of aerial photographs is an understanding of the conversion of analogue imagery to digital imagery, typically through the use of scanning densiometers, video digitizers, or CCDs (Jensen, 2000). The idea is that the very high spatial resolution of the analogue photographic product (a function of the film density and processing chemistry) can be adequately captured if there is a relationship between dye exposure and output gray tone. Due to the presence of bidirectional reflectance, pixel values will be affected by their location within the photo (Holopainen and Wang, 1998); this problem can be more sharply defined if topographic effects are pronounced (Dymond, 1992) or if radial displacement is severe. In general, analogue aerial photographs taken under the same exposure and flying conditions will have higher spatial resolution than their digitized counterparts. In satisfying the modest general mapping requirements in many forestry applications, particularly at the stand level, this may not be a limiting factor in the development of digital aerial photography applications.

Airborne Digital Sensors

A wide range of digital sensor systems have been developed and deployed to support forest science, operations, and management applications. Most digital sensor systems are characterized as research or near-operational, and can be considered in most markets at this time to be merely complementary to, rather than fiercely competitive with, aerial photography and field work. One possible exception is the fast-developing digital frame camera (King, 1995), increasingly thought to be a likely replacement for conventional aerial photographic cameras in the near future — in the view of some practitioners, as soon, perhaps, as a decade (Caylor, 2000). Examples of digital camera and videography imagery are contained in Chapter 3, Color Figure 2.

Multispectral Imaging

Multispectral scanners have been generating imagery for use in environmental applications for several decades and have been continuously improving. Early systems operated with sweeping mirrors, followed by the development of pushbroom instruments using linear arrays. Recently, Wewel et al. (1999) described the world's first fully automated digital multispectral scanner system; the High Resolution Stereo Camera (HRSC), originally designed for the exploration of Mars, has been modified for terrestrial applications. Multispectral scanners, digital frame cameras (King, 1992), and multispectral video systems (Roberts, 1995) operate on principles of solid-state imaging techniques, can be mounted in aerial photography platforms, and with the exception of data storage, conceptually can be considered virtually identical to analogue cameras in operation. Digital systems simply replace the analogue film emulsion in a camera with an array of photosites embedded in a substrate material such as silicon. Incident photons excite electrons in each photosite. This charge can be converted to an analog signal, such as the NTSC or HDTV standards, in direct linear proportion to the incident radiation. The signal is digitized within the system and output to some media (depending on data rate).

Jensen (1996) and Vincent (1997) reviewed various types of scanners, videographic and frame camera sensors and the different technologies that have been incorporated into operational systems as the field has matured. While these systems do not yet replace aerial photography (Hegyi et al., 1992), they can reduce the need to conduct intensive field sampling and large-scale resource aerial photography, and they can produce near-photo-quality analogues from the original digital image data with similar spatial resolution and contrast. While they cannot match aerial photo

TABLE 3.5
Principal Advantages and Disadvantages of Digital Systems Compared to Aerial Photography

Disadvantages
Generally small view angle which does not allow cost-effective large-area mapping
Extremely high data rates which can overwhelm most recording media
Multidimensional radiometric and geometric calibration which requires significant investment
Advantages
Digital formats, providing multiple analytical functions
Greater spectral range and sensitivity (than photography)
Near-real (or real enough) time capability

Source: Adapted from King (1995) and Roberts (1995).

quality with similar resolutions, differences in light conditions and geometric registration of frames are no longer huge problems, paving the way for the acquisition of sequential multispectral video imagery (Bobbe et al., 1993). An example of available equipment in this category is the four-camera system called the ADAR (Benkleman et al., 1992). This system was used to simulate AVHRR bands with 0.5m pixel resolution (Hardy and Burgan, 1999); after band-to-band registration, solar zenith angle corrections, and a disabled gain and pre-set aperture, comparisons of imagery acquired on four different dates were used to monitor live moisture content of different forest canopies and understories.

Principal advantages and disadvantages of airborne digital systems compared to aerial photography are outlined in Table 3.5. The main challenges in the operations side of employing these types of sensor systems in forestry and other applications include (Roberts, 1995; King, 1995):

- Multispectral information can be obtained most easily if multiple cameras are deployed; but intercamera registration difficulties may be created depending on the configuration used and the stability (vibration) of the platform (Nixon et al., 1985; Everitt et al., 1991; Neale and Crowther, 1994);
- Gain and automatic exposure control should almost always be disabled to prevent voltage saturation (overexposure or "blooming") and underexposure, and to allow comparisons of imagery at different times and places (Franklin et al., 1995b); the most common solution is to adjust camera exposure settings during an initial test of the system in-flight;
- Filtering of image data can be accomplished with optical or gel filters or computation filters; these filters can be used effectively to select spectral properties (e.g., remove haze) and reduce or eliminate geometric and vignetting effects (Pellikka, 1996);
- A range of deployment issues (such as system availability and dedication to the mission, reliability, complexity, and system component integration)

increase uncertainty in any kind of airborne remote sensing (Wulder et al., 1996); of these, system complexity can quickly overwhelm new users as new components (such as GPS/INS integration) are developed.

Some of these operational issues, when considered as challenges to the possible replacement of aerial photography as the principal data source in forest mapping, will no doubt be addressed as computer hardware and software improve, and as experience in applying this type of remote sensing becomes more widespread. "There is little doubt that at some point film will become obsolete" (Greenfield, 2000: p. 23). There are two very good reasons for this situation:

1. The reduction in waiting time between digital and film processing (Linden, 2000), and
2. The increased manipulation capability with digital data (Stow et al., 2000).

The remaining obstacles to operational use of airborne videography are thought to be related principally to computer storage and power (Um and Wright, 1999). Since computer processing speed continues to double approximately every 18 months, there may not be much more time to wait. Nevertheless, the limits on the digital frame camera sensor, and airborne platform technology in particular, have led King (1995: p. 266) to comment that "Users should not be falsely led to believe that accurate automated thematic mapping of precise land cover types can be conducted on an operational basis." Five years later, this situation has not yet changed, but perhaps it should be noted that the critical limits are no longer imposed during the data collection, but rather by the methods (or lack thereof) to convert the data to information of value to the users of these technologies.

HYPERSPECTRAL IMAGING

Imaging spectrometry (also occasionally referred to as imaging spectroscopy or spectrographic imaging) is a relatively new category of remote sensing instruments. These sensors can sample the electromagnetic spectrum in many, very narrow spectral intervals, creating hyperspectral imagery. Airborne spectrometry has been carried out extensively using instruments such as the Compact Airborne Spectrographic Imager (CASI) (Babey and Anger, 1989; Babey et al., 1999), and the Airborne Visible/Infrared Imaging Spectrometer (AVIRIS) (Porter and Enmark, 1987). In one analysis of canopy gaps created by the death of individual trees, Blackburn and Milton (1997) acquired hyperspectral imagery; one of the main findings again emphasized the importance of the flexibility of an airborne sensor — no other way of collecting data for this application appeared even remotely possible. Real value was provided by the high sensitivity of the spectrograph, which could be operated under clouds and variable skies.

Another example of a hyperspectral sensor is the MEIS-FM imager (called the Multispectral Electro-Optical Imaging Spectrometer for Forestry and Mapping). This system was developed by Neville and Till (1991) from an earlier multispectral pushbroom sensor. After a few acquisition trials, the MEIS-FM was considered a superior source of information in forest management when compared to aerial photographs because of the additional spectral information. However, no real tests of the information content or the ability to extract the information have yet been conducted — a critical issue is the cost of such careful comparative studies. Hybrid systems, incorporating principles of solid-state imaging and spectroscopy, have been developed. The typical result of these developments has been a proliferation of inexpensive sensors that can provide hyperspectral imagery at data rates that can choke the typical image analysis computers almost as fast as they can be built (Sun and Anderson, 1993). Software, as usual, lags far behind hardware capability.

Synthetic Aperture Radar

Radar imagery can be used as a multipurpose data source for forestry and vegetation applications, especially in areas of severe cloud cover limitations (Bush and Ulaby, 1978) or where a historical record of radar data exists. More commonly, however, the use of airborne radar can be considered a specialized application in forestry. Probably the most significant impediment to wider use of the data has been that, despite significant technological advances made with airborne, and more recently, free-flying satellite or Shuttle SAR imagery, the digital data, and the means to handle them effectively, are not widely available (Quegan, 1995). The free-flying satellite SAR data record began in 1991 with single wavelength and polarization SAR imagery from the ERS-1 satellite. Data acquired from Seasat in 1978 were a short-lived sensation; but the forestry potential was quickly recognized as low or moderate due to the influence of topography on the signal (Rauste, 1990). Subsequently, SAR data continued to be underutilized in forestry applications. Another reason for the relatively low general use of radar data in forestry is the short wavelengths, not ideally suited for penetration of forest canopies (Dobson, 2000).

Although digital applications are not yet fully developed, manual radar image analysis has been used in many forests around the world with some considerable success. One commercial program of airborne radar image analysis included the provision of inputs to tropical forest operations and management planning activities through forest cover type mapping, monitoring of logged/cutover areas, plantation mapping, and operations planning (Thompson and Macdonald, 1995). The airborne system, called the STAR-1, was a stereo X-band, HH polarized SAR integrated with an INS and GPS system that provided three-dimensional positional accuracy to less than 10 m. Three basic digital products were generated routinely: (1) stereo flightline strips, (2) image mosaics, and (3) topographic maps/DEMs. The radar image interpretation for forest management was conducted on image mosaics using standard photomorphic techniques, and at scales ranging from 1:20,000 to 1:100,000. Earlier applications and systems focused on the unique characteristics of airborne SAR image data that were shown to be sensitive to moisture variations (e.g., Bradley and Ulaby, 1981).

Airborne programs in recent years have served to spur research and increase interest in radar in forestry. For example, the multiband, multipolarization AIRSAR operated by NASA/JPL has been flown over many forest sites in the U.S. and Europe, and a second program called GLOBESAR has been operated by the Canada Centre for Remote Sensing since 1993. Forestry applications, only one of several target disciplines, were considered in need of data and method development so that full use of spaceborne SAR data could be made. These programs typically have a focus on biomass or timber volume estimation (Le Toan et al., 1992; Ranson et al., 1996; Liao and Guo, 1998). The basic idea is that the radar beam is attenuated by the forest canopy, and this attenuation can be directly related to the characteristics of that canopy. The radar image is dominated by volume scattering except in instances where the target presents an angular reflector. Here the double-bounce portion of the signal may be enhanced; then, the correlation between the double-bounce signal and the angular target (typically the stem or bole, but perhaps also the branching characteristics) may be the basis of volume extraction. Much of the information content of SAR imagery is contained in the texture patterns rather than the pixel tone or backscatter coefficients (Weishampel et al., 1994; Treitz et al., 2000).

A promising new development in radar remote sensing is a long-wavelength VHF SAR known as the CARABAS system (Israelsson et al., 1997; Fransson et al., 2000). Operating in the 20- to 90-MHz range (radio frequencies) the system was originally designed to detect military installations under forest canopies; at these wavelengths, the foliage and most branches are essentially invisible, providing almost no signal attenuation. A very strong dihedral bounce return is obtained from the main stem of trees leading to the distinct possibility of accurate estimation of biomass and volume across a wide range of forest types and conditions.

LIDAR

New developments in lidar remote sensing have been reported that dramatically increase the potential of these data in forest management (Dubayah and Drake, 2000). Lidar remote sensing has the potential to provide forest canopy height and structure measurements through spot measurements or newly developed scanning systems that provide unprecedented precision in estimating spatially variable forest canopy structure. Accurate digital forest height measurements, in particular, have been notoriously difficult — virtually impossible to generate by means of any other airborne or satellite remote sensing approach. Lidar devices can be developed in any part of the visual or near-infrared portion of the electromagnetic spectrum. In practice, the design of these devices is limited by the availability and relative efficiencies of both lasers and optical detectors at the desired wavelengths, and the constraint of eye safety. Efficient lasers are available for a number of discrete bands between 500 and 1600 nm. However, the most efficient optical detectors tend to operate between 800 and 1000 nm, and as a consequence, most terrestrial lidar systems tend to operate in this range. Eye safety is a key concern in the 400- to 1200-nm range; some designs use longer wavelengths (where the eye is less sensitive) to achieve higher power. Lidars that use laser light are typically distinguished from lidars that use other types of light (e.g., xenon, flash) (Wehr and Lohr, 1999).

The operation of one of the earliest lidar systems deployed in forestry, the NASA AOL (Airborne Oceanographic Lidar), originally designed and used for near-shore ocean bottom mapping, was described by Maclean and Krabill (1986: p. 9):

A short pulse of laser light approximately 7 nanoseconds (ns) in duration is emitted, and reflected out of the aircraft through a series of mirrors. When the laser pulse is intercepted by the forest canopy, a portion of the pulse is reflected back to the aircraft. Of the remaining energy, some proceeds through the canopy and is reflected off the forest floor and back to the aircraft as a secondary return pulse. The time difference between the initial return from the tree canopy and the secondary return can be converted to a height measurement using the known value of the speed of light.

The resulting data were analyzed as individual "hits" on the canopy and the ground in a profile beneath the aircraft at a rate of 500 laser pulses per second. Up to five measurements per meter were taken when the aircraft was moving over the target area at 100 m/s at an altitude of 150 m above ground. The footprint (or pixel resolution) of the laser measurement was approximately 0.7 m, and integrated INS information was used to position the profile on the ground.

A new scanning lidar sensor known as SLICER (Scanning Lidar Imager of Canopies by Echo Recovery) has been deployed in studies of forest structure, biomass, and canopy volume (Lefsky et al., 1999a,b; Means et al., 1999). SLICER differs from the earlier-generation lidars in that the entire laser return signal is digitized. From an altitude 5000 m above ground, a waveform is generated from multiple canopy elements and the ground over a 10-m footprint with an 11-cm vertical resolution. This broad footprint is a key difference. As the pulse reaches and proceeds through the canopy, reflectances from all canopy elements (foliar and woody) and the ground are measured. The resulting waveform provides a top-to-bottom view of forest canopies unlike the individual hits of the spot lidars; when coupled with multispectral imagery, an image of tree heights and spectral properties can be obtained (Chapter 3, Color Figure 3). No commercial operators of the airborne large-footprint scanning lidar yet exist (Means et al., 1999), but a small-footprint lidar with the potential to record entire waveforms has been tested in measuring tree heights and stand volume (Nilsson, 1996). The Vegetation Canopy Lidar (VCL), a proposed satellite lidar similar to SLICER, may be launched soon (2001); an airborne lidar that can simulate VCL data was described by Blair et al. (1999) and used by Weishampel et al. (2000) in mapping canopy structure of a tropical rainforest in Costa Rica.

SATELLITE PLATFORMS AND SENSORS

The principal satellite platforms and sensors appropriate as an information source for the sustainable forest management problems discussed in this book are listed in Table 3.6. Most systems provide observations from a single sensor payload in either the optical (e.g., Landsat, SPOT, IRS) or microwave (e.g., Radarsat, ERS-1) portion of the spectrum; occasionally, a satellite will carry both types of sensors (e.g., ALMAZ). A few systems have thermal detectors or other sensors as part of the package but, in essence, the choice of a platform/sensor package in support of forestry applications has been limited in the past to a few satellites with quite similar characteristics in the optical or microwave portions of the spectrum.

TABLE 3.6
Characteristics of Selected Existing and Proposed Satellite Platforms and Sensors for Forestry

Identification	Sensor	Number of Bands	Spatial Resolution (m)
Current Operational Satellites (Year 2000)			
Landsat-5	TM	7	30–120
	MSS	4	82
Landsat-7	ETM+	7	15–30
SPOT-2	HRV	4	10–20
SPOT-4	HRV	5	10–20
	VI	4	1150
RESURS-01-3	MSU-KV	5	170–600
IRS-1B	LISS	4	36–72
IRS-1C, -1D	LISS	4	23–70
	PAN	1	5.8
IRS-P4 (Oceansat)	OCM	8	360
JERS-1	VNIR,SWIR	8	20
	SAR	1	18
Almaz	SAR	3	4–40
Radarsat	SAR	1	9–100
ERS-1, -2	AMI (SAR)	1	26
	ATSR	4	1000
Space Imaging	IKONOS-2	5	1–4
NOAA-15	AVHRR	5	1100
NOAA-14	AVHRR	5	1100
NOAA-L	AVHRR	5	1100
Orbview-2 (Seastar)	SeaWiFS	8	1130
CBERS-1	CCD	5	20
	IRMSS	4	80–160
	WFI	2	260
Terra (EOS AM-1)	ASTER	14	15, 30, 90
	MODIS	36	250, 500, 1000
	MISR	4	275
Proposed Satellites (Launch Window 2000–2007)			
Earthwatch	Quickbird[a]	5	0.82–3.2
Orbview-3	Orbview	5	1–4
Orbview-4	Orbview	5	1–4
	Hyperspectral	200	8
IRS P5 (Cartosat)	Pan	1	2.5
IRS P6	LISS	7	6–23.5
	AWiFS	3	80
SPOT-5	HRV, VI	5	5–1150
KVR-100	Camera	1	1.5
TK350	Camera	1	10

TABLE 3.6 *(Continued)*
Characteristics of Selected Existing and Proposed Satellite Platforms and Sensors for Forestry

Identification	Sensor	Number of Bands	Spatial Resolution (m)
EO-1	Hyperion	220	30
	LAC	256	250
	ALI	10	10–30
WIS	EROS	1	1
CBERS-2	CCD	5	20
	IRMSS	4	80–160
	WFI	2	260
Resource21	A,B,C,D	5	20
ADEOS-II	GLI	36	250–1000
Kompsat	CCD	5	10
ARIES	ARIES-1	97	10–30
ALOS	VSAR	1	10
	AVNIR-2	5	2.5–10
Envisat-1	ASAR	1	30,150
	MERIS	15	300, 1200
Radarsat-2	SAR	1	6.25–500
LightSAR	SAR	4	3–100
XSTAR	XSTAR	10+	20
NEMO(HRST)	AVIRIS	211	5–30
EROS-A1, -A2	Pan	1	1.5
EROS-B1	Pan	1	0.82
Aqua (EOS PM-1)	MODIS	36	250–1000
Resource21	CIRRUS	6	10–100
MTI	MTI	15	5
NOAA-M, -N	AVHRR	5	1100

[a] Launch failed November 20, 2000.

Source: Adapted from Morain, 1998; Chen, 1998; Glackin, 1998; Stoney and Hughes, 1998; Dowman, 1999; Barnsley, 1999; Estes and Truelove, 1999, http://rs320h.ersc.wisc.edu/ERSC/Resources/EOSF.html.

In the near future, the satellite remote sensing scene will change dramatically (Glackin, 1998). A series of tremendous changes is forecast to take place in virtually every aspect of international spaceborne satellite remote sensing — sensor, platform, delivery, and product. The most significant change is the projected increase in the number of satellites of all types; space reconnaissance (military intelligence) (Richelson, 1991) and of principal interest here, Earth observation (Robertson and Cvetkovic, 1991; Hyman, 1996; Wynne and Oderwald, 1998). This will create new competition in the provision of imagery. As many as 99 satellites of all types were scheduled for launch between 1996 and 2006 (Fritz, 1996); Stoney and Hughes

(1998) discussed characteristics of 31 new Earth-viewing remote sensing satellites planned, or in various stages of planning, construction, or launch in a 2-year window to the year 2000. Clearly, these numbers suggest the beginning of widespread international availability of remote sensing data; increasingly through commercial developments, and continued growth in remote sensing methodology and technology (e.g., increased spatial and spectral resolution). However, launch and satellite tracking failures in 1997 and 2000 (Quickbird), and 1999 (IKONOS 1), resulted in a delayed timetable for some of the new hardware. Some of these new or proposed satellites and sensors are also listed in Table 3.6.

The existing and proposed satellite sensors generally belong to one of four groups of instruments (Stoney and Hughes, 1998):

1. Landsat-like,
2. Hyperspectral,
3. High spatial resolution, or
4. Radar.

In the first category are the Landsat, SPOT, IRS series of satellites, each of which is scheduled to be expanded and extended with some sensor improvements. For example, future Indian Remote Sensing (IRS) P5 and P6 platforms will carry an enhanced Linear Imaging Self-Scanning System (LISS) with one 2.5-m panchromatic channel and as many as seven 6-m multispectral channels. Existing data are used in updates for roads and harvesting patterns in several areas of the world; these new data will likely increase the acceptance of the new methods (Hall et al., 2000b).

The hyperspectral satellites include the Australian ARIES and a number of EOS platforms; much of this hardware will remain experimental with a research focus aimed at unlocking the wealth of information in the huge data streams to be generated. The high spatial resolution group of satellites are primarily funded and operated by private corporations; Chapter 3, Color Figure 4 contains an example of IKONOS-2 imagery acquired in British Columbia for forest mapping purposes. The goal of the high spatial resolution image providers is a high level of penetration into the aerial photography market (Glackin, 1998); early suggestions (Fritz, 1996: p. 44) were that "well over 50% of the imaging provided by the aerial survey market will be replaced by this high-resolution satellite imagery." Additional development of new markets for satellite remote sensing products, such as real estate planning and property assessment, were thought likely.

A number of radar satellites are planned following the success of the European Space Agency ERS-1 and Canadian Space Agency Radarsat programs (van der Sanden et al., 2000). These data will continue to be used in some applications in forestry, often in an integrated format with optical data. Satellite radar data are particularly well suited in ocean and ice applications. The focus for radar systems will continue to be on providing all-weather day/night sensing (Quegan, 1995) and on radargrammetric (or interferometric) applications, including topographic mapping (Leberl, 1983; Zebker and Goldstein, 1986). The research community eagerly awaits consistent availability of multipolarization and multifrequency SAR data since data with these characteristics hold such great promise in forestry applications (Dobson, 2000).

GENERAL LIMITS IN ACQUISITION OF AIRBORNE AND SATELLITE REMOTE SENSING DATA

In this chapter, some data collection and some preprocessing issues were reviewed. The idea was to help provide a feel for the range of data and data types that can be obtained in a remote sensing approach to forest management questions. This was not an exhaustive survey, as many sensors and data issues were not considered. While there may not be a data collection tool for every data collection problem, remote sensing *is* a very flexible data source and opportunities are increasing as new platforms and sensors are developed and deployed. Despite this, it is certainly obvious that more than a few remote sensing application projects have failed, or at least returned disappointing results. Now, it is appropriate to consider the general limits that exist in using remote sensing data.

Often, remote sensing applications have been less than impressive because of a lack of clarity in the statement of objectives. What exactly is the information requirement that remote sensing is trying to satisfy? One fundamental problem has been that a key feature of the remote sensing experimental design was implemented without regard for the end user, sometimes years or decades before the particular application project of interest was conceived and attempted — that is, the platform and sensor characteristics were developed long before the application project was initiated. Even if the project was envisioned before the sensors were built and the platforms deployed, the imagery that is obtained often represents fundamental trade-offs in design that are a result of compromises in original mission execution (e.g., orbital characteristics for satellites, or time-of-day for image acquisitions for airborne sensors).

Perhaps the best example is the continued lack of a multipolarization, multifrequency satellite SAR system. Despite research since the 1970s highlighting the critical need for multifrequency, multipolarization SAR data for effective imaging of forests, no such satellite system has yet been launched. Planned launches of a polarimetric L-band sensor on ALOS Palsar, and multipolarized C-band sensors on Envisat and Radarsat-2, will soon correct this deficiency. In the meantime, continued use has been made of single frequency, same polarization satellite SAR sensors, or by combining data from different platforms (e.g., Radarsat, JERS-1, Shuttle missions). The forestry applications have been less successful than would otherwise have been possible.

A possible limitation on the use of remote sensing in forestry might exist in the form of institutional inertia in management agencies; no doubt, resistance to technological change has been experienced. This will likely continue to occur. For example, cloud cover has been raised reliably over the past 30 years as a constraint preventing the more widespread adoption of a remote sensing approach in forestry. Cloud cover can be a major problem in many areas of the world. But for many areas, several hundred thousand images exist in various satellite remote sensing archives; it was possible to create a generally cloud-free Landsat data set of Brazilian Amazonia, for example (Skole and Tucker, 1993). Kuntz and Siegert (1996), working in East Kalimantan, found that no Landsat or SPOT imagery had been acquired of their study area without major cloud problems — ever! Instead, they used a series of

KFA1000 photographs taken from the MIR space station and ERS-1 SAR data. By avoiding steep topography they could recognize five forest classes suitable to develop an initial forest inventory: undisturbed lowland *Dipterocarp* forest, undisturbed forest, selectively logged forest, clearcuts, and secondary forests.

In such regions, it may be operationally practical to use available microwave satellite imagery when optical data are not available, or when cloud conditions prevent their use (Fransson et al., 1999). ERS-1, JERS-1, and Canada's Radarsat — operational radar satellite programs since the early 1990s — routinely acquire imagery useful in forestry applications in all weather conditions and at night. A more accurate interpretation of the current situation may be to recognize that as forest management and planning itself has adapted, a further step in evolution is needed to take advantage of the full range of possible remote sensing data.

There are some real limits to remote sensing data, limits that go beyond data acquisition, data pricing, and sensor design questions (Landgrebe, 1997), limits that flow largely from the infrastructural problems users have had when actually acquiring and using the data. For example, the failure to commercialize the Landsat program in the early 1990s may be partly attributable to the lack of infrastructure in place to support users (Williamson, 1997). Apparently, few were positioned to make effective use of the data on a commercial basis; even ordering satellite data was a challenge! What data were available? What cloud cover conditions were acceptable? What level of preprocessing was desirable? What format was the digital tape? There were few user-selected options available as data suppliers often acquired data under strict orbital conditions not particularly amenable to the desired data characteristics.

A realistic assessment of the limit that the lack of infrastructure imposed on early attempts to apply remote sensing technology to forestry might have more to do with the crisis of quality and overpromotion, the overselling of the technology (Wynne et al., 2000); there were real problems with the ability of the data to handle the tasks that resource management professionals needed to have done by remote sensing. There were real misunderstandings about what the data could and could not do with appropriate processing and interpretation. Many such limits remain beyond the control of the typical data user — if the data are not acquired under the conditions suitable for the application there are more options, but still far fewer than are available in many data collection exercises. Apart from the continuation of such infrastructural limits, of interest are the limits that transcend a particular sensor package or method of analysis, but which relate to the fundamental properties of remote sensing data.

The first and most obvious limit experienced in any remote sensing project relates to the phenomenological unit of data collection: the pixel (a function of the sensor instantaneous field of view, electronics, and mission conditions). There are always limits to what can be accomplished with data; no data set, remote sensing or otherwise, can possibly satisfy all possible needs, hopes, or desires. But it is the very nature of the remote sensing pixel that can lead to frustration, however, because although objects smaller than the pixel can often be seen, there is a fundamental limit suggested by pixel size in any data set. The pixel is imposed on the landscape but is not likely to match the contents of that landscape (Fisher, 1997). There is no single correct pixel size for use in forestry applications. There has always been tension between acquiring data with sufficient detail (grain or resolution) and covering the area of interest (extent); this tension may complicate data storage and data processing efficiency, but more significantly, there are the analysis problems created by pixels being larger than desired for a given application (Cracknell, 1998). Related to the pixel size and scale problem are the real limits caused by the other image resolutions: spectral resolution (enough bands in the right places?), radiometric resolution (sensitivity to the features of interest — forests?), and temporal resolution (repeat coverage controllable?). Too often the following situation has arisen: wrong data, right application.

The second major limit that is imposed by remote sensing data relates to the relatively poor information extraction methods that exist. The following two chapters consider this issue in some detail. Remote sensing has always been capable of generating an overwhelming amount of data, but from the beginning of the satellite programs such as Landsat, it has been obvious that even if exactly the right pixel size were available for all applications, even if spectral sensitivity to ground conditions of interest were available, a weak link is the ability to use "rigorous, broadly applicable, and user-acceptable analysis procedures" (Landgrebe, 1997: p. 866). H-resolution remote sensing data are a case in point; airborne data with less than 1 m pixel size in multiple wavelengths have been available for several decades, yet the image processing capability for such data is pathetic relative to information content. Another problem: algorithms that work at one pixel resolution may not be transportable to imagery at other resolutions (and scales) (Chen, 1999).The general problem will be addressed throughout the course of this book and in the many references provided, but is worth repeating: the ability to extract information from image data, information that is known to exist in the image data, is limited by understanding and methodology.

The proliferation of existing and new satellite platforms and sensor designs, and the continued improvements in airborne sensor technology, guarantee the continuation of the data-rich environment to which remote sensing practitioners are accustomed. But this helps create the third general limit in the acquisition of remote sensing data: the availability of trained and experienced remote sensing data analysts. The nature of remote sensing data must be understood; the many methods of remote sensing data collection and analysis must be mastered; and this knowledge must be applied to a disciplinary field, such as forestry, in an ever-changing technological milieu. The key to the future will no doubt lie in continuing improvement of the capability to process useful information from these data — and humans are needed to facilitate this process and to use the information in forestry applications. Probably the most constraining limit in remote sensing continues to be the creation and maintenance of a user community that can understand the data and the methods of remote sensing as well as they understand the problems of forest management, and vice versa for the remote sensing community.

There may always be limits to what can be done with the pixel as the phenomenological unit of analysis (Fisher, 1997). There may always be limits created by the information extraction methods, which can seem both astonishingly complex (compared to earlier methods) and stunningly primitive; but still unable to complete tasks the user would like to be able to do (Townshend and Justice, 1981). There may always be a mismatch between the available data and the ideal data for a given application. As more of the user community become familiar with these immediate issues of relevance to remote sensing forestry applications, and work more closely with imagery and image analysts (Oderwald and Wynne, 2000), then limits imposed by the conceptual difficulty of organizing the landscape on a pixel-by-pixel basis, the data themselves, and the methods of image analysis, will likely be overcome more quickly.

4 Image Calibration and Processing

This revolutionary new technology (one might almost say black art) of remote sensing is providing scientists with all kinds of valuable new information to feed their computers

— K. F. Weaver, 1969

GEORADIOMETRIC EFFECTS AND SPECTRAL RESPONSE

A generic term, spectral response, is typically used to refer to the detected energy recorded as digital measurements in remote sensing imagery (Lillesand and Kiefer, 1994). Since different sensors collect measurements at different wavelengths and with widely varying characteristics, spectral response is used to refer to the measurements without signifying a precise physical term such as backscatter, radiance, or reflectance. In the optical/infrared portion of the spectrum there are five terms representing radiometric quantities (radiant energy, radiant density, radiant flux, radiant exitance, and irradiance). These are used to describe the radiation budget of a surface and are related to the remote sensing spectral response (Curran, 1985). When discussing image data, the term spectral response suggests that image measurements are not absolute, but are relative in the same way that photographic tone refers to the relative differences in exposure or density on aerial photographs. Digital image spectral response differs fundamentally from photographic tones, though, in that spectral response can be calibrated or converted to an absolute measurement to the extent that spectral response depends on the particular characteristics of the sensor and the conditions under which it was deployed. When all factors affecting spectral response have been considered, the resulting physical measurement — such as radiance (in $W/m^2/\mu m/sr$), spectral reflectance (in percentage), or scattering coefficient (in decibels) is used. Consideration of the geometric part of the image analysis procedure typically follows; here the task is the correct placement of each image observation on the ground in terms of Earth or map coordinates.

It is well known that spectral response data acquired by field (Ranson et al., 1991; Gu and Guyot, 1993; Taylor, 1993), aerial (King, 1991), and satellite (Teillet, 1986) sensors are influenced by a variety of sensor-dependent and scene-related georadiometric factors. A brief discussion of these factors affecting spectral response is included in this section, but for more detail on the derivations the reader is referred to more complete treatments in textbooks by Jensen (1996, 2000); Lillesand and Kiefer (1994); and Vincent (1997). If more detail is required, the reader is advised to consult papers on the various calibration/validation issues for specific sensors (Yang and Vidal, 1990; Richter, 1990; Muller, 1993; Kennedy et al., 1997) and platforms (Ouaidrari and Vermote, 1999; Edirisinghe et al., 1999).

There are three general georadiometric issues (Teillet, 1986):

1. The influence of radiometric terms (e.g., sensor response functions) or calibration,
2. The atmospheric component, usually approximated by models, and
3. Target reflectance properties.

Chapter 3 presented the general approach to convert raw image DN to at-sensor radiance or backscattering using the internal sensor calibration coefficients. To summarize, the first processing step is the calibration of the raw imagery to obtain physical measurements of electromagnetic energy (as opposed to relative digital numbers, or DNs, see Equation 3.1) that match an existing map or database in a specific projection system. In SAR image applications, the raw image data are often expressed as a slant-range DN which must be corrected to the ground range backscattering coefficient (a physical property of the target, see Equation 3.3). These corrections, or more properly calibrations, together with the precise georeferencing of the data to true locations, are a part of the georadiometric correction procedures used to create or derive imagery for subsequent analysis. Of interest now are those additional radiometric and geometric processing steps necessary to help move the image analyst from working with imagery that is completely internally referenced (standardized digital numbers, radiance, or backscatter on an internal image pixel/line grid) to imagery that has removed the most obvious distortions, such as view-angle brightness gradients and atmospheric or topographic effects. The results are then georeferenced to Earth or map coordinates (Dowman, 1999).

In optical imagery, the three major georadiometric influences of interest include the atmosphere, the illumination geometry (including topography and the view angle), and the sensor characteristics (including noise and point-spread function effects) (Duggin, 1985). In SAR imagery the dominant georadiometric effects are the sensor characteristics and the topography. When considering the individual pixel spectral response as the main piece of information in an image analysis procedure, a difference in illumination caused by atmospheric, view-angle, or topographic influences may lead to error in identifying surface spectral properties such as vegetation cover or leaf area index. The reason is that areas of identical vegetation cover, or with the same leaf area index, can have different spectral response as measured by a remote sensing device solely, for example, because of the differences in atmosphere or illumination geometry on either side of a topographic ridge.

In general, in digital analysis, failure to account for a whole host of georadiometric influences may lead to inaccurate image analysis (Duggin and Robinove, 1990) and incomplete, or inaccurate remote sensing output products (Yang and Vidal,

1990). In some situations, uncorrected image data may be virtually useless because they may be difficult to classify reliably or be used to derive physical parameters of the surface. But not all imagery must be corrected for all these influences in all applications. In many cases, imagery can be used off-the-shelf with only internally consistent calibration, for example, to at-sensor radiances (e.g., Wilson et al., 1994; Wolter et al., 1995). Almost as frequently, raw image DNs have been used successfully in remote sensing applications, particularly classification, where no comparison to other image data or to reference conditions has been made or is necessary (Robinove, 1982). Use of at-sensor radiance or DNs is exactly equivalent in most classification and statistical estimation studies; rescaling the data by linear coefficients will not alter the outcome. Even in multitemporal studies, when the differences in spectral response expected in the classes can be assumed to dominate the image data (for example, in clearcut mapping using Landsat data), there may be no need to perform any radiometric calibration (Cohen et al., 1998).

General correction techniques are referred to as radiometric and geometric image processing — in essence, radiometric processing attempts to reduce or remove internal and external influences on the measured remote sensing data so that the image data are as closely related to the spectral properties of the target as is possible. Geometric processing is concerned with providing the ability to relate the internal image geometry measurements (pixel locations) to Earth coordinates in a particular map projection space. All of the techniques designed to accomplish these tasks are subject to continual improvement. In no case has any algorithm been developed that resolves the issue for all sensors, all georadiometric effects, and all applications. This part of the remote sensing infrastructure is truly a work in progress.

Radiometric Processing of Imagery

Some sensor-induced distortions, including variations in the sensor point-spread response function, cannot be removed without complete recalibration of the sensor. For airborne sensors, this means demobilization and return to the lab. For satellites, this has rarely been an option, and only relative calibration to some previous state has been possible. Some environmentally based distortions cannot be removed without resorting to modeling based on first principles (Duggin, 1985; Woodham and Lee, 1985); for example, variations in atmospheric transmittance across a scene or over time during the acquisition of imagery. Often, it is likely that such effects are small relative to the first-order differences caused by the atmospheric and topographic effects. Typically, these are the more obvious radiometric and geometric distortions. Image processing systems often contain algorithms designed to remove or reduce these influences. Experience has shown that atmospheric, topographic, and view-angle illumination effects can be corrected well enough empirically to reduce their confounding effects on subsequent analysis procedures such as image classifications, cluster analysis, scene segmentation, forest attribute estimation, and so on. The idea is to develop empirical corrections to remove sensor-based (e.g., view-angle variations) and environmentally based (e.g., illumination differences due to topographic effects, atmospheric absorption, and scattering) errors.

In the optical/infrared portion of the spectrum, raw remote sensing measurements are observations of radiance. This measurement is a property of the environment under which the sensor system was deployed. Radiometric corrections typically involve adjustments to the pixel value to convert radiance to reflectance using atmosphere and illumination models (Teillet, 1997; Teillet et al., 1997). The purpose of a scene-based radiometric correction is to derive internally consistent spectral reflectance measurements in each band from the observed radiances in the optical portion of the spectrum (Smith and Milton, 1999).

The simplest atmospheric correction is to relate image information to pseudo-invariant reflectors, such as deep, dark lakes, or dark asphalt/rooftops (Teillet and Fedosejevs, 1995). For the dark-object subtraction procedure (Campbell and Ran, 1993), the analyst checks the visible band radiances over the lakes or other dark objects, then correspondingly adjusts the observed values to more closely match the expected reflectance (which would be very low, close to zero). The difference between the observed value and the expected value is attributed to the atmospheric influences at the time of image acquisitions; the other bands are adjusted accordingly (i.e., according to the dominant atmospheric effect in those wavelengths such as scattering or absorption). This procedure removes only the additive component of the effect of the atmosphere. The dark-target approach (Teillet and Fedosejevs, 1995) uses measurements over lakes with radiative transfer models to correct for both path radiance and atmospheric attenuation by deriving the optical depth internally.

These pseudo-invariant objects — deep, dark, clear lakes or asphalt parking lots (Milton et al., 1997) — should have low or minimally varying reflectance patterns over time, which can be used to adjust for illumination differences and atmospherically induced variance in multitemporal images. An alternative to such scene-based corrections relies on ancillary data such as measurements from incident light sensors and field-deployed calibration targets. In precise remote sensing experiments, such measurements are an indispensable data source for more complex atmospheric and illumination corrections.

A large project now being planned by the Committee on Earth Observation Satellites (CEOS) (Ahern et al., 1998; Shaffer, 1996, 1997; Cihlar et al., 1997) to produce high-quality, multiresolution, multitemporal global data sets of forest cover and attributes, called Global Observation of Forest Cover (GOFC), contains several different "levels" of products based on raw, corrected, and derived (classified or modeled) imagery (GOFC Design Team, 1998).

1. Level 1 data — raw image data
2. Level 2 data — calibrated data in satellite projection
3. Level 3 data — spatially/temporally resampled to *true* physical values
4. Level 4 data — model or classification output

Existing methods of radiometric processing are considered sufficient for the general applications of such data, and users with more detailed needs can develop products from these levels for specific applications. For example, in studies of high-relief terrain with different (usually more detailed) mapping objectives, it has clearly been demonstrated that raw DN data cannot be used with sufficient confidence; more complex radiometric and atmospheric adjustments must be applied to obtain the maximum forest classification and parameter estimation accuracy (Itten and Meyer, 1993; Sandmeier and Itten, 1997).

Such atmospheric corrections are now much more commonly available in commercial image processing systems. For example, a version of the Richter (1990) atmospheric correction model is a separate module within the PCI Easi/Pace system. The model is built on the principle of a lookup table; first, an estimate of the visibility in the scene is required, perhaps derived from the imagery or an ancillary source, from which a standard atmosphere is selected that is likely to approximate the type of atmosphere through which the energy passed during the image acquisition. Second, the analyst is asked to match iteratively some standard surface reflectances (such as golf courses, roads, mature conifer forests) to the modeled atmosphere and the image data. An image correction is computed based on these training data. When coded this way, with additional simplifications built in, the corrections are not difficult, costly, or overly complex to apply (Franklin and Giles, 1995). However, it is important to be aware of the assumptions that such simplified models use, since the resulting corrections may not always be helpful in image analysis. Thin or invisible clouds, smoke, or haze, for example, will confound the algorithm because these atmospheric influences are not modeled in the standard atmosphere approach.

Topographic corrections are even more difficult and the results even less certain; the physics involved in radiant transfers in mountainous areas are incompletely understood and daunting to model, to say the least (Smith et al., 1980; Kimes and Kirchner, 1981; Dymond, 1992; Dubayah and Rich, 1995). This complexity, coupled with the obvious (though not universal) deleterious effect that topography can have on image analysis, has given rise to a number of empirical approaches to reduce the topographic effect well enough to allow subsequent image analysis to proceed (Richter, 1997). The topographic effect is defined as the variation in radiance from inclined surfaces, compared with radiance from a horizontal surface, as a function of the orientation surface relative to the light source and sensor position (Holben and Justice, 1980). Corrections for this effect have been developed, together with attempts at building methods of incorporating the topographic effect into image analysis to better extract the required forestry or vegetation information from the imagery. Neither of these two ideas — correcting for topography, or using topographic information to help make decisions — has attained the status of an accepted standard method in remote sensing image analysis.

Unfortunately, while the various georadiometric factors are all interrelated to some extent (Teillet, 1986), it is clear that the effects of topography and bidirectional reflectance properties of vegetation cover are inextricably linked. These effects are difficult to address, and may require substantial ancillary information (such as coincident field observations or complex model outputs). Clearly, due only to topography and the position of the sun, north-facing slopes would appear darker and south-facing slopes would appear lighter, even if the vegetation characteristics were similar. The difference in topography causes a masking of the information content with an unwanted georadiometric influence (Holben and Justice, 1980). In reality, some of these influences are actually aids in manual and automated image interpretation; for example, the subtle shading created by different illumination conditions on either side of a topographic ridge can be a useful aid in identifying a geological pattern, in developing training statistics, and in applying image analysis techniques. In automated pattern recognition and image understanding this topographic shading can lead to higher levels of information extraction from digital imagery. The use of stereoscopic satellite imagery to create a DEM is largely based on the presence of a different topographic effect in two images acquired of the same area from different sensor positions (Cooper et al., 1987).

The complexity of atmospheric and topographic effects is increased by the non-Lambertian reflectance behavior of many surfaces depending on the view and illumination geometry (Burgess et al., 1995; Richter, 1997). Surfaces are assumed to be equally bright from all viewing directions. But since vegetated surfaces are rough it is clear that there will be strong directional reflectances; forests are brighter when viewed from certain positions. This has given rise to a tautology: to identify the surface cover a topographic correction must be applied; to apply a topographic correction the surface cover must be known. In the early 1980s, the problem was considered intractable and computationally impossible to model precisely using radiation physics (Hugli and Frei, 1983); this situation has not yet changed; the Lambertian assumption is still widely used (Woodham, 1989; Richter, 1997; Sandmeier and Itten, 1997).

Empirical topographic corrections have proven only marginally successful. Most perform best when restricted to areas where canopy complexity and altitudinal zonation are low to moderate (Allen, 2000). In one comparison of topographic correction approaches, only small improvement in forest vegetation classification accuracy was obtained using any one of four commercially available techniques (Franklin, 1991). In another study with airborne video data, Pellikka (1996) found that uncorrected data provided 74% classification accuracy compared with 66% or less for various illumination corrected data. The topographic correction decreased classification accuracy. After an empirical postcorrection increased the diffuse radiation component on certain slopes, a significant increase in accuracy was obtained. The tautology! These authors emphasized the uncertain nature of the topographic corrections using simple sun sensor-target geometric principles, and with empirical and iterative processing were able to provide data that were only marginally, if at all, more closely related to the target forestry features of interest. But for many image analysts, even these corrections are difficult to understand and apply in routine image analysis.

Although there have been attempts to provide internally referenced corrections (i.e., relying solely on the image data to separate topographically induced variations from target spectral differences) (Eliason et al., 1981; Pouch and Compagna, 1990), most empirical corrections use a digital elevation model to calculate the illumination difference between sloped and flat surfaces (Civco, 1989; Colby, 1991). These early approaches typically assumed that the illumination effects depended mainly on the solar incident angle cosine of each pixel (i.e., angle between the surface normal and the solar beam) (Leprieur et al., 1988); but this assumption is not valid for all covertypes, and not just because of the non-Lambertian nature of most forested surfaces. In particular, forests contain trees which are geotropic (Gu and Gillespie, 1998). In forests, the main illumination difference between trees growing

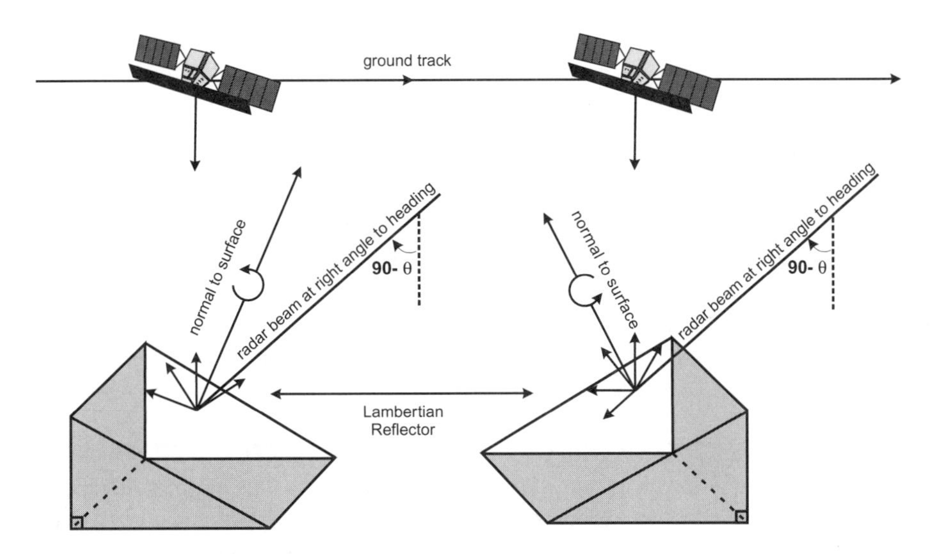

FIGURE 4.1 An initial correction geometry employed to reduce the topographic effect on airborne SAR data. The dominant effect in SAR imagery over rugged terrain is caused by the slope. This influence can be reduced by correcting the data for the observer position by comparing to the normalized cosine of the incidence angle. The correction assumes a Lambertian reflectance surface and does not consider that forest canopies are "rough." A cover-specific correction may be necessary to allow the SAR data to be related to the characteristics of the vegetation rather than the terrain roughness and slope. (Modified from Franklin, S. E., M. B. Lavigne, B. A. Wilson, et al. 1995a. *Comput. Geosci.,* 21, 521–532.)

on slopes and on flat surfaces is in the amount of sunlit tree crown and shadows that is visible to the sensor, rather than the differences in illumination predicted by the underlying slopes.

In the microwave portion of the spectrum, radiometric corrections are needed to derive backscatter coefficients from the slant-range power density. For environmental effects, SAR image calibration and correction require calibration target deployment (Waring et al., 1995b). By far, the strongest georadiometric effects on SAR imagery are caused by azimuth (flight orientation) and incidence angles (defined as the angle between the radar beam and the local surface normal) (Domik et al., 1988). The influence of local topography can be dramatic as high elevations are displaced toward the sensor and the backscattering on slopes is either brightened or foreshortened. Simple image corrections using DEM-derived slopes and aspects do not completely restore the thematic information content of the imagery. The wavelength-dependent energy interactions are too complex to be well represented by simple cosine models (Domik et al., 1988; Van Zyl, 1993); however, cosine-corrected imagery will likely be more useful (Hinse et al., 1988; Wu, 1990; Bayer et al., 1991). Figure 4.1 shows the initial correction geometry that has been employed to reduce the topographic effect on airborne SAR data (Franklin et al., 1995a).

Table 4.1 contains examples of original and corrected values for some example pixels extracted from Landsat and SAR imagery. Examples of the cosine and modified cosine corrections are shown for three pixel values extracted from earlier work

TABLE 4.1
Example Original Uncorrected and Corrected Pixel Values for SAR and Landsat Sensors Based on Relatively Simple Correction Routines Available in Commercial and Public Image Processing Systems

Original DN	Incidence Angle	Azimuth	Surface Slope	Surface Aspect	Corrected Value	Sensor and Type of Correction	Ref.
23	117.5	270	60	315	38	SAR Cosine	Franklin et al., 1995
62	67	151	6	180	49	Landsat Cosine	Franklin, 1991
69	57	151	6	180	57	Landsat Modified Cosine	Civco, 1989

(Franklin, 1991; Franklin et al., 1995a). The table shows the original DN value collected by a west-looking airborne SAR sensor over a steeply sloping north aspect. This geometry produced an image DN value much lower than the DN on a flat surface without any topographic effect; the purpose of the correction is to estimate how much brightness to add to the pixel value. The opposite effect is shown in the two Landsat pixel examples. Here, the surface was gently sloping into the direction of the sun, and the result was that the surface appeared brighter than a flat surface would under the same illumination conditions. The purpose of the cosine correction is to reduce the brightness; the first correction reduced the brightness based solely on the illumination and target topography (Franklin, 1991). A second correction applied to slightly different image illumination conditions was based on the modification of the cosine by an estimate of the average conditions for that image (Civco, 1989).

These corrections are shown to indicate the types of corrections that are widely available. Such corrections must often be used in highly variable terrain or areas in which the precise differences in spectral reflectance on different slopes are not of interest — classification studies, for example. These corrections do not adequately account for all aspects of radiative transfer in mountain areas (Duguay, 1993); they are first-order approximations only, ignoring diffuse and adjacency effects, for example, and as such may or may not be useful depending on data characteristics, the level of processing, and the purpose of the image application. Because these corrections may not work, one of the more powerful methods to deal with the topographic effect has been to use the DEM data together with the spectral data in the analysis; Carlotto (1998: p. 905), for example, built a multispectral shape classifier, "instead of correcting for terrain and atmospheric effects." This idea of avoiding or incorporating unwanted georadiometric effects such as topography into the decision-making classification or estimation process is discussed in more detail in later sections.

View-angle effects can reduce the effectiveness of airborne sensor data because of the wide range of viewing positions that airborne sensors can accommodate during a remote sensing mission (King, 1991; Yuan, 1993). Wide-angle and off-nadir views will introduce variable atmospheric path lengths in an image scene, thereby introducing different atmospheric thicknesses that need to be corrected during the atmospheric processing (Pellikka, 1996). Such differences in atmospheric path length are usually minor, particularly if the sensor is operated below the bulk of the atmosphere; instead, the bidirectional effect is the main difficulty. Ranson et al. (1994) described several experiments with the Advanced Solid-State Array Spectroradiometer (ASAS), an instrument designed to view forests in multiangle (off-nadir) positions (Irons et al., 1991). The idea was to reconstruct the bidirectional reflectance factors over forest canopies. As expected, higher observed reflectances were recorded in or near the solar principal plane at viewing geometries approaching the antisolar direction (Ranson et al., 1994). Others, using multiple passes over a single site with wide-view-angle sensors, observed similar effects (Kriebel, 1978; Franklin et al., 1991; Diner et al., 1999). The view angle will also determine the projected area of each pixel and introduce a more complex geometric correction (Barnsley, 1984). Pixel geometry is constant across-track for linear arrays, but variable for single-detector configurations.

View-angle effects are typically much smaller in most satellite systems compared to those in airborne data, but are sometimes apparent in wide-angle or pointable satellite systems such as the SPOT (Muller, 1993), AVHRR (Cihlar et al., 1994), SPOT VEGETATION (Donoghue, 1999), or EOS MODIS sensors (Running et al., 2000). For satellites, the view-angle effect can "mask" or hinder the extraction of information as is typically the case with single-pass airborne data. This situation will deteriorate with still larger view angles and higher spatial detail satellite imagery. The importance of the view-angle effect will depend on (Barnsley and Kay, 1990).

1. The geometry of the sensor — i.e., the sizes of the pixels and their overlap relative to the illumination sources
2. The geometry of the target — i.e., the variability of the different surface features visible to the sensor

No systematic approach for correcting these two effects has been reported although systems that deal simultaneously with geometric, topographic, and atmospheric corrections are now more common (Itten and Meyer, 1993). But experiments with multiple incidence angle high spatial resolution data are relatively rare. As with topographic corrections, there is the parallel attempt not to simply correct view-angle effects in imagery (Irons et al., 1991), but instead to use the variable imaging conditions to extract the maximum amount of information in the imagery that is attributable to the different viewing geometry. Sometimes referred to as an "angular signature" (Gerstl, 1990; Diner et al., 1999), this approach has provided some improved analytical results. For example, at the Boreas site in northern Canada (Cihlar et al., 1997a), when BRDF data were extracted from multiple view-angle hyperspectral imagery, higher classification accuracies of species and structural characteristics of boreal forest stands were possible (Sandmeier and Deering, 1999). Off-nadir viewing improved the forest information content and the performance of several different multispectral band ratios in discriminating forest cover and LAI (Gemmell and McDonald, 2000).

The more general interpretation of view-angle effects, especially in single-pass imagery or in compositing and mosaicking tasks, is that the effect is an impediment to image analysis and to image classification (Foody, 1988). Fortunately, in many cases the view-angle effect is approximately additive in different bands and therefore can be cancelled out by simple image processing; for example, image band ratioing (Kennedy et al., 1997). Another approach is to apply a profile correction based on the observed deviation from nadir data measurements (Royer et al., 1985). Each profile value is based on averaging many lines for a given pixel column at a constant view angle or distance from nadir. The resultant values are fitted with a low-order polynomial to smooth out variations which result from localized scene content. The polynomial is used to remove view-angle dependence by predicting a new pixel value relative to the nadir position and replacing or correcting the actual value proportionally. The overall effectiveness of the view-angle corrections in reducing variance unrelated to vegetation and soil surfaces has been confirmed under numerous different remote sensing conditions, particularly in the presence of a brightness gradient that is clearly visible in the imagery. But these corrections are inexact. In one comparison of four different empirical methods of view-angle correction for AVIRIS data, Kennedy et al. (1997: p. 290) found at least one method provided "blatantly inappropriate brightness compensation" thereby masking true information content more severely than in the uncorrected imagery.

GEOMETRIC PROCESSING OF IMAGERY

The accuracy of spatial data — including imagery — can be considered as comprised of two components.

1. Spatial or locational accuracy
2. Thematic accuracy

Thematic accuracy has often been a major concern in remote sensing (Hord and Brooner, 1976). Validation of thematic accuracy, at least in classifications, has recently attained the status of a standardized procedure in remote sensing (Congalton and Green, 1999). Accuracy assessment procedures now exist as an integral part of virtually every commercially available image processing system, and accuracy assessment can be considered an essential element in any remote sensing application. The idea of thematic accuracy is intricately tied to the issue of validation of remote sensing data products, discussed more fully in later sections.

Spatial or locational accuracy has long been of interest because of the promise that remote sensing contained to satisfy mapping needs; from the collection of the earliest images, there was concern with the capability to locate accurately on the Earth's surface the results of the image analysis (Hayes and Cracknell, 1987). Geometric corrections are applied to provide spatial or locational accuracy (Burkholder, 1999). Geometric distortions are related not only to the sensor and imaging geometry, but also to the topography (Itten and Meyer, 1993; Fogel and Tinney, 1996); corrections, then, are applied to account for known geometric distortions based on the topography or sensor/platform characteristics and to bring the imagery to map coordinates. This latter exercise is also commonly known as geocoding.

Working with digitized aerial photographs, Steiner (1974) outlined the typical sequence of steps in registration of digital images to a map base. These steps are illustrated in Chapter 4, Color Figure 1*, which contains an example rectification and resampling procedure for an airborne image and satellite image dataset with map coordinates.

1. Perform a theoretical analysis of possible geometrical errors so that an appropriate form of transformation can be selected.
2. Locate corresponding ground control points in the reference (map) and image (pixel/line) coordinate systems.
3. Formulate a mathematical transformation for the image based on the georeferencing information.
4. Implement the transformation and subsequently resample the image data to match the new projection/georeference.

Such corrections can be relative (i.e., to another image, map, or an arbitrary coordinate system) or absolute (i.e., to a global georeferencing system in Earth coordinates). The availability of GPS has rendered subpixel geometric corrections tractable in remote sensing. During Step 2 above, the analyst would typically either identify GCPs in map data or use a GPS unit on the ground to collect GCPs visible in the imagery. Step 3 requires an understanding of the types of geometric errors that must be modeled by the transformation; the order of the polynomial increases as more errors are introduced to the correction. Particularly in mountainous terrain, image points may be shifted due to scan line perspective displacement, a random characteristic of the orbital parameters and the terrrain. This effect is not normally dealt with during polynomial transformations, even if higher-order polynomials are defined (Cheng et al., 2000). Instead, users concerned with the relief displacement and geometric distortions caused by topographic shifting of pixels must consider more complex orthorectification procedures. The ready availability of high-quality DEMs — or the ability to derive these DEMs directly from stereocorrelated digital imagery (e.g., Chen and Rau, 1993) — has provided a foundation for the orthorectification of digital satellite and aerial imagery, at least at the resolution of the DEM (usually a medium scale such as 1:20,000).

In Step 4 a decision must be made on the type of resampling algorithm to use; little has been reported in the literature to guide users in this choice (Hyde and Vesper, 1983). A general preference for the nearest-neighbor resampling algorithm exists, apparently because this algorithm is thought to minimize the radiometric modification to the original image data that are introduced by area (mean) operators, such as the cubic convolution or bilinear interpolation algorithms. However, even nearest-neighbor resampled data differ from original imagery since some pixels and scan lines may be duplicated and individual pixels can be skipped, depending on the resolution of the output grid.

A fine adjustment after the main correction could be based directly on a comparison of image detail (Steiner, 1974); such an adjustment would be based on feature or area comparisons (Dai and Khorram, 1999). Feature-based registration implies

* Color figures follow page 176.

that distinct entities such as roads and drainage networks can be automatically extracted and used to match imagery over time. Area-based registration usually works on the correlation of image tone within small windows of image data and therefore works best with multiple images from the same sensor with only small geometric misalignment. Few studies have attempted these procedures (Shlien, 1979), and the processing software is not widely available. Because of the complexity of the processing, current approaches to image registration are largely constrained by the tools which have been made available by commercial image processing vendors (Fogel and Tinney, 1996). Typically, the fine adjustment is simply another application of the same four processing steps over a smaller area. For example, most satellite images can be obtained from providers who will supply a standard georeferenced image product. The four geometric processing steps are applied before delivery. In the case of airborne data, it is possible to geocode the imagery in flight; certainly, immediately following acquisition. However, many users find that these global geometric corrections do not match the local geometry in their GIS — possibly because the original aerial photography on which their GIS data layers are based do not meet the geometric accuracy now possible from satellites and airborne systems. The imagery can be corrected to differentially corrected GPS (and, in the case of airborne imagery, INS) precision, and this will likely exceed the accuracy and precision of most archived aerial photography which underly the base maps from which GCPs are typically selected.

Improved techniques are needed to support the analysis of multiple sets of imagery and the integration of remote sensing and GIS. Geometric corrections are typically easier in satellite imagery because of lower relief effects and higher sensor stability (Salvador and Pons, 1998b). As GIS and remote sensing data integration becomes more common and the tools are improved, it seems likely that manual identification of GCPs must soon be replaced by fully automated methods of georeferencing (Ehlers, 1997). As well, improvements are needed in reporting the characteristics of the geometric correction, including improved error analysis that considers not only geometric accuracy but geometric uncertainty in spatial data locations.

IMAGE PROCESSING SYSTEMS AND FUNCTIONALITY

An image processing system is a key component of the infrastructure required to support remote sensing applications. In the past few decades the evolution of image processing systems has been nothing short of astonishing. Early systems were based on mainframe computers and featured batch processing and command line interfaces. In the absence of continuous-tone plotters, photographs, or cathode-ray tubes, output was to a line printer; if a lab was fortunate and well-equipped, a program was available or could be written to provide the printer with character overstrike capability. Imagine pinning strips of line printer output to the boardroom or classroom end wall, stepping back 15 or 20 paces, and interpreting the image! Thankfully, output considerations have changed drastically; then, considerations included the closeness of print spacing, the maximum number of overprint lines the paper could withstand, the blackest black possible with any combination of print characters, and textural effects (Henderson and Tanimoto, 1974). Now, the issue of screen real estate and what-you-see-is-what-you-get (WYSIWYG) continues to create an inefficiency; but plotters and printers have revolutionized output. Concerns regarding effective use of disk space and memory, efficiency, programming language, and machine dependence, have remained fairly constant.

Increasingly, image processing systems with camera-ready output are found on the desktop, with interactive near-real-time algorithms and a graphical user interface (GUI). The number of functions available has increased enormously — now, image processing systems can feature many tens or even hundreds of separate image processing tasks. But a new tension has emerged between the simplicity of use of these systems — point and click — and mastery of the actual functionality necessary to provide advanced applications results. The feel of the system (Goodchild, 1999) may be as important to the user as the actual way in which tasks are accomplished.

At one time, it appeared inevitable that the increasing complexity of image processing systems, in order to be comprehensible to users (Wickland, 1991) or even experienced image analysts, would lead to a situation in which image processors could only be operated in conjunction with a plethora of expert systems (Goldberg et al., 1983, 1985; Estes et al., 1986; Fabbri et al., 1986; Nandhakumar and Aggarwal, 1985; Yatabe and Fabbri, 1989). Many efforts have been made to build such systems to guide, direct, and even complete remote sensing image analysis. A key stimulus has been the desire to better integrate remote sensing output with GIS data (McKeown, 1987). Progress has been slow; success is most apparent in automation and expert systems where the algorithms are not data dependent, and the tasks are simple enough that human talents are not really needed (Landgrebe, 1978b) when choosing data characteristics, calibration, database queries, software selection, software sequencing, and archive, for example (Goodenough et al., 1994). The principal need in forestry remote sensing for automation and expert systems in the near term may be in the maintenance and construction of large databases and complex analytical operations involving multiple computer platforms, groups of tasks, and well-known sequences of individual procedures — rule-based image understanding (Guindon, 2000), for example.

Now, as in the larger world of GIS, increasing emphasis on expert systems in the analysis of remote sensing imagery in key decision making within an analytical process "seems to fly directly in the face of the view that computers empower people" (Longley et al., 1999: p. 1010). Few people willingly subscribe to multiple black boxes. In any event, complete or even partial automation of image analysis functions is not yet a realistic goal for many, if not most, forestry remote sensing applications. Instead, human participation in image processing is likely to continue to require a full range of computer assistance, from completely manual data analysis techniques along the lines of conventional aerial photointerpretation to human-aided machine processing. Image processing systems have evolved to accommodate this range of computing needs, but it is apparent that this theme will continue to preoccupy many remote sensing specialists and image processing system developers.

Different strategies have prevailed in terms of image processing functionality as the field has dealt with certain issues, and then moved on to others in response to the user community and the rapidly developing remote sensing and computer technology. Today, it is apparent the focus has shifted from exploratory studies to perfecting and standardizing techniques and protocols — a renewed commitment to building methods of radiometric correction, image transformation, nonstatistical classification, texture analysis, and object/feature extraction seems to be emerging in the literature. Congalton and Green (1999) noted strikingly different epochs in the development of the methods of classification accuracy assessment, ranging from widespread early neglect of the issue to concerted efforts to provide standardized methods and software as the field matured. The first stage of image processing development occurred in the early 1970s; the need was to develop the tools to ensure the new field of remote sensing was not inadvertently slowed by a lack of analytical techniques. Wisely, the main emphasis was on building information extraction tools and applying the quantitative approach in new applications, such as crop identification and acreage estimation (Swain and Davis, 1978). In the early days of digital image processing and remote sensing, scientists and engineers were focused on building classifiers and object recognition tools, image enhancements and feature selection procedures, and automating some of the (now) simpler image processing tasks such as lineament detection, spectral clustering, and geometric error estimation.

The main focus was on engineering computer systems and applications that would immediately make the benefits of remote sensing available to users; so, with "little time for contemplation" (Curran, 1985: p. 243) scientists began developing and testing ways of extracting information from the new image data. Multispectral classification and texture analysis, detection of regions and edges, processing multitemporal image datasets, and other tasks that are reasonably straightforward today, appeared nearly insurmountable given the available imagery and the computers and software capabilities. However, the fundamental algorithms in such everyday tasks as geometric registration (Steiner, 1974), multispectral pattern recognition (Duda and Hart, 1973; Tou and Gonzalez, 1977; Fu, 1976), per-pixel classification (Anuta, 1977; Jensen, 1978; Robinove, 1981), object detection (Kettig and Landgrebe, 1976), feature selection (Goodenough et al., 1978), and image texture processing (Weszka et al., 1976; Haralick et al., 1973) were established in that early push to develop the field. These algorithms can still be discerned beneath the GUI surfaces of microcomputer-based image processing systems today (Jensen, 1996; Richards and Jia, 1999). Like a veneer over these fundamental image processing algorithms, a series of procedures or protocols — ways of doing things — has emerged in a growing standardization of image analysis tasks (Lillesand, 1996). As systems have matured, users are less concerned with the technical complexities of image processing (Fritz, 1996).

The general direction and thrust over the past few decades has been to provide increasingly sophisticated image processing systems commercially and through the public domain or government-sponsored developments. Most public domain packages are not multipurpose in the sense that they do not support a full range of image analysis, are not reliably upgraded, and are periodically threatened with discontinuity, perhaps because of budget cuts or shifting priorities in public institutions. The situation may not be much different in the private sector! Some commercial systems were designed with a particular image processing focus in mind and are not partic-

TABLE 4.2
Main Tasks Supported by Commercially Available Image Analysis Systems

Processing Module	Approximate Number of Tasks
Data Format, Import, Export, Support	13
Graphics Display	4
Radiometric Correction	2
Enhancement	13
Registration/Rectification	6
Mosaicking	1
Terrain Analysis (DEM)	1
Classification	9
Map Production	3
Customization (programming)	2

Note: A total of 10 modules and more than 50 individual tasks.

Source: Modified from Graham and Gallion, 1996.

ularly robust; they may perform extremely well, even optimally in certain tasks, but may not support the full range of necessary functionality. Jensen (1996: p. 69) listed more than 30 commercial and public domain digital image analysis systems, suggesting that more than 10 of these had significant capability across a wide range of tasks in image preprocessing, display and enhancements, information extraction, image/map cartography, GIS, and IP/GIS. Of these ten, five or six are commercially available in full-function versions.

These commercial systems appear to have established market acceptance in the remote sensing community, and are marketed to that audience and the applications disciplines with promises of wide-ranging image processing functionality and linkages to GIS, cartographic, and statistical packages. From the perspective of the user, it appears that the dominant systems have only slightly differing operating and architectural philosophies. All systems will have a few key hardware components necessary for image display (monitor and color graphics card), fast processing of raster data (CPU and wide bus), and various supporting devices for storage (hard drive, backup, and compression drives). Table 4.2 is a summary of the main tasks supported by virtually all of the five or six commercially available image processing systems (see Graham and Gallion, 1996: p. 39). Within a general class of industrial-strength image processing systems there may be reasonable comparability (Limp, 1999). Some systems have good SAR processing modules, others have good DEM capability, still others offer custom programming languages. None is purpose-designed for forestry applications.

In recent reviews, Graham and Gallion (1996) and Limp (1999) compared a range of image processing systems focusing on the main commercial packages. The reviews keyed on such features as interoperability with GIS packages, multiple data formats and CAD operations, visual display and enhancement, classification methods, rectification and registration, orthophotography, radar capabilities, hyperspectral data analysis, user interface, and value. Such reviews are helpful in generating a sense of the functionality in any given image processing system relative to its competitors. For those aiming to acquire image analysis functionality, such reviews are most useful when preceded or accompanied by a user needs analysis. For example, in sustainable forest management applications it is probable that the image processing system would need to provide.

1. A high level of processing support for high and medium spatial detail optical/infrared imagery (airborne, IKONOS, and Landsat type data sets).
2. A high degree of interoperability with both raster- and vector-based GIS capability.
3. A good, solid output facility (note that maps are expected to be a prominent remote sensing output, but in many situations the existing GIS system can provide that functionality, reducing the demands on the image processing system).

In probably the most important respect for forestry, that of image analysis functionality for forestry applications, the commercial and publicly available image processing systems in many ways remain primitive and unwieldy; "Earth observation technology ... has not yet managed to provide whole products that are readily available, easy to use, consistent in quality, and backed by sound customer support" (Teillet, 1997: p. 291). For example, compared to the rapid, manual interpretation of imagery by trained human interpreters, computer systems are relatively poor pattern recognizers, poor object detectors, and poor contextual interpreters. Computers obviously excel in tedious counting tasks requiring little high-level understanding, such as in classifying simple landcover categories based on the statistical differences among a limited set of spectral bands. This is fine; humans always have much better things to do! But most systems do not provide extra tools to help in training large-area classifiers (Bucheim and Lillesand, 1989; Bolstad and Lillesand, 1991; McCaffrey and Franklin, 1993); most do not have a comprehensive set of image understanding tools (Gahegan and Flack, 1996; Guindon, 1997); most will not provide contextual classifiers, complex rule-based systems, or several different classifiers based on different classification logic (Franklin and Wilson, 1992; Peddle, 1995b); or shape (or tree crown) recognition algorithms (Gougeon, 1995); high spatial detail selection key-type classifiers (Fournier et al., 1995); multiple texture algorithms (Hay et al., 1996); customized geographic window sizes (Franklin et al., 1996); advanced DEM analysis, e.g., hillslope profiles (Giles and Franklin, 1998); atmosphere, view-angle and topographic correction software (Teillet, 1986); evidential reasoning and knowledge acquisition modules (Peddle, 1995a,b); multiple data fusion options (Solberg, 1999); and so on.

It is important to note that extracting information about forests from imagery will range from the simple to the complex; from analogue interpretation of screen displays and map-like products to multispectral/hyperspectral classification and regression; to advanced modules for texture processing, linear spectral mixture analysis, fuzzy classifiers, neural networks, geometrical/optical modeling, and automated tree recognition capability. There are presently few good guidelines to offer users on choices among different image processing approaches — this comment, originally made in 1981 by Townshend and Justice, suggests that the complexity of remote sensing image processing continues to outpace the accumulation of experience and understanding. In the recent development phase of such systems, a focus appears to have been on ease-of-use (GUIs), interoperability with GIS, and particularly, the increased availability of algorithms for automated processing for mapping (e.g., orthorectification and cartographic options). Continued improvements in the ease-of-use of image processing systems, supporting tasks, classification and prediction algorithms, and image understanding provide new opportunities for remote sensing in sustainable forest management applications.

What follows is a presentation of some of the issues and decisions that users will face in execution of remote sensing applications in sustainable forest management. The discussion will not exhaustively document the strengths or deficiencies of any public or commercial image analysis systems, but instead will focus on the need for continued algorithm development in certain areas, continued research to keep pace with new applications, and a continued commitment to aim for more functional and integrated systems for spatial analysis. For example, interpreting a remote sensing image in normal or false color on a computer display is quite simple, even easy, once the relationship between image features and ground features is completely understood; but this understanding is dependent on the display itself, the screen resolution, size of monitor (screen real estate), speed of refresh, the capability of the display software to generate imagery, and options to suit the individual interpeter. Can the user zoom the image quickly? Can the imagery be enhanced on-the-fly? Can different data sets be fused or merged on-screen? These issues, while important to the user in the sense that they can make life easier, are not as critical as the analytical functionality of the system — the ability of the system to respond to the extraction of information in a flexible way.

Understanding those options, in addition to having access to new, faster, more dynamic ways of displaying imagery, may lead to greater insight into the role and applications of remote sensing in forest management, forest science, and forest operations. In essence, it should be possible for those interested in using remote sensing for sustainable forest management applications to specify the main types of sensor data that must be handled, the level of image processing required or expected, and the number and type of output products that will be generated for a given management unit or forest area. Only then would it be appropriate to consider the available software systems in light of these needs.

IMAGE ANALYSIS SUPPORT FUNCTIONS

The basic support functions common to all image processing systems and required in virtually any remote sensing application are data input, sampling, image display, visualization, simple image transformations, basic statistics, and data compression (storage). Data input issues include reading various image data formats, conversions, and ancillary data functions. Many of the problems with data input could be considered the domain of the GIS, within which remote sensing image analysis may be increasingly conducted; most GIS and image analysis systems come with a wide array of conversion routines. As noted in the previous section, georeferencing is a key to successful GIS and image data input, but data conversion may be a decisive issue because of the time and cost involved (Weibel, 1997; Molenaar, 1998; Estes and Star, 1997). For users of image analysis systems and geographical information systems, deciphering several spatial data formats can represent a formidable barrier to be overcome before the real battle — the analysis of the data — begins (Piwowar and LeDrew, 1990). Some estimates for data conversion range as high as 50% of the cost and effort in a GIS project (Hohl, 1998).

Building data layers is another preliminary task that can consume resources. After converting all the data formats, Green (1999) pointed out that, typically, considerable additional resources are used in many large area resource management projects in building GIS data layers. The remaining budget can be used to comprehensively develop only one or maybe two analysis questions. Building and georeferencing data layers aside, the real task of image analysis begins with correct image sampling and the derivation of simple image transformations for use in subsequent remote sensing analysis in support of forest management applications. The generation of appropriate image displays and data visualization products revolve around issues such as computer graphics capability, color transformations, and output options; these, and data storage issues, may be largely dictated by the hardware environment in which the remote sensing software resides.

It is not the intention in this book to review extensively the basic image analysis and image processing environment; instead, an understanding of the range and types of tasks in the technological infrastructure is provided such that a more complete background in specific areas of interest can be acquired by further reading. A selection and some examples of particularly important tasks in forestry applications are discussed.

Image Sampling

In remote sensing applications, sampling generally consists of:

1. The creation of image databases from scenes either by "cookie-cutting" or mosaicking
2. The generation of pixel coordinate lists for use in various image analysis tasks

Sub-area creation procedures are widely available in commercial image processing systems, which might include options for variable area extraction and mosaicking across image edges to remove image differences caused by different illumination conditions or sensor packages. Masking the original image data with physical limits or arbitrary boundaries such as political or socioeconomic units is a common task; perhaps the mask is a boundary or polygonal coverage read-in from a GIS where different vector files are stored.

The large volumes of remotely sensed and other geospatial data used in natural resource applications such as forest mapping and classification have created the need for multiple sampling schemes in support of image analysis (Franklin et al., 1991). Later, as different applications are reviewed, considerations emerge concerning the design of a sampling scheme for the collection of ground data to support remote sensing applications (Curran and Williamson, 1985). Typically, it is possible to assume that the ground-based sampling issues have been dealt with by the use of conventional forest sampling schemes, perhaps modified for remote sensing; the multistage sampling and multiphase sampling strategies, for example (Czaplewski, 1999; Oderwald and Wynne, 2000). These must be sensitive to the spatial variability of the area, the minimum plot size, the number of plots that are feasible with the available resources, the type of analysis that is contemplated, and the desired level of confidence in the results. In all sampling, a plot on the image must correspond precisely with the plot on the ground (Oderwald and Wynne, 2000).

Pixel sampling in the form of coordinate lists is required in support of other image analysis tasks such as the creation of image displays and histograms, principal components analysis, image classification, and other image transformations (Jensen, 1996). Sampling can be used in support of the selection of mapping or classification variables (Mausel et al., 1990), assessment of preprocessing functions such as atmospheric or topographic corrections (Franklin, 1991), field-site selection for training areas (Warren et al., 1990), and classification accuracy assessment (Congalton and Green, 1999). The samples can be random, systematic, stratified, or purposive (Justice and Townshend, 1981), depending on the purpose of the sampling. The output of pixel sampling is usually an attribute table which is a compilation of image values referenced by location (Table 4.3). The idea is that once the image data have been georeferenced, the individual pixel spectral response can be associated with the field or GIS data in statistical or deterministic analysis routines.

Image Transformations

Simple image transformations may be very useful in understanding image data; image ratios and multitemporal image displays may be key in understanding and enhancing differences between features in a scene and over time. A few basic image transformations have been used frequently in forestry applications, although many different image transformations have been designed for specific applications. For example, in mapping biomass in northern forests, Ranson and Sun (1994a,b) created multifrequency and multipolarization SAR image ratios as a way of maximizing the information content of the airborne SAR imagery. Each frequency or polarization appeared best correlated with a different feature of the forest; ratioing allowed the information content of the many different images to be captured in a smaller data set.

Typically, the ideas behind image transformations are

1. To reduce the number of information channels that must be considered
2. To attempt to concentrate the information content of interest into the reduced number of bands

The normalized vegetation difference index (NDVI) is a common image transformation in vegetation studies (Tucker, 1979). The NDVI may be the single most

TABLE 4.3
Example Attribute Table Created by Pixel Sampling

Point Coordinate[a]	Spectral Values					DEM Data			GIS Data	
(Row, Column)	TM1	2	3	4	5 …	Elevation	Slope	Aspect …	Polyid	Species Code …
123, 267	98	42	28	129	75 …	1341	21	187	53990	27
945, 1903	81	65	23	101	79	1209	11	341	768904	12
4312, 5672	109	87	57	184	121	987	5	98	456219	21
… …	…	…	…	…	…	…	…	…	…	…
… …	…	…	…	…	…	…	…	…	…	…
… …	…	…	…	…	…	…	…	…	…	…
… …	…	…	…	…	…	…	…	…	…	…

[a] Expressed in geographic coordinates.

successful remote sensing idea responsible for wider use of remote sensing data in ecology and forestry (Dale, 1998). The development of NDVI (which more strongly relates reflectance as measured in the image to forest conditions), was instrumental in showing that useful information could be extracted from remote sensing imagery, and once the forest information content of the NDVI was determined it became more obvious which applications would be worthwhile. NDVI is based on the use of a near-infrared (IR) band and a red (R) band:

$$NDVI = (IR-R)/(IR+R) \tag{4.1}$$

The NDVI will range between –1 and +1; while the extraction of NDVI from imagery is straightforward, the interpretation of NDVI values for different forest types has sometimes been problematic. Normally, one would expect that high NDVI would be found in areas of high leaf area. Foliage reflects little energy in the red portion of the spectrum because most of it is absorbed by photosynthetic pigments, whereas much of the near-infrared is reflected by foliage (Gausman, 1977). The normalized difference would emphasize, in a single measure, the effect of these two trends (Tucker, 1979). However, it has been shown (Bonan, 1993) that the NDVI is an indicator of vegetation amount, but is related to LAI only to the extent that LAI is a function of absorbed photosynthetically active radiation (APAR); remotely sensed reflectance data are actually related to the fraction of incident photosynthetically active radiation absorbed by the canopy (FPAR) (Chen and Cihlar, 1996). The relationship between NDVI and FPAR, discussed more fully later in the book, varies for different vegetation and forest types (Chen, 1996).

Corrections to NDVI values and the use of various other indices have been reported; for example, the soil-adjusted vegetation index (SAVI) accounts for soil effects (Huete, 1988). Others have used mixture models to first eliminate the shadow effects within coarse pixels; then NDVI derived from shadow-fraction indices can be used (Peddle et al., 1999). Generally, different indices should be considered depending on the circumstances under which the image transformation is to be used. Fourteen different indices were summarized by Jensen (2000) with some suggestions for their use in different types of image analysis. The main issues appear to be

1. The extent to which the atmosphere (or more generally, image noise) has been corrected prior to calculation of an index
2. The range of forest conditions that are of interest (i.e., from areas with sparse vegetation to areas with full canopy coverage, or perhaps only a limited range of forest conditions)

The Tasseled Cap Transformation (Kauth and Thomas, 1976; Crist and Cicone, 1984) has been used to reduce MSS and TM image dimensionality to fewer, more easily displayed and interpreted dimensions; the result is two (in the case of MSS data) or three (in the case of TM data) statistically significant orthogonal indices that are linear combinations of the original spectral data. The original reason for developing the Tasseled Cap Transformation was to capture the variability in spectral characteristics of various agriculture crops over time in indices that were primarily related to soil brightness, greenness, yellowness, and otherness — as crops emerged in the spring the relative differences in the growth and phenology could be summarized. Since then, the transformation has been thought of as a simple way of creating a physical explanation for changes in other surface conditions. Few physical studies have been reported relating that explanation to different terrain features; nevertheless, the idea has considerable merit.

These linear combinations of TM bands 1 through 5 and band 7, can emphasize structures in the spectral data which arise as a result of particular physical characteristics of the scene. A different set of coefficients must be used depending on the imagery and the extent of earlier processing (Crist, 1985; Jensen, 1996); for example, shown here are the coefficients for Landsat-4 TM imagery:

$$\text{Brightness} = 0.3037(TM1) + 0.2793(TM2) + 0.4743(TM3) + 0.5582(TM4) + 0.5082(TM5) + 0.1863(TM7) \quad (4.2)$$

$$\text{Greenness} = (-0.2848(TM1)) + (-0.2435(TM2)) + (-0.5436(TM3)) + 0.7243(TM4) + 0.0840(TM5) + (-0.1800(TM7)) \quad (4.3)$$

$$\text{Wetness} = 0.1509(TM1) + 0.1973(TM2) + 0.3279(TM3) + 0.3406(TM4) + (-0.7112(TM5)) + (-0.4572(TM7)) \quad (4.4)$$

Broad interpretations of these indices have been reported. For example, Collins and Woodcock (1996) have suggested that single-date Landsat TM data are dispersed mainly in this three-dimensional space called brightness/greenness/wetness, and measurements of differences in these quantities over time can be a good indicator of vegetation change — forest mortality — caused by insect activity. Brightness is a positive linear combination of all six reflective TM bands, and responds primarily to changes in features that reflect strongly in all bands (such as soil reflectance). Greenness contrasts the visible bands with two infrared bands (4 and 5) and is thought to be directly related to the amount of green vegetation in the pixel. The wetness index is dominated by a contrast between bands 5 and 7 and the other bands. Generally, reflectance in the middle-infrared portion of the spectrum is dominated by the optical depth of water in leaves (Horler and Ahern, 1986; Hunt, 1991). A more appropriate name for the wetness component might be maturity index (Cohen and Spies, 1992) or structure wetness (Cohen et al., 1995a), since "it appears to be the interaction of electromagnetic energy with the structure of the scene component and its water content that are responsible for the response of the wetness axis" (Cohen et al., 1995a: p. 744). This suggests that not only the total water content, but its distribution within the pixel, is important.

Global transformation coefficients such as these can be used, but if training data are available then a local transformation can be created that is more sensitive to the actual distribution of features in the scene. The scene dependence of the Tasseled Cap Transformation has resulted in some difficulty in interpretation of these indices, in the same way that principal components analysis sometimes can be problematic (Fung and LeDrew, 1987). Sometimes it is not at all clear what information the new components or indices contain. Regardless, interpretation of the new brightness/greenness/wetness image space often can be simplified compared to interpretation of the six original reflectance bands. These transforms, and others, represent one possible approach to the data reduction and feature selection problem in remote sensing; with many bands to choose from and many redundancies and multicollinearity in linear models to deal with, such image transformations can provide an exploratory tool to better understand the data, and also generate input variables that are more closely related to the features of interest for other more advanced image processing tasks, such as classification and change detection. As hyperspectral data become more common, it seems likely that individual indices and image transformations such as NDVI, second derivatives of the red-edge, green peak reflectance, and Tasseled Cap Transformations will become more valuable as data reduction and analysis techniques.

Data Fusion and Visualization

A third reason to conduct image transformations is to merge different data sets. There may be a need to create more graphically pleasing and informative image displays that emphasize certain features or spectral response values. More generally, this is one of the main objectives of data fusion techniques. One common display transformation that can also be used in data fusion is known as the intensity-hue-saturation (IHS) transform. Three separate bands of data are displayed or mapped into IHS color coordinates, rather than the traditional red-green-blue (RGB) color space (Figure 4.2). A traditional color image is comprised of three bands of visible light (blue, green, red) projected through their corresponding color guns; a false color image is shown with a green band projected through the blue color gun, a red band projected through the green color gun, and a near-infrared band projected through the red color gun.

Hue is a measure of the average wavelength of light reflected by an object, saturation is a measure of the spread of colors, and intensity is the achromatic component of perceived color. During the IHS transform, the RGB data are reprojected to the new coordinates (Daily, 1983). This is a powerful technique to view different aspects of a single data set (e.g., multiple bands and spatial frequency information) or to merge different data sets with unique properties. For example, if a new band of data (say, a SAR image) was inserted in place of the intensity channel during an IHS transformation of TM data and the reverse transform implemented back to RGB space, a new type of display would be created from the fusion of the two data sets. While striking enhancements for manual interpretation can be created this way, the digital use of these transformed data may be more difficult to justify. The reason is that it may no longer be apparent how to relate the new color space to the features that provided the original spectral response (Franklin and Blodgett, 1993).

Data fusion techniques have emerged as key visualization tools as well as providing improvements in classification accuracy, image sharpening, missing data substitution, change detection, and geometric correction (Solberg, 1999). In visualization, images are interpreted following special enhancements perhaps designed to reveal specific features in maximum contrast (Ahern and Sirois, 1989; Young and White, 1994). Visualization techniques can be used with data from different satellite

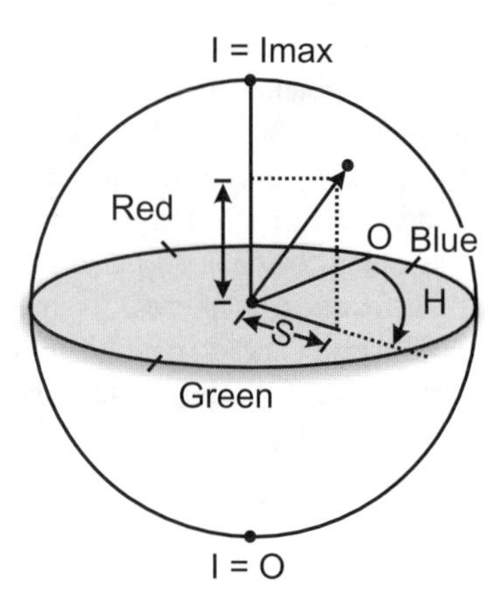

FIGURE 4.2 The hue-intensity-saturation color coordinate system shown in relation to the red-green-blue system normally employed in image displays. The HIS transform is used in data visualization and fusion by converting data from normal RGB color space to the HIS coordinates, substituting image data (e.g., a higher spatial detail panchromatic channel) or manipulating the individual band ranges and retransforming the data for display or analysis. (From Daily, M. 1983. *Photogramm. Eng. Rem. Sensing,* 49: 349–355. With permission.)

or airborne sensors — or from different sources of information such as imagery and map products. The idea is to display these data in original RGB format or after some initial processing step. Generally the use of multiple data-set visualization data can be shown to provide advantages over the use of individual data sets alone. For example, Leckie (1990b: p. 1246) used SAR and optical data together in a forest type discrimination study in northern Ontario that was aimed at separating general species classes, and concluded that "There is a definite synergistic relationship between visible/infrared data and radar data that provides significant benefits to forest mapping." This synergy has been used to combine multitemporal ERS-1 SAR data and multispectral Landsat TM data, and thereby increase the classification accuracy of Swedish landcover maps (Michelson et al., 2000).

Robinson et al. (2000) used spectral mixture analysis in image fusion; they showed that the choice of fusion method depends on the purpose of the analysis. For example, if the desire was to improve overall accuracy, or if a specific feature in the image was of interest, then different approaches would provide imagery with optimal characteristics. The key issue in data fusion is the quality of the resulting imagery for the purpose of analysis (Wald et al., 1997). But how to address or measure quality? Few guidelines exist. In one study, Solberg (1999) provided a Markov random field model to assess the effect of using different sensor data, spatial context, and existing map data in classification of forests. Different levels of data fusion were developed. The results could be considered a warning against the indiscriminant combining of data in black-box algorithms; the dominant influence by the existing map products could be traced with effects related to the fusion of data and features, but the influence at a third (high) level of data fusion, decision-level fusion, was less easily understood, even with relatively simple classes (e.g., soil, shrub, grassland, young and old conifer stands, deciduous).

Principal components analysis (PCA) is a well-known statistical transformation that has many uses in remote sensing: reduction of the number of variables, fusion of different data sets, multitemporal analysis, feature selection, and so on. Many different possible image processing options have been employed with PCA. Decorrelation stretching is based on the generation of principal components from a combined, georeferenced multiple image data set. The components are stretched, and then transformed back to the original axes; the resulting imagery is often amazingly different. Using data from the Landsat TM and an experimental high spatial resolution shuttle sensor called the Modular Opto-electronic Multispectral Scanner (MOMS), Rothery and Francis (1987) set the first principal component to a uniform intensity before applying the inverse transformation. The color relations of the original color composite were retained, but with albedo and topographically induced brightness variations removed. These and other transformations (perhaps based on regression analysis or filtering) can be used to merge multiresolution and multispectral imagery from different sensors (Chavez, 1986; Price, 1987; Chavez et al., 1991; Franklin and Blodgett, 1993; Toutin and Rivard, 1995, 1997). Chapter 4, Color Figure 2 contains an example data fusion procedure using Landsat and Radarsat imagery of an agricultural area in southern Argentina.

Basic statistics, such as band means and standard deviations, are necessary for display of remote sensing data (Jensen, 1996) and a whole host of statistics may be employed directly or indirectly in analysis of the data. The link between remote sensing and statistics is strong; clearly, remote sensing can be considered a multivariate problem (Kershaw, 1987) and probabilistic methods constitute one of the most powerful approaches to the analysis of multivariate problems. Remote sensing image analysts must be conversant in multivariate statistics in order to complete many image processing tasks. But image analysis systems are not equally versatile in their provision and use of statistics; two approaches in recent years have been to write the required statistical modules in an external database or programming language (Franklin et al., 1991), or to build an interface between the image analysis system and an existing statistical package (Wulder, 1997). The intent is to provide a larger range of statistical tools to the image analyst. As will be shown in later examples in this book, sometimes this linkage has proved to be extremely valuable in concluding a remote sensing experiment, study, or operational application.

IMAGE INFORMATION EXTRACTION

The goal of remote sensing in forestry is the provision, based on available or purposely acquired remote sensing data, of information that foresters need to accomplish the various activities that comprise sustainable forest management. Unless remote sensing has been relegated to only pretty pictures mounted on the wall (but recall, this phase of remote sensing has been declared over!), in every case the first step must be the extraction of information from remote sensing data by manual interpretation or computer — that is, from the spectral response patterns — in an appropriate format. A few of the more obvious ways to extract information rely on visual analysis, data visualization, spatial algorithms that extract specific features such as edges or textures, object detection routines, change detection and change trajectory comparison methods, and multispectral classification tools which attempt to identify homogeneous classes and create generalized areas for mapping. In sustainable forest management applications, it appears that three general but different kinds or types of digital remote sensing information are of interest.

1. Continuous forest variable information (e.g., spectral response estimation of crown closure or LAI)
2. Forest classification information (e.g., spectral response categorization of forest covertypes)
3. Forest change or difference information (e.g., differences in spectral response, crown closure, or class over time)

Each of these types of information can be used by foresters in a multitude of ways, but first the spectral response data must be extracted from the imagery and, by using image analysis techniques, converted to one of the three types of information. No one single image analysis approach has the potential to optimize the extraction of image information, but a suite of image analysis tasks exist that when used together can facilitate the process of converting the remote sensing data into the necessary information products. Different types of image analysis have emerged that use different aspects of the imagery. Each of these has spawned numerous options, and will likely continue to evolve as foresters increasingly look to extract from remote sensing the information needed to support sustainable forest management applications in the future.

Image analysis is a dynamic field in which new ideas and methods, and improvements and refinements of early techniques, have emerged over the past 25 years almost continuously. A major trend in these developments has been the search for increasingly automated procedures that can extract information from imagery; however, it is still the case that many image processing tasks require human intervention, human judgment, and human guidance in order to operate successfully. In continuous variable estimation, as in classification and change detection, the results are largely dependent on the quality and comprehensiveness of the input training data (Salvador and Pons, 1998b). Training data can be acquired using the manual selection of pixels by class or through some statistical approach (e.g., spectral clustering); that is, training samples can be obtained using a strategy based on human knowledge (e.g., select certain stands known or thought to represent the desired ground condition) or can be obtained using some statistically based strategy, or perhaps a combination of these two. Regardless, the degree to which training data relate to the desired information product can often be the difference between the success or failure of a remote sensing project.

Continuous Variable Estimation

A continuous variable, such as LAI, might be required as input to a model of productivity; other continuous variables might be suitably presented as either continuous variables or as categorical variables. For example, biomass, volume, crown closure, and height estimates are thought of as continuous variables, but can often be generalized without significant loss of detail or value into a discrete number of classes. Continuous variable estimation occurs primarily by one of a few common forms of inversion modeling, including regression analysis, neural networks, reflectance modeling, or radiative transfer modeling (Strahler et al., 1986). The objectives of these modeling approaches are virtually identical; the differences are found in the methods and the degree to which the results are robust. Can the results be used in conditions different from those under which they were generated?

Obviously, the most robust approach relies on modeling based on first principles of radiative transfer. By accounting for all possible interactions between the source of energy, the target, the sensor, and the media, complete model inversion can be achieved. Typically, estimating the value of a continuous variable, such as LAI or crown closure, is only one (and probably not the most important one) of the possible objectives in using such models (Nilson and Ross, 1997: p. 56):

1. Recognition of how a reflected signal is formed;
2. Identification of the primary factors that determine a reflected signal and its temporal and spatial variability;
3. Simulation of the effects on reflectance of various scenarios of ecosystem development, including successional changes and management effects;
4. Determination of various ecosystem parameters from remotely sensed data by means of inversion;
5. Interpretation and normalization of remotely sensed data (for example, extending the measured data to another solar elevation or phenological stage).

Remote sensing research scientists are primarily concerned with making progress in understanding spectral response and its applications (objectives 1, 2, and 3); users of remote sensing data are primarily interested in obtaining information from remote sensing data (objectives 4 and 5). Objective 4 could be restated to represent the goal of all those interested in how well the measured variable (reflectance or backscattering coefficient) can be used to predict a biophysical variable such as LAI, canopy closure, or stand volume. In the words of Kuusk and Nilson (2000: p. 245), "Can a forest reflectance model act as an interface between imagery and forestry databases?"

What is of interest is the relationship between the measured variable and the surface condition and the error and statistical significance of the relationship (Gemmell, 1995, 1998; Trotter et al., 1997). However, the use of radiative transfer models is not yet commonplace in either the remote sensing or forestry user communities; none of the major commercial image analysis systems contains even a simple radiative transfer model, and specialized code for modeling is both hard to find and difficult to use. The number of input variables, and the high level of understanding that these models demand, suggest that it will be some time before applications specialists regularly access this approach in their efforts to increase the value of remote sensing data in forestry.

Continuous variable estimation in remote sensing has been accomplished much more frequently through an empirical search for relationships, typically using a regression analysis. These studies follow the traditional statistical probability design;

relate two sets of variables — one field set of variables and one remote sensing set of variables — derive the least-squares fit, invert the relationship, test the relationship independently, and examine the residuals and standard error of the prediction. In remote sensing, the development of regression equations sometimes follows a classification. The classes are used to stratify the landscape and reduce the variance to an acceptable degree that can be modeled linearly with the available spectral response patterns (Franklin et al., 1997a,b).

A large number of empirical studies have been completed in this vein and are reviewed in later chapters of this book and in other sources. For example, in one treatment, Stellingwerf and Hussin (1997) presented regression predictors derived from optical and microwave remote sensing measurements for:

1. Forest area
2. Number of trees
3. Tree height
4. Dbh
5. Age
6. Site
7. Volume
8. Biomass

Their work was an attempt to document a set of actual prediction equations for remote estimation of each of these variables of interest; however, the actual predictors documented are unlikely to be useful in any particular forest region, being heavily dependent on the type of forest and data characteristics involved. The way in which the predictors were obtained and the general form of the prediction, on the other hand, is likely to be a helpful guide to those developing specific prediction relationships elsewhere in the world. As always, the general interpretation of the predictors (coefficients) and the relationships must be based on an understanding of the physical relationships that exist and which constrain remote sensing applications.

An alternative — or a complementary method — to such data-driven regression studies of the relationships between a continuous forest variable and remote sensing spectral response is the canopy reflectance model. An important class of such models useful in forestry applications is the geometric-optical (GO) model (Li and Strahler, 1985; Jupp and Walker, 1997; Gemmell, 2000). A GO model is based on the understanding provided by more detailed radiosity and radiative transfer models (Myneni and Ross, 1991; Nilson, 1992), but mechanistically and statistically portrayed using the size, orientation, and shape of cones, disks, and spheres to represent tree structures. Such models occupy a position somewhere between the wholly theoretical radiative transfer equations and the completely data-driven regression approach. Some models use both geometric-optical and radiative transfer components (Nilson and Peterson, 1991) and have reached an amazing degree of complexity, able to closely mimic the more powerful radiosity models across a wide range of remote sensing conditions (Gerstl and Borel-Donohue, 1992). Nilson and Ross (1997) described the subcomponents of one such model; the final output relating a field variable such as crown diameter to spectral reflectance observed by an airborne sensor was provided by considering four model components:

1. The optical model of a needle
2. The optical model of a shoot

3. The optical model of a tree crown
4. The optical model of forest community (a forest stand)

The Li-Strahler model (Li and Strahler, 1985) is a GO model designed for inversion to provide estimates of the size and density of trees contributing reflectance in relatively coarse resolution (L-resolution) remotely sensed images (i.e., where pixel size exceeds average tree crown size). The reflectance observed by the sensor is modeled as consisting of two forest components, each with a shadowed and sunlit form: (1) the tree crowns and (2) the background.

Reflectance is then considered to be a linear combination of four terms and their areal proportions (Woodcock et al., 1997):

$$S = K_g G + K_c C + K_t T + K_z Z \tag{4.5}$$

where S is the brightness value of a pixel; K_g, K_c, K_t, and K_z stand for the areal proportions of sunlit background, sunlit crown, shadowed crown, and shadowed background, respectively. G, C, T, and Z are the spectral signatures of the respective components and must be acquired using ground or aerial-based measurements (end-member spectra), a spectral library, a simulation, or the image data themselves (e.g., training data from test stands). Typically, these signatures are defined and collected with a recognition of the assumption that such signatures can be relative or normative; it has long been understood that spectral signatures are subject to the particular geographic area and measurement technology and that "Unique, unchanging spectral signatures do not exist in the natural world" (Hoffer, 1978: p. 270). The Li-Strahler model can predict the density of trees within a TM pixel in real forest settings, but is unable to predict tree size because of the influence of variable understory reflectance (Gemmell, 2000), and clumping of trees in stands (Woodcock et al., 1997).

The spectral response of trees on a background becomes a mixture of the four components, and thus, when inverted, this form of the GO model equation is equivalent to the simple version of a spectral mixture analysis, or an SMA equation (Hall et al., 1996). More complex mixture modeling is used to represent the spectral mixture of components as a linear combination of the response of each component (Shimabukuro and Smith, 1991). Because each pixel is assumed to contain information about the proportion and spectral response of each component, it may be possible to unmix the pixel spectral response to determine the likely contributions of individual components; for example, two pixels may contain the same proportion of two different species of trees which reflect differently, or perhaps different proportions of the same tree species created by age differences.

A parallel interest in the development of radiative transfer models has occurred in the radar portion of the spectrum. The Michigan Microwave Canopy Scattering Model, or MIMICS, approach is one of the better-known and more widely available tools (Ulaby et al., 1990). MIMICS is a first-order vector radiative transfer model that simulates the microwave interaction within closed canopy forest areas comprised of forest crowns and trunks on flat terrain, with an underlying soil surface with known properties. Different components of the forest are modeled statistically as cylinders (e.g., trunks) and thin disks (e.g., deciduous leaves). Each element is described by its orientation, size, density and dielectric properties; crown depth and stand density can be varied. The purpose of this model is to predict the radar backscattering properties of forests. By inversion, then, the backscatter coefficients observed by remote sensing can be used to predict the type of forest conditions that created the observation. The MIMICS model has been used in calibration (Beauchemin et al., 1995), interpretation of imagery (Dobson et al., 1992), and in laboratory and field studies aimed at acquiring a better understanding of actual SAR image characteristics over forests (Chauhan et al., 1991; Sun et al., 1991; Sun and Ranson, 1998). In reviewing accomplishments in this field later in the book, it will become apparent that, like their optical/infrared counterparts, such models are likely to continue as research tools rather than practical image analysis tools (Skelly, 1990; Kasischke and Christensen, 1990; Ahern et al., 1991).

Image Classification

Image classification occurs primarily by employing categorical (and increasingly fuzzy) decision rules. The objective of classification is to generate spatially explicit generalizations that show individual classes selected to represent different scales of forest organization. Of all the available ways of extracting information from remote sensing data, image classification in particular can be considered a prime candidate for a standard methodology with the potential to be distilled into a protocol that can be extended spatially and temporally (Lillesand, 1996). However, a large number of factors will influence the success of a classification project. These factors have hindered the development of a standard protocol (Gong and Howarth, 1990), but the basic steps are reasonably well known (Chapter 4, Color Figure 3).

- The development of a classification system comprised of individual and hierarchical classes across the landscape
- The derivation and implementation of methods and algorithms that can be applied with understandable error patterns and identified uncertainty in decision making
- The application of a statistically sound accuracy assessment and validation of mapping products

The first and most important step is the development of the classification system — identifying the classes that are to be mapped. Conventionally, the system would contain classes that are exhaustive and mutually exclusive; in fuzzy systems, this requirement can be relaxed, allowing intergradations of classes and mixed communities (Townshend, 2000). All other decisions in a remote sensing classification project flow from this decision. Classification units may not always coincide with mapping units (Driscoll et al., 1978), but the link between classification and mapping is very strong, as will be evident in the review of forest covertype mapping applications in Chapter 6. Here, it is important to note the dramatic effect that the classification system can have on the procedure of classification.

The classes in the scheme must be defined by their ground characteristics for the classification products to be useful. It is usually not difficult to separate forest from other land covertypes; but is an area with few trees (very low crown closure) still a forest? Typically, the desired classification scheme is comprised of a list of hierarchical classes, mutually exclusive and exhaustive of the landscape of interest. As reviewed in Chapter 6, a typical starting point is the Anderson et al. (1976) USGS land use and land cover classification system, although innumerable specialized classifications have been proposed and used almost since the beginning of remote sensing applications. A common supporting mechanism for these classifications is provided by current understanding (Mabbutt, 1968; Driscoll et al., 1978; Bailey et al., 1978; Bailey, 1996) and recent exercises in ecological land classification (Sims et al., 1996). The goal of ecological land classification is to map important functional or compositional differences at the scale of interest by using information on topography, vegetation, soils, and climate (Lacate, 1969; Klijn, 1994; Bailey, 1996); this is done so the map boundaries express the mapper's belief that important functional or compositional differences exist on either side (Rowe, 1996). The close, often unclear relationships between classes, forest stands, and forest ecosystems will already be evident.

Regardless of the types of classes selected, experience has shown that the classification scheme must be developed with full involvement of at least two collaborative interests: (1) those generating the classification products and (2) those using the classification products. Occasionally these two groups are one and the same; but too frequently, the users are not directly involved in the classification process. Acquiring user input into the classification scheme at the beginning of the process will almost certainly help generate more usable classification products from remote sensing imagery. Asking for user input only to evaluate the finished product is not a wise option.

Once the critical decision on the nature of the classes has been made, the first and most important effect is that one of two different methods must be employed to structure the classification process (Duda and Hart, 1973; Lillesand and Kiefer, 1994). Initially, the choice is between (1) supervised classification or (2) unsupervised classification. In a supervised approach, the project begins with a set of classes. In an unsupervised approach, a set of classes emerges from the image analysis. The implications of using a supervised or an unsupervised approach, and differences between them, are discussed in this section. The modifications that have been devised to alleviate some of the constraints in each approach are reviewed briefly in the following section.

Supervised classification begins with a known set of classes; these classes are then separated or combined based on the statistical or other properties of the image data. The question in a supervised classification is: How well do the data separate these classes? That is, the classes are not in doubt; the concern is only with the ability of the data to capture their differences, and with the performance of the classifier in determining the correct distinctions. If the classification system is one which has been imposed on the data, perhaps using a classification structure adopted or devised elsewhere, the methods of classifying remote sensing data will be supervised. If there are classes that must be mapped, regardless of how well the data represent them, the method will be supervised. In a supervised approach the accuracy of the classification is a good indication of its utility; if the classes of interest are classified well by using the remote sensing data, then the maps produced will be very useful.

If, on the other hand, the classes are not known but are to be decided during the remote sensing project, an unsupervised approach is suggested. Unsupervised classification usually refers to classes which are not predetermined; the statistical or clustering properties of the image data are used to find a set of natural classes (Thomas et al., 1987). The question in an unsupervised classification is: What classes do the data provide? In an unsupervised approach, the classes may be very accurate, but they may not coincide with the classes that are desired, making the classification products less useful. In Minnesota, Bauer et al. (1994) found that only about 50% of the Landsat-derived unsupervised classes could be named or labeled as distinct entities on the ground useful for forest inventory; on the other hand, in that same study the supervised approach was determined to be inadequate, primarily because of extreme forest complexity.

It is clear that when a forest area is mapped for forest inventory purposes using aerial photography, essentially an unsupervised approach is taken. The idea is to map the forest stand conditions with a preconceived idea of what those conditions must be (e.g., different species composition, crown closure, height class, and density), but with no real predefined classification scheme; the interpreter simply recognizes areas that have the right characteristics to be labeled a stand. When recognizing stands — or rather erecting the boundaries between stands (Lowell et al., 1996) — on aerial photographs, the interpreter has some idea of what is a sufficient difference that must be used to separate one stand from another. The role of aerial photography has long been central to this forest inventory and classification idea. It is rare for the data to be considered unequal to the task; instead it is assumed that the differences recognized on the photograph must be a result of the differences that exist among the stands. The production of usable forest classifications, resource assessments, and terrain analyses (Avery, 1968; Christian and Stewart, 1968; Townshend, 1981b) without the perspective of the aerial photograph "from above" is difficult to imagine.

Initially, in remote sensing the idea is to recognize areas that are distinct because they have a different spectral response pattern for each pixel. Digital remote sensing has been slow to adopt the aerial photographic classification philosophy, and hence has rarely been able to generate classifications at the level of an operational forest management unit — in other words, remote sensing cannot produce classifications of forest stands that are identical to those produced by the interpretation of aerial photography by experienced interpreters, because a totally different premise is involved in what represents a significant difference to be recognized. Photointerpreters are delineating spatially connected homogeneous units; per-pixel classification is simply assigning pixels to categories regardless of their spatial location. Thus a digital classification of (digitized) aerial photographs would not produce the identical stands as would the experienced photointerpreters working with analogue versions of those photographs. Completely different ways of classifying the data are involved. However, even experienced photointerpreters cannot produce the stands identical to those produced by other experienced photointerpreters working with the same photographs. Actually, no technology or process yet known can produce forest stands identical to those delineated by experienced photointerpreters! Not surprisingly, digital per-pixel classifications that compare well with current forest inventory classifications are rare. For example, the increased spatial resolution of some remote sensing data has simply introduced new challenges; it is already abundantly clear that image analysis tasks are not well designed to handle these new data.

The remote sensing difficulties are easy to enumerate. The per-pixel classification of aerial and satellite data will not generate forest stands because of the constraints caused by the different image resolutions; for example, the relatively low spatial resolution prevents adequate distinction of species (from historical satellites, although not always with airborne data and now with new satellite data). But even high spatial resolution imagery in which individual species can be discerned, generally cannot be used to identify the same stands as are interpreted on aerial photos. The most significant constraint is likely to be the heavy dependence on spectral response at the individual pixel level. No photointerpreter would restrict interpretation to a single image attribute: tone. Instead, a full range of photodescriptors are used to classify and map stands. While computer analysis is far more thorough in analyzing spectral response — tone — information than human interpreters (Cihlar et al., 1998), in remote sensing the tools to perform adequate analysis of the other descriptors have been lacking (Green, 2000). Remote sensing suffers from poor texture analysis, poor pattern analysis, poor shape analysis, poor object (size) analysis, poor shadow analysis, poor topographic analysis, and very poor location-association analysis. The lack of stereoscopic coverage was noted early (Story et al., 1976); because of poor base/height ratios, the topographic effect in satellite imagery was not usable in the same way that topography could be analyzed on aerial photographs in stereopairs. This capability, missing from coarse L-resolution large area satellite imagery, can provide the photointerpreter with virtually random access to three-dimensional landscapes, a powerful advantage in recognizing forest stand differences.

Few digital analysis tools have been developed to more carefully mimic this capability of aerial photography, but Toutin (1997) reported progress in understanding the concept of chromo-stereography; physiological and psychological aspects of color vision and depth perception enable synthetic generation of depth from any independent source of data (elevation, gravimetry, GIS polygonal data, and more) for use with image data. A good-quality DEM is not required if one of these other data sources is used to synthesize depth in the remote sensing image through manipulation of the IHS transformation within the radiometric content of the integrated data set. The result is a normal flat color image which can be viewed without glasses, but when viewed through special chromo-stereoscopic lenses jumps into 3D in much the same way that good-quality stereo-airphotographs do when viewed under stereoscopes.

Still, all of these aspects of digital classification using remote sensing have changed; there is better spatial resolution, better use of ancillary information such as DEMs, better access to other image features such as texture, and better ability to relate spectral response to the features of interest in the stand (e.g., species composition and crown closure). Much has changed, but much remains the same — there is not yet a complete digital remote sensing forest stand mapping and classification system, but it is only a matter of time until such systems are developed based on early promise (Hagner, 1990). Access to better GIS tools and data, provision of hyperspectral data, and greater use of other image information such as image context, shape, and texture are required to stimulate higher levels of success in automated classification for forest stand mapping. The current capability is slowly improving, and the future is wide open (Green, 2000: p. 39):

> *Automated classification of high-resolution imagery can create forest ecosystem maps that are comparable in accuracy and resolution to traditional photo maps and more consistent, quicker to produce, richer in content, and easier to use to monitor change.*

Once such a system has been developed, the goal of producing traditional stand maps will perhaps become rather less important than adjusting the mapping requirements of foresters to include the logic of using digital image classification for forestry purposes. Existing mapping products have not evolved enough to allow the new remote sensing digital data to emerge as valuable. Apparently, foresters need convincing evidence that the information provided by remote sensing classifications can meet or exceed their requirements; and how better to determine if this is the case than by comparing them to their current classification products, generated in traditional, well-understood methods? Is it likely that building new algorithms that mimic human interpretation is simpler than adjusting stand mapping paradigms?

In 1981, Robinove discussed the logic of multispectral classification with coarse-resolution satellite imagery, noting that since image pixels are comprised of spectral response contributed by all of the components of the landscape — the vegetation, soils, landform, bedrock, and water — they can best be classified by referring to those original components of the landscape. There has always been this difficulty of identifying a single forest attribute (such as species composition) upon which the classification will be built. Even with better spatial resolution and these other improvements, new challenges are created that reduce the likelihood of generating consistent air photo-like forest stand classifications — but this is the same problem experienced in manual aerial photointerpretation! In many ways, the analogue and digital ways are still a world apart in the type of classification products that are generated routinely. Probably, it is a relatively simple matter of opportunity and cost before a larger degree of acceptance of the differences in the classification of digital imagery compared to traditional methods is recognized, accepted: normalized. One possibility might be the adoption of a dynamic forest inventory structure based on individual trees resolved in multiple spatial resolution remote sensing data. Once the technology is understood and mastered, new forms of organizing the landscape will emerge to allow those characteristics to move to the forefront.

The next important decision in the digital classification protocol involves selecting the variables to be classified. Which bands and image transformations are most useful in classification? Numerous studies have reported on band selection among a wide variety of image transforms, vegetation indices, reduced bandsets, augmented sets, ancillary data sets, and simulated bands. First, there must be some reason to suspect that a band or transform will contain the information needed for it to be selected. Users must guard against the tendency to throw everything at hand at a classifier, expecting the classifier to make sense of the data. This is one of the dangers of the hyperspectral era — so many bands, so much temptation! It is always wise to keep in mind the old computer adage of garbage in, garbage out. This applies in remote sensing classification studies as in most other data-rich applications.

Many classifiers, including the popular maximum likelihood (ML), can suffer from the Hughes phenomenon (Kim and Swain, 1990), in which classification accuracy decreases as additional data layers are introduced, even though these additional data layers provide new information that should aid the analysis. This phenomenon can be attributed to a sensitivity to greater levels of multidimensionality (e.g., hyperspectral airborne imagery, hypertemporal image sequences, and large GIS inventory data) (Peddle, 1995a) or ancillary data channels which violate statistical assumptions of parametric classifiers (Strahler, 1980), or both. The decision on variables is followed by the choice of the decision rule to apply in the process of classifying those variables. Actually, these decisions should be considered together since many classifiers can only use certain types of data. Others, such as evidential reasoning (ER) (Peddle, 1995a,b) or neural networks (Benediktsson et al., 1990), do not display sensitivity to increasing numbers of data dimensions.

The final step in the classification protocol is an assessment of classifier performance and map output accuracy. Performance can be assessed in training areas if the relative accuracy of the classifier is of principal interest; this type of accuracy assessment is valid only if the training data are truly representative of the study area (Swain, 1978), something that is rare and difficult to verify. Independent test areas not used in training the classifier should be examined to report on classification accuracy, usually in the form of a contingency table or confusion matrix (Congalton and Green, 1999). These test areas should be identified as belonging to individual classes in the classification scheme unambiguously using field observations; alternatively, aerial photographs or other remote sensing data can be used to verify that the image classification has correctly labeled the test area. These other accuracy assessments typically generate less confidence in the image classification product.

Errors in a remote sensing classification, when compared to a reference such as field or aerial photointerpretation, are typically expressed as errors of omission and errors of commission in a contingency table (an example is shown in Chapter 8, Table 8.1), also sometimes known as the classification confusion matrix (Lillesand and Kiefer, 1994). If the remote sensing classification and the reference data are in agreement, there are no errors and 100% agreement between the two data sources can be recorded. If the two are not in agreement, one of the data sets is considered to be in error. The resulting error can be considered in two ways. A pixel in the remote sensing classification shown to be a member of a class other than the one shown in the reference data is said to be omitted from the correct class and committed to the incorrect class. Then, the omission errors for a class are the pixels that should be shown on the map as that class, but which are shown as members of the other classes. The commission errors (or false positives) for a class are the pixels shown as that class but which actually are pixels from other classes. What is important here is to note two different types of errors may be interpreted in the same classification problem; two faces of the same coin. These error compilations reflect the difference in importance that the two types of classification error can have on users and producers of remote sensing classifications. Pixel errors in contingency tables are sometimes expressed as the user's accuracy (the number of correctly classified pixels divided by the total number of pixels in that class), and the producer's accuracy (the number of correctly classified pixels in that class divided by the total number of training pixels).

Improvements in virtually every step of the classification process — the classes, the classifier, the variables, the collection of training data, the ways of assessing accuracy — are all possible and continue to rate a high priority on the research agenda of many remote sensing analysts (Bolstad and Lillesand, 1991; McCaffrey and Franklin, 1993). To a large extent, in many projects a few simple actions can generate most of the desired increases in accuracy that are needed to increase the value of the remote sensing classification products. One of the most obvious developments that has the potential to revolutionize the image data is the presence, in many managed forests, of a comprehensive GIS which may contain ancillary data such as DEMs, but also detailed forest stand descriptions. The integration of remote sensing into this GIS (and vice versa) has sometimes been problematic; the data formats are quite different, for example, necessitating much data conversion and manipulation. Chapter 4, Color Figure 4 contains examples of these three integration activities.

The significant improvements that can be obtained appear to flow from relatively straightforward actions, such as the inclusion of ancillary digital data and the careful use of image texture. These are discussed in separate sections below. Improvements in the overall process of classification are less common and are discussed in the following section.

Modified Classification Approaches

Significant changes have been made to the classification process as a result of continued efforts at building the optimal classification protocol for forestry applications. Modified versions of the two main classification methods — supervised and unsupervised — have been around almost as long as the original pure methods, and have been shown to outperform either the supervised or unsupervised methods under certain conditions (Fleming and Hoffer, 1975). In fact, pure versions of the supervised and unsupervised approaches are actually relatively rare, since users and analysts have often been interested in devising ways to extract more value from the imagery with less investment (in computer time, in training data collection, in assessing the results, and so on).

One of the earliest hopes was that a method could be devised that could combine greater control over the classes (characteristic of the supervised classification approach) with increased statistical validity and reduced demands on training data (characteristic of the unsupervised classification approach). An early modification that has since become almost standard procedure, particularly when classifying large areas with limited field resources, is to cluster the image data before collecting training data (Table 4.4). The initial unsupervised clustering can be used to allocate resources more efficiently; for example, to train classes in the supervised approach. This is typically an iterative process that can encompass the collection of reference information, the compilation of local expertise, visits to the study site, and the initial analysis of the available satellite and airborne image data (Marsh et al., 1994).

TABLE 4.4
A Typical Procedure for Quantitative Analysis of Remote Sensing Data

Step 1. Use unsupervised analysis and image transformations to enhance raw image data by deriving and displaying spectral classes.
Step 2. Use reference data to associate informational classes with spectral classes. Select training areas.
Step 3. Apply clustering and spectral separability analysis to the training areas to derive unimodal classes.
Step 4. Use feature selection or feature extraction to determine appropriate variables for classification.
Step 5. Apply decision-rule of choice to the area of interest.
Step 6. Evaluate results by considering (a) classifier accuracy on training data, (b) classifier accuracy on test data.
Step 7. Refine the analysis or prepare the results for the user.

Source: Modified from Swain, P. H. and S. M. Davis, Eds. 1978. *Remote Sensing: The Quantitative Approach,* McGraw-Hill, NY. With permission.

Spectral classes generated in an unsupervised way can be labeled in a systematic process. For example, Debinski et al. (1999: p. 3284) generated 50 ISODATA clusters from Landsat TM data to discriminate the gross land cover types (forest vs. meadow), and among a gradient of meadow types (xeric to hydric):

> *... each spectral class was then identified using aerial photography and personal knowledge of the study region and assigned to an information class representing a vegetation type to create a final map of spectrally distinct vegetation classes.*

There were five forest classes mapped in their Greater Yellowstone Ecosystem study area; two pure conifer and three mixed conifer classes based on density (sparse, medium, and dense stands). The purpose of the initial clustering was to provide remote sensing input to habitat class characterization that was used to structure a biodiversity sampling scheme.

This process of training the classifier or labeling clusters is heavily dependent on the skill and experience of the analyst (Jensen, 2000). In a typical supervised image classification project, hundreds of training area samples must be acquired, usually with great care and justification, but sometimes with little more than passing reference to the field. What value can be placed on a skilled airphoto interpreter *cum* image analyst *cum* GIS technician who can generate suitable training areas for a satellite image classification, with minimal field work, and a few old and faded aerial photos? Adams et al. (1995: p. 145–147) noted that:

> *few observers have a mental image of a TM spectrum of a forest, whereas most know what a forest looks like in the field.*

Good image analysts, like the good aerial photointerpreters before them, are those who can hold and use that mental image — that abstraction of a forest. Perhaps this ability is simply accumulated experience or an innate spatial facility. The deeper understanding it connotes might be used in numerous though serendipitous ways to produce higher-quality, more accurate classification products from remote sensing images.

Clustering before collecting training data can help avoid a later step in which classes are spectrally combined because they are not separable with the necessary degree of accuracy. Another possible modification has been to collect training data, then cluster those data before deciding on a final training data set to provide to the classification decision rule. Mixing and matching unsupervised and supervised techniques to yield an optimal training data set or to refine classes already mapped can be a simple and effective way of improving the overall classification process. Changing the order or the purity of the supervised or unsupervised process may make the process more efficient and sometimes can produce an information product with higher accuracy.

An ideal image classification approach does not yet exist; all methods at some point must handle a three-way compromise between:

1. The information classes that are desired
2. The spectral information content of the imagery
3. The method of making class decisions

The information classes and the spectral classes are almost never in complete agreement (Swain and Davis, 1978), and every decision rule yet devised will make less than optimal choices in the presence of real noise or various levels of ambiguity. These problems have led many to consider the classification process — not only the decision rules, but the entire set of procedures that comprise the classification protocol — as a special circumstance of fuzzy logic. Fuzzy classification methods have recently attracted a great deal of attention because of their ability to soften the decision for each pixel; instead of only a single class assignment, each pixel would belong to every class but with different degrees of membership. A fuzzy approach could even include a collection of fuzzy training areas; fuzzy accuracy assessments, as well as fuzzy variables, are all part of the fuzzy methodology. A continuum of fuzziness in classifications may ultimately provide the optimal set of compromises (Foody, 1999).

An ideal classification method would satisfy these five criteria (Cihlar et al., 1998).

1. Accurate
2. Reproducible
3. Robust (and able to exploit fully the information content)
4. Applicable uniformly
5. Objective (i.e., not dependent on the analyst's decisions)

These design criteria were selected by Cihlar et al. (1998) to measure the performance of a new hybrid classifier called classification by progressive generalization (CPG). The development of the CPG hybrid classifier attempted to overcome the deficiencies of the supervised (mainly in the reproducibility criteria) and unsupervised methods without creating a completely new and awkward procedure. At the same time, CPG was thought of as an effective bridge of the gap between visual analysis of imagery and digital image analysis (Beaubien et al., 1999).

At the heart of the CPG is a method of using enhanced visual image analysis to identify training areas that are then subjected to rigorous statistical analysis. After many years of experimentation and image interpretation, the enhancement-classification method (ECM) procedure was devised (Beaubien, 1994). The idea originated in visual image analysis and has evolved to now support a fully digital method within CPG. The ECM enhancements are standardized functions of the original Landsat TM image bands; transformations that are similar to the weighted indices discussed earlier, such as Tasseled Cap indices, now optimized for the forest types of interest (Beaubien, 1994). The procedure resolves the difficulty noted by many image analysts (e.g., Ahern and Sirois, 1989: p. 61) "of the ability of the interpreter to distinguish real differences between two images which are artifacts of the enhancement process." This difficulty is based on the fact that scene enhancements are derived using image statistics that can change from scene to scene and with changing objects in any given scene (e.g., clouds, number of large lakes, differing distributions of forests and grasslands, and so on). By using standardized enhancements, the subtle differences between enhanced image products that might lead to different interpretations are avoided.

Three basic steps are required in a larger flow of the CPG procedure tasks (Figure 4.3).

- Digital contrast enhancement of the image channels to maximize visual discrimination among classes (using ECM)
- Selection of training values which represent various surfaces (colors) present in the image and together encompass the range of land covertypes in the image
- Classification of each pixel with a minimum distance function, and agglomeration of the resulting spectral classes into thematic classes

This CPG approach is one of several modifications to the traditional supervised and unsupervised approaches that appear to have moved the classification process into a closer alignment with both user needs and user capabilities. This hybrid classification attempts to capitalize on the particular strengths of the supervised method by using a known set of classes and human interpretations of spectral differences. The classification also attempts to retain the strengths of the unsupervised method by spectral clustering. Finally, the classification is doable with off-the-shelf commercial and publicly available software capability.

Another reasonably simple modification used more and more frequently in forestry remote sensing classification applications has been to employ the existing forest inventory data in some fashion during a classification procedure. This is usually done in an attempt to improve the training data collection or the ability to separate certain classes known to be spectrally similar (Goodenough, 1988). With increasing amounts of forest inventory data available in digital formats in a GIS, it has become much easier to develop better methods of using them in a classification procedure. For example, the forest inventory data can be used to guide training area selection or to sort out clustering results within known forest stands — simple sort, merge, and mask GIS operations (e.g., Boresjö Bronge, 1999).

Chalifoux et al. (1998) considered Landsat TM spectral response values of conifer forests associated with defoliating insect activity within GIS polygons; the goal was to use the existing forest inventory classification and report only those

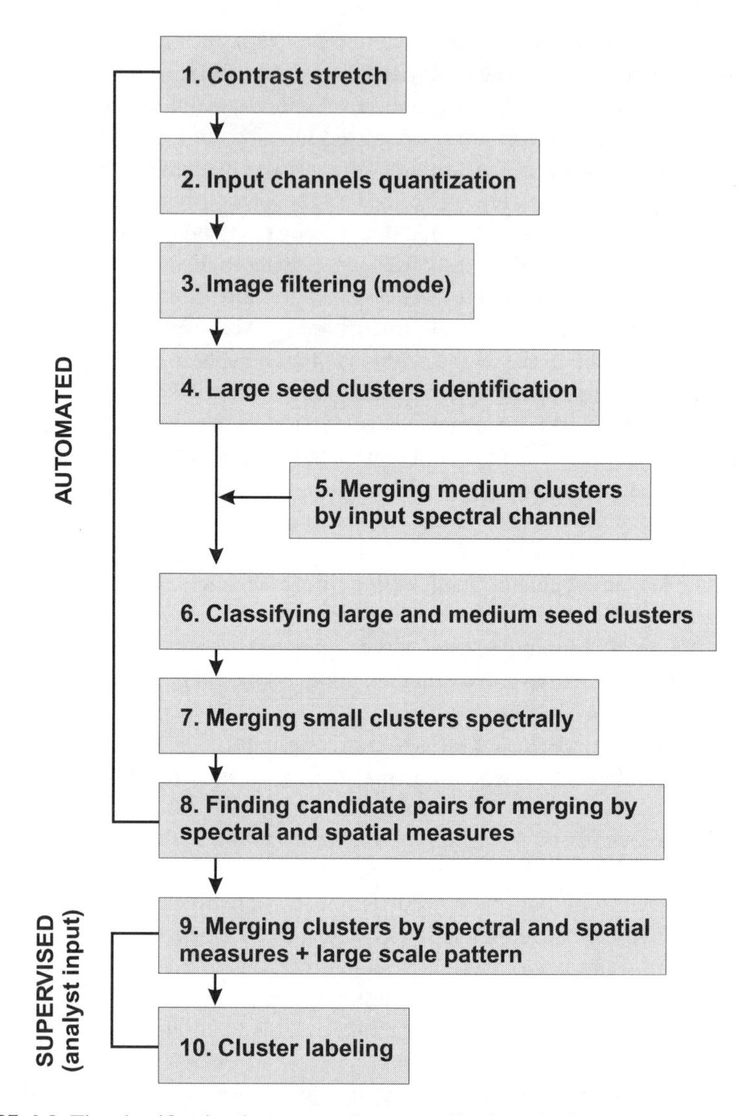

FIGURE 4.3 The classification by progressive generalization (CPG) process flowchart. The idea is to minimize the weaknesses and incorporate the strengths of the supervised classification method (e.g., analyst ability to select training features) and the unsupervised classification method (e.g., clustering techniques) to generate an optimal forest classification map product of large areas. An important component of the CPG process is to use an enhancement procedure during the training stage. This enhancement is locally designed to maximize the spectral distinctiveness of known forest covertypes in a standardized way and thereby reduce or eliminate analyst subjectivity. (From Cihlar et al. 1998. *Int. J. Rem. Sensing,* 19: 2648–2704. With permission.)

existing classes (i.e., polygons) which have likely been disturbed by the defoliators. A map-guided classification procedure was designed to answer the question: In each forest inventory polygon, what is the class of eastern spruce budworm (*Choristoneura fumiferana* Clem.) damage severity? The procedure disregards the classification of individual pixels, and instead classifies the entire polygon based on the distribution of spectral response patterns within the polygonal boundary as follows:

1. Superimpose forest inventory polygons on satellite imagery.
2. Compute distribution of spectral response patterns within polygons.
3. Differentiate four categories (healthy stand, lightly defoliated stand, highly defoliated stand, heavy mortality stand) based on mean spectral response within each polygon.
4. Compare to independent surveys (and a conventional classification procedure) of spruce budworm defoliation.

Development of careful ways of stratifying the image data by ancillary data and forest inventory polygons will usually improve the process of classifying the feature of interest (Strahler, 1981); Chalifoux et al. (1998) reported that the map-guided classification was on the order of 85% correct, compared to 50% accuracy achieved with a conventional supervised approach. Use of this stratification idea is related to the layered classification approach discussed in the following section, and has been used to facilitate the use of DEM data (Fleming and Hoffer, 1979; Hutchinson, 1982) and forest inventory data (Ekstrand, 1994b; Franklin and Raske, 1994; Wulder, 1997) in image transformations and in classifications. The purpose of stratification is to reduce the variance and the size of the area that must be handled; the process of classification is not significantly altered, essentially retaining the same steps, and only the area examined and the training data are modified.

In all of these modifications of the classification process, there remain three essential requirements in order to achieve successful classification results (Landgrebe, 1978b: p. 340).

1. The classes must be spectrally distinguishable.
2. The list of classes must be exhaustive.
3. The classes must be of interest to the user.

The first of these is critical from the perspective of the remote sensing analyst, and the third is obviously of critical importance to the user. Generally, it all comes back to the classification scheme, from which a valid list of classes must be produced. No classification process, modified or otherwise, can make much headway if the classes that must be mapped are not radiometrically distinct in the data that have been acquired to map them.

Increasing Classification Accuracy

Most classifications are iterative, trial-and-error exercises in which several runs are completed, and then modifications are implemented depending on the results and the available resources. Step 7 in Table 4.4 contains this essential truth; at the conclusion of the process, either present the map to the user or go back and try something else to improve it! Some improvements can be found in a better training area selection, the use of different variables (e.g., transformed data) in the classifier, perhaps even a different choice of classification algorithm. Fortunately, there are numerous ways to increase classification accuracy (Strahler, 1980, 1981; Hutchinson, 1982).

One of the most obvious ways to increase classification accuracy is to deploy and use improved airborne or satellite sensors. In other words, collect different remote sensing data better suited for the classification task. For example, if the classification task was aimed at discriminating smaller forested wetlands rather than providing details within broad landcover types, perhaps the SPOT satellite or an airborne sensor with fewer bands but greater spatial resolution would provide a more accurate classification than would be obtained using the Landsat TM sensor (Franklin et al., 1994). Multiple observations by the same satellite or another sensor system are useful to detect phenological changes of vegetation (Wolter et al., 1995); if understanding vegetation changes over time would help identify particular classes, then choosing new imagery that captures those changes would likely result in increased classification accuracy. Similarly, a combination of SAR and optical/infrared data has been used to increase vegetation discrimination (Leckie, 1990b); if the classes differ not only in their spectral response in the optical portion of the spectrum, but also in the microwave portion of the spectrum, then providing both types of spectral response to the classifier will likely result in increased classification accuracy.

Acquiring multiple imagery for a study site, or airborne imagery instead of satellite imagery, can lead to other difficulties, not the least of which could be the additional cost to the project generated by the acquisition, and the new requirement for precise geometric registration. A higher standard of geometric and radiometric processing might be necessary. Another approach to improve classification results is to combine remote sensing imagery with digital ancillary data, perhaps derived from a GIS or from different types or combinations of image information (Fleming and Hoffer, 1979; Jensen, 1978; Strahler, 1981; Franklin and Wilson, 1992; Chalifoux et al., 1998). In many areas, for example, a DEM can be easily generated, or may already be available. Digital topographic data can be a source of immediate improvements in classifier performance (Anuta, 1977; Strahler et al., 1978; Hutchinson, 1982). Depending on the classes that are of interest, a powerful method of improving classification accuracy is to incorporate additional variables derived from the DEM, including elevation, slope, and aspect directly in the classification. If geomorphometric data (Evans, 1972, 1980) are related to the distribution of classes in the study area, then adding geomorphometric variables to the classification decision rule can provide increased accuracy (Franklin, 1987, 1994).

The use of ancillary data in classification is not without problems; for example, adding new DEM variables often increases the demands for training data, and can require the use of different classifiers if the new data have very different characteristics from the spectral response patterns. Slope/aspect artifacts — such as stepped vs. smooth slopes between digitized and rasterized contours, or significant breaks in areas of discontinuous elevation data (e.g., lakes, shadows in imagery) — can create difficulties in classification maps that use both remotely sensed and DEM data (Franklin and Wilson, 1992; Franklin et al., 1994). Before considering adding DEM data to the classification, another idea, discussed earlier, is to use the DEM data to correct remote sensing spectral response patterns for the topographic effect — those variations in slope and aspect which can mask the true information content of the image data (Teillet et al., 1982; Ekstrand, 1994a; Gu and Gillespie, 1998; Dymond and Shepherd, 1999). Radiometric corrections are not always simple to implement, and removing the topographic effect can sometimes be detrimental, particularly if there is a close association between the classes of interest and topography — as in, for example, ecologically based classifications that attempt to map landscape units and land systems comprised of recurring patterns of vegetation, soils, and topography. But radiometrically corrected imagery will often provide data that are more closely related to the actual spectral properties of the target.

Traditional statistical classifiers used in remote sensing, such as the maximum likelihood algorithm, the discriminant function, the minimum distance to means algorthim, the parallelepiped classifier, all operate in Euclidean space. Most early digital multispectral remote sensing data, for which these classifiers were built, were ratio-level data that have characteristics easily handled in such probabilistic-based decision-rules formulations. These classifiers, and their assumptions, are well-described in the pattern recognition (Duda and Hart, 1973; Tou and Gonzalez, 1977), statistical, and remote sensing literature (Swain and Davis, 1978; Lillesand and Kiefer, 1994; Jensen, 1996). But much of the new data that may be available to a classification project might not be ratio-level data, but rather data provided on different data types and scales — perhaps including a mix of ordinal (e.g., ranked soil fertility classes), interval (e.g., dominant species), or nominal (e.g., class 1 of 12) data. A classifier that is nonparametric and thus can use a range of data types, including ratio-level data, would be needed in these situations.

To handle these new data types, classifier decision rules based on fuzzy logic (Bezdek et al., 1984; Cannon et al., 1986; Wang, 1990a,b), evidential reasoning (Lee et al., 1987; Peddle, 1995a), and neural networks (Benediktsson et al., 1990; Carpenter et al., 1997, 1999; Trichon et al., 1999; Jensen et al., 1999), have been developed or modified for remote sensing applications. These new decision rules are increasingly available in commercial image processing systems or as add-on packages. Decisions are made, not on probabilistic rules, but by using different mathematical theory and logic. A few of these classifiers operate in the same way that conventional classification algorithms, such as maximum likelihood, operate. For example, they require the same preparation and iterative steps; only the actual decision rule is different (Step 5 in Table 4.4). For others, the entire process of classification must be adjusted to take into account the demands of the classifier.

The complexity of the nonstatistical classification decision rule, and ultimately of the entire classification process (including interpreting the results), may be formidable. For example, Pinz et al. (1996) described "active fusion," a computer procedure to combine information from multiple sources on the basis of three different mathematical theories: (1) probability theory, (2) the Dempster-Shafer theory of evidence, and (3) fuzzy set logic. Experimentally, they showed a significant reduction in the number of information sources required for a reliable decision on classification of Landsat data for agricultural crops. Accuracy may be increased, but it is unlikely that such a process will lead to the classification of significantly different classes, given the same data set and objectives, than a conventional statistical algorithm. From the systems viewpoint, however, it is significant that the final classification accuracy obtained can be significantly increased with little or no additional investment in training area data collection or input variable manipulation.

Desachy et al. (1996) integrated remote sensing data and expert knowledge in a classifier based on fuzzy neural networks. The approach recognized that expert knowledge — similar to the knowledge that air photointerpreters use when classifying vegetation on aerial photographs with manual methods — can be an ideal source of information to rectify errors in previously run classifications. The ICARE (Image CARtography Expert) system stored expert knowledge as a set of production rules with certainty factors — in fact, these rules tended to resemble similar compilations that could be generated within most GIS forest inventories. For example, an expert may express knowledge about pine stands in the following way:

"Pines are mainly located on south slopes from 800 to 1500 m above sea level."

This was translated into a production rule (using a rule compiler written in PROLOG) as an if-then statement with a specific weight (heuristically determined) attributed to the use of the qualifier 'mainly':

if class 'pines'
then (south slopes) and (800 < elevation < 1500m)
rule = 0.8 corresponding to 'mainly'

The number of rules and the levels of different rules can be large (Desachy et al., 1996), but are best kept to some manageable number since initially all pixels to be classified must be considered with each rule. In an example of a southern India tropical forest vegetation classification using Landsat TM data and a DEM, 11 rules were invoked. The improvements in classification accuracy ranged up to 14% when compared to the average result for supervised maximum likelihood classification. By reconsidering this classification product with subsequent fuzzy neural networks built through a separate learning process, an additional 11% increase in accuracy was achieved, with the final map accuracy determined to be 83% in agreement with field observations.

Obviously, this classification approach can require considerably different setup work than a conventional maximum likelihood classification. The process may be excessively demanding of training data and analyst computer skills (Benediktsson et al., 1990; Peddle et al., 1994; Binaghi et al., 1997). In general, new classifiers, notably neural network routines, have not adequately balanced the need for the analyst to understand how decisions are reached with the assumption of a black-box philosophy. For this reason, it appears that, although many of these algorithms can produce superior decisions and subsequent classification results, when compared to traditional statistical classifiers they are not yet in high demand in operational settings.

One promising development, based on the fact that neural networks and statistical classifiers operate in very different ways, is the use of an integrated approach (Wilkinson, 1996). Using different classifiers at different points in the classification process holds the promise of achieving maximum classification accuracy through selective decision making; in other words, the algorithm is designed to invoke a particular classifier decision rule only for a decision to which it is optimally suited. One issue has been problematic: deciding on when to use a particular decision rule at different points in the image analysis process, based on the available data and the type of decision required.

An early example of this approach was provided by Franklin and Wilson (1992) in the form of a layered, three-stage classifier that used spectral data, DEM data, and spectral-DEM data in areas for which the data were ideally suited to provide the right decision. In valley reaches, where DEM data were not contributing to discrimination of vegetation classes, the classification was based solely on the spectral response data and a minimum distance to means rule. In more complex terrain, in which different slopes and landforms were of interest, the DEM data were more powerful in separating out classes. A maximum likelihood classifier was used in those situations. In other areas, both spectral and DEM data were needed. The layering occurred with the data (i.e., spectral, DEM, or combined), the methods (minimum distance or maximum likelihood), or both.

Layered classifications, or more generally, classification trees (Hansen et al., 1996), can provide the same type of increases in accuracy that have been reported in more complex decision rule processes provided the data are not too complex (i.e., they are restricted to one or two data types). But a new demand is for classifiers that can handle multisource digital data. This is especially evident in classification projects in which remote sensing data are only one of the input variables necessary to create acceptable classification products. One method that has been developed to handle the multisource digital data sets in a classification task is the evidential reasoning (ER) classifier (Peddle, 1995a,b). Based on the Dempster-Shafer theory of evidence (also known as Belief Theory), the classifier is able to incorporate all four types of geospatial data (nominal, interval, ratio, and ordinal) (Dempster, 1967). Much of this would be difficult to incorporate into any new classification based on satellite remote sensing data if only statistically based decision rules were available. One of the few options available to the image analyst using a conventional statistical classifier is to attempt to use these types of data to help guide a classification procedure (Chalifoux et al., 1998), or to augment training data collection (Goodenough, 1988).

The ER classifier, on the other hand, imposes no constraints on data type and has two other distinct advantages: (1) the classifier makes no statistical assumptions about the data and (2) the classifier outputs not only the classification map (integrating satellite imagery and GIS data together in a single, complex decision rule), but several interpretive measures (e.g., measures of support, conflict, plausibility, and consensus). After the evidence from each data source has been derived for each class, this information is compiled into a mass function, or evidential vector (magnitude of support and plausibility for each class), from which it must be combined to identify the class with the overall greatest magnitude of integrated evidence with respect to the support, plausibility, and uncertainty measures (Peddle, 1995a).

This is achieved by source-specific orthogonal summation (Dempster, 1967). For source 1 (with mass μ_1 over a set of labels α) and source 2 (with mass μ_2 over a set of labels β), the orthogonal sum ($\mu_1 \oplus \mu_2$) to determine the mass μ' assigned to a labeling proposition χ is computed as:

$$\mu'(\chi) = \left[1 - \sum_{\alpha_i \cap \beta_j = \phi} \mu_1(\alpha_i)\mu_2(\beta_j)\right]^{-1} \sum_{\alpha_i \cap \beta_j = \chi} \mu_1(\alpha_i)\mu_2(\beta_j) \qquad (4.6)$$

From this formulation, the extent of conflict between the two sources can also be computed. This process is repeated sequentially for each source in the data set, after which all mass functions have been reduced to a singular evidential vector for which a decision rule can be invoked to determine a final pixel label classification. In the task of forest resource mapping, there may be significant advantages in using the ER approach — or any fuzzy or neural network classifier — rather than a statistically based classifier, because of this ability to provide "soft" in addition to "hard" classification maps (Foody, 1999).

Image Context and Texture Analysis

One of the better-known weaknesses of all of the remote sensing image classifications discussed to this point is the almost sole reliance on the spectral response pattern at individual pixel locations in the classification decision. Many other types of information could be made available to the classifier if the methods of extracting that information from the imagery were not so fragmented and poorly developed. One of the more promising alternatives has been to consider the classification of image data in a spatial context. The premise is that a pixel's most probable classification, when viewed in isolation, may change when viewed in some context (Haralick and Joo, 1986). Context classifiers operate spatially as well as spectrally (Gurney, 1981; Haralick, 1986). The simplest context classifiers use neighboring pixels to help decide, confirm, or change the classification or labeling of a center pixel; later attempts to broaden the context classifier algorithm required work with the spatial correlation function between pixels (Khazenie and Crawford, 1990), map context (Solberg, 1999), and contextural parameters (Chen, 1999). These concepts are closely related to spectral mixture and image texture analysis.

Per-pixel classifiers can be outperformed in certain classification tasks in which the objective is to detect objects or homogeneous regions by another class of image processing routines generally known as image segmentation. Forest inventory polygons — forest stands — would appear to be an ideal segmentation target (Woodcock and Harward, 1992). Instead of deciding pixel membership based solely on spectral response patterns, image segmentation typically involves the use of edge and region analysis to find spatially identifiable features. The idea is that texture, shape, and context can be exploited during the classification decision, in addition to spectral response patterns at a point (Kettig and Landgrebe, 1976; Cross et al., 1988; Lobo, 1997). Image texture is a quantification of the spatial variation of image tone values that defies precise definition because of its perceptual character (Hay et al., 1996). In aerial photos, experienced air photointerpreters use texture to identify changes in the spatial distribution of forest vegetation.

This use of texture flows naturally from the powerful innate ability that humans have in recognizing textural differences, although the complex neural and psychological processes by which this is accomplished have so far evaded detailed scientific explanation. Insight into how texture might be analyzed by computer has focused on the structural and statistical properties of textures (Haralick, 1986). The hope is that by a combination of per-pixel and area-based texture processing more accurate classifications of remote sensing imagery can be generated (Connors and Harlow, 1980; He and Wang, 1992; Lark, 1996; Ryherd and Woodcock, 1997). Parallel to this use of texture in classification, interest has developed in texture itself as a variable in forest applications (Coops and Culvenor, 2000). Texture can be directly related to different aspects of forest stand structure, including age, density and leaf area index (Cohen and Spies, 1992; St-Onge and Cavayas, 1995, 1997; Wulder et al., 1996). Chapter 4, Color Figure 5 contains a graphical example of the inherent texture quality of high spatial detail imagery. Also shown is a simplified representation of the potential power of texture in augmenting spectral data, such as mean and standard deviation values (or signatures), in small windows.

Texture variables have been suggested based on first-order statistics (e.g., standard deviation or variance), second-order statistics, frequency domain or Fourier power spectrum, spatial autocorrelation functions (e.g., semivariance), and structural image features. By far the most common approach has been to use second-order statistics derived from image spatial co-occurrence (Haralick et al., 1973). The assumption is that texture information on an image is contained in the overall or average spatial relationship which the gray tones in the image have to one another. Those relationships are specified in spatial (or gray-level) co-occurrence matrices which are computed for four directions between neighboring pixels within a specified (moving) window on the image. The co-occurrence matrix is a summary of the way in which pixel values occur next to each other in a small window divided by the number of pixels in the window. This basic procedure has repeatedly proven its value on a wide variety of imagery and in a wide variety of applications (Franklin and Peddle, 1987).

In well-defined areas texture can be highly discriminating; when the processing window is applied in very heterogeneous areas, or crosses boundaries of homogeneous units, the resulting texture values tend to vary widely depending on the chance location of the window (Townshend, 1981). Since it is often the texture differences themselves that define where boundaries are placed, operational procedures are needed to constrain texture calculations; violating or straddling stand boundaries, for example, reduces the ability of the texture measures to be related to within-stand variability. Normally, image co-occurrence texture analysis procedures require the user to identify five different control variables.

1. Window size
2. The texture derivative(s)
3. Input channel (i.e., spectral channel to measure the texture of)
4. Quantization level of output channel (8-bit, 16-bit, or 32-bit)
5. The spatial component (i.e., the interpixel distance and angle during co-occurrence computation)

Of these, window size is perhaps most critical (Marceau et al., 1990; Franklin et al., 1996) and least understood; for example, if pixels that occur next to each other are used in the compilation of the co-occurrence matrix, what is the effect of spatial autocorrelation (Foglein and Kittler, 1983)? Apart from this problem, the spatial co-occurrence approach generates a lot of data. For example, assuming that seven derivatives of the co-occurrence matrix are available (actually there may be even more), six different spectral channels (often there are more), window sizes ranging from 3 × 3 to 21 × 21 (ten different sizes, but why stop there?), three quantization levels (could be worse), and four possible directions (spatial component), the result would be more than 5000 different texture channels for a single application. This output would overwhelm even the most sophisticated classifier. What can it all mean?

Multiscale texture is an open-ended way of generating awesome amounts of data. Choosing a set of texture variables to use has been problematic. In general, texture appears to be a scene-specific image variable; successful texture analysis in one application does not necessarily imply global applicability. Therefore, selection of texture variables should probably be based on an iterative study of the particular image data set and forest conditions of interest. One approach might be to use feature selection statistics and attempt to identify the optimal variables. Statistical methods, such as Bhattacharyya distance measures do not work well with this type of data volume, however. In earlier work, visual analysis of texture displays was used to understand the way in which texture represented differences in the imagery (e.g., Franklin and Peddle, 1990). This approach obviously has limitations — who wants to look at 5000 different texture images? In another study, 208 wavelet (multiscale) texture features were reduced to about 50 using statistical methods (Wu and Linders, 2000); but 50 texture features are still a serious amount of data for many classifiers to handle. In the end, no universally best feature set of textures (from SAR data) for mapping clearcuts and burns was found.

Many improvements in the co-occurrence procedure have been suggested. These include the elimination of directional counting (Sun and Wee, 1983) and the use of geographic windows rather than fixed geometrical windows (Franklin et al., 1996). An optimal window size can be predicted using a semivariance procedure. If no real information about the size of the objects can be obtained, then a multiscale texture analysis is indicated based on the use of several different window sizes. But, if the size of the window can be related to the feature of interest, then obviously only that window size should be selected. For example, Coops and Culvenor (2000) suggested that if *a priori* crown size estimates were available, then the spatial pattern in high spatial detail forest imagery could be discerned over reasonably large windows that distinguished different forest stands. In one of the few recent comparative texture studies, Carr and Pellon de Miranda (1998) found that semivariance textures produced higher accuracies than co-occurrence when classifying microwave images, but that co-occurrence texture measures produced higher accuracies when classifying optical imagery.

This finding may highlight one specific weakness of spatial co-occurrence that has restricted even more widespread use (Wezka et al., 1976; Franklin and Peddle, 1987; Sali and Wolfson, 1992; Treitz et al., 2000); when imagery is highly variable (as in SAR imagery, with speckle and coherent noise), spatial co-occurrence may not adequately capture the entire spatial autocorrelation function, but instead responds disproportionately to minor variations over smaller areas. The texture derivatives in these cases seem unrelated to the main features of interest. On the other hand, co-occurrence texture has the decided advantage of widespread availability; most commercial image processing systems rely on spatial co-occurrence for texture analysis functionality, perhaps together with a selection of simple first-order statistics (such as standard deviation or gradients) in variable windows. More complex texture analyses, perhaps designed optimally for a particular type of image data (e.g., high spatial detail imagery or SAR images), are not yet widely available (Hay et al., 1996; Wilson, 1996). However, few purpose-designed texture analyses (i.e., using texture formulated specifically as a result of the study of the forest feature of interest) have provided results that are demonstrably superior to the general texture methods based on available procedures such as spatial co-occurrence.

Change Detection Image Analysis

A series of images acquired over time with radiometric and geometric fidelity can be subjected to a different type of image analysis aimed at identifying anomalies and confirming patterns over time. Change detection can be accomplished by visual analysis; perhaps by loading corresponding bands from multidate images into different computer display channels. No change, positive change, and negative change appear as different colors in the image, useful for a general examination of landcover change. However, the composition and quantity of change cannot be identified or calculated readily from the results of visual change detection; an intermediate (manual) mapping step would be necessary.

Using digital image analysis, change detection can be automated. Two approaches, illustrated in Chapter 4, Color Figure 6, have shown promise.

1. Image or map pixel-to-pixel analysis such as trend analysis, classification, and differencing (Singh, 1989)
2. Image or map area-to-area analysis in which change can be generalized within larger structures, such as GIS-based forest stand polygons (Wulder, 1997)

Change detection methods work best when imagery is selected to detect optimally the type of change that has occurred in the landscape (Olsson, 1994; Adams et al., 1995; Varjö, 1996; Yool et al., 1997; McCay, 1998; Cohen and Fiorella, 1999). For example, to detect rapid changes to the environment, such as clearcuts, a short time span sequence of images is required, whereas to detect trends and to forecast a change in forest growth, a longer time span and larger number of images might be needed (Häme et al., 1998). In general, to reduce the need for image calibration in imagery that is to be compared, image parameters should remain constant, i.e., the same time of year, time of day, spectral bands, sensor, sensor look angle, spatial resolution, and so on. Imagery with obvious features that can be confused with change, such as clouds or extreme soil moisture conditions, should be avoided. Images to be compared must be carefully georectified and registered to the same map projection to avoid mistaking misregistration for change. An assumption for virtually all change detection techniques is that the areal extent of the changes to be detected is larger than the spatial resolution of the imagery. An exception might be the newly developed temporal mixture models that extend the linear unmixing of pixels over time.

There are also differences in the way algorithms to detect change will work on different types of imagery, and in the detection of different types of change. For example, numerous studies have shown that annual detection of change in forest or land cover is possible by classification techniques (Cohen et al., 1998; Häme et al., 1998), but an annual classification approach would not likely be very sensitive to changes in leaf area caused by insect defoliation (Franklin and Raske, 1994). Insect defoliation can cause a change in class, but often the changes are less distinct unless the class scheme is very precise; instead, an image-differencing algorithm would likely have more success, since the differences in leaf area caused by the defoliation would be more apparent in the original spectral response than in a class-by-class compilation.

The direct multidate classification approach involves independent classifications of imagery from different dates being compared to detect changes in the landscape. Traditional classification of a reference image to develop base classes can be combined with spectral cluster information derived from a second image, now classified to show only changes that have occurred since the first image was acquired (Häme et al., 1998). Constraints on the types of changes that are of interest can be generated by considering different thresholds for class-to-class change; for example, the aim might be to detect change that differs from normal vegetative succession that is fairly sudden, and for which the areal extent is minor compared to the area covered by vegetative succession. In one approach, the characteristics of the spectral classes from the first image are used to classify the second image. If the two images are radiometrically exact, this eliminates one major source of classification differences in two dates that might lead to false identification of change.

Change detection methods can use both multispectral classification and image segmentation (Bruzzone and Fernandez Prieto, 2000); the algorithm could select homogeneous groups of pixels (parcels) from both images and compare them, focusing solely on areas that do not match according to the classification. By limiting the second classification to areas of change, the classification requirements are not as complicated or demanding. One advantage of the classification method of change detection is that absolute calibration of the imagery is not required if the changes are prominent or if good ground data for training sets are available. Disadvantages are that errors in classification may be compounded in the change detection analysis, resulting in a misinterpretation of change. Class change detection requires a complete change of class before the change will be found. And, if classification methods are used, the accuracy of classification in each image date and over time must be evaluated (Franklin and Wilson, 1991a; Congalton and Brennan, 1998).

When using image differencing techniques, the digital spectral response from one image date can be subtracted from, or otherwise compared to, the spectral response of another date. The magnitude of the change is the Euclidean distance between the two points. Areas with change will have large differences in value, while those with little change will have small differences (Johnson and Kasischke, 1998). A threshold for deciding which pixel values belong to the change class must be selected; this can be followed by the production of a binary change/no change mask in which all areas below the threshold (no change) are masked out. The direction of the vector relates to whether the change is positive or negative. For example, a negative change might be a loss of vegetation, whereas a positive change might represent vegetative regrowth (Cohen and Fiorella, 1999). Alternatively, a ratio of the band of one image date to the band of another image date can be generated; for areas with no change, the value of the ratio will be close to one. Areas of change will deviate to higher or lower ratio values. How much deviation is required for change is based on selected threshold values which, ideally, would be derived from field observations or other training data collection procedure.

There is a clear distinction between image data transformation — to detect change in spectral response — and labeling or identification of change. Image differencing, PCA, and change vector analysis are all linear combinations of spectral data acquired at different times such that the data space is rotated in a way perceived as most useful to extract change information from the data. The actual labeling of pixels and areas of change requires the use of a set of rules, perhaps derived through thresholding, classification, regression, or image modeling.

IMAGE UNDERSTANDING

Image understanding is a different approach to the analysis of remote sensing imagery based on a combination of cartography, computer vision technology, and knowledge issues (McKeown, 1984; Haralick and Shapiro, 1992). Image understanding is considered the digital or computer equivalent of human image or scene interpretation (Guindon, 1997). The highest goal of image understanding is to generate the digital equivalent of human spatial reasoning applied to images and other spatial data (Wang et al., 1983; Papadias and Egenhofer, 1996); for example, McKeown (1984: p. 92) wrote that the goal is "to understand how knowledge can be used in the image interpretation process to produce systems that are capable of detailed analysis of complex scenes." Spatial and temporal reasoning within a GIS environment has become more powerful as theoretical advances in behavioral geography, cognitive science, and environmental psychology have converged and been incorporated in a formalized framework (Egenhofer and Golledge, 1998).

Image understanding has been defined as "the development of techniques and computational systems for the *automated extraction of scene properties* from satellite or aerial imagery for specialist domains" (Muller, 1988: p. 85, italics added). From this perspective, specific criticisms aimed at the multispectral classification approach, in particular, seem formidable. In that approach, only modest amounts of automation seem possible, and the scene properties that can be extracted are limited to those associated with the list of classes. Statistical methods — continuous variable estimation, change detection, multispectral image classification, even the more advanced forms of modeling — are thought to be wholly inadequate to overcome fundamental problems, which range from the presence of mixed pixels, lack of spectral definition for most classes of interest, and temporal instability in multitemporal data sets. Manual interpretation methods are considered inadequate on the grounds of speed, accuracy, and cost (Muller, 1988). Only an image understanding approach based on spatial knowledge engineering seems likely to resolve the issues and create the necessary objective and quantitative information extraction methods.

Two factors have combined to encourage the growth of image understanding tools as part of the remote sensing infrastructure.

1. Increasing availability of digital data of all kinds, but particularly high spatial detail imagery
2. Lack of success in analyzing such data with existing (largely statistical) approaches to image analysis

As this new imagery with different characteristics — such as increased spatial resolution and hyperspectral bandsets — becomes more common, the image processing field will continue to expand to include techniques with a focus on feature elements rather than solely relying on statistical analysis of pixels. The emerging consensus is that these features can be extracted by analysis of remotely sensed images based on shape descriptions, rather than spectral properties (Haralick et al., 1987). The logic has been extended to include object-specific characteristics of interest to foresters, such as tree-crown outlines, which might be recognizable for different species (Gougeon, 1995). Typically, image understanding methods rely on rules of generic structural characteristics for features or objects within a scene, and image models which elaborate on the expected characteristics and functional relationships. The task of identifying and elaborating on rules for the interpretation of image objects has been greatly simplified through advances in knowledge engineering, though the endeavor is still vastly complex (McKeown et al., 1999).

By far, the greatest progress has been made in understanding aerial photographs in which the objects that are automatically extracted are buildings, airport structures, and roads (McKeown, 1984; Nicolin and Gabler, 1987; Matsuyama, 1987). Specialized software is required. For example, Guindon (1997) has described three different specialized computer software systems in which work has focused on constructing explicit definitions of roads, airports, and built-up areas for automated interpretation of high spatial detail satellite and aerial imagery. In natural resources applications, some progress has been reported in recognizing surface mineral properties from airborne and space-borne imaging spectrometer data (Chiou, 1985), and in Landsat-type satellite imagery in automatically extracting ridges and streams (Wang et al., 1983), forest stands, and toposequences. Actually, very little progress in image understanding applications in forestry has been reported. But one promising application of image understanding procedures has focused on extraction of geomorphological or hydrological objects from digital elevation models and satellite imagery: objects such as drainage networks (Wang et al., 1983; Smith et al., 1990), land components (Dymond et al., 1995), and valley features (Tribe, 1992).

It is worthwhile examining this application in more detail; while much simpler than spatial reasoning to create forest stands from spectral response patterns, for example, the complexity of even this simple geomorphic data extraction from DEMs can be illustrative of the direction of the approach. A second example is provided of another difficult spatial reasoning task: the recognition and extraction of individual tree crowns in very high spatial detail imagery.

Drainage landscape features are visible as patterns on raw or classified satellite imagery, and rules to describe their behavior on the landscape are relatively simple; for example, drainage accumulates downslope. While seemingly simple, almost trivial, such reasoning applied to regional or even local landscapes can become exceedingly complex, and can overwhelm even the most powerful computers available. Noise, data quantization, and the grid spacing in the sampling of topography in the DEM can create ambiguities, including artificial pits and ridges; in narrow gorges, streams can appear to cross ridgelines; and in smooth terrain in the presence of lakes, features such as streams can lose their coherence (Qian et al., 1990). Obviously, methods of extracting information from DEMs will only be accurate to the degree that the data are accurate (Blaszcynski, 1997). When extracted drainage networks are examined, it might become apparent that streams are broken, merge with others, or flow in two or more directions. No human interpreter would make such errors, but building the high-level computer programs to handle this type of spatial reasoning has not been easy. Here again, a new application of evidential reasoning (ER) is found to provide inference in the case of uncertain information (Qian et al., 1990).

The hillslope profile provides the fundamental unit of analysis for the study of geomorphic processes; a slope unit is defined as a section of the profile having relatively homogeneous form, process, and lithology, with upper and lower boundaries located at breaks of slope (McDermid and Franklin, 1995; Giles and Franklin, 1998). The significance of breaks of slope in the quantitative analysis of landforms has long been recognized (Scheidegger, 1986). The objects which are input to the geomorphological analysis of the DEM, then, are comprised of these slope units. A rule-based system encoding this knowledge (for example, drainage accumulation downslope) can be used to guide and enhance the extraction of geomorphic information from DEMs (Leighty, 1987) and, in time, spatial image data sets. This approach underlies the use of topography in partitioning the predictions of ecosystem process models to landscape units and applied to large landscapes, as described in Chapter 5.

In high spatial detail multispectral imagery, the objects of interest could be individual tree crowns and inter-tree crown features such as understory and shadows. Such objects must be separated from the background and other objects, perhaps using a classical classification rule such as maximum likelihood, mathematical morphology, or a segmentation routine. It is well known that a single image contains information at different scales or frequencies (Ahearn, 1988), but that digital methods to extract these different types of information are lacking. One idea is to develop a

TABLE 4.5
A Generic Catalogue of Spatial Discriminators to Recognize Tree Crowns in Digital Imagery

Tree Crown Visual Descriptor	
Crown Outline	**Crown Radiometric Profiles**
Contamination	Contamination
Translucence	Longitudinal form
Size	Longitudinal symmetry
Contour shape	Lateral form
Boundary elements	Lateral symmetry
	Contrast between pixels
Bright Areas	**Tree Shadows**
Contamination	Contamination
Number of regions	Description
General description	Translucence
Definition	Uniformity
Variability of intensities (max.)	Relative size vs. outline
Average size	
Shape	
Spatial arrangement	

Source: Modified from Fournier, R. A., G. E. Edwards, and N. R. Eldridge, *Can. J. Remote Sensing,* 21, 285, 1995. With permission.

catalogue of spatial discriminators that could be used together with the single-pixel information. Using high spatial detail (<40 cm per pixel) airborne multispectral imagery of forests, this type of visual catalogue was designed by Fournier et al. (1995). The interpretation process was broken down into actions on quantifiable individual features (Table 4.5), such as the crown outline, radiometric profiles, bright areas within the crown, and shadow regions. Crown outlines are usually part of any photointerpretation key for tree species recognition, and visual descriptors might include such terms as billowy, star-shaped, circular, symmetrical, rounded, and so on (Zsilinszky, 1964).

Digital counterparts to these visual descriptors for use in image understanding can be constructed using such terms as contour shape, size, boundary elements, and sharpness. For example, the bright areas embedded within individual crowns provide information on the separability of the individual tree crown feature from the background (Dralle and Rudemo, 1997) and may also contribute to the identification of the tree species. The occurrence of a single point of maximum radiance often results from a uniform foliage distribution and a smooth crown perimeter; Wulder et al. (2000) demonstrated a critical relationship between crown size and pixel size for this maxima filter. The algorithm generated image-based tree stem counts similar to field estimates. At a spatial image resolution of 1 m, a tree crown radius of 1.5 m appeared to be the minimum size for reliable identification of tree locations. The approach is not yet able to provide high levels of accuracy for mixed forests.

In another example, Gerylo et al. (1998) used the single bright pixel or maxima feature to isolate individual white spruce (*Picea glauca*) and lodgepole pine (*Pinus contorta*) crowns on 25-cm multispectral video imagery; the effectiveness of the single maxima search was reduced when hardwood (aspen — *Populus tremuloides*) crowns were examined, because these crowns were much less likely to have a single maxima pixel at this spatial resolution. Rules were devised based on the hypothesized size of aspen crowns in the area. In other areas, multiple maxima and crown fuzziness might suggest inherent crown structure irregularities; aspen, or jack pine (*Pinus banksianna*). Large branches could suggest white pine (*Pinus strobus*) (Fournier et al., 1995). The isolated objects are identified through knowledge resident in the combined human interpretation and machine recognition system; the ideal approach would see a rule-base which could be invoked whereby the shape or association of an object with other objects can be used to resolve ambiguous identification. In the case of the digital tree crowns (Fournier et al., 1995), white spruce could be characterized as single and compact, red pine (*Pinus resinosa*) as multiple and compact, and jack pine as multiple and extended. If individual crowns can be extracted from imagery, then such an approach would work very well.

Such descriptions accord well with what is known to be important in the photointerpretation of these species. Much effort has been made to optimize the knowledge acquisition process. Sorting and quantifying the knowledge used by tree species photointerpreters has not been a simple task; the end of this process is not yet in sight. The early problem, not yet overcome, was noticed by Wang et al. (1983: p. 94):

> *Because humans have a great innate facility for extracting information from visual shapes, forms, and textures, photointerpreters often do not devote much conscious thought to their analyses of the detailed relationships between light and dark that convey information about the content of a visual scene. The reasoning process may be implemented through a series of implicit steps that are not immediately obvious even to those who conduct the interpretation.*

In forest applications the catalogue approach to digital tree species identification with spatial discriminators has been augmented by a host of rule-based systems and proof-of-concept studies in single-tree isolation, regeneration surveys and forest health assessment, stand structure, crown closure, gaps, and stand volume mapping (Hill and Leckie, 1999). But what developments are ready for use now? One of the more promising directions has been to facilitate the integration of many different aspects of remote sensing research with high spatial detail imagery (Gong et al., 1999).

1. Three-dimensional photogrammetric analysis
2. Automated image understanding based on aerial photointerpretation rules
3. Hyperspectral image classification, texture, and contextural techniques

This approach has been introduced as photo-ecometrics, a subfield of a new interdisciplinary endeavor defined as the science and technology of obtaining reliable ecological measurements over large landscapes. Photo-ecometrics suggests a new level of automation in forest information extraction from remote sensing may be just over the horizon based on the integration of advances in the different fields of measurement; photogrammetry, photointerpretation, field spectroscopy, fisheye photography, and digital remote sensing applied to hyperspectral and multispectral data (Gong et al., 1999). By combining spatial morphological information from very high spatial detail stereo-imagery with traditional multispectral, textural, and contextural image processing it may be possible to overcome some of the problems that each approach (i.e., photogrammetric, image processing, image modeling) suffers in isolation. Examples of interest to foresters that may be possible by integrating these different approachs have been suggested in automated tree species recognition, tree crown delineation, closure estimates, and height measurements.

Few of these image understanding tools — including systems for spatial reasoning, basic image understanding, tree crown delination, stem-counting, or photoecometrics — exist in the commercially available functionality of image processing systems. In some cases, approaches and software tools have been under development for 15 or more years (McKeown, 1984; Gougeon, 1995). Are these applications really a form of image understanding? The primitive nature of the spatial operators, compared to human capability, suggests that rule-based image understanding for tree species recognition, let alone for stand recognition, is far from becoming reality for most remote sensing image analysts. It is expected that this situation may change. Methods of knowledge engineering continue to improve and methods, data, and applications may soon begin to converge.

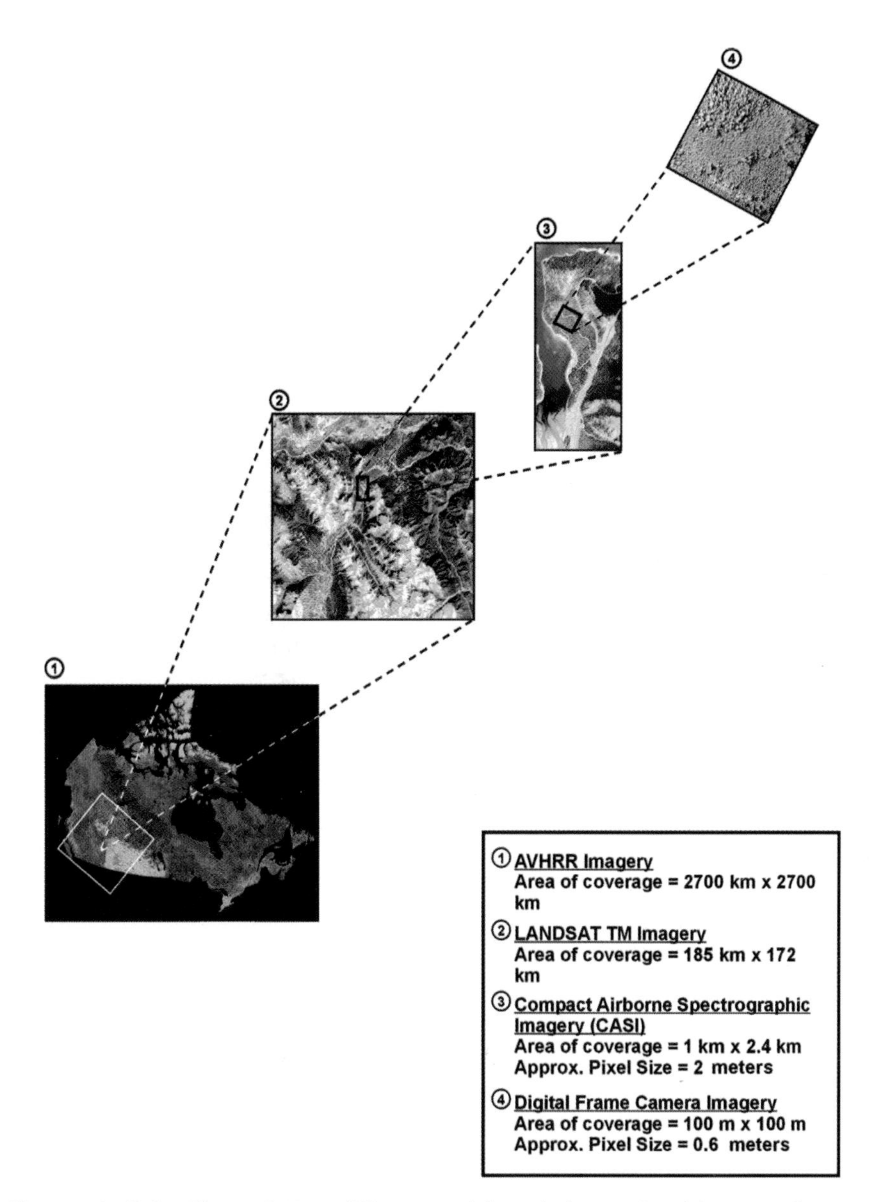

Chapter 3, Color Figure 1 Four different spatial resolutions and aerial extent of remote sensing imagery suitable for observation of forest processes operating over different scales. (1) Low spatial resolution imagery for mapping of forest cover by broad community type (coniferous, deciduous, mixed wood); abiotic/biotic characteristics; Level I physiographic and climatic classifications; AVHRR map of Canada (From Cihlar et al., 1999). (2) Medium spatial resolution imagery for mapping forest stand, patch, or ecosystem level characteristics and dynamics; tree species; crown diameters; tree density; the number of stems; stand-level LAI; Level II forest cover type and vegetation type classification. (3 and 4) High spatial resolution imagery for mapping individual trees and other discrete ground objects (understory assemblages); forest structure; forest cover (crown diameters, closure); LAI; understory composition or rare species detection.

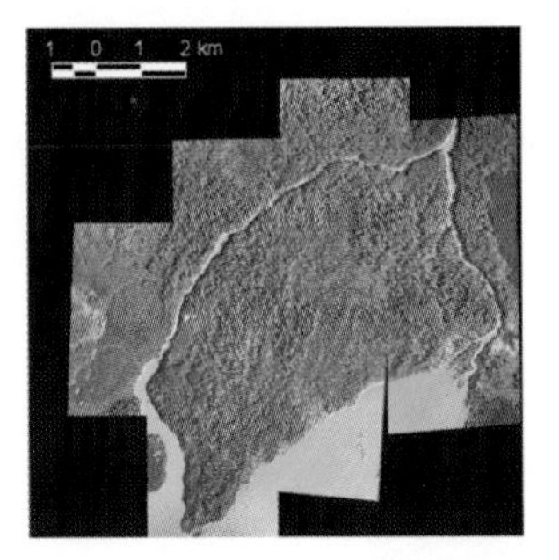

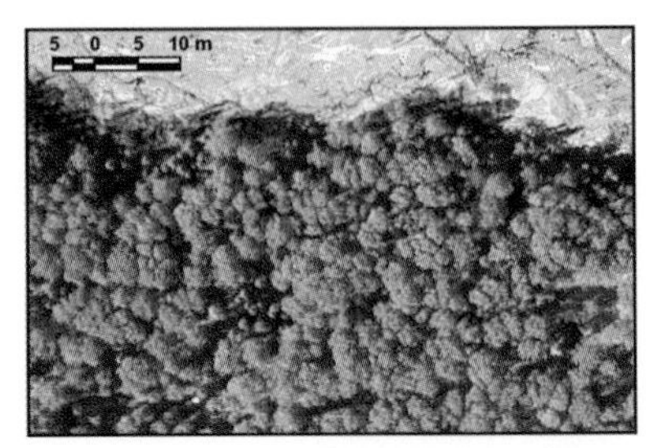

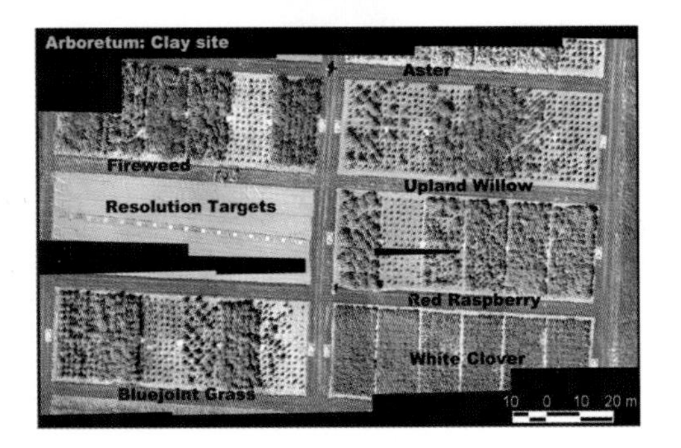

Chapter 3, Color Figure 2 Examples of airborne imagery collected in support of forest mapping and detection work. Clockwise from top left: mosaic of 50 cm pixel, multispectral, digital frame camera; imagery (10 nm bands at 550, 670 and 900 nm) of a damaged forest at a mine site in northern Ontario; 25 cm Kodak DCS420CIR imagery of individual aspen tree crowns at the same mine site; 57 cm Kodak DCS460CIR image of a temperate hardwood forest damaged by a severe ice storm in Quebec; two 57 cm Kodak DCS460CIR images of managed forests in eastern Ontario damaged by the same ice storm; mosaic of 25 cm leaf-on Kodak DCS420CIR imagery of 2-year-old regeneration at the Sault Ste. Marie Arboretum; (All imagery courtesy of Dr. Doug King, Carleton University.)

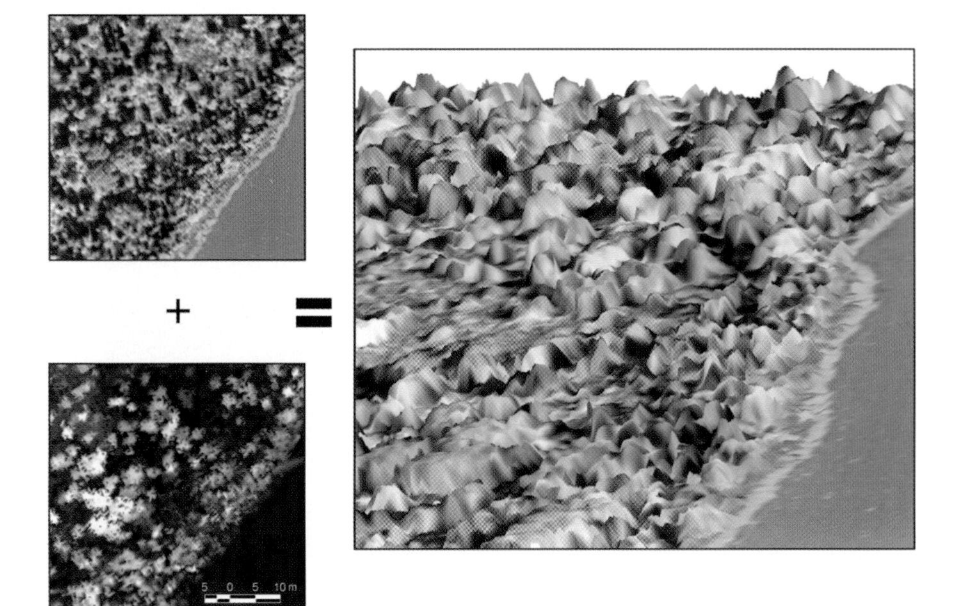

Chapter 3, Color Figure 3 Multispectral image (top) with scanning lidar (height) data (bottom) combined to produce a digital canopy map with spectral and height properties enhanced. (Lidar data acquired by LaserMap Image Plus and courtesy of Dr. Benoît St-Onge, Université du Québec à Montréal.)

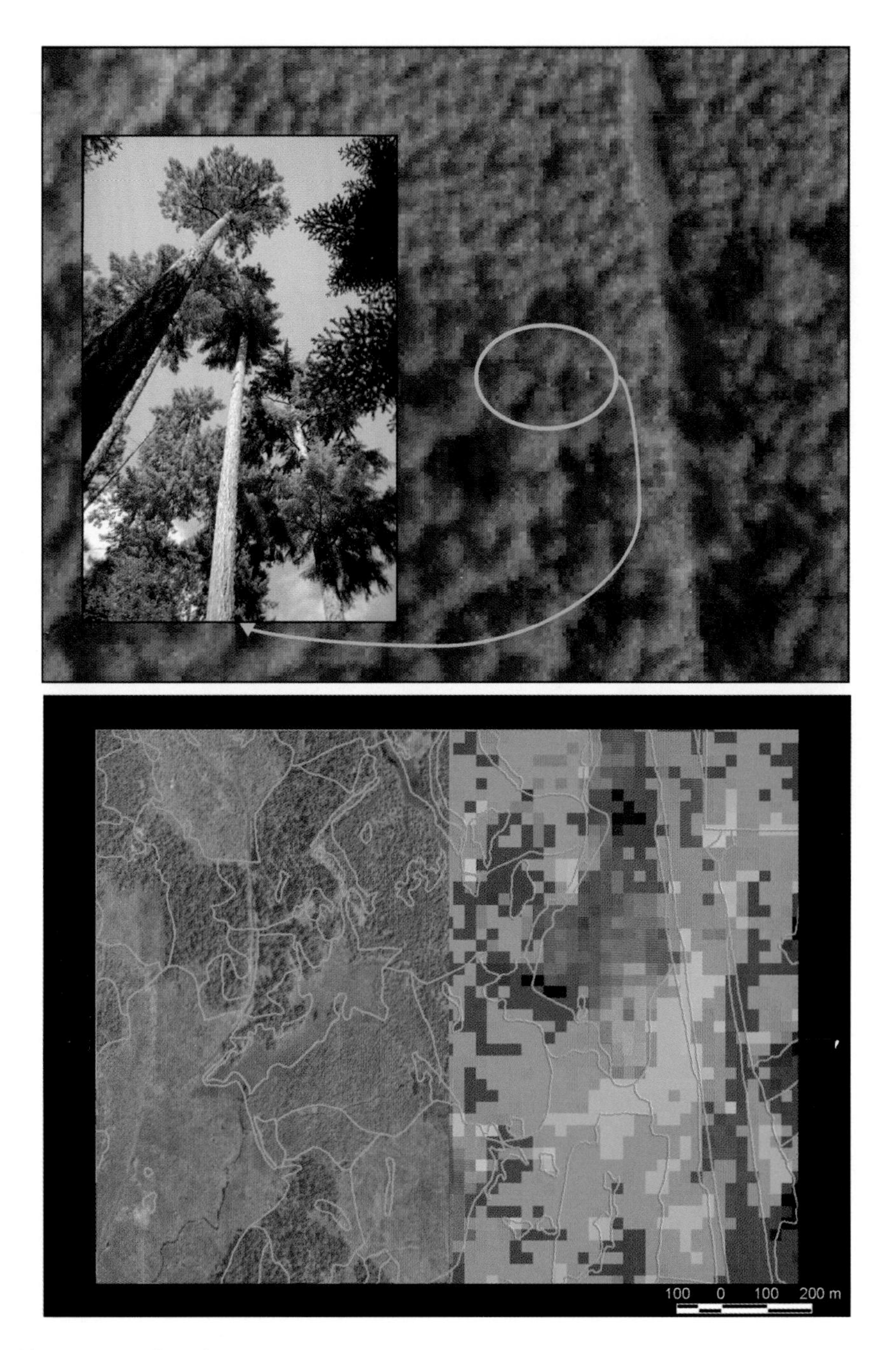

Chapter 3, Color Figure 4 IKONOS-2 panchromatic imagery of Douglas-fir forest stands in British Columbia. Individual crowns of large Douglas-fir trees can be distinguished (top); forest stand polygons originally interpreted in aerial photographs overlaid on the IKONOS and TM data (bottom). (Example courtesy of Dr. M. A. Wulder, Canadian Forest Service, Natural Resources Canada. IKONOS-2 © Space Imaging.)

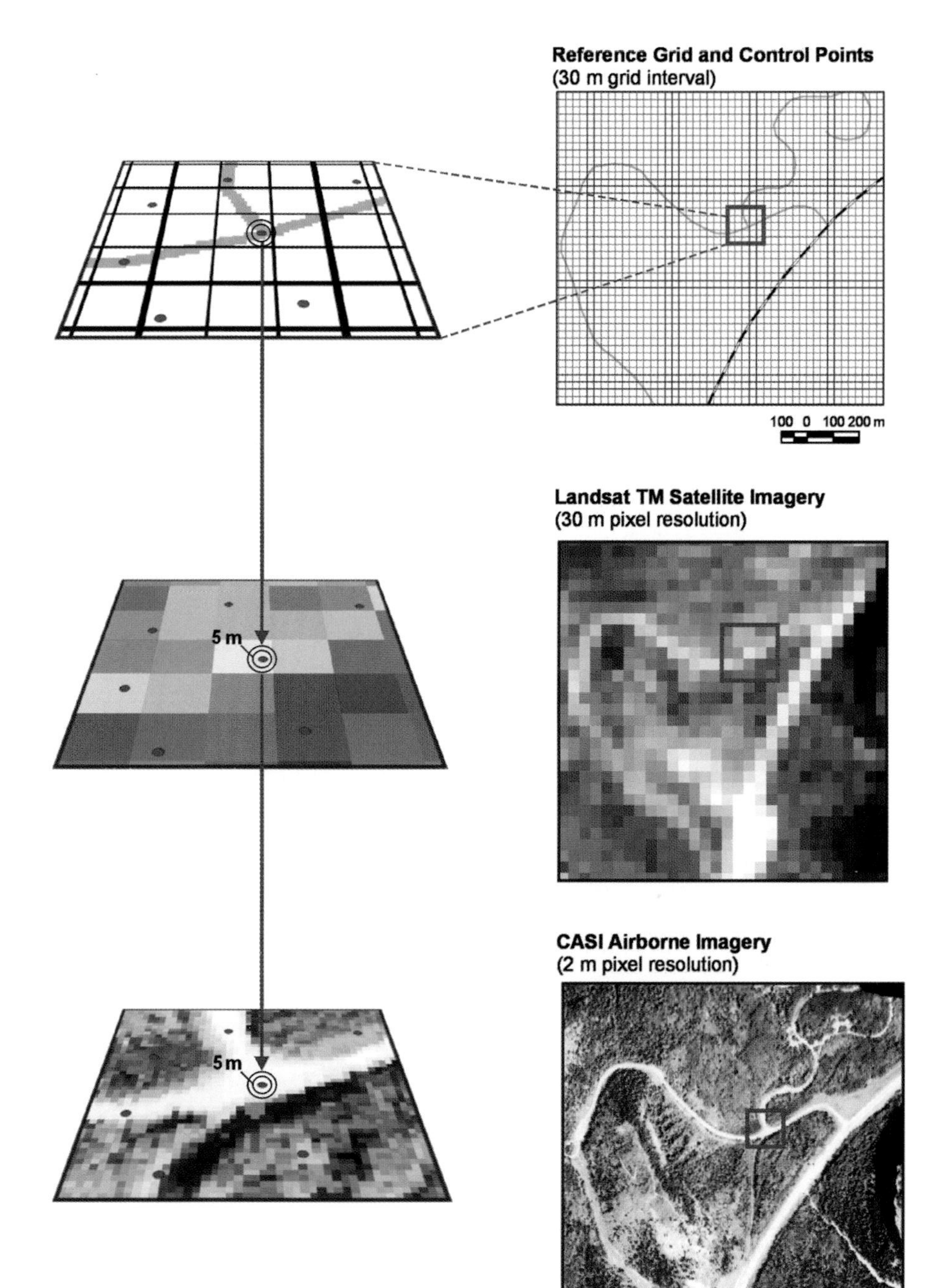

Chapter 4, Color Figure 1 Connection of image data to map data and absolute coordinates. First, ground control points must be identified; typically, each control point is located using Global Positioning System technology on the ground, but with some degree of error; subsequent identification of the image pixel is used to build a transformation between the map (top), satellite image (middle), or airborne image (bottom) coordinate systems.

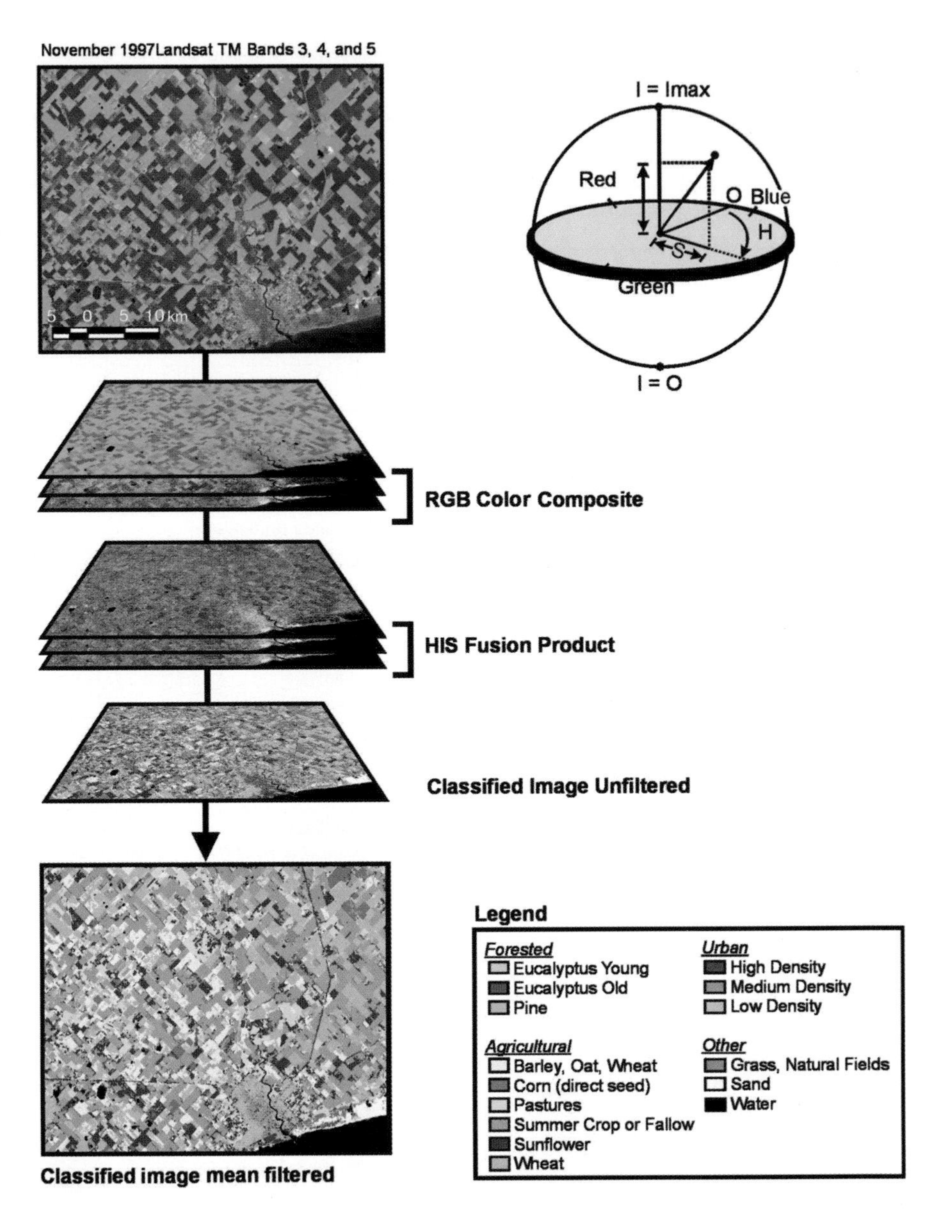

Chapter 4, Color Figure 2 Example of satellite data fusion using Radarsat and Landsat TM data of southeastern Buenos Aires, Argentina, results in a new display of image data, an integration of the spectral properties in the microwave and optical/infrared parts of the spectrum, and higher classification accuracy for certain classes. The HIS color coordinate system (top right) was used. (From Daily. 1983. ASPRS. With permission. Modified from Presutti et al., 2000. Radarsat imagery processed by RSI, © CSA 1997.)

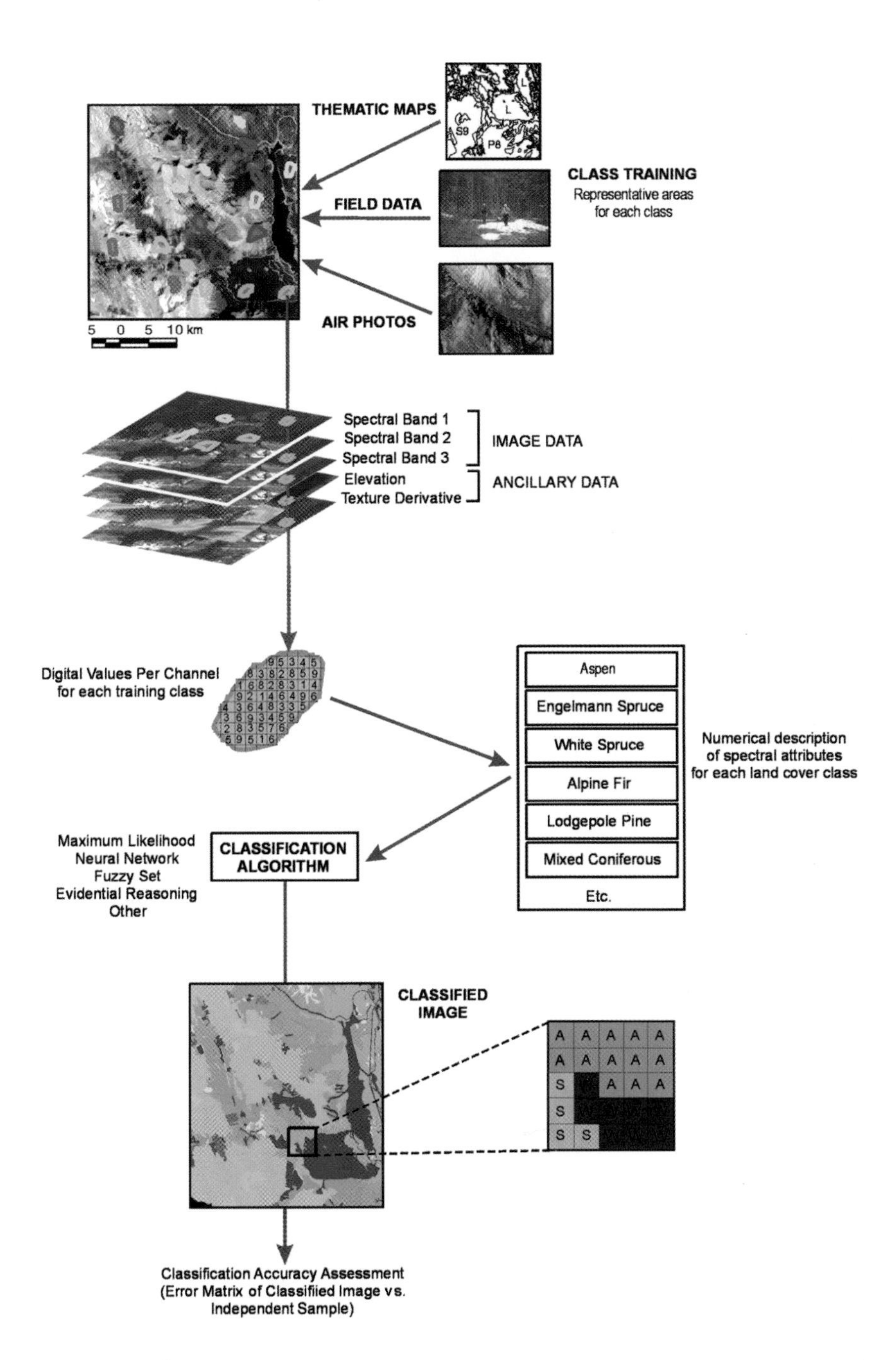

Chapter 4, Color Figure 3 Basic steps in an image classification procedure applied to satellite imagery. Existing maps, aerial photos, and field data are used to identify training areas (top); selection of image data and ancillary data such as DEM and texture results in a training signature for each of the desired classes (middle); choice and application to entire image of decision-rule results in classification map that must be assessed for accuracy (bottom).

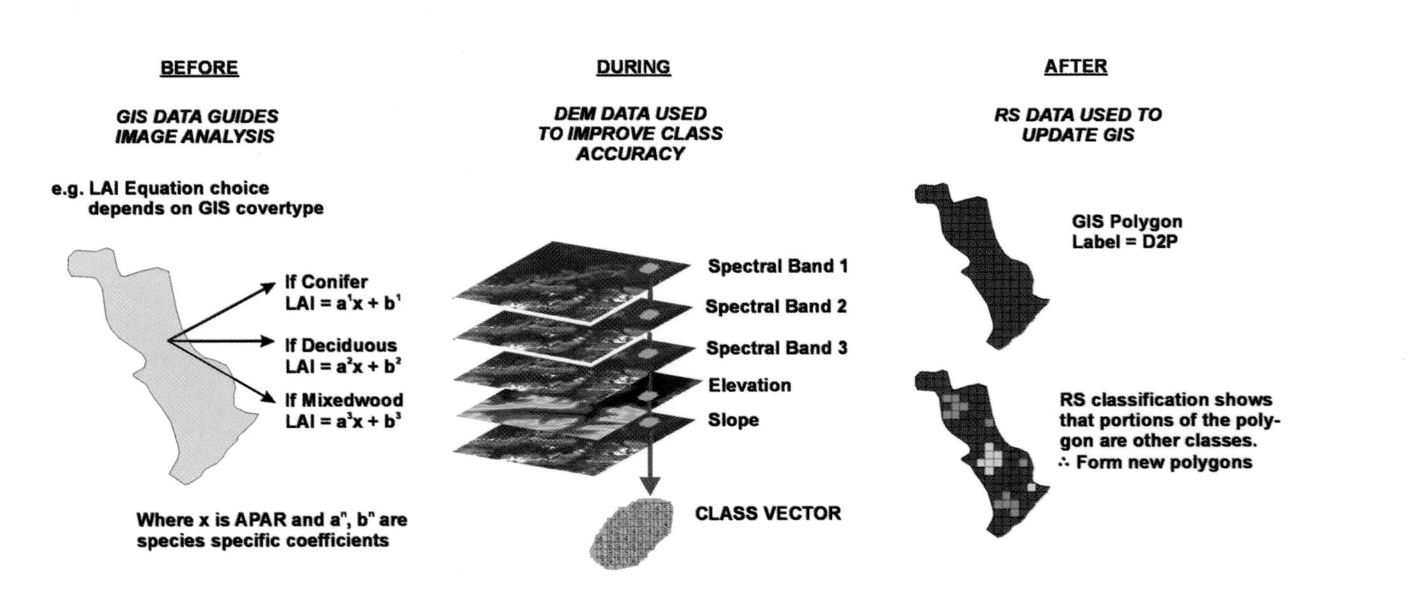

Chapter 4, Color Figure 4 Integration of remote sensing data into GIS can occur before, during and after image processing designed to extract information from the imagery. Before: information on stand type within a GIS polygon database is used to select the appropriate equation to convert remotely sensed APAR to LAI (species-specific). During: a DEM is used to augment a decision-rule to increase the likelihood of correct per-pixel classification. After: a GIS polygon forest stand (labeled D2P or D-class density, age class 2 pine) is updated with new information on variability within that stand.

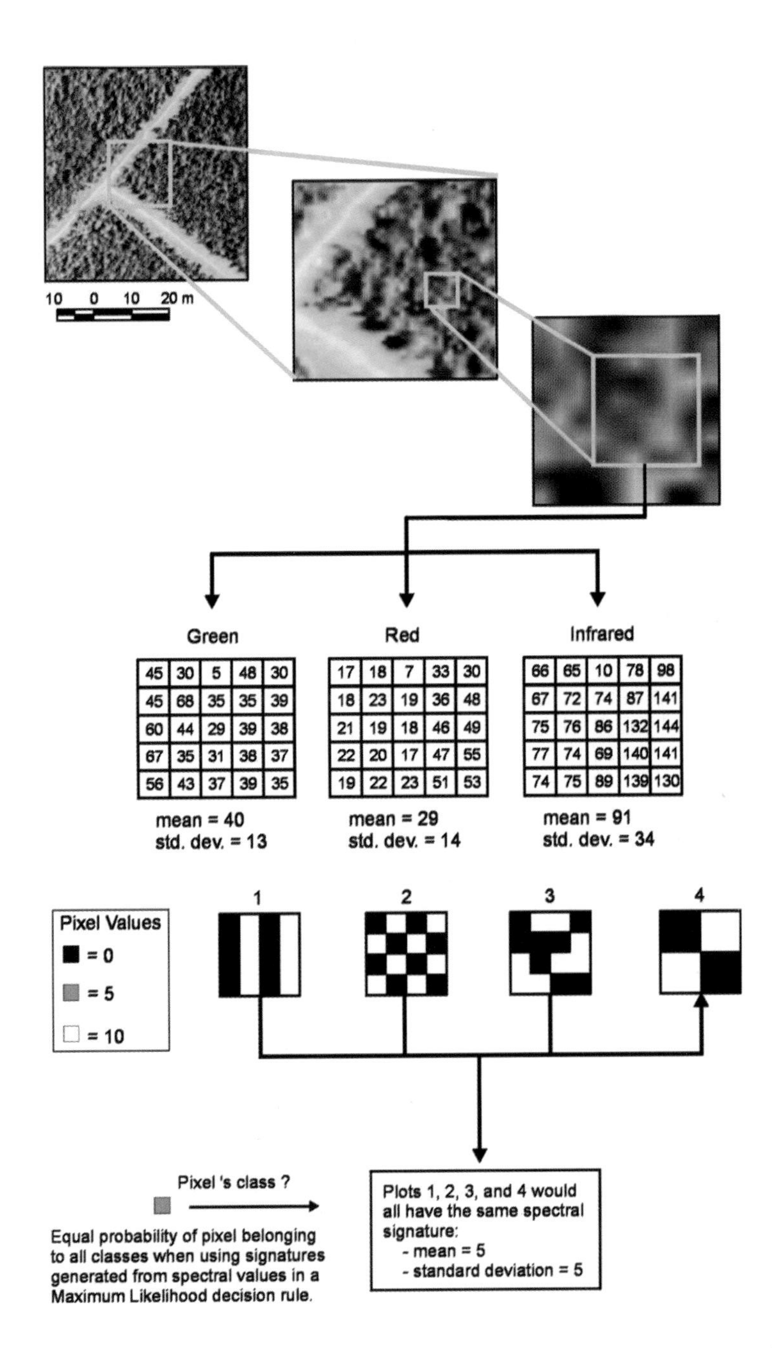

Chapter 4, Color Figure 5 Generation of image signatures leads to the use of spatial variability or image texture analysis (in high spatial detail CASI imagery, the image mean and standard deviation values do not accurately reflect stand characteristics) (top); simplified pixel windows emphasize this key problem (bottom). Pixel windows 2 and 3 are highly textured; pixel windows 1 and 4 are patterned. Image spatial co-occurrence can be used to differentiate these windows and the real image data at different scales.

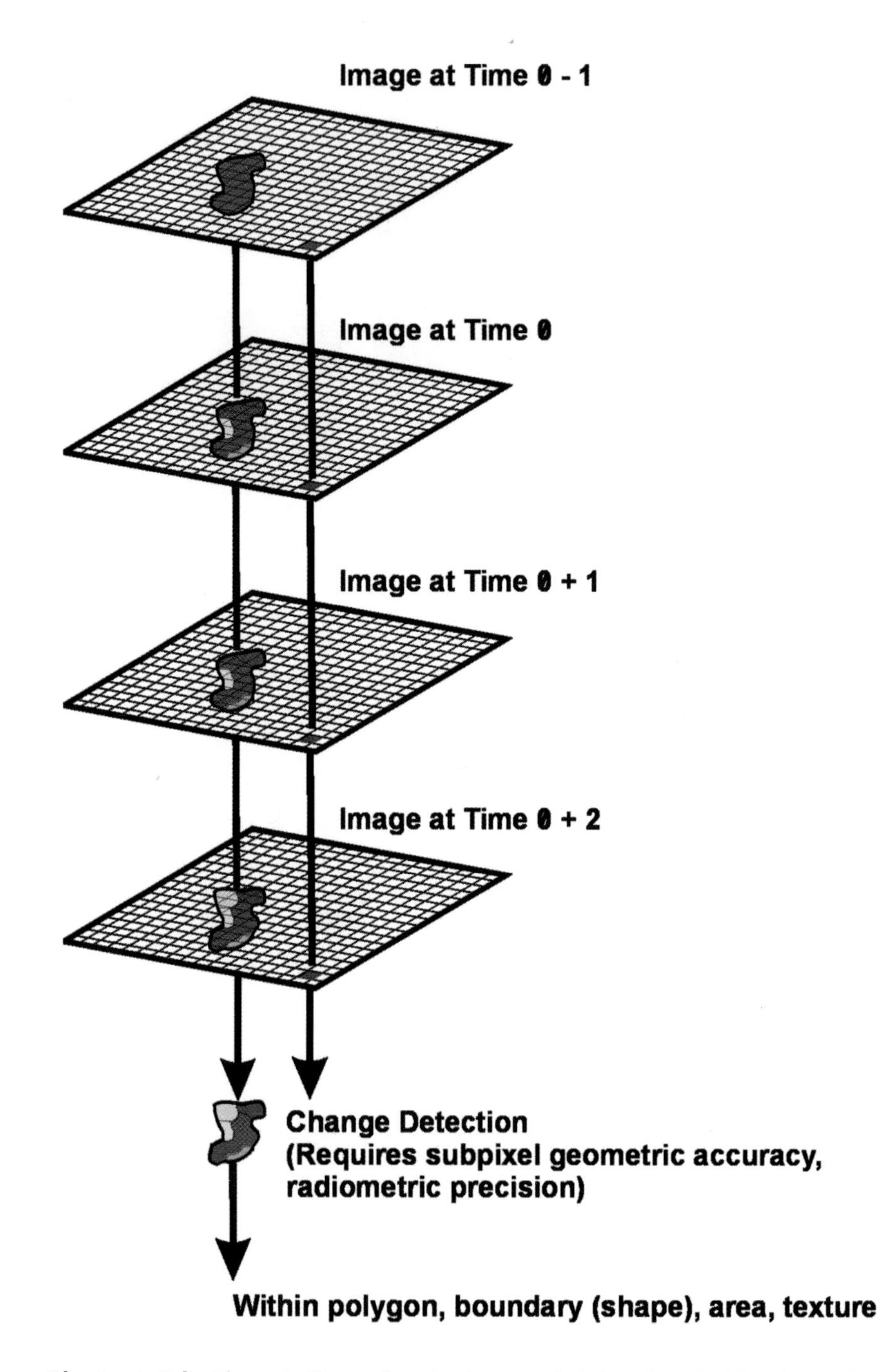

Chapter 4, Color Figure 6 Change detection image analysis based on pixel-by-pixel analysis or area-based operations within existing forest structures. Key decision points are the type of pre-processing required (georadiometric factors) and the choice of change detection procedure that can detect and identify changes of interest in forest conditions.

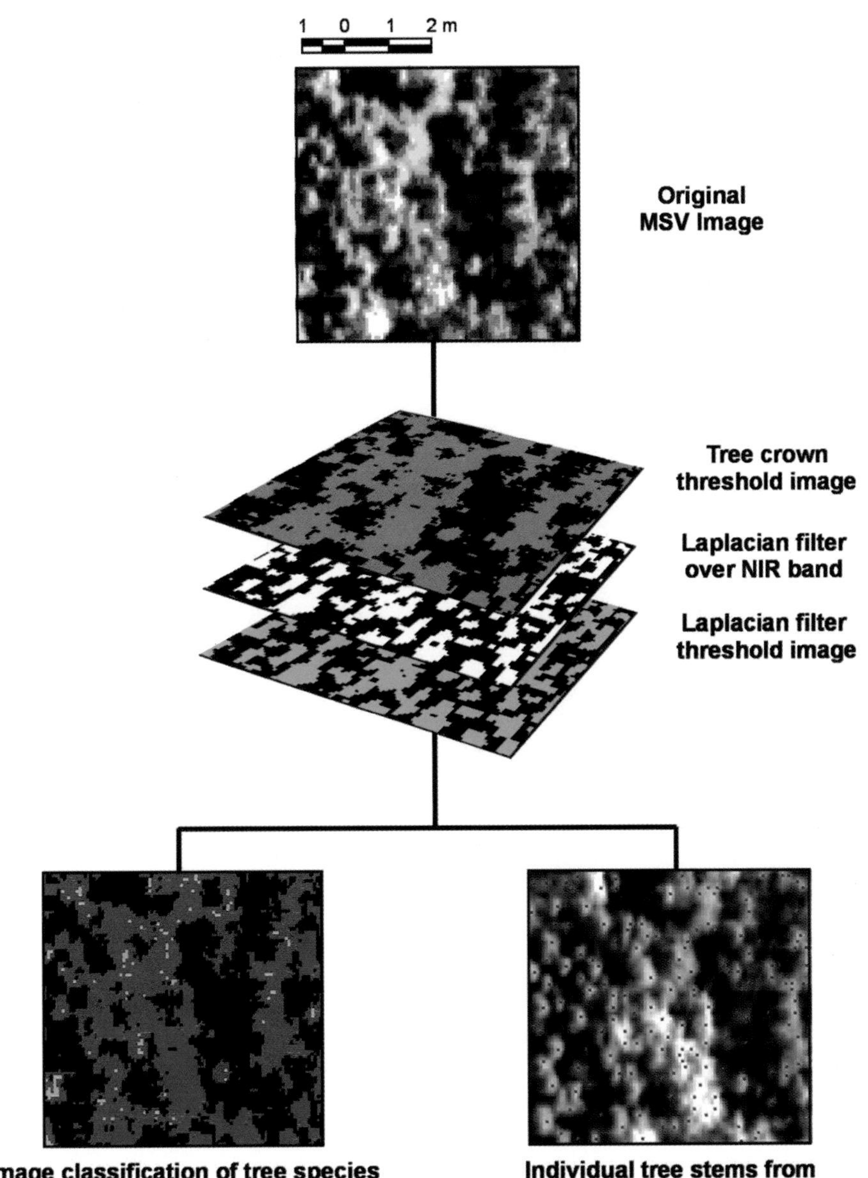

Chapter 6, Color Figure 1 Image classification of tree species, tree crown threshold image, and individual tree stems in high spatial detail videographic imagery (25 cm pixels). By identifying individual tree crowns, crown areas, and species, a simple maxima filter can be used to count stems and build basal area and volume estimates. (Modified from Gerylo et al., 1998.)

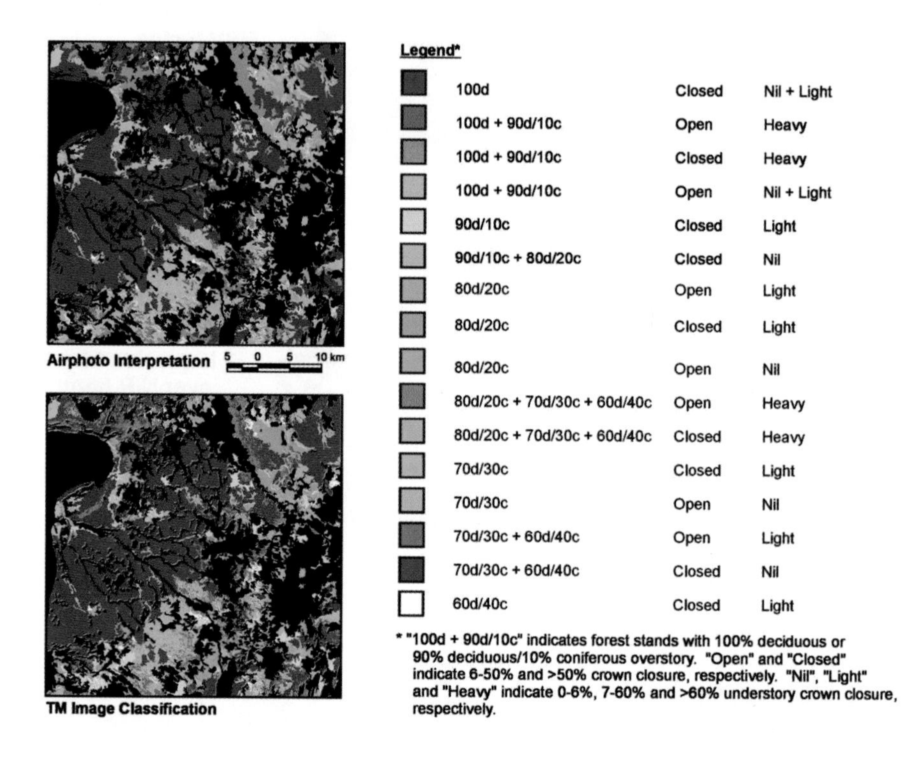

Chapter 6, Color Figure 2 Comparison of understory maps produced from aerial photointerpretation of 1:10,000 scale color infrared photography acquired during early spring (leaf-off conditions) and a leaf-on/leaf-off multitemporal Landsat TM classification (From Hall et al., 2000a. With permission. Example courtesy of Dr. R. J. Hall, Canadian Forest Service, Natural Resources Canada.)

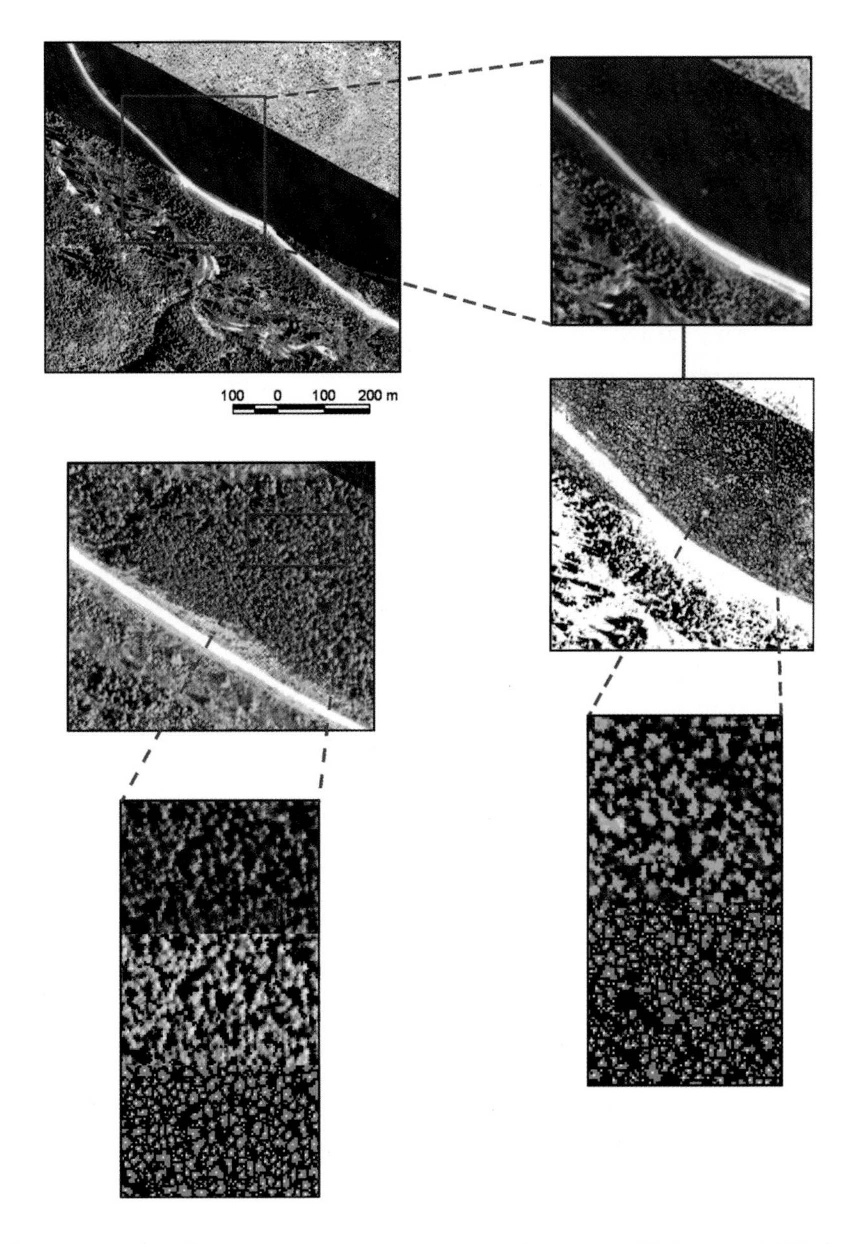

Chapter 7, Color Figure 1 Use of Compact Airborne Spectrographic Imager (CASI) data in identification of species and stem counting in support of forest inventory in British Columbia conifer forests. The large image contained cloud shadows and brightly lit areas; enhancements suitable for each were designed (top). Stem segment maps outlining individual tree crowns, a conifer mask, and an understory mask were applied (middle). Individual tree stems were identified based on rules for crown size, maximum brightness of tree crown pixels, and spectral reflectance information (bottom). Note identification of crowns (white points), individual trees (red points) and singletons (yellow points) in detailed imagery. (Example courtesy of Dr. Doug Davison, Itres Research Ltd., Calgary, Alberta.)

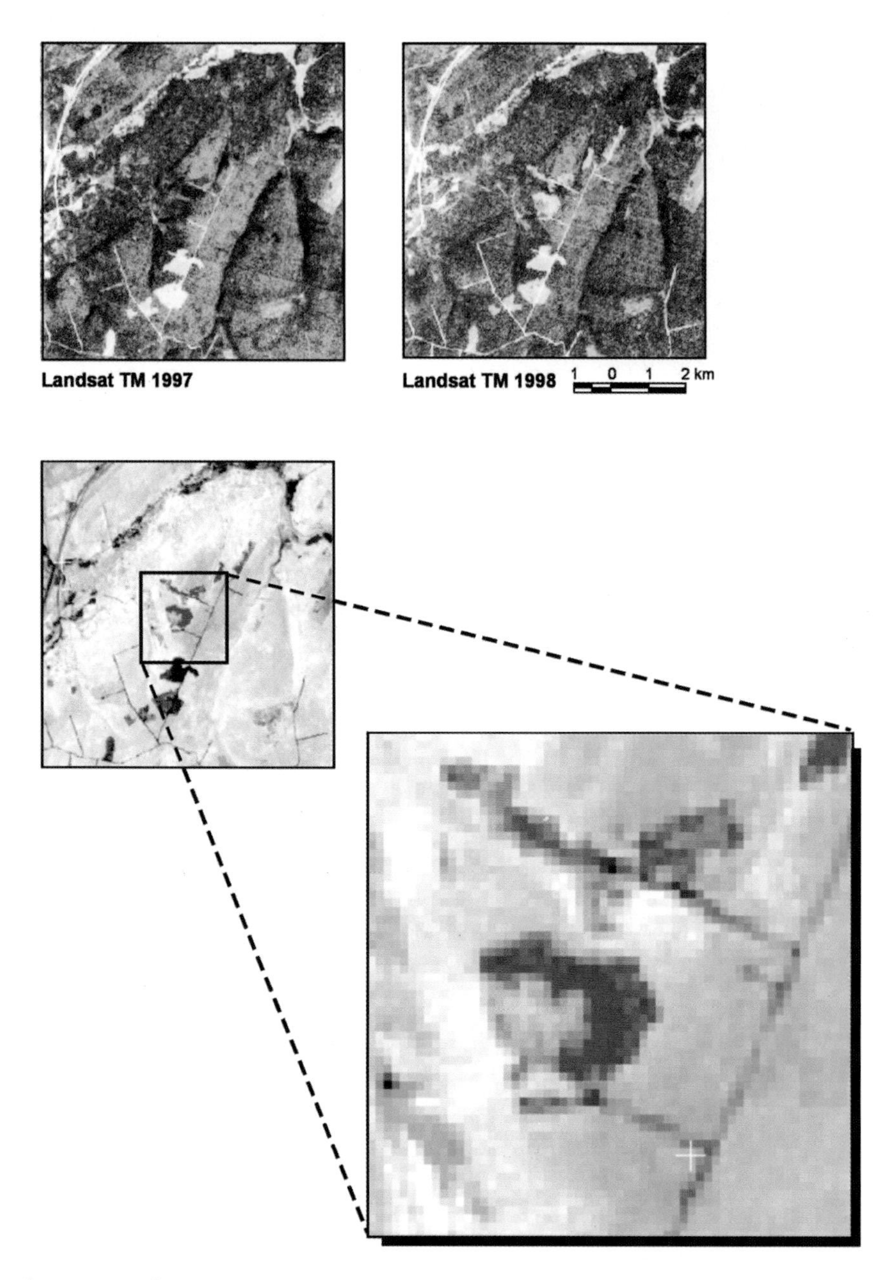

Chapter 8, Color Figure 1 Annual change detection of partial harvesting in a red pine stand in New Brunswick using the enhanced Landsat TM wetness difference image. In the original imagery (top), areas that were harvested appear brighter in the 1998 image. Harvesting results in an increase in brightness, a decrease in greenness, and a decrease in wetness, which was enhanced after simple image subtraction. In the wetness index, areas that are bright red were clear-cut; pink tones represent areas in which the wetness index was reduced proportionally to the reduction in canopy leaf area (commercial thinning operation, about 30% of the basal area removed). (TM image processed by RSI, © CSA 1997, 1998.)

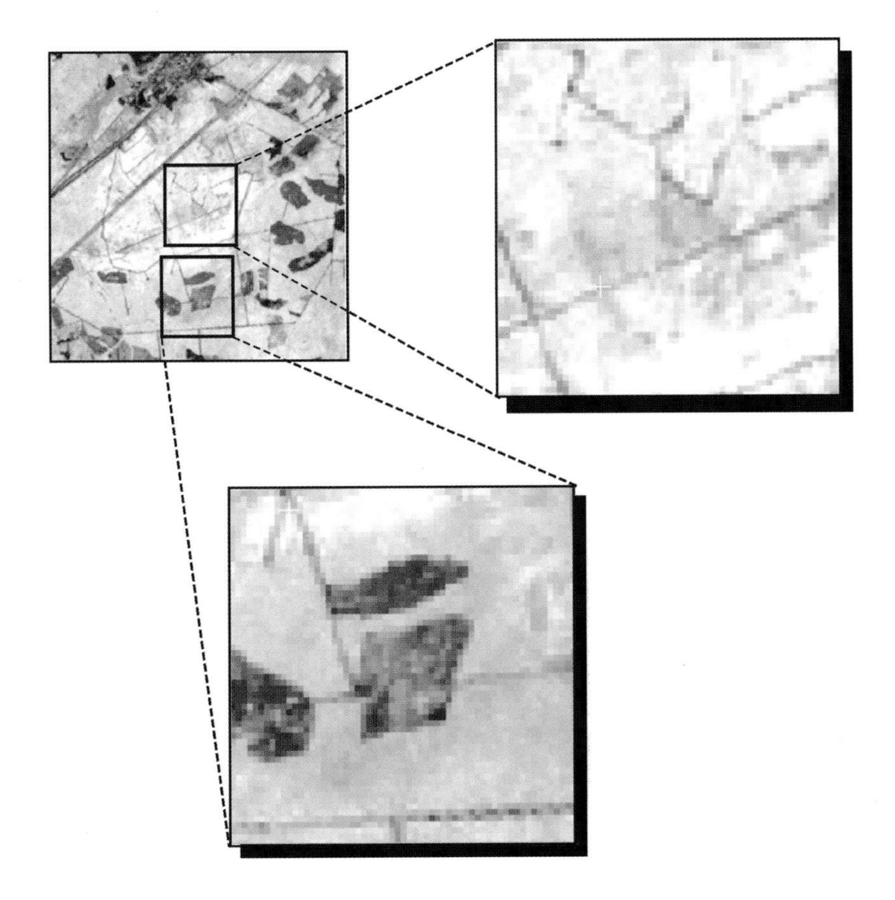

Chapter 8, Color Figure 2 Annual change detection of harvesting and silviculture treatments in conifer stands in New Brunswick using the enhanced Landsat TM wetness difference image. Clearcuts appear bright red, seed tree cuttings are less red, with more white and pink pixels; wetness in the thinned plantation changed only a small amount, resulting in a light pink tone. (TM image processed by RSI, © CSA 1997, 1998.)

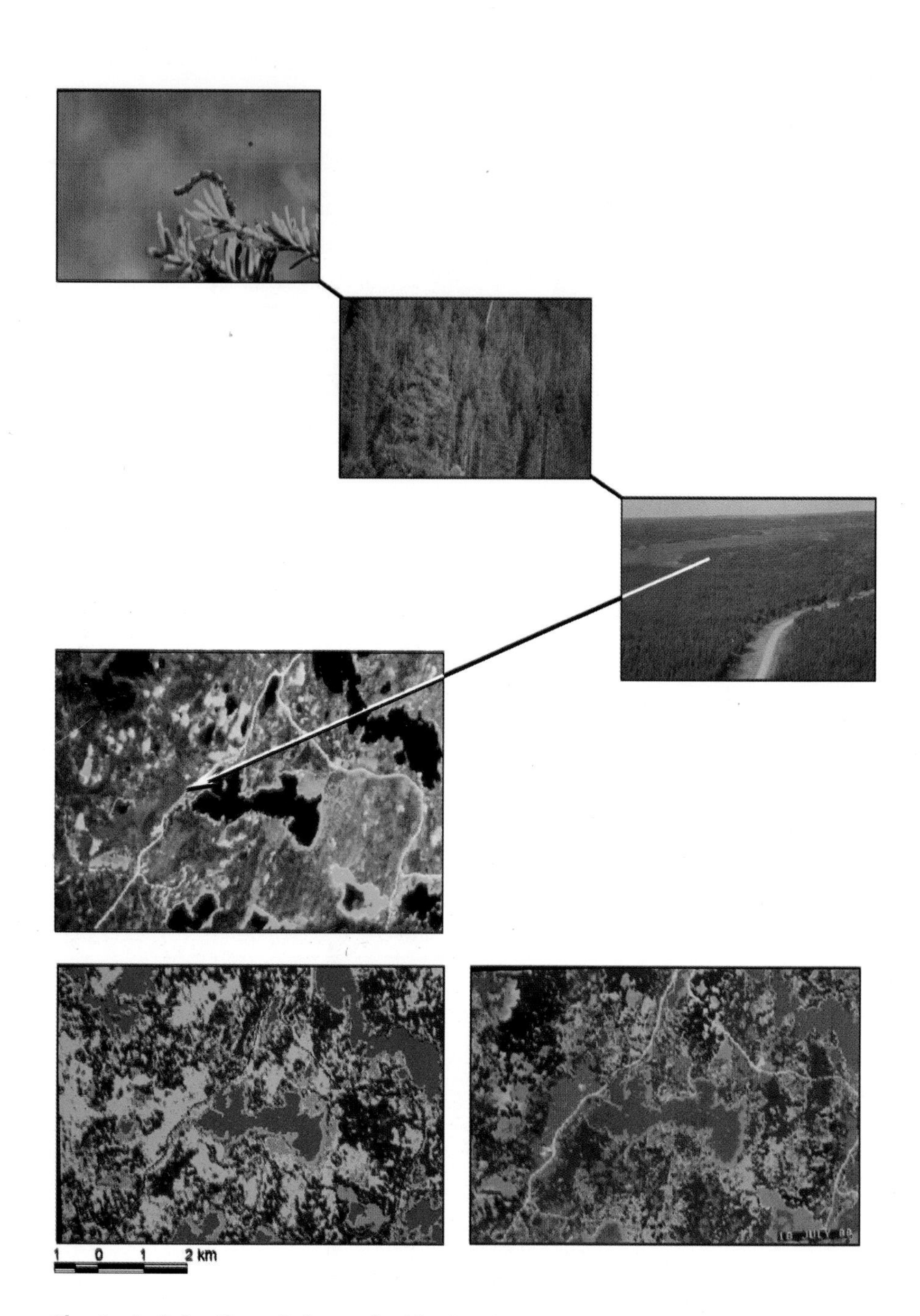

Chapter 8, Color Figure 3 Image classification change detection using SPOT HRV imagery in Newfoundland during two years of hemlock looper defoliation. The insect feeds on balsam fir foliage causing necrosis and a red stage that can be clearly separated into light, moderate, and severe defoliation classes. As the damage accumulated, mortality was classified in imagery acquired the following growing season. (SPOT image © CNES 1987, 1988.)

5 Forest Modeling and GIS

... of the future developments in the handling of remote sensing data, none is likely to be more important than their integration with other data sources, to produce a comprehensive geographic information system.

— J. R. G. Townshend, 1981

GEOGRAPHICAL INFORMATION SCIENCE

Geographical information systems (GIS) are computer-based systems that are used to store and manipulate geographic information (Aronoff, 1989). Like remote sensing, GIS have emerged as a fully functional support for resource management following a series of intensive, synergistic, technologically driven activities over the last four decades. Developments have been built on the strengths of successive revolutions in computer technology and geography. GIS have their modern origins in the 1960s and 1970s, but conceptually can be traced much farther back to the earliest requirements to assess land capability using multiple criteria, and the need to perform map overlays. The potential contribution of GIS to sustainable forest management appears enormous; here is the ideal tool with which forest management issues can be addressed — simply, the relevant tasks are

1. To assemble a spatially referenced database across all relevant scales, and then
2. Put multiple analytical tools in the hands of the users so that the accumulated information can be made to provide answers that are needed.

The simplicity of these statements and the general, casual, attitude toward geographical information and mapping sometimes found in forestry, are deceptive; GIS is no simple process! A great deal of complexity has become subsumed under the GIS label (Longley et al., 1999). GIS, like remote sensing, appears ill-defined and very broadly based. It is comprised of geographic objects (polygons, lines, points) and their attributes, with or even without reference to spatial components and complicated topology. Currently, it is defined more by what is done under the banner of GIS than by any coherent definition of the field. GIS in forestry tends to be comprised of two major endeavors:

1. Geographic data management, including data collection, database development, and archiving, and
2. Geographic data analysis, including modeling and information extraction.

In natural resources management, the time and effort devoted to the first task, geographical information management, is enormous. For those businesses and governments with substantial lands to manage, managing the vast array of spatially referenced information on those lands has emerged as an onerous responsibility, and can consume vast amounts of human and capital resources (Green, 1999). In recent years, many of the significant problems in this activity have been resolved — for example, database development, storage, output, and processing speed bottlenecks. Now, a trend to increasing emphasis on the latter set of tasks — that of geographic data analysis — is becoming apparent in the GIS research and applications literature.

A prognosis on the final form of GIS and its contributions to sustainable forest management is premature and, because of the many known and unknown factors influencing the development and applications of GIS, would likely be unconvincing. The evolution of GIS is not yet complete (Longley et al., 1999). Instead, it is instructive to consider that during the last 10 years, a transition has taken place in GIS related to fundamental issues of geographic information, methods, and practical implementation of GIS in applications. The original concepts and tools of geographical information systems continue to develop into a geographical information science (GIScience) (Goodchild, 1992). Comprised of concerns with the technical and scientific issues surrounding the use of geographical data in natural science and social science applications, GIScience appears well on the way to acceptance as a separate field with a unique focus and research agenda (Goodchild and Proctor, 1997). Practically speaking, GIScience appears to be rapidly replacing a GIS technological agenda with a mapping/functional analysis agenda. In the future, there will be increasing emphasis on using GIScience to satisfy user needs (Albrecht, 1998; Gibson, 1999) as the technological problems which have preoccupied GIS developers appear to be in recession — solved, for the most part, or at least understood.

A new GIScience mandate: providing the scientific basis for increased use of the new tool of GIS in real-world applications. In forestry, GIScience geographic data analysis is already making a substantive contribution to sustainable forest management in at least three ways:

1. Integration of multiple data sources, including remote sensing data,
2. Provision of input to models and the appropriate environment to run, validate, and generate model output, and
3. Mapping and database development.

The first two contributions focus on the role of remote sensing and models within the infrastructure provided by a forestry GIS (Landsberg and Coops, 1999). These two components are a critical development to facilitate flexible and innovative operational, tactical, and strategic forest management planning.

Remote Sensing and GIScience

Is remote sensing actually a part of GIScience? Uncertainty over whether remote sensing and GIS are actually different aspects of the same science has been common (Estes, 1985), but a growing consensus is emerging. The relationship between remote sensing and GIS is so strong that some have suggested that the potential contribution of each cannot be realized without continued, and finally, complete integration of the two endeavors (Ehlers et al., 1993; Estes and Star, 1997). There may be some resistance to this idea as GIS and remote sensing evolved at different rates, and tended to remain separate (Aronoff, 1989). Each field is serviced by separate journals and societies, but there are many common points of contact including meetings in which the other technology is heavily featured. Perhaps only a change in attitude or perspective is needed to further the goals of integration (Edwards, 1993). As Goodchild (1992: p. 35) has suggested, "Ultimately it matters little to which of the many pigeon holes we assign each topic … one person's remote sensing may well be another's geographical information science."

The reality today is that almost every usable remote sensing image and image product will reside and find application at some point in its lifetime in a GIS environment. Obviously, a key methodological focus in remote sensing has been the extraction of forestry information from imagery using tasks in the image processing system. Again stating the obvious, much of the information produced by the analysis of imagery is geographic information. Increasingly, that information must be managed, together with other forestry information, in the GIS. The image processing system can be seen as one part of a larger GIS; to users, this makes great sense, simplifying some of the data issues, and methodology within the technological approach (Landsberg and Gower, 1996; Treweek, 1999). In turn, the GIS can be seen as one part of the larger, emerging world of GIScience, encompassing all issues of spatial data analysis and mapping (Haines-Young et al., 1993; Atkinson and Tate, 1999; Longley et al., 1999). One task of the new GIScience paradigm is to enable smooth integration of all the assembled technologies in support of the disciplinary tasks set before it.

A quick glance at the literature of the past few decades reveals a symbiosis which can be seen to exist from the earliest, tentative first steps in remote sensing and GIS. An early concern was to use the GIS to manage the raw images as a spatial archive (Tomlinson, 1972). A suite of tools and techniques to provide image display and data exchange was built into most early GIS. Practically speaking, modern GIS contain the descendants of these tools, sometimes in the form of still more powerful tasks (such as the creation of polygons from image classification output, polygon decomposition, cleaning, and dissolve). GIS users and developers have long understood that much of the data required as input to their emerging systems would be obtained by remote sensing (Burroughs, 1986; Aronoff, 1989). Updating a GIS with remote sensing information continues to be an important and complex application area (Wulder, 1997; Smits and Annoni, 1999). It is now widely understood that GIS and remote sensing integration goes both ways.

In the late 1970s and early 1980s, for example, remote sensing scientists began to recognize that many image analysis tasks could be improved with access to other digital spatial data. These data — DEMs, soils maps, ecological land classifications, geophysical surfaces, and others — were increasingly held within a supporting GIS or relational database/computer cartography environment. Landgrebe (1978b) listed five key limits on the extraction of useful information from remote sensing data: the four types of image resolution (spectral, spatial, radiometric, and temporal), and the quality of ancillary data. On this level alone it seems likely that the dependency between the science and technology of remote sensing and the science and technology of geographical information will continue to strengthen. This strength will be based on the fact that rarely will the analysis of remote sensing or GIS data alone provide an advantage over the analysis of both together; one obvious exception exists in areas where the existing remote sensing or GIS data are unsuitable or untrustworthy for a given mapping application, perhaps derived through some now obviously deficient but previously acceptable methodology.

Using GIS data to generate or supplement training data for image classifiers is increasingly common, as are combinations of GIS and remote sensing data in a single classification process. The effect of using remote sensing data from different sensors, the effect of image spatial context, the effect of existing map data in remote sensing forest classification, are all more readily addressed within the GIS environment (Solberg, 1999).

Despite these developments, there still may be a strong tendency to consider GIS simply as a useful way to generate remote sensing output products — principally, forestry maps. No doubt a primary focus in remote sensing and GIS integration will continue to be maps and time-series of maps to support forest monitoring. Obviously, one of the primary ways in which forest managers access and present data is through the use of maps. A completely seamless digital environment that results in good, understandable maps based on the unique benefits of digital data is predicted to follow the largely paper-oriented era just passing (Davis and Keller, 1997). Remote sensing and GIS are moving rapidly to quantitative digital maps which tie the tremendous, but finite, complexity of landscape models to the infinite complexity of reality. An issue is to maintain or increase user accessibility to the science behind the maps. The capability of the GIS to determine the underlying uncertainty in the remote sensing data structures and maps and to document error propagation in spatial data are critical components of the analysis of remote sensing imagery with other digital data (Joy et al., 1994; Zhu, 1997).

The complementarity of GIS and remote sensing (Wilkinson, 1996) can lead to increased capability for many types of environmental modeling and analysis. Increased GIS and remote sensing integration gives rise to a new concern: GIS and image processing system interoperability (Limp, 1999). Available commercial image processing systems differ only slightly in their ability to link to GIS, to handle ancillary data, to be used with field data, and to assist with sampling problems. All of these tasks, long recognized as critical in forestry, need to be documented carefully in any application. All are supported to some degree by virtually all of the commercially available remote sensing image analysis and GIS systems — separately. The key issue is how to move quickly between the two systems, taking advantage of functionality that might exist in one system, but not in the other. There is concern over reducing the amount of data conversion that must take place (Hohl, 1998). But even within the GIS community interoperability is a major issue — how to ensure different GIS can talk to each other, share data, repeat analyses, provide comparable output? "Interoperability between computing infrastructures needs — much like every information exchange — a set of common rules and concepts that define a common understanding of the information and operations available in every cooperating system" (Vckovski, 1999: p. 31). For those relying heavily on the remote sensing information as a primary input to the GIS, or requiring GIS information to analyze imagery, what features are needed to make the interface smooth?

A common language and an instruction set providing seamless transfer of data would be a premium advantage. The current marketplace appears to be responding to this issue. Vckovski (1998) has gone further; users need to be provided with an environment in which they use a virtual data set. The system would feature transparent data access, web-based interoperable tools, geolibraries of objects and tools, adaptive query processing, and quick datum and projection changes. The key new development is a set of interfaces which provide data access methods. The virtual data set is not a standardized structure of physical data format, but a set of interfaces facilitating the ability to exchange and integrate information that is meaningful. Against this measure, current interoperability among GIS and image processing systems, and between the two, is practically zero.

But increasingly, GIS functionality and image processing functionality are interchangeable; some key examples now exist where a GIS system has been used to interpret or process imagery in ways that just a few short years ago seemed exclusively the domain of proprietary image processing systems (Verbyla and Chang, 1997). Unsupervised classification, supervised classification, accuracy assessment, filtering and enhancements, removing noise — typically these functions were the reason to have an image processing system; now, all can be completed within a single GIS package without reference to a separate image processing system. Since the GIS typically has a large mandate within a resource management organization (Worboys, 1995; Burroughs and McDonnell, 1998; Goodchild, 1999), larger by far than the mandate enjoyed by most remote sensing, this trend might lead one to conclude that a separate image analysis system may be redundant in some situations.

Since the systems are developing so quickly, with new functionality emerging almost overnight, the emphasis shifts to the GIS/remote sensing field personnel. A new position — a spatial data analyst — sometimes assumes greater responsibility and importance within the organization. One of the most valuable skills of any spatial data analyst is the ability to get something done that seemingly was not possible with the existing system. However, the complexity of some of the operations in remote sensing and GIS can be underestimated. Frustration can occur when analysts use a remote sensing image analysis system as if it were a GIS, or a GIS as if it were an image analysis system beyond the fairly simple processing mentioned above (classification or image enhancement). Typically, a GIS will contain many hundreds of individual tasks based on as many as 20 functional (universal) operations (Albrecht, 1999) which can be grouped into four main analytical functions (Aronoff, 1989):

1. Maintenance and analysis of the spatial data — common GIS and image processing tasks would include data conversions, geometric transformations, and mosaicking;
2. Maintenance and analysis of the attribute data — none of these individual tasks would overlap between GIS and image processing systems;
3. Integrated analysis of spatial and attribute data — common GIS and image processing tasks would include classifications and neighborhood operations; and,
4. Output formatting — many of these individual tasks would be common to GIS and image processing systems.

Image processing systems, as we have seen, can also contain many tens or even hundreds of tasks in broad areas (Chapter 4, Table 4.2; Graham and Gallion, 1996). Having such a variety and number of individual tasks in one computer system alone may create problems in training and upgrading skills. For example, it may take more than one year to learn most GIS systems (Albrecht, 1999). Individual user-interface design, the language of commands, and numerous aspects of system look and feel help create a steep learning curve for users (Goodchild, 1999).

A probable outcome of these conflicting pressures is that there will be, at some point in time, one single (monolithic) GIScience environment comprised of these many tasks in several, perhaps tens, of functional groups. Perhaps through vertical integration remote sensing image analysis will be one or two functional groups within this large system. Presently, though, the situation is much less integrated; if there is a stand-alone need to do image analysis, then likely a stand-alone image analysis system is required. If there is a need to do GIS analysis — and in forestry, based on the dominance of the inventory as an information source, this seems obligatory, then a stand-alone GIS is required together with the appropriate training and support.

GIS and Models

Forest models represent a key piece of infrastructure required in support of sustainable forest management. Models allow generalizations from sites to regions and can be used to predict, investigate, or simulate effects over a wide range of conditions and scales. Ecological models have developed "as tools for projecting the consequences of observations or theories about how ecosystems may change over time" (Shugart, 1998: p. 7). Substitute "stands" for "ecosystems," and the value of this new tool is quite apparent under any forest management strategy; but under sustainable forest management with its pressing need to better understand ecosystems, models may be an indispensible information resource. Models facilitate experimental design and interpretation of results, the testing of current hypotheses and the generation of new ones; models form a framework around which empirical observations can be organized (Laurenroth et al., 1998). By recognizing the cultural aspects of data management and modeling, a three-way relationship designed to alleviate the problems that flow from the enormous accumulation of scientific data, is emerging between (Olson et al., 1999):

1. Empirical data collection,
2. Multidisciplinary data analysis, and
3. Computer modeling.

Obviously, GIS and remote sensing are wonderful ways of accumulating enormous collections of empirical observations, but this creates the need for better, more powerful tools to help make sense of these data. Models represent one such powerful tool.

A wide variety of forest models exist, ranging from the individual tree growth and mortality models, to gap or stand models of competition and structure, to global models of productivity (Shugart, 1998). The proliferation of models threatens to overwhelm their promising role as a helpful tool in forest management. For example, Landsberg and Coops (1999) list three types of models that have been developed to deal with aspects of, or approaches to, forest productivity: (1) standard growth and yield models, (2) gap models, and (3) carbon balance or biomass models. Battaglia and Sands (1998) and Shugart (1998) provide more comprehensive listings, but only a few of these models are expected to emerge as *bona fide* management tools.

In the past, some forest management questions were resolved primarily by using descriptive empirical models, usually known as traditional growth and yield models. But this view appears to be changing. Other types of models are reaching new levels of sophistication at the same time that they are increasingly able to answer questions posed by managers (Battaglia and Sands, 1998). Here, the promise appears to be in those carbon balance or ecosystem process models; at least in some forests, such models appear to have a greater likelihood of current or near-future use as tools by managers. Their use in operational settings has been made more likely by virtue of the wider use of remote sensing and, especially, GIS technology (Bateman and Lorett, 1998). In fact, the availability of GIS data and the design of models that require GIS data to run appear to have been instrumental in bringing these models into a more mainstream position in forest management. For example, the prevalent perception that process-based models are suited only for research applications, since their original design was to help explain theoretical ecosystem functioning questions (Waring and Running, 1998) appears to be quickly fading.

GIS and modeling have been and will continue to be used alone in forestry, but each has benefited from key developments in the other. These developments have facilitated new insights and applications. The demands of ecosystem process models for spatially explicit data often, but not always, obtained by remote sensing can only be addressed for large areas within the framework of a GIS. Increased interest in the results of forest ecosystem (Leblon et al., 1993) and grassland modeling (Burke et al., 1990) has spurred wider availability of various types of GIS data — biogeographical data: DEMs, forest inventory covertype maps, spatially explicit meteorological data, and finally, biophysical remote sensing information. On the other hand, database development to serve forest modeling applications has stimulated progress in using and refining forest models.

The availability of GIS data and ecological models has created a number of new analytical possibilities, including a new emphasis on ecological impact assessment (Treweek, 1999). Typically, an ecological impact assessment is a more focused environmental impact assessment. The greater focus on ecosystem processes is made possible by improvements in ecosystem science and ecological theories. When the data are compiled to support such assessments, ecological concepts can be explored at different temporal and spatial scales with the help of models. For example, the influence of human disturbances can be examined within the context of the natural disturbance and successional patterns across watersheds rather than in small artificial management units (Dale, 1998). Obviously, field approaches to ecosystem ecology are highly variable and differ in regional settings, but with GIS and modeling approaches it is possible to simulate empirical or natural history and to devise experimental and comparative ecosystem studies (Likens, 1998). GIS applications of this type might include cumulative effects models, regional habitat studies, land use planning, ecological mitigation planning, and landscape level monitoring.

A recent emphasis on the provision of landscape metrics from remote sensing imagery within a GIS environment is an indication of a trend to map quantification and landscape modeling (O'Neill et al., 1988; McGarigal and Marks, 1995; Frohn, 1998; Elkie et al., 1999). Those efforts are accompanied by exhortations to the user community to increase awareness and understanding of the science behind the tool; as always in computer applications, users perhaps need to be reminded: garbage in, garbage out. Further discussion of the landscape models occurs in a later section of this chapter.

ECOSYSTEM PROCESS MODELS

One type of forest model — the ecosystem process model — has recently emerged and is intricately linked to remote sensing technology with its multiscale applications and numerous kinds of output potentially useful in forest management decision making. Waring and Running (1999) dubbed this kind of model the integrative model. The integration occurs with remote sensing, climate, ecophysiology data, and understanding of ecological processes. In one review, Battaglia and Sands (1998) referred to these models as APAR (absorbed photosynthetically active radiation) or hybrid APAR-process models, suggesting best uses of such models would be found in global carbon modeling. In fact, understanding global carbon cycles through modeling is part of the C&I of sustainable forest management, and has been suggested as a sufficient justification for development of regional carbon flux models based on remote sensing inputs. For example, Cohen et al. (1996a) developed a carbon flux model of the U.S. Pacific Northwest precisely because of the need to document the contribution of these forests in managing global forest resources to enhance carbon sequestration. This issue has emerged in many areas around the world as an important regional goal which is dependent on local (ownership) forest management practices.

Despite improvements in the models and the potential of synergy in coupling modeling and remote sensing technologies, surprisingly few examples exist of successful simulation of forest ecosystem processes (Ong and Kleine, 1996; Lucas and

Curran, 1999). The Boreal Ecosystem Productivity Simulator (BEPS) model represents a combination of ecophysiology, remote sensing and climate models which are linked to estimate NPP, to help natural resource managers in Canada achieve sustainable development of forests (Liu et al., 1997). The critical inputs to BEPS include LAI (ten-day composites from AVHRR, EOS MODIS, or SPOT VEGETATION satellite imagery), available water capacity of soil (from the Soil Landscapes of Canada (SLC) database, a national soils database similar to the U.S. STATSGO but compiled at 1:1 million scale), and gridded daily meteorological variables (shortwave radiation, maximum and minimum temperatures, humidity, and precipitation). To obtain NPP, BEPS runs in five steps:

1. Soil water content is modeled by considering the soil water balance (using the soil bucket concept) and calculations of rainfall input, snowmelt, canopy interception, evapotranspiration, and overflow;
2. Mesophyll conductance is calculated as a function of radiation, air temperature, and leaf nitrogen concentration; canopy stomatal conductance is calculated as a function of radiation, air temperature, vapor pressure deficit, and leaf water potential (which, in turn, is a function of soil water content modeled in Step 1);
3. Daily photosynthesis is calculated as a function of mesophyll conductance and canopy stomatal conductance constrained by LAI and daylength; maintenance respiration is estimated for each vegetation type and biomass class using nighttime average air temperature (for aboveground components), and soil temperatures (for belowground components);
4. Daily maintenance respiration is subtracted from daily gross photosynthesis, which is summed for the annual time step;
5. Growth respiration (assumed to be a constant fraction of gross photosynthesis) is subtracted to yield NPP estimate.

The NPP estimates are spatially explicit at the scale of the biome or ecoregion. Currently, ecosystem models such as BEPS may be most useful at the global, regional, or biome scale (Ruimy et al., 1994), but concerns related to global carbon budgets are a part of sustainable forest management at the local level. It would be useful for forest managers working at the stand, ecosystem, or landscape level, to be able to embed their NPP estimates in these smaller-scale strata: ecoregions, biomes, and natural regions. The goal is to create and run models which can scale between the different features — biomes to landscapes to local stands, and back again.

The concept of the landscape level or landscape scale has been tarnished somewhat (Allen, 1998); this terminology, landscape level or ecosystem scale, is thought to be imprecise and potentially misleading, but it may not be as critical here to deal with the semantics and meaning of these terms. What is generally meant by the intermediate step of the nested or hierarchical NPP models such as BEPS or DIPSIM (Ong and Kleine, 1996) is the ecosystem scale with an understandable spatial extent of a few hundred hectares, a drainage area, or a watershed. Stand-level modeling is similarly imprecise in theory, but in practice it is generally understood what is meant. It may be important to stress again the linkage in remote sensing between pixel size and spatial extent, as presented in the general hierarchy of image scale in Chapter 3. When reference is made to the intermediate- or medium-scale image data, the implication is that these data are well suited to the landscape scale of analysis. This is simply to provide an idea of the relative amount of detail that can be extracted from the different types of imagery; similarly, the term landscape-level gives a general idea of the spatial extent of viable modeling estimates of key processes.

The use of remotely sensed data with a purely physiological model so that it can be applied at a landscape scale — over a few drainage areas or the area of a Landsat TM image, for example — has had a significant effect on the applicability of the model for landscape managers where it has been used in real management situations. The use of models in management of individual stands is increasing, and will likely improve with access to better remote sensing data (Coops and Waring, 2000). Currently, the principal benefits of using remote sensing in the modeling exercise can be summarized as (Coops, 1999):

1. Allowing details of management and disturbances to be incorporated into the climate-driven estimates of growth (e.g., thinning, insect infestation), and
2. Extrapolating spatially across the landscape.

These process models represent an effective way of providing estimates of important variables that are difficult to measure directly (Peterson, 1997). The mechanics are reasonably straightforward, though not usually simple. Remote sensing data are used to generate initial conditions (e.g., covertype) and driving variables (e.g., LAI) for such models, and to validate (or reparameterize) model output (Peterson and Waring, 1994; Lucas and Curran, 1999). Resource managers can use ecosystem process models to describe the forest stand conditions at a point in time relative to a range of potential management treatments and an historical database, and they can generate projections of future growth and stand development. Some models include the ability to model forest disturbance and management actions such as thinning (Landsberg and Coops, 2000). Applications in a wide variety of areas, including wildife habitat assessment, biodiversity, and growth assessment, are now possible.

However, the input needs of these models can be very demanding — some are designed to run with near-continuous remote sensing input (e.g., global-scale AVHRR, SPOT VEGETATION, or MODIS composites). At the landscape scale, a key simplification is the use of a single satellite image obtained during summer (full leaf conditions); a single estimate of LAI can be used to approximate the photosynthetic capacity of the forest for the entire growing season (Franklin et al., 1997b; Coops and Waring, 2000). In this way, the models can be used to estimate stand or site net primary production with certain critical information on land cover, soils, topography, and climate (Bonan, 1993; Hall et al., 1995).

In the future, improved ecosystem process models may replace empirical stand growth and yield models (Landsberg and Coops, 1999). These field-based models suffer from the potentially fatal limitation of not being robust under conditions of climate change, because they are based on past data. Initially, it is expected that complex process-based models which do not suffer from this limitation — that is, can provide reliable predictions of ecosystem behavior and structure under future, new atmospheric conditions (Friend et al., 1997) — will be used in combination with growth and yield models. For example, Ollinger et al. (1998: p. 324) position their model of forest productivity at the regional level "because they provide an important intermediate between detailed plot-level information and coarse-scale modeling of global fluxes." Critical to successful application of their model is the provision of a satellite-derived landcover map to represent actual vegetation cover, rather than only potential vegetation.

Figure 5.1 contains a block diagram of the essential components in one ecosystem process model, BGC ++ (Hunt et al., 1999). The model was derived from an earlier model called BIOME-BGC (Running and Hunt, 1993), which in turn, was derived from the earlier mechanistic conifer forest ecosystem model called Forest-BGC (Running and Coughlan, 1988; Running and Gower, 1991; Running, 1990, 1994). BGC ++ was designed to generalize ecosystem biogeochemical and hydrological cycles across a wide range of lifeforms and climate. The model uses two time-steps, daily and annual, and requires (1) climate station records (air temperature, radiation, precipitation, humidity, atmospheric CO_2), and (2) GIS site data such as soil texture, coarse fragment content, and depth, to estimate soil water-holding capacity for use in the daily water balance. Modeled carbon dynamics include daily canopy net photosynthesis and maintenance respiration, annual photosynthate allocation, tissue growth, growth respiration, litterfall, and decomposition. Table 5.1 contains a listing of the major parameters of BGC ++ used in one model run for a study of balsam fir stands in western Newfoundland (Hunt et al., 1999).

Hydrologic Budget and Climate Data

The daily time step in BGC ++ simulates the hydrologic budget, including estimation of soil water content and stomatal conductance. These are strongly determined by LAI. The important driving variables in the hydrologic budget are the minimum and maximum air temperatures, relative humidity, solar radiation, and precipitation. Climate data can represent a real challenge to modelers. Such data are often sparse and of dubious quality in representing regional patterns, particularly in mountainous areas (Running et al., 1987). Without multiple weather/met stations, mean monthly minimum and maximum temperature and precipitation surfaces are generally interpolated from available weather/met stations, often located in valleys, far from the slopes that are of interest. Climate model parameters such as incoming solar radiation are usually modified from sunshine estimates at airports for different slopes and aspects and daylength. If the spatial variation of meteorological conditions can be quantified, this information can be used to improve estimates of site hydrologic balance and evapotranspiration (Nemani and Running, 1989; Price, 1990), snowmelt and water discharge, and ultimately, terrestrial vegetation productivity (Running, 1990). In at least one study (Unger and Ulliman 2000), Landsat TM thermal band data were found to relate better to coincidental mean maximum daily forest ecosystem ambient air temperature than several estimates derived by modeling.

An accurate soils map is a tremendous asset in forest ecosystem productivity modeling. Soils and topographic data tend to improve landscape productivity models

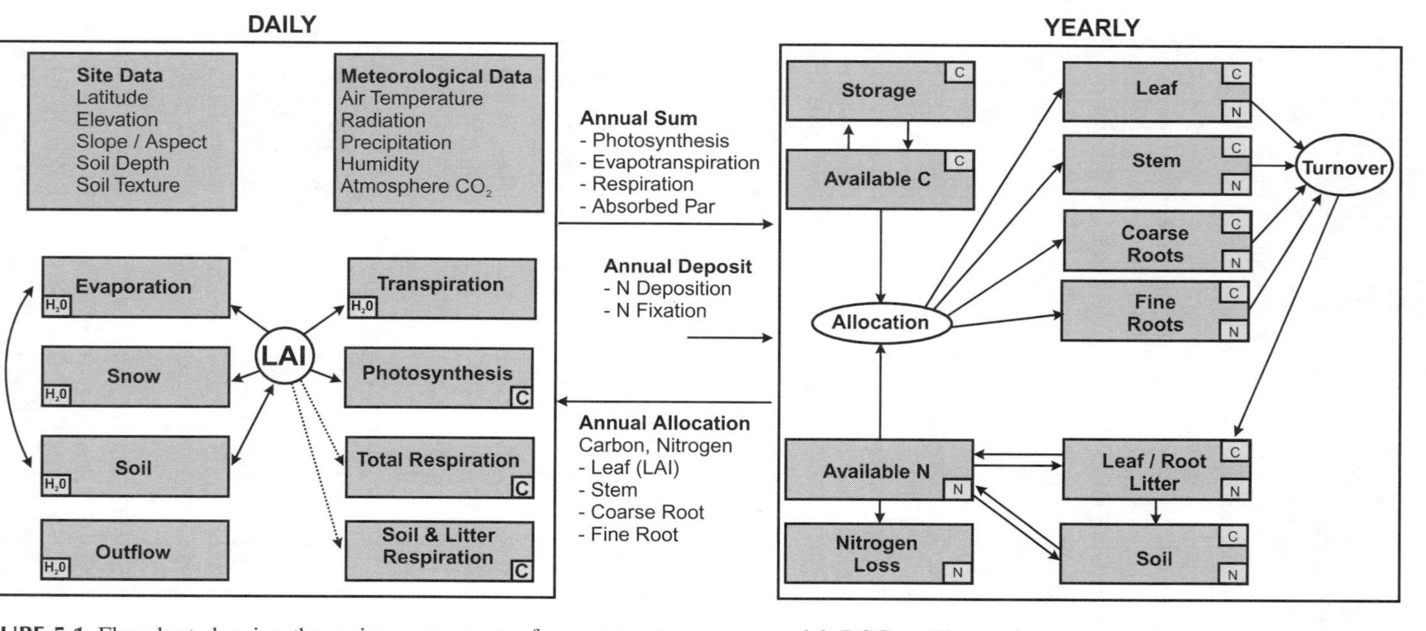

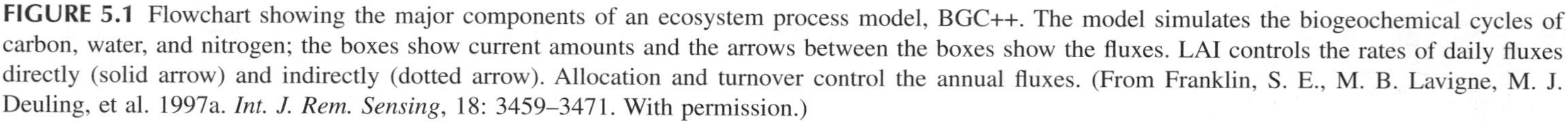
FIGURE 5.1 Flowchart showing the major components of an ecosystem process model, BGC++. The model simulates the biogeochemical cycles of carbon, water, and nitrogen; the boxes show current amounts and the arrows between the boxes show the fluxes. LAI controls the rates of daily fluxes directly (solid arrow) and indirectly (dotted arrow). Allocation and turnover control the annual fluxes. (From Franklin, S. E., M. B. Lavigne, M. J. Deuling, et al. 1997a. *Int. J. Rem. Sensing*, 18: 3459–3471. With permission.)

TABLE 5.1
List of Ecosystem Process Model Requirements Includes Information on Climate, Site, and Physiological Status. Shown Here Are the BGC ++ Model Parameters and Variables

Variable or Parameter
Required Daily Inputs
Day of year
Air temperature, maximum
Air temperature, minimum
Precipitation (water equivalent)
Calculated and Optional Inputs
Daylength
Total solar radiation
Total photosynthetically active radiation
Dew point temperature
Soil temperature (20 cm depth)
Atmospheric CO_2
Site Variables
Latitude
Slope and aspect
Elevation
Albedo
Soil water content at field capacity
Soil water content at 1.5 MPa
Initial water: soil, snowpack
Daily Outputs
Water fluxes: transpiration, evaporation, runoff
Soil water content
Pre-dawn leaf/soil water potential
Daily net photosynthesis
Daily maintenance respiration
Annual Outputs
Carbon increments: leaf, stem, root, litter
Total growth respiration
Major Physiological Parameters
LAI
Specific leaf area
Fraction of carbon in dry matter
Maximum stomatal conductance
Boundary layer conductance
Maximum photosynthetic rate
CO_2 compensation point
Critical soil/leaf water potential

TABLE 5.1 *(Continued)*
List of Ecosystem Process Model Requirements Includes Information on Climate, Site, and Physiological Status. Shown Here Are the BGC ++ Model Parameters and Variables

Q10 for maintenance respiration
Maintenance respiration: leaf, stem, root
Growth respiration: leaf, stem, root
Carbon allocation: leaf, stem, root
Precipitation interception coefficient
Light extinction coefficient
Leaf turnover coefficient
Stem turnover coefficient
Fine root turnover coefficient
Initial carbon: leaf, stem, root, soil, litter

Source: Modified from Running and Coughlan (1988) and Hunt et al. (1999).

(Mummery et al., 1999). However, such maps are rare, and if not rare, often incomplete or at an inappropriate scale (Payn et al., 1999). In the U.S., the State Soil Geographic Database (STATSGO) has been compiled at 1:250,000 scale from a combination of soil survey data and information on geology, topography, climate, and vegetation, supplemented with remote sensing imagery. STATSGO data have been used in forest growth capacity model development and testing; for example, Coops and Waring (2000) used these data to infer soil fertility and soil water holding capacity in Oregon. By focusing on growth capacity rather than forest growth, the model could be greatly simplified. But even using the model to predict growth capacity at a 200-m spatial resolution, however, the inadequacies of the STATSGO database became apparent. Difficulties were experienced in modeling N processes (annual mineralization, deposition, uptake and allocation to canopy, and losses).

In other studies, because of the scarcity of reliable soil information a digital elevation model (DEM) has been used to estimate soil depth and other soil characteristics, by assuming a relationship between the position of the stand on the slope and soil development (Moore et al., 1993a). In many environments the soil-landscape relationships can be predicted by geomorphometrics such as slope steepness, curvature, wetness indices, stream-power, and local relief (Pike, 1999); for example, in hydric soils in the glaciated landscape of Minnesota, Thompson et al. (1997) found that these variables explained much of the variation in a soil color index. Zheng et al. (1996) created a compound topographic index as the function of the contributing area upslope and the slope. According to Coops and Waring (2000), higher values of this index tend to be found in the lower parts of watersheds and in convergent hollow areas associated with soils of low hydraulic conductivity, or areas with more gentle slope than average (Clerke et al., 1983; Beven and Wood, 1983). Soil depth and silt and clay content tend to increase from ridge tops to the valley bottoms (Singer and Munns, 1987) even though few hillslopes have a single parent material (Hammer, 1998). The underlying principle is based on the fact that landforms significantly affect site productivity and the distribution of forest ecosystems (Clerke et al., 1983; McNab, 1989, 1993; Host et al., 1987; Moore et al., 1993b; Swanson et al., 1988); if this influence can be understood, simplified, and quantified by automated landform delineation (Blaszcynski, 1997), the process of modeling productivity can be made more accurate.

The issue of capturing the essential variability in slopes related to soil characteristics with the correct DEM resolution has not yet received much attention (Isard, 1989; Mitasova et al., 1996; Pike, 1999), but in many jurisdictions using a DEM in this way is likely to provide a simpler, more accurate modeling solution quicker than waiting for better soils maps to be produced. A DEM can be used to partition the landscape into homogeneous hillslope units that can then be modeled individually (Band et al., 1991; McDonnell, 1996). The terminology must be clarified; a hillslope is the drainage area contributing flow to a stream link from one bank, and a stream link is a stretch of stream channel along which no tributaries enter (Band et al., 1991). The first step is to extract the stream network from the DEM; as has already been suggested, this is no trivial task (Qian et al., 1990). The second step is to determine the drainage area upslope of each pixel in the DEM. A suitable threshold must be used to ensure that hillslope units make "geomorphic sense" and are not too numerous. Again, no simple task. Each hillslope unit is then parameterized with mean slope and aspect, and the spatially referenced GIS data (such as the soils layer, if available) are called in to run the productivity model. This approach reduced the spatial aggregation error that can accumulate with arbitrary pixel sizes as basic modeling units. For example, Pierce and Running (1995) simulated NPP for a landscape at four successive levels of landscape complexity and grid cell sizes; estimates could vary by as much as 30%.

The issue of spatial variability in topography and soils is coupled with the issue of spatial variability in vegetation and reflectance as measured by remote sensing (Landsberg and Coops, 1999). Using a combined remote sensing/DEM approach, it is possible to establish a landscape modeling approach that would be based on homogeneous units that are similar to the photomorphic units used by management as forest covertypes. Such units — defined in a Rhode Island deciduous forest ecosystem study as areas of high geomorphological heterogeneity on the basis of soils and topographic indices — can be highly related to biodiversity at the plot scale (Burnett et al., 1998) and the landscape scale (Nichols et al., 1998). A region without any soils database was studied by Giles and Franklin (1998). They presented the "landform logic" in partitioning a mountainous area in southwest Yukon into homogeneous geomorphic units based on the idea of a geomorphic signature. Combining DEM and spectral response data obtained from a single image source — stereoscopic analyses of satellite imagery — simplified the number of separate data layers that had to be acquired. It is expected that this approach can at least create basic modeling units in which the assumptions of homogeneity can be more confidently applied (Band et al., 1991; Coughlan and Dungan, 1997).

Forest Covertype and LAI

In the ecosystem process models, such as BEPS and BGC ++ , a daily time step procedure is used to determine the rates of photosynthesis, autotrophic respiration (sum of growth and maintenance respiration), and nitrogen transformation. These depend heavily on LAI and forest covertype assumptions. LAI is one of the most important variables in many process models; step back, and consider that "The terrestrial biosphere is like a chlorophyll sponge blanketing the Earth with a thickness proportional to LAI" (Running, 1994). LAI determines the APAR and stomatal area, and hence strongly influences CO_2 uptake, evapotranspiration, and nutrient cycles (Bonan, 1993). The BEPS developers (Liu et al., 1997: p. 174) stated that reliable and accurate LAI data were a prerequisite for regional application of a process model because "LAI strongly affects all components of the model, including radiation absorption, transpiration, photosynthesis, respiration, rainfall interception and soil water balance." Individual process models either assume a maximum LAI based on species, climate, and soil constraints, or require an estimate of LAI from remote sensing. Spectral response patterns can be used to produce LAI estimates by radiation modeling or by empirical indices such as the NDVI. This application is described in some detail in Chapter 7.

Photosynthesis is strongly dependent on LAI and leaf nitrogen (Farquhar et al., 1980); the assumption has usually been made that the forest canopy behaves as a single big leaf. Obviously, leaves have very different characteristics depending on their position in the canopy and their orientation; some improvements to the big leaf estimates of photosynthetic rates have been reported if consideration is provided for differing leaf morphology and age (Chen et al., 1999b). Typically, growth respiration is assumed to be a constant fraction of photosynthesis for each type of forest cover, while heterotrophic respiration and maintenance respiration of roots are determined by soil temperature.

Forest covertype may be one of the most important variables in controlling assimilation rates, carbon allocation, nutrient use and litter, decomposition, and productivity (Bonan, 1993; Waring and Running, 1998). In the BEPS modeling effort, the significance of forest covertype information was illustrated in comparison with five other models; three had higher NPP in broadleaf deciduous forest than in evergreen needleleaf forest, but two had the opposite. The differences could be avoided if covertypes were used to pre-stratify the model runs. In another study, Coops and Waring (2000) confirmed the importance of stratifying forest productivity models by covertype data; in their case, the covertype information was obtained by access to the state GIS forest inventory database. The problem of internal polygon — or stand — homogeneity again arises (Franklin et al., 1997b). The internal variability in covertypes within GIS forest inventory polygons can lead to erroneous model assumptions, and hence, model output can be seriously biased. In New Brunswick, stand estimates of NPP, derived by assuming the GIS label of dominant species for a stand was correct, could differ by as much as 30% from the estimates obtained by first classifying Landsat TM imagery into a few general forest covertypes (softwood, hardwood, mixedwood) and using the classes obtained in this way to call the model parameters (Franklin et al., 1997b). This approach circumvents the well-known polygon variability problem; even if the polygon is labeled a softwood stand in the GIS, it is more or less likely that some of the stand area would be hardwood. Remote sensing classification can reveal this internal homogeneity at the same time as the original LAI estimates are generated from NDVI or other vegetation index (Franklin et al., 1997b).

The seasonal (or annual) time step of BGC ++ uses the available nitrogen (net N mineralized and N inputs), and the available carbon (NPP), to allocate C and N to leaf, stem, and coarse and fine roots. Stand water and nitrogen limitations are used to alter the internal dynamics of the leaf/root/stem carbon allocation fraction (Running and Gower, 1991). In one study (Lucas et al., 2000), spatial estimates of leaf nitrogen concentration were derived through their relationship with LAI, and this improved the functioning of the model in estimates of stem carbon production.

Model Implementation and Validation

One of the most important questions for the use of ecosystem simulation models in examining hypotheses of forest processes is (Hunt et al., 1999: p. 159), "What kind of data are important for model testing?" Generally, models are most credible if they come with a long history of development and testing, such as expressed in the continuing development of Forest-BGC (Running, 1984; Running and Coughlan, 1988; Running and Gower, 1991), then BIOME-BGC (Running and Hunt, 1993), and now BGC ++ (Hunt et al., 1996; Hunt et al., 1999). Model lineage or history is certainly one type of data required by managers to help understand and assess model results. However, the actual use of the model in forest management, rather than solely as a research tool, would also likely be considered an essential piece of information when assessing model formulations and results; i.e., has the model been used successfully in helping make forest management decisions? Does the model output fit with the types of questions asked by forest managers? Are quality assessments of the model output available?

Some of the reasons advanced to explain why process models have not been used extensively in forest management are that they are overly complex, they require too many input parameters, they are difficult if not impossible to validate at scales of interest (Running, 1994), and their output cannot be readily understood as helping to answer specific questions of interest and concern to forest managers (Battaglia and Sands, 1998; Landsberg and Coops, 1999). Kasischke and Christensen (1990) were concerned with building connections between forest ecosystem process models and microwave backscatter models; in many ways, 10 years later not much progress has been achieved in dealing with the diverse model subcompartments and their linkage to provide meaningful forest management information. In general, models must improve in their ability to move from description to explanation, from just predicting harvestable products to understanding the limits and constraints to growth (Battaglia and Sands, 1998). Based on the large number of indicators associated with productivity in sustainable forest management, it appears that this is what managers must be able in achieve in the short term.

Agreement among different models — for example, a process model such as BGC ++ vs. a simpler, lumped-parameter model — has been suggested as another type of validation test (Battaglia and Sands, 1998). If results of one model are used to validate or provide input to another model, there are potential validation tests that can be performed among nested submodels (Jupp and Walker, 1997; Nilson and Ross, 1997). Validation of internal logic and variables can be used to increase confidence in model predictions of the response of trees, stands, or ecosystems to

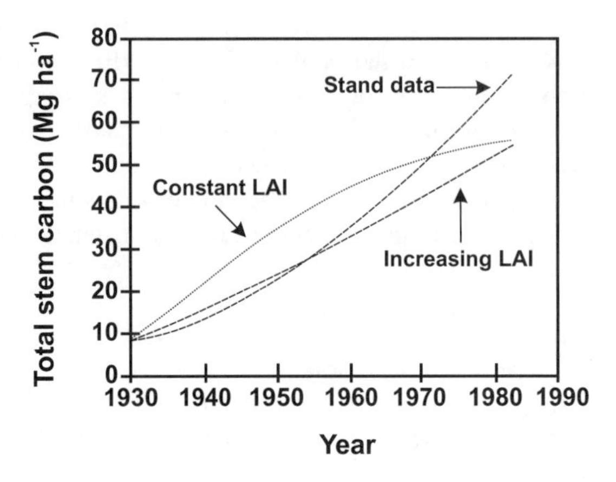

FIGURE 5.2 Measured total accumulation of stem carbon for a ponderosa pine stand and simulated accumulation of stem carbon using an ecosystem process model. Shown are the simulations assuming either a constant LAI or an LAI increasing from 2.0 to the maximum of approximately 5.0 for this stand. Actual accumulation was measured by tree ring analysis, which can accurately depict seasonal variation in climate. The difference between the stand data and the simulations were thought to be largely a result of the relatively poor climate data used in the model. (From Hunt, E. R., Jr., F. C. Marin, and S. W. Running. 1991. *Tree Physiol.*, 9: 161–171. With permission.)

changed environmental conditions. A second common validation approach is to consider smaller regions or local types within the larger modeling scale, then compare aggregated or nested results for reasonableness.

"Probably the most accepted form of model validation is to compare predicted model output directly to observed behavior" (Running, 1994: p. 238). Typically, to validate the carbon cycle components, model estimates of leaf and stem growth are presented as increments in aboveground biomass. These estimates can be compared to field observations or empirical growth models for individual sites. In one test, Hunt et al. (1991) found that the ecosystem model could better integrate the effects of interannual climate variability on stem carbon gain than could multiple linear regression models (the traditional growth and yield approach) (see Figure 5.2). The differences are not large, but over time and over a large management unit such as a drainage basin, they can accumulate quickly. Dendochronological models, however, can outperform ecosystem process models when tree ring data are available. Tree ring analysis can better reflect the actual LAI of the stand during early growth stages.

The hydrologic cycle may be validated with stream runoff and soil moisture measurements. The nutrient cycle may be validated with leaf litter bioassays. All of these validation exercises must confront the issues of scale and expense. In one of the most comprehensive validation exercises, Running (1994) used the Forest-BGC model at seven sites in Oregon, and evaluated predicted and observed pre-dawn leaf water potential (hydrologic cycle), aboveground net primary production (ANPP), equilibrium LAI and 100-year stem biomass (carbon cycle), and leaf nitrogen concentration (N-cycle). Defining the water-holding capacity of the rooting zone and the maximum surface conductance for photosynthesis and transpiration rates were the most critical system variables that defied routine field measurement. To validate the model accurately and completely in the field would have required many more resources than were available for even this dedicated study; few forestry agencies have the resources or the mandate to engage in such recommended validation exercises.

The criticisms that most models are too complex, and require too many input variables for routine use, have been accepted as a challenge in the modeling community (Landsberg and Coops, 2000). One response has been to create more user-friendly model interfaces; modeling can be accomplished off-line, in much the same way that data from the thousands and thousands of PSPs are used to assess growth and yield by the compilation and fitting of curves. To reduce computational constraints a look-up table approach may be used for simulations; this alleviates the need to run the model for each pixel in the scene (Lucas and Curran, 1999). Remote sensing data and GIS data are critical in using the model by stratifying the area by variables of interest, for example, ecoregions or subregions. An example structure for an ecosystem process model look-up table is contained in Table 5.2.

The appropriate mix of remote sensing and modeling has not yet been determined for best application in forestry. The use of standard sites as measures of normal community development appears essential (Jupp and Walker, 1997); remote sensing can then be used to initialize model runs, and to make observations that can be compared to model output in known conditions and sites. The role of remote sensing is to confirm that the model is working well, to document that the expected changes in forest conditions have occurred as predicted, and to identify when the forest conditions and model predictions are not in agreement (Peterson and Waring, 1994; Nilson and Peterson, 1994). These observations, when coupled with the appropriate model outputs, will allow for efficient scaling of ecosystem processes from the instrumented site to larger and larger areas, eventually encompassing an entire biome, and potentially the globe (Running and Hunt, 1993; Waring et al., 1995a,b).

Finding the difference between the actual value and potential value of some remotely sensed biophysical condition is a powerful approach in future monitoring of regional and global ecosystem functioning, based on regular acquisition and analysis of NDVI derived from AVHRR imagery (Stoms and Hargrove, 2000). The approach is to find training sites where actual NDVI approximates baseline values, formulate a model that best predicts these values, and apply that model to biophysical predictors to map potential NDVI. Then, actual NDVI are examined to determine areas where productivity may be reduced (drier sites), or increased (wetter sites, perhaps agricultural irrigation) compared to predicted values. A detailed version of this logic was used by Franklin et al. (1997b) in Landsat TM NDVI prediction of forest stand LAI; actual LAI (from remote sensing) and predicted LAI (from climate and soils conditions predicted by the model BIOME-BGC) in some areas were significantly different. Field checks determined that, in the areas where actual NPP was lower than predicted by the model, plantations had been established in which the species was inappropriate for the site.

SPATIAL PATTERN MODELING

Sustainable forest management recognizes the importance of spatial forest structure, defined as the mosaic of forest patches varying in composition and size, altered by

TABLE 5.2
Example Ecosystem Process Model Look-up Table Structure by Remote Sensing and GIS Data. This Example Has Four Dominant Species Types, Up to Ten Density Classes, Up to Seven Age Classes, Up to Four Management Types, and LAI Expressed in 0.5 Increments between 0.5 and 10 for a Total of 14,560 Table Entries

Dominant Species Code	Density Class	Age Class	Management Type	LAI
01 (Red spruce)	1			
	2			
	3			
	4	1		
	5	2		
	6	3		
	7	4	1	
	8	5	2	
	9	6	3	
	10	7	4	0.5
		—	—	1.0
		—	—	1.5
		—	—	2.0
		—	—	—
02 (Aspen)	1			
	2			
	3			
	4	1		
	5	2		
	6	3	1	0.5
		—	—	1.0
		—	—	1.5
		—	—	2.0
		—	—	—
03 (Jack pine)	—	—	—	—
04 (Balsam fir)	—	—	—	—
—	—	—	—	—
—	—	—	—	—

natural events (geomorphological and ecological processes) and by human intervention (Eng, 1998; Baskent, 1997, 1999). The management of the forest landscape is driven more by the need to regulate changes than by any specific distributions (e.g., age class distributions for timber supply or maximum mean annual increment). It is the spatial distribution of features in the landscape that indicates whether goals are being met for timber, wildlife, water regulation, recreation, and visual quality. Spatial forest structure can be quantified from measures of the composition and structure of landscape patches mapped from elements obtained by remote sensing. For example, from classifications of covertype, LAI, or other defined elements of the surface, patch relationships that affect landscape dynamics, i.e., diversity, complexity, association, and connectivity can be calculated. The variety and relative abundance of patch types are measures of composition and include patch richness, patch diversity, and diversity indices. The size distribution of patches, the dispersion of patch types throughout the landscape, the contrast among patches, the patch shape complexity, the contagion or clumping of patch types, and the corridors between patches are all structural components of the landscape that can be quantified (McGarigal and Marks, 1995).

For monitoring landscape structure via remote sensing, the first question is, What constitutes a homogeneous patch at the ecosystem level? One approach is to separate patches in the same way that forest stands have been mapped: using aerial photointerpretation. Patches, such as clearcuts or other disturbance features, are then embedded in the background mosaic — the forest stands. Homogeneous patches are those areas of the forest that can be recognized as homogeneous on aerial photographs. Or, the stands are generalized, perhaps by species dominance, to create covertypes; the background mosaic, then, becomes the covertype. Another approach is to employ image classification using digital remote sensing data (Slaymaker et al., 1996). Patches are therefore the areas of the forest that can be recognized as homogeneous in their spectral response patterns (and perhaps textures). Clearly, subsequent measures of landscape pattern and patch diversity are not only dependent on the definition of a patch and the mosaic, but also on the consistency of the techniques used to identify and map patches using the specified elements and the homogeneity of other attributes that might be included within the patch. When using remote sensing imagery it is always necessary to distinguish between the spectral classes (based on reflectance of energy) and the information classes (based on human perceptions of what constitutes a community of plants) which will be used to create the patch/mosaic landscape. There is an interaction between the number and type of mapping categories and the resulting complexity observed in the landscape structure.

O'Neill et al. (1996) now ask, At what spatial scale is it relevant to monitor, report, and assess landscape patterns? One of the applications of spatial modeling is the quantification of spatial relationships and patterns to allow comparisons between different areas, for example, adjacent watersheds with different forest management operating plans. Such comparisons may provide clarification of the impact of land management decisions on landscape structure, and hence on biodiversity and productivity. By monitoring patch relationships over time, important changes which influence ecological phenomena such as animal movements, hydrology, spread of disturbance, and net primary productivity may be detected (O'Neill et al., 1991). The transition zones between patches, the patch boundaries, perform important ecological functions, allowing passage of biotic and inorganic factors between patches and contacts between core- and edge-dwelling species (Metzger and Muller, 1996). Boundaries resulting from anthropogenic activities (roads, fields, clearcuts) are generally sharper and less complex than those generated by natural processes, i.e., transitions between meadow and forest, or conifer and mixed conifer. Points where the boundaries of three or more landscape elements congregate may be important centers of resources and corridors for wildlife (Forman and Godron, 1986).

Issues such as nature reserve design, adjacency considerations in harvest scheduling, road access, and integration of production and conservation are all possible applications of spatial pattern modeling. The approach requires careful evaluation of the trade-offs associated with forest management options at the forest ecosystem and landscape level. Here, the focus is on the various measurement and monitoring issues that must be addressed in spatial pattern modeling with remote sensing (Frohn, 1998).

Remote Sensing and Landscape Metrics

A review of remote sensing issues affecting landscape metrics is presented in this section together with an introductory comment on the role of landscape metrics in forest management; by necessity, the review is not exhaustive, but is presented to serve two purposes:

1. To help introduce users to the immense literature that has developed and continues to grow, based on the integration of remote sensing and landscape ecological concepts and practices, and
2. As a caution against possible measurement errors that can result if landscape metrics are used haphazardly.

Software to calculate landscape metrics — from input data in vector or raster formats — has been available in stand-alone packages for some time (McGarigal and Marks, 1995). Recently, software modules for patch and landscape analysis have been integrated into a widely used commercial GIS system (Elkie et al., 1999). The result is that large numbers of metrics are now available to those with spatially explicit data sets interested in quantifying landscape composition and structure (Table 5.3).

Some of these metrics are modeled directly on ecological theory and observations. For example, certain metrics, such as core area, require information concerning habitat requirements for the species of interest. Many other metrics are independent of underlying ecological process or habitat requirements, relying strictly on the geometric and spatial relationships of patches (O'Neill et al., 1988; McGarigal and Marks, 1995; Metzger and Muller, 1996; Riitters et al., 1995). Given the many metric options, deciding on a set suitable for a particular study has been problematic. Understanding the influence of image resolution, pixel size, number of patch classes, patch size, patch shape, and raster orientation on metrics of landscape pattern is critical when analyzing the types and quantities of change in the landscape over time. Clearly, the metrics chosen should offer unique information and have ecological relevance (Griffiths et al., 2000).

Foresters have long been aware of many of the issues surrounding patch/mosaic dynamics in managed forests (Franklin and Forman, 1987), but interpreting the ecological relevance of an individual metric, let alone the overall landscape composition and structure, has been problematic (Davidson, 1998). However, even without a complete understanding of how landscape patterns affect the complex biotic/abiotic dynamics within and among ecosystems, interpretations of landscape composition and pattern, such as assessments of fragmentation and connectivity, are considered important indicators of sustainable forest management practices, perhaps leading to greater understanding of biodiversity and species richness. Landscape metrics and

TABLE 5.3
Landscape Metrics Organized by Area and Type of Measurement Illustrate the Complexity of Structural, Compositional, and Boundary Quantification

Index Type	Index Description/Definition
	Area Metrics
Total landscape area	
Largest patch index (%)	Percentage of area accounted for by the largest patch
Number of patches	
Patch density	Number of patches per unit area
Number of classes	
Mean patch size	
Patch size standard deviation	Absolute measure of patch size variability
Standard deviation of mean patch size	Percentage variation (relative)
Dominance	The degree to which proportions of each patch type on the landscape predominates
Permeability	Area of unsuitable patches (for transmission) divided by the total area
	Edge Metrics
Total edge	Total length of all patch edges
Edge density	Length of patch edge per area
Contrast-weighted edge	Length of patch edge per area, weighted by edge contrast
Total edge contrast index	The degree of contrast between a patch and its immediate neighborhood
Mean edge contrast index	The average contrast for patches of a particular class
Area-weighted MECI	Patches are weighted by their size
Isolation	% Edge adjoining similar patch types
	Shape Metrics
Landscape shape index	Measures of landscape compared to a standard
Mean shape index	Average patch shape (perimeter/area) for a patch class
Area-weighted mean shape index	Patches are weighted by their size, then mean shape calculated for class and landscape
2 × log fractal dimension	Departure of landscape mosaic from Euclidean geometry
Fractal dimension	The complexity of patch shape on a landscape
Mass fractal dimension	The total complexity of the map matrix
Mean patch fractal dimension	Based on the fractal dimension of each patch
Area-weighted mean patch fractal dimension	Patches are weighted by their size, then fractal dimension calculated for class and landscape
Elongation	Diagonal of smallest enclosing box divided by the average main skeleton width
Square pixel (SqP)	The shape complexity of patches on a landscape
	Core Area Metrics
Core area	Area of interior habitat defined by specified edge buffer width

TABLE 5.3 *(Continued)*
Landscape Metrics Organized by Area and Type of Measurement Illustrate the Complexity of Structural, Compositional, and Boundary Quantification

Index Type	Index Description/Definition
Number of core areas	
Core area density	Number of core areas per unit area
Mean core per patch	
Core area standard deviation	Absolute measure of core area variability
Disjunct core	Within a patch, 2 or more disjunct core areas
Total core area index	The percentage of a patch comprised of the core area
Nearest Neighbor Metrics	
Nearest-neighbor distance	The distance of a patch to the nearest neighboring patch of the same type based on edge to edge distance
Proximity index	The size and proximity distance of all patches whose edges are within a specified radius of the focal patch
Mean nearest-neighbor distance	For a class or for the landscape as a whole
Nearest-neighbor distance standard deviation	A measure of patch dispersion
Spatial autocorrelation	Patch type spatial correlation; patch type distribution
Mean proximity index	For a class or for the landscape as a whole
Interpatch Distance	
Diversity, Richness, and Evenness Metrics	
Shannon's diversity index	A single number that captures both abundance and variety. The amount of information per patch
Simpson's diversity index	A single number that captures both abundance and variety. The probability that any types selected at random would be different types.
Patch richness	Number of different patch types
Patch richness density	Patch richness standardized to per area
Relative richness density	Richness as a percentage of the maximum potential richness
Shannon's evenness	Relative abundance of different patch types
Simpson's evenness	Relative abundance of different patch types
Interspersion/Juxtaposition, Contagion, and Configuration Metrics	
Contagion	The tendency of landcovers to clump within a landscape
Dispersion	Degree of fragmentation/complexity of patch boundaries
Association	Concentration of spatially distributed attribute variables
Interspersion	The number of pixels in a 3 × 3 pixel square that are of a different habitat than the central pixel
Juxtaposition	Habitat edges are weighted by their habitat quality for each organism and those surrounding the central pixel in a moving window are summed.
Fragmentation	The tendency of landcovers to break into small pieces within a landscape
Patch per unit area (PPU)	The degree of fragmentation of patches on a landscape

TABLE 5.3 *(Continued)*
Landscape Metrics Organized by Area and Type of Measurement Illustrate the Complexity of Structural, Compositional, and Boundary Quantification

Index Type	Index Description/Definition
Connectivity and Circuitry	
Connectivity	Number of links in a class network divided by the maximum number of links
Circuitry	Number of circuits in a class network divided by the maximum number of circuits

the view "from above" are the essential tools in achieving the necessary insights in building and supporting these landscape interpretations. Practically speaking, however, landscape metrics are not very well understood; no single metric is sufficient for quantifying spatial pattern or the distribution of spatial pattern (Hargis et al., 1998). The choice of metrics will depend on experience, the questions to be addressed, and the process at hand (Spies and Turner, 1999).

Two important difficulties in metric interpretation related to the information content (or unique information) are highlighted here:

1. Metrics are sensitive to the data characteristics, and
2. Metrics are highly interrelated.

First, individual metrics may be sensitive to map scale, number of classes, size and shape of patches, spatial distribution of patches, and many other factors. Using sensitive metrics for landscape comparisons could result in misinterpretation if conditions are not held constant. For example, landscape metrics calculated using different satellite imagery may not be comparable because the pixel size affects the types of patches, and also can influence the computation of individual landscape metrics. In one study, classifications of SPOT HRV, Landsat TM, and AVHRR images of the same area in northern Wisconsin showed very different patch areas, shapes, and locations (Benson and MacKenzie, 1995). Small bodies of water that were detected in the SPOT and TM data were not recorded in the AVHRR data. The percent water and number of lakes decreased as the spatial resolution increased while the contrast, the number of patches, the average lake area, perimeter, and fractal dimension increased. Estimates of some metrics (e.g., homogeneity and entropy) were relatively invariant across the images.

The pixel size of a digital image may affect certain landscape measures. For instance, the number of pixels that are adjacent to one another governs the metric contagion, a measure of patch aggregation (O'Neill et al., 1988; Li and Reynolds, 1993). A smaller pixel size does not change patch aggregation, but the value of contagion increases due to the increased number of pixels that are adjacent to one another (Frohn, 1998). Raster orientation changes the proportions of pixel adjacency. By shifting an image 45°, the straight edges of a rectilinear patch become serrated as the corners of edge pixels jut into the adjacent patch; this shift effectively increases the proportion of adjacent pixels (Figure 5.3). Because contagion is determined by calculating pixel adjacencies, raster orientation affects the values of contagion. In addition to measuring patch clumping, contagion takes into account the proportional representation of patch types in the landscape. Consequently, the number of patch classes can influence contagion, even though no change in spatial pattern has occurred. Pixel size also alters fractal dimension, a measure of patch complexity that is estimated using a linear regression of the patch area and patch perimeter in pixel units (rather than metric units). Different sized squares resulted in various measures of fractal dimension, caused by diverging rates of change for the area (exponential) and perimeter (linear) with increasing size (Frohn, 1998).

Nine landscapes with patterns of increasing fragmentation were simulated while controlling the size and shape of patches and the type of growth (enlarging patches, abutting patches, and buffered patches) (Hargis et al., 1998). Patch size and shape showed significant effects on measures of edge density, contagion, mean nearest neighbor distance (for thinly distributed patches), the proximity index, perimeter-area fractal dimension (for abutting patches), and mass fractal dimension (for enlarging patches with increased disturbance). Some metrics — such as contagion, mean nearest neighbor distance, mean proximity index, edge density, perimeter-area fractal dimension, and mass fractal dimension — were relatively insensitive to the spatial arrangement or composition of patches (Hargis et al., 1998). There is not a single landscape metric that quantifies the spatial distribution of patches, which can have an important impact on certain ecological processes that depend on connectivity including the flow of organisms, pollen, and seeds across the landscape. Finding measures to quantify the spatial distribution of patches is important; for example, for detecting changes that affect the basic biological processes of distribution and migration.

Second, and perhaps equally important, many landscape metrics are highly interrelated. Several metrics share fundamental measures of patch size, shape, perimeter-area ratio, and inter-patch distance (Cain et al., 1997; Hargis et al., 1998; Li et al., 1993; Riitters et al., 1995). To find a set of uncorrelated landscape metrics, Riitters et al. (1995) performed a multivariate factor analysis on 26 metrics calculated for 85 land use and land cover maps; 87% of the metric variation was explained by the first six factors (Table 5.4). These factors were interpreted as composites of correlated measures representing:

1. Average patch compaction,
2. Overall image texture,
3. Average patch shape,
4. Patch perimeter-area scaling,
5. Number of attribute classes, and
6. Large-patch density-area scaling.

The implications are that in order to avoid erroneous interpretations and redundancy in analysis of landscape metrics derived from remote sensing, comparisons of landscape patterns should be made at the same scale and image spatial resolution,

FIGURE 5.3 The landscape metric contagion — a measure of landscape structure — is influenced by the orientation of raster pixels used to represent landscape patches. In this example, contagion decreased from 0.21 to 0.03 solely as a result of the increased pixel edges associated with an image rotation of 45°. (From Frohn, E. 1998. *Remote Sensing for Landscape Ecology*, CRC Press, Boca Raton, FL.)

TABLE 5.4
A Factor Analysis of Landscape Metrics Can Be Used to Reduce a Large Number of Individual Metrics to a Few Orthogonal Composite Measures. Here Are Two Examples in which Multiple Metrics Obtained from Maps or Imagery Acquired Over Time Are Reduced to No More than Six Factors

Factor	Group of Metrics (Riitters et al., 1995)	Metric Best Representing Group (Riitters et al., 1995)	Factor	Group of Metrics (Cain et al., 1995)
1	Average patch compaction	Average patch perimeter ratio	1	Texture
2	Image texture	Shannon contagion	2	Patch shape and compaction
3	Average patch shape	Average patch area normalized to the area of a square with the same perimeter	3	Patch shape and compaction
4	Patch perimeter-area scaling (fractal measures)	Patch perimeter-area scaling	4	Perimeter-area scaling
5	Number of attribute classes	Number of attribute classes	5	Perimeter-area scaling
6	Large-patch density-area scaling	Not considered relevant	6	Number of attribute classes

Source: Adapted from Riitters et al. (1995) and Cain et al. (1997).

and through a set of metrics that are ecologically understandable and statistically independent. Both of these suggestions have been very difficult to implement in real world applications in which historical and variable data sets have been used, and in which limited resources to validate and test landscape metrics have been made available.

6 Forest Classification

Land, considered in the broadest sense, has an extremely large number of attributes that may be used for classification and description, depending on the purpose of the classification and the needs of the classifier.

— C. J. Robinove, 1981

INFORMATION ON FOREST CLASSES

Remote sensing can provide information on forests through classification of spectral response patterns. Of interest is a summary of the distribution of classes, and map products that depict the spatial arrangement of the classes. The process of mapping the results of classification must necessarily follow the rules of logic, which express formally the philosophy and criteria by which maps for various management applications will be created and assessed (Robinove, 1981). In addition, classification and mapping are always done for some purpose; it is this purpose, and the skill of the analyst, which exert perhaps the strongest influence on the accuracy and utility of the final products. In this world of limited resources, computer support, and personnel, there are only a few practical ways in which the optimal remote sensing classification, from which usable maps can be obtained for sustainable forest management, can be accomplished.

The many issues and approaches to forest and land classification and mapping have generated a rich and specialized literature and language; what follows is an attempt to sort out some of the larger issues, particularly from the perspective of the producer and user of remote sensing classifications and maps in sustainable forest management. Of specific interest are the insights sought by users, who may need to understand and appreciate the role that unique forest classifications and maps obtained from remote sensing data can have in the process of forest management. For example, it is expected that remote sensing will continue to be the technology of choice in the creation of classifications and maps that are timely, synoptic, and at a particular level of detail that supplements the many map products available from the forest inventory GIS. Are maps produced from the classification of remotely sensed data fundamentally different from maps generated through GIS database queries? One expectation is that remote sensing will continue to be used to create maps that cannot be obtained readily or effectively in any other way. What are the unique aspects of remote sensing classifications?

Three themes or broad-scale issues affecting the implementation and use of a regional classification hierarchy to map forest vegetation are used to structure this discussion (Franklin and Woodcock, 1997):

1. Vegetation mapping requires a conceptual model of vegetation as a geographic phenomenon (gradients or patches mapped as fields or entities on the basis of vegetation attributes alone, or vegetation and environmental attributes).
2. Vegetation mapping is generally carried out within the context of spatial, temporal, or taxonomic hierarchies.
3. Taxonomic and process hierarchies are not necessarily spatially nested, e.g., different vegetation formations occur on the same landscape, and cover types occur discontinuously across different landscape units.

These three issues are discussed in the following sections. First, the process of classification and mapping is briefly introduced with a view to understanding the niche that remote sensing can occupy in mapping forests. This is followed by a discussion of the prevailing classification philosophies, and illustrative lists of classes and hierarchies that might be used. This discussion is followed by a brief recap of issues associated with remote sensing data and methods, covered more fully in earlier chapters. Then the chapter focuses on some highlights from the applications literature on using remote sensing at the various levels, or scales, of forest classification.

Mapping, Classification, and Remote Sensing

A map is a product of three operations (Robinove, 1981):

1. The definition of a hierarchical set of classes,
2. Assignment of each individual to a class — or the use of the decision-rule, and
3. Placement of the classified individual in its correct geographic position — the actual creation of the map.

The objective of image classification and mapping, then, is to use a decision-rule to generalize or group objects (pixels) according to the list of classes defined in Step 1 by examining their attributes — their spectral response patterns. Mapping is the completion of Step 3, the process of extending the classification to cover the spatial extent of the (georeferenced) area of interest. The list of classes defines in many ways the best way to develop the decision-rules and create the maps — but recall that the list of classes requires a conceptual model of vegetation as a geographic phenomenon (Franklin and Woodcock, 1997). As will be seen, not just any class list will be appropriate for use with remote sensing data.

Classification is used to determine the differences in attributes among the classes that will be mapped, or to allocate individuals to the classes based on these differences. Therefore, it is hoped, different landscape units will exist on either side of the line drawn on the map and on the ground between two classes. A landscape unit is homogeneous or acceptably heterogeneous with respect to an attribute or set of attributes of the forest used in the classification, such as plant lifeform, species composition, or tree density. Hierarchical forest classification is aimed at organizing the forested landscape into successively smaller units — roughly, forest covertypes, forest ecosystems, and forest stands — that can be managed uniformly (Bailey et al., 1978). The expectation in forestry is that the smallest landscape units, forests stands, will respond to a given management treatment in a coherent, predictable manner. Stands can be aggregated to represent forest ecosystems which, in turn, can be aggregated into forest covertypes at a particular scale useful to managers. Increasingly, information on the spatial extent and arrangement of forest covertypes, forest ecosystems, and forest stands are required for effective management. It should be clear that categorical resolution is defined by the definition of the unit and the cartographic resolution is defined by the map scale.

Note that this is a simplification of the true complexity of forest classification for management purposes, but this may be as good a structure as any from which to consider the wide variety of classifications and mapping products necessary to accomplish the goals of forest management. It seems unlikely that there will be a one-to-one correspondence between spectral response patterns, forest covertypes, forest ecosystems and forest stands; the different levels of classification provide an opportunity to consider the appropriate methods that must be used to convert the spectral response into the desired groupings of forest conditions on the ground.

What is meant by forest covertype can be understood by referring to the differences in classes that are to be mapped, and considering the more general case of vegetation types. There are, perhaps, as many ways of creating vegetation or forest types as there are attributes to divide them. Realistically, only a few ways of dividing one area from another area, and calling them different vegetation types, are of practical use. One approach — which goes by many different names, including the physiognomic approach — conforms to the general notion of vegetation types understood and used by most biologists, ecologists, foresters, and other resource management professionals (Whittaker, 1975). Vegetation classes are selected and described based on specific structural features, such as the percent cover by species in different strata (canopy, shrub, herb, moss layers). These structural features are simple to measure and record in the field using visual estimates, line intercepts, or crown cover photo models; although great care must be taken to ensure the sample is large enough, sites are selected according to a valid sample design, and reliable estimation or measurement procedures are followed (Curran and Williamson, 1985; Zhou et al., 1998). Vegetation types are usually considered equivalent to remotely sensed vegetation classes when these classes are carefully constructed and described using field or aerial photographic data.

Another way to think of vegetation or forest covertype is to consider the categorical resolution of the classification exercise. Each vegetation class within a single level, and at each successive level of the hierarchy, is different from the other classes in the way in which it is comprised of layers of vegetation. The layers can be described by considering a simple structural aspect of the class, such as the dominant species or amount or density of vegetation in each layer. The uppermost layer is often the most important in defining the class (Spies, 1997). The lower layers may be modifiers of the canopy layer description; this approach differs from the detailed floristic classifications and integrated classifications described in subsequent sections, although classes defined in this way can be a hierarchical component of either an ecological or more detailed floristic system. When vegetation types are not sharply defined, transitional classes may be required (Foody and Boyd, 1999).

The use of remote sensing in this process is based on the fact that the differences on the ground between vegetation types can be isolated or separated as differences in the image characteristics. When different vegetation structures define the classes, and these classes correspond with recognizable vegetation types on the ground, there is good reason to believe that the types can then be mapped with digital remote sensing data and methods (Merchant, 1981). The number of vegetation types described as part of a structural system that can be classified on satellite remote sensing imagery is large, and not yet fully known for a range of environmental conditions at a variety of scales and different sensor data (Graetz, 1990; Kalliola and Syrjanen, 1991; Franklin et al., 1994).

A simple example of the classification using remotely sensed data of common vegetation types that are known to differ on the ground can illustrate this ideal situation. Mangrove vegetation communities (or types) are known to differ in their structural features, particularly with respect to the density of dominant species (Davis and Jensen, 1998; Gao, 1999). Satellite and aerial remote sensing imagery acquired by optical/infrared and microwave sensors are known to be influenced by the amount of vegetation cover. In Mexico, Ramirez-Garcia et al. (1998) used this knowledge to map 10 classes, including 2 mangrove communities, with over 90% accuracy using a Landsat TM image, a supervised maximum likelihood classifier, and approximately 80 field plots. In French Guiana, Proisy et al. (2000) interpreted airborne SAR multipolarization and multifrequency imagery in 12 stands representing different mangrove communities, and successfully determined different levels of forest biomass representing different successional stages of mangrove forest dynamics. These studies illustrate the ideal case for the selection of remote sensing data and a classification approach; vegetation types are known to differ on the ground in ways that are amenable to a remote sensing measurement.

Sometimes, vegetation types are defined using structural attributes that are not amenable to remote sensing. Vegetation types defined on the basis of understory characteristics alone, for example, will not likely be spectrally distinct because the differences between the classes — perhaps the presence or absence of certain understory species — cannot often be detected reliably in full leaf-out with multispectral or microwave remote sensing data (Ghitter et al., 1995). The ability to classify such vegetation types with these remote sensing data would be near minimal, and would be restricted by the ability of what is remotely sensed — the canopy layer and gap structure — to predict what occurs beneath. Sometimes, image characteristics are known to be only poorly correlated with vegetation types, and ancillary data are used to help in the classification; even this may not be enough to provide high classification accuracy.

No doubt this simple way of considering the process of classification and deriving classification hierarchies by considering the characteristics of vegetation is already confusing enough, but the structural approach is only part of the classification problem. Many classifications are driven by reference not only to vegetation structure, but to a whole host of environmental factors (Frank, 1988; Franklin and Woodcock, 1997). In some areas of the world, vegetation is classified on the basis of site characteristics rather than the actual vegetation structure (Beauchesne et al., 1996). Since the resulting vegetation types are not based on observed vegetation structure, or even successional stages, they are not likely to be reliably determined from satellite imagery (Kalliola and Srjanen, 1991). The biophysical inventories of many of Canada's National Parks were constructed in this way (Lacate, 1969; Bastedo et al., 1983); homogeneous units were outlined on aerial photographs, but then named or labeled not primarily for the vegetation they contained, but rather for the interpreted site characteristics based more confidently on the hydrological regime and soil conditions than the existing vegetation.

Pure forms of the ecological land classification approach may have limited spectral distinctiveness — but it is worthwhile considering the broader classification literature to understand better the different types of classes that can arise when implementing a remote sensing classification using vegetation structure and environmental factors. In a broader sense, these latter classifications are more likely to generate the increased understanding that is needed of forest communities and ecosystems. It may be useful to examine this type of classification to determine how remote sensing can best contribute.

Roughly speaking, there are three quite different (yet linked) philosophical positions from which the list of classes for use with remote sensing data can be designed. The choice of the list of classes helps define the distinctiveness of the maps and the units that will be portrayed:

1. The genetic approach — landscape units are described by classes that differ on the basis of causal environmental factors (Mabbutt, 1968);
2. The parametric approach — landscape units are described by classes that differ on the basis of quantitative parameters (Blaszcynski, 1997); and
3. The integrated (or landscape) approach — landscape units are described by classes that differ on the basis of multiple criteria that describe recurring patterns of topography, soils, and vegetation (Mabbut, 1968; Christian and Stewart, 1968; Robinove, 1979, 1981).

These approaches are not pure, but rather represent ways in which three separate maps could be generated for the exact same piece of forest; all three can be used to generate map products of great interest and use in sustainable forest management for a variety of different applications. The forest stand maps of particular interest in forest management are an example of a mixed approach — typically, parametric and landscape criteria are used in their creation. The vegetation typing based on structure discussed above is a form of the parametric approach. Vegetation typing based on environmental factors, the ecological or biophysical land classification maps (Lacate, 1969), typically represent an almost pure form of the landscape approach.

Geomorphological or surficial geology maps are good examples of land classified according to the genetic approach. Classes might include depositional differences (McDermid and Franklin, 1995): alluvium, colluvium, eolian, and stable.

There is a long and valuable tradition of using remote sensing data in such mapping — more so in geology than in geomorphology (Young and White, 1994). Classification is not usually the main image processing approach used. The relationship between spectral response and the genetic attributes of interest is often weak or masked by marginally related or completely unrelated factors, such as in areas of dense vegetation or glaciated terrain. Geobotanical applications tend not to be based principally on the classification of spectral response, but rather on the interpretation of spectral differences (Vincent, 1997). Genetic land classifications are not used extensively in forest management, except perhaps as an ancillary source of information. Such maps can be useful in understanding soils and hydrology and in productivity modeling, for example. However, another example of a genetic classification, the stand origin map, has great value in forest management.

The parametric approach requires the description of terrain in physical, chemical, or engineering terms (Robinove, 1981). Geochemical and geophysical mapping are pure examples of the parametric approach, but for obvious reasons are not used extensively in vegetation mapping. A pure form of this approach to land classification based on vegetation data does not exist in forestry, but Kimmins (1997) referred to a version of this type of classification as the vegetative approach. The most common parametric classifications of interest in forestry use vegetation structure data; the quantitative structural features of vegetation such as percent cover in different layers. Maps constructed from this perspective have a major role in many forestry mapping projects and are amenable to remote sensing. A second parametric classification may be based on digital elevation model data. The many attempts to automate terrain analysis based on slope morphometry (Evans, 1972, 1980; Zevenbergen and Thorne, 1987; McDermid and Franklin, 1995), and to generate quantitative taxonomic schemes for terrain types and landforms based on geomorphometric data extracted from DEMs (Pike, 1988, 1999; Dikau, 1989; Blaszcynski, 1997), attest to the power of this classificatory approach.

Classifications of remotely sensed data based solely on spectral response patterns, as are most unsupervised clustering maps, qualify as parametric classifications. But rarely will a map constructed only with reference to spectral classes prove useful in application. Typically, the spectral classes are related in some way to the informational classes of interest to foresters, and those informational classes are more often constructed with reference to vegetation structure, floristics, or physiography. When other data are used, such as DEMs, or the clusters are modified to consider other attributes (merging clusters to create new class labels), a remote sensing classification may resemble more pure forms of the genetic or landscape classifications. Earlier, Robinove (1979, 1981) argued that since the spectral response of individual pixels was comprised of the total environment contribution reflectance (including vegetation, soils, and topography), then image classification was more similar to classification according to the integrated or landscape approach.

The landscape approach is sometimes called a biophysical or ecosystematic approach (Kimmins, 1997). Here, the classifier considers each parcel of land unique and classifies each on the basis of a complex of attributes — usually soils, topography (or landform), and vegetation — that are applicable to the purpose of the map (Robinove, 1981). Such classes when mapped over a landscape create the homogeneous units that are the phenomenological unit of management, sometimes called land facets, terrain units, or perhaps ecosites. The generic term for land classification results, landscape units, is preferred here to avoid confusion with these more specialized classifications.

It makes sense to say that all of these approaches generate classifications that are useful in sustainable forest management. To a large degree the approaches are interrelated, using many of the same variables and differing only in the scale at which they seem to work best. In fact vegetative (parametric) and ecosystematic (integrated) approaches tend to nest within the climatic and physiographic schemes (genetic), and are actually best considered as simply more detailed versions of the same procedures. How can understanding these ideas help in building a successful remote sensing classification project?

Purpose and Process of Classification

The purpose of the classification influences the desired end product and will help shape the actual process of mapping. Forest covertype, ecological classifications, stand maps, in fact all forest classifications, are designed to help answer two specific questions about the land (Sauer, 1921; Robinove, 1981):

- For a given area of land, what are its (forest) attributes?
- For a given use of land, which areas have the proper (forest) attributes?

Since there may be an infinite number of attributes, the first question typically reverts to a query aimed more at understanding which are the attributes of interest. In classification, the attributes of interest become the criteria upon which classes will differ: species composition, density, age, productivity, and so on. If the purpose of the map is to allow contiguous areas to be depicted in their natural state, then a single classification scheme will be needed for all areas to be mapped. That class scheme may be an imposed, generic classification structure — such as the Anderson et al. (1976) scheme discussed below. But rarely will a general purpose classification serve several specialized purposes equally well (Robinove, 1981; Bailey, 1996).

If the purpose of the classification is well-defined locally, then perhaps the class structure can be local as well. The optimal data and methods to achieve the desired product will be more obvious, but the use of such a map elsewhere (in adjacent forests, for example) will be less certain. If the purpose is not well defined, or subject to variability (perhaps shifting budgetary conditions), then the data and methods will be less certain; it will not be obvious which are the better data to use and which are the best methods. One likely outcome is that compromises may enter into the construction of the map. An obvious point at which this compromise can occur is the scale of the map. If the purpose of the map was not well defined, then it is likely that the appropriate map scale will not be particularly obvious. There is greater likelihood that the map will be constructed using source data that may turn out to be too fine or too coarse in resolution, rendering the final product less useful. The point is this: a remote sensing derived classification can be printed at any map scale, but the resolution of the source data are the critical factors in whether a useful map is produced. Often the question of scale and source data resolution are combined in the concept of the minimum mapping unit (MMU) — the smallest coherent object (e.g., polygon) expressed individually on the final map product.

Typically, the purpose of any general forest covertype classification is to provide an overview, a reconnaissance, an order-of-magnitude assessment of the forest condition and extent, the first or second level in the hierarchy of mapping products which might contain many levels, often culminating in the ecological community map (Beauchesne et al., 1996). Detailed forest covertype maps are required by managers in planning field work, preliminary stand assessment, the construction of covertype volume tables, forest community assessment, and a myriad of other uses. Identifying these uses will possibly help avoid the production of a map from remote sensing in which the spectral and spatial characteristics of the image classes are not completely compatible with the land-cover classes identified on the ground (Marsh et al., 1994). The difficulty of relating classifications to human use of the classification relates to the fact that remote sensing can reveal the spatial distribution of cover and species, but human users often interact with vegetation on the basis of its physical structure (in fairly small areas) and genetic properties (Smith et al., 1999).

In many ways, the methodological design (Curran, 1987) is an important issue to consider when reviewing remote sensing covertype classifications or when contemplating the initiation of a new classification project. While statistical results will vary from place to place, the way in which those classification products were generated has often proven equally valid in producing usable classification products under a wide range of forest and landscape conditions in many diverse places of the world. Classifications are essentially empirical creations, however, generally speaking, the fact that three classes of forest covertypes (softwood, hardwood, mixedwood) can be classified with approximately 85% accuracy in New Brunswick, Canada (Franklin et al., 1997a) suggests that approximately that level of accuracy can be achieved in a classification using these data and methods virtually anywhere in the world that a similar forest condition exists. Ranson and Sun (1994a: p. 152) put it this way:

> *... identifying different forest stands is possible, but not easy when the biomass of these stands are high. The principal components analysis we employed represents a 'best case' for separating the classes in our study area. The combination of channels may change with the landscape and should be determined from training data. However, the classification accuracies reported should be similar for similar sensors and forest types.*

Many factors may influence the success of a remote sensing classification and the performance of the image analyst; consider the effect that the comprehensiveness of the backgrounds of those on the project team (Robinove, 1979, 1981) and the degree to which the array of human resources assembled matches the size of the task to be completed (Green, 1999) might have on the final results. The complexity of the area for which a remote sensing covertype map must be produced will influence decisions. If the area is highly variable, then there will likely be more classes, rather than few — more variables, rather than few. If the area is not very well mapped or known, there will likely be more emphasis on field data collection.

Classification is an inherently multidisciplinary effort, benefiting greatly when people from different disciplines come together and view the landscape with their different perspectives.

There are remote sensing forest classification precedents in virtually all the major biomes of the world. However, some areas are better understood than others because of extensive prior work or the presence of long-term research initiatives (e.g., Shoshany, 2000). For example, some temperate, Mediterranean, and boreal conifer forest community types have been of interest to remote sensing scientists for several decades. A number of studies have been built up that enable any new classification project to benefit from what has been learned in that environment. The existence of these earlier studies can influence the design and outcomes of any new classification exercise.

CLASSIFICATION SYSTEMS FOR USE WITH REMOTE SENSING DATA

A glance at a listing of some classes used in the classification of Landsat type satellite imagery over the past 30 years for the purposes of general vegetation typing or land cover mapping provides a general idea of the kind of detail that is possible (Table 6.1). Digital classification of vegetation always begins with (1) an image and (2) a list of desired or expected classes. The process, typically, then considers the selection of the input data to be classified, the algorithm to be applied in the decision-rule, and the assessment of accuracy (Pettinger, 1982). Since all such classifications are applied on the basis of rules that conform to an internal logic that can be described, documented, and repeated, the results often depend on the purpose of the classification, the environmental context, and the skill of the analyst.

A good example of a hierarchical vegetative classification system is the Anderson et al. (1976) Land Use and Land Cover Classification System comprised of four Levels (I, II, III, IV). This classification scheme was published for use in the U.S. (the forest classes are shown in the first part of Table 6.1), but the logic can be applied almost anywhere. The system, designed for use with remote sensing data, assumes that no one ideal classification of land use and land cover can be developed, but flexible classes and an open-ended structure can be used to accommodate many of the different uses that such classification maps are intended to serve. The system has its origins in the mapping of land associations by aerial photographs, and is therefore not a pure parametric approach, but is linked to the landscape approach. The list of classes, and the general approach suggested by Anderson et al. (1976), has found wide acceptance as the basis for digital classification using remote sensing (Jensen, 2000). Numerous regional examples exist of this type of nested, hierarchical, standardized, and comprehensive classification approach.

A good example of a hierarchical ecosystematic classification system is described by Bailey (1996). The hierarchy of ecosystem units is based on almost a century of ecosystem research and land mapping applications around the world. As managers in many countries struggled with the need to recognize linkages between parcels of land based on energy and material exchanges, an integrated view of land

TABLE 6.1
Examples of Forest Classes and Levels Used in Landsat Sensor Image Classification

Level I	Level II	Level III	Level IV
Anderson et al. (1976) North America — Classification: General/Vegetative			
Forest land ------	Deciduous forest -----	Species levels	
	Evergreen forest		
	Mixed forest		
	Forested wetlands		
Beaubien (1979) Eastern Canadian Boreal Forest — Classification: General/Vegetative			
Forest	Softwood ---------	Very dense mature Bf	
	Hardwood	Mature Bf	
	Mixedwood	Young Bf	
		Overmature Bf	
		Overmature Bs with Bf	
		Overmature Bs (low density)	
		Open Bs	
		Ws regeneration	
		Defoliated Bf (hemlock looper)	
		Dead Bf (looper kill)	
Beaubien et al. (1999) Western Canadian Boreal Forest — Classification: General/Vegetative			
	Coniferous Forest-----	High crown density	
		High crown density, younger	
		Medium crown density	
		Medium crown density, lichen cover	
		Low crown density	
		Low crown density, lichen cover	
		Very low crown density	
	Deciduous forest -----	High crown density	
		Low crown density	
	Mixed forest -------	Mixed coniferous forest	
		Mixed deciduous forest	
		Mixed open forest	
		Mixed with shrubs	
	Open land---------	Wetlands	
		Burns ------------	Recent (black)
			Older (green)

TABLE 6.1 *(Continued)*
Examples of Forest Classes and Levels Used in Landsat Sensor Image Classification

Level I	Level II	Level III	Level IV
Pettinger (1982) Southern Idaho, U.S. — Classification: General/Vegetative			
Forest land	Conifer		
	Hardwood	Aspen	
	Mixed	Conifer and Aspen	
	Forested wetland	Riparian hardwoods	
Franklin (1987) Northern Canada — Classification: Ecosystematic, Integrated or Ecological Community			
Forest land	Conifer forest		
	Woodland (open forest)		
Skidmore (1989) Southeast Australia — Classification: General/Vegetative			
Forest land	Silvertop Ash		
	Yertchuk		
	Stringybark Gum		
	Blueleaved Stringybark		
	Tea tree		
	Black Oak		
	Silvertop Ash-Gum		
Davis and Dozier (1990) Southern California — Classification: Ecosystematic, Integrated or Ecological Community			
Forest land	Conifer forest		
	Oak forest		
	Oak Chaparral		
	Chaparral		
	Coastal Scrub		
	Grassland		
	Riparian woodland		
Marsh et al. (1994) Brazilian Amazon — Classification: General/Vegetative			
Forest	Gallery		
	Secondary	Semideciduous (broadleaf mesophytic)	
		Tall semideciduous	
		Riparian	
Wolter et al. (1995) Northern Midwest U.S. — Classification: Floristic/Species			
	Conifer	Red pine	
		Jack pine	
		Black spruce	
		White spruce	
		Mixed swamp conifer	
		Tamarack	
		Northern white cedar	

TABLE 6.1 *(Continued)*
Examples of Forest Classes and Levels Used in Landsat Sensor Image Classification

Level I	Level II	Level III	Level IV
	Hardwood ---------	Black ash	
		Northern red oak	
		Northern pin oak	
		Sugar maple	
		Trembling aspen	
		Mixed aspen	
	Mixedwood --------	Balsam fir — aspen	
		E. white pine — hardwood	
		Paper birch — conifer	
		E. hemlock — yellow birch	
		Black ash — lowland conifer	
		Northern pin oak — pine	
		Jack pine — oak	
Jakubauskas (1996) Yellowstone National Park, U.S. — Classification: Ecosystematic, Integrated or Ecological Community			
		Lodgepole pine	Successional stages (5)
			Postfire regeneration
			Dense, small dbh
			Mature
			Mesic, mixed
			Xeric
			Pine beetle infest.
Hall and Knapp (1994a,b) and Cihlar et al. (1997) Northern Saskatchewan, Canada — Classification: General/Vegetative			
	Evergreen needleleaf ---	Wet conifer --------	Crown density classes (4)
			High (>60%)
			Medium (40–60%)
			Low (25–40%)
			Very low (10–25%)
		Dry conifer	
		Deciduous broadleaf ----	60–80% broadleaf trees
			40–60% broadleaf trees
		Mixed	
Ramirez-Garcia et al. (1998) Nararit, Mexico — Classification: Floristic/Species			
	Low deciduous forest		
		L. racemosa (Mangrove community)	
		A. germinans (Mangrove community)	

classification developed that could accommodate the holistic approach of recognizing units at different scales by their common attributes. The site or ecosite is the smallest (a few hectares) homogeneous ecosystem recognized by foresters and range scientists; it is comprised of not only key criteria in the vegetation layer but an understanding of the functioning relationships between components in the vegetation, soils, topography, geology, and climate.

The ecosystematic approach is based largely on the definition (or philosophical understanding) of a landscape unit as a homogeneous area of soils, topography and vegetation easily recognizable on aerial photographs. Originally, this way of viewing the landscape was applied over large, unmapped areas in Australian land systems (Christian and Stewart, 1968) and Canadian ecological land classifications (Lacate, 1969). Refining these concepts, Bailey (1996) refers to the lowest level of landscape units as microecosystems. Linked sites create a landscape mosaic (mesoecosystem, or land system, or ecosection) that from above resembles a patchwork largely defined by landforms (Swanson et al., 1997). Landscape mosaics combine to form macroecosystems that are consistent with broad physiographic regions, for example, the lowland plains of the western U.S., and are principally separated by climatic criteria. In Canada, the Ecological Land Classification process culminated in the following hierarchical ecological land classification terminology and associated mapping scales (Rubec, 1983):

- Ecoregion (1:3,000,000 to 1:1,000,000),
- Ecodistrict (1:500,000 to 1:125,000),
- Ecosection (1:250,000 to 1:50,000),
- Ecosite (1:50,000 to 1:10,000), and
- Ecoelement (1:10,000 to 1:2,500).

A scaling relationship between different remote sensing data (e.g., different spatial resolution and areal extent of imagery from different sensor/platforms) and these ecological land classification units was first discussed by Murtha (1977), who provided a hierarchical cross-listing of the relationships between scale, categorical detail, biogeoclimatic ecosystem classification, and management activity. Understanding these co-relationships may create greater understanding of the source and resolution of resource conflicts that help generate the demand for multiscale information in the first place.

LEVEL I CLASSES

Climatic and Physiographic Classifications

The climatic and physiographic classifications are generally broad mapping systems that cover large areas, such as continents, usually with little spatial detail. Physiography is the comprehensive study of surface form, geology, climate soils, water, and vegetation, and their interrelationships (Townshend, 1981c); clearly, only very general differences can be interpreted and classified using the influences of this broad set of properties. Simple physiographic class descriptions include water, forest, cultivated lands, urban development, grassland, and alpine, but at larger and larger scales finer and finer divisions are introduced, and the physiographic approach smoothly integrates with the landscape approach (Mabbutt, 1968).

Many such general classifications exist based on climate and physiographic features in which more detailed forestry classifications are embedded. For example, in Canada all forest, vegetation, and resource classifications are organized into ecological regions (Ecological Stratification Working Group, 1996); in Alberta, a similar regionally sensitive function was performed by the classification of Natural Regions and Subregions (Strong, 1992). The landscape units are usually defined at the scale of mapping below (i.e., smaller than) about 1:500,000. At this mapping scale, the resulting physiographic maps largely resemble climatic classifications and have their greatest impact as regional and global information resources. Their utility in sustainable forest management is as the first layer, or step, in the classification hierarchy — at the strategic level of information — for example, in climate change modeling, prediction of carbon flux for countries and continents (Gaston et al., 1997; Cihlar et al., 2000), and in calculating areal extent of the broad physiographic features for a region (Vogelmann et al., 1998; Lunetta et al., 1998). In Alberta, for example, a certain percentage of land in each of the natural subregions is targeted for preservation in a natural state; in British Columbia, the set-aside target is 12% of all ecosystems (ecosections are used to define the terrain and the biogeoclimatic ecosystem classes are used to define the ecology) (Murtha et al., 1996).

Traditionally, when using climate or physiographic mapping criteria, potential vegetation is considered rather than actual vegetation. With this approach, the individual plants and communities that comprise a landscape unit are less important than the broad patterns of growth constrained by climate. As remote sensing information products such as the continental NDVI data sets with global coverage at low spatial resolution became available, it was possible to consider the actual vegetation within physiographic provinces or climate zones. Classifications of vegetation produced directly from large-pixel satellite reflectance data such as acquired by the AVHRR, SPOT VEGETATION, or MODIS sensors are good examples of this updated physiographic classification approach (Cihlar et al., 1997; Foody and Boyd, 1999). This is a more useful classification structure in global modeling studies, as well as in providing information that can be used in broad planning exercises. Such physiographic maps are likely to be produced as part of the organizational infrastructure of a country or region rather than within individual forest management units (Loveland et al., 1991).

One example is described in more detail here. In considering the global scale, Running et al. (1995) suggested one remote sensing approach for these small-scale (i.e., large area) climatic and physiographic classification systems for mapping. A classification system was based on classes distinguishable in the coarsest resolution satellite imagery for which global converage was practical (e.g., AVHRR, SPOT VEGETATION, or MODIS data). Six fundamental vegetation classes that differ in three fundamental attributes resulted:

1. Permanence of aboveground biomass,
2. Leaf longevity, and
3. Leaf type.

The first criterion separated areas with a permanent respiring biomass from annual crops and grasses (Running et al., 1994). The second criterion separated evergreen from deciduous canopies, a critical distinction for carbon-cycle dynamics of vegetation. The third criterion created classes based on needle-, broad-, and grass-leaf types. Once these classes were mapped, regional climate data — precipitation amounts, for example — could be used to create subclasses at lower levels of a hierarchy. The simplicity of these classes compares favorably to the more complicated floristic logic used in earlier continental-scale physiographic classifications, also based largely on remote sensing (e.g., Loveland et al., 1991). Such a system is clearly designed less for forestry than global ecology and carbon budget modeling. The maps are primarily useful in forestry as a way of organizing the more detailed mapping that must be done on a regional and local scale. The minimum mapping units are quite large (e.g., several to many square kilometers).

LARGE AREA LANDSCAPE CLASSIFICATIONS

The physiographic and climatic classifications discussed in the previous section often have nested hierarchical subclasses, or subzones, that continue division but with more precise criteria. Many of these successful land classification systems are based to a large degree on the landscape approach; it is the integration of several different land attributes that constitutes a difference of interest to the classifier. Obviously, even a continental scale physiographic and climatic classification is an integrated classification, but at such a small scale (large area extent) as to be of little interest to forest managers in operational settings. Here, classes defined by landscape methods tend to work best at larger scales (smaller area covered), and when local conditions are accommodated. This means that detailed classifications in one area will not often be transferrable; the classes may not be transferred, but the methods of recognizing and classifying them certainly can be. The aim is to facilitate the logical and repeatable separation of large areas of land into increasingly smaller landscape units that suit the needs of the user.

Many of the early land cover classification, land systems, soil assessment, and forest resources mapping projects grew out of the photomorphic tradition that had been the dominant land mapping paradigm following the widespread adoption of aerial photography as a base mapping tool in the 1950s (Stellingwerf, 1966; Christian and Stewart, 1968; Townshend, 1981c). For example, Webster and Beckett (1970: p. 52) commented that "a procedure for predicting soil or other terrain attributes over large areas with limited access was seen to depend on terrain classes within each of which the terrain was of the same kind *and which could be consistently recognized from air photographs*" (italics added). These surveys were designed to indicate (usually in a comprehensive interpretive map legend), but not actually to map the detailed terrain characteristics (Christian, 1958) and to allow a stratification such that additional detailed surveys (for the purpose of mapping soils, hydrological features, slopes, or vegetation) could be planned and embedded within the larger context (Lacate, 1969).

One brief example can serve to illustrate the role of aerial photography in this integrated landscape classification paradigm. Paijmans (1970) recognized major vegetation groups in New Guinea based on dominant life forms. The vegetation groups were readily distinguishable on air photos, providing the logical framework for the final vegetation classification which was based on detailed photointerpretation and field work on structure and floristics (Paijmans, 1966). First, the interpreters worked to separate out grassland, mixed herbaceous vegetation, palm and pandan vegetation, scrub and thicket, savanna, woodland, and forest. Second, relief features were used to determine hydrological and soils conditions (coastal saline and brackish environments, beach ridges and swales, coastal back plain, floodplains, hills and mountains, undissected plateau). Third, some assessments of land capability were made based on agricultural and forestry resource uses. The strategy was "to first delineate as many different photo patterns as one can, and then to determine, by field investigation, which patterns are significant in terms of land capability" (Paijmans, 1970: p. 99).

This is precisely the same strategy employed in many digital satellite remote sensing projects; first, find as many spectrally distinct features as possible (unsupervised clustering) and second, label or otherwise train the individual classes in a supervised classification. The classification paradigm that was used to guide the use of aerial photographs throughout the 1940s to 1970s was immediately extended in digital remote sensing classifications. The aim was to map Level I forest and landscape classes from satellite data. The first step was often a manual interpretation of satellite image hardcopy products (Rubec, 1979, 1983; Gregory and Moore, 1986) or simple computer displays of band ratios, density slices, and stretches (Clark et al., 1985). The photomorphic approach was seen as an interim method to manually explore the new digital satellite and airborne imagery data (and, almost incidentally, generate usable maps) until automated classification techniques were more fully developed and available. Early remote sensing practitioners sometimes felt that the best approach was to enhance the image and leave it in the hands of a competent interpreter (Story et al., 1976; Jobin and Beaubien, 1974; Ringrose and Large, 1983; Rubec, 1983; Ryerson, 1989). This reduced the amount of training that would be required to generate significant map products to a few hours or days, rather than the lengthy learning times required for a digital approach to be implemented.

Experience in photointerpretation was not always an asset in this process; it was found that experienced photointerpreters were soon bored with the process of outlining photomorphic units (Kreig, 1970) and were more interested in higher-order cognition and deductive reasoning (Colwell, 1968). Unskilled interpreters could be expected to bring higher energy and enthusiasm to the task, and the task did not require high levels of training or skill. Another advantage to manual interpretation of imagery was that no sophisticated equipment was required (Oswald, 1976). This meant that in many areas of the world in which a technological infrastructure could not be supported, only manual methods were contemplated as feasible.

But elsewhere, the drive to create objective methods of classification sometimes created an atmosphere in which the expertise of the interpreter was considered unnecessarily subjective. This is still an important underlying rationale for continued development of automated methods and expert systems. Now, it is more or less understood that all classificatory methods are subjective, differing only in degree of subjectivity and an understanding of the influence of this subjectivity on the actual map results. It is important to continue development of increased automation in classification so that the digital nature of the data can be fully exploited, but this will likely succeed only when the process is fully integrated with the recognized power of human image analysis (Swain and Davis, 1978). The human mind is perhaps the finest available tool for synthesis and analysis of image patterns; instead of discrediting human skill in interpretation, a more appropriate strategy is to utilize, as much as is possible, the expertise of the interpreter. The concept of visual interpretation of remote sensing imagery for classification is far from obsolete.

As image spatial resolution continues to improve (e.g., IRS-1D with 5.8 m panchromatic, IKONOS-2 with 1 m panchromatic and 4 m multispectral data) and photo-quality imagery becomes more common from satellite altitudes and improved airborne systems, a resurgence in manual image interpretation can be expected using the principles of the photomorphic approach. On-screen digitizing of forest roads using SPOT 10 m panchromatic imagery, for example, has been used in areas where a high contrast between roads and surrounding features can be expected (Jazouli et al., 1994). In Canada, Alberta Environment (Dutchak, 2000) initiated a 3-year, $3,000,000 program to update access features (roads, seismic cuts, and depletions) in forested areas of the province using manual interpretation of orthorectified IRS 5.8 m spatial resolution images. The approach is time-efficient and is more likely to be adopted by operational forest management units than the automated extraction of roads and other access features from multispectral imagery. Even in urban areas with high road densities and highly structured patterns, automated approaches to road detection and mapping are barely considered feasible (Karimi et al., 1999; Guindon, 2000). Optimal tools (and data) are not yet readily available (Wang and Liu, 1994).

Several different digital approaches have been used in classification of Level I class mapping applications, based on an analysis of the spectral differences among the classes of interest in different regions of the world. For example, in temperate and boreal regions, forest areas exhibited tonal differences on early false-color composite Landsat images which indicated variations in stands or successional stages (Heath, 1974; Beaubien and Jobin, 1974; Fleming et al., 1975; Oswald, 1976). Dark tones were produced by dense stands of old-growth trees. Mature and older stands of white spruce, western hemlock (*Tsuga heterophylla*), mountain hemlock (*Tsuga mertensiana*), subalpine fir (*Abies lasiocarpa*), and western redcedar (*Thuja plicata*) showed darker tones than did lodgepole pine (*Pinus contorta)* or Douglas-fir stands. Subsequent studies noted that variations in image interpretation could be caused by spectral bands, spatial resolution, temporal resolution (seasons), and processing (atmospheric conditions and photo quality) (Beaubien, 1979).

In areas of flat terrain, such as Anticosti Island in Quebec (Beaubien and Jobin, 1974), the forest classes visible in normal and false-color composite Landsat satellite images were thought to be formed principally by species differences, stand age, and density. The younger and/or denser a stand, the higher its spectral response, especially in the near infrared; growth rate appeared to generate a similar effect. In more rugged terrain, such as on the Laurentian Plateau in Quebec (Beaubien, 1979), the forest classes (comprised of the same species) visible in the imagery were more influenced by slope. Stands with a greater exposure to sunlight contained more black spruce (*Picea mariana*), and those south-facing stands were older and had larger diameters and lower densities than those with more northerly exposures. The reflectance, therefore, expressed a balancing of factors … "old stands with a fair proportion of black spruce will have a higher reflectance because they are exposed to sunlight, and vice versa for younger stands growing on slopes with a northern exposure" (Beaubien, 1979: p. 1142). Younger stands typically were brighter and more variable than older stands, which tended to be darker and more smoothly textured in Landsat imagery (Walsh, 1980, 1987; Franklin, 1987). Cutovers were very bright, burned areas were dark, and forest defoliation was bright, but not as bright as cleared areas.

Level I classification studies all over the world proceeded (and still do) from this type of basic observation of the spectral and physical differences between adjacent areas that differ physiographically or structurally on the ground. The concern is to translate these general image patterns into Level I categories useful in:

1. Estimating the regional extent of forest cover (Markon, 1992; Prins and Kikula, 1996),
2. Reconnaissance mapping in areas for which more detailed maps do not yet exist (Talbot and Markon, 1988; Wilson et al., 1994),
3. Global and regional forest inventory (Loveland et al., 1991; Ahern, 1997; Homer et al., 1997), and
4. Creating a base for landcover, climate, and carbon budget change and modeling studies (Foody and Boyd, 1999; Cihlar et al., 2000).

At Level I, the principle is that a forest covertype class must be part of a system which is clearly based on a physiographic or structural attribute, such as vegetation cover. In reality, such classes may be defined almost without regard to the data that will ultimately be used to map them — almost any source of spatially explicit information (aerial photographs, satellite imagery, even DEMs) can be used to produce such general classes at the coarse scale of the hierarchy. Because of the general nature of the classes, the maps will be quite accurate (Pettinger, 1982). After all, with appropriate spatial resolution in any of these data sources it is hard to confuse forest and water, meadow and rock.

While such maps are not simple to validate (Thomlinson et al., 1999), validation ensures that derived products meet claimed specifications. Validation of classification maps can be considered as part of the general difficulty in validating the products of remote sensing data analysis (Cihlar et al., 1997b):

1. Initial product validation — the process of establishing the quality of an algorithm by assessing the product generated by the algorithm; and

2. Continuing (process) validation — the process of establishing how well the algorithm performs if the area of interest, time, and data are changed (e.g., new satellite sensor in a different year in a different forest type).

As in validation of modeling results, image classification results can be validated by comparison to some independent assessment, perhaps field-based observations. In large-area classifications this may be difficult; how does one observe classes over many hectares corresponding to individual 1-km pixels? Another image classification product generated using different data can be used (Biging et al., 1995; Moody and Woodcock, 1995; Kloditz et al., 1998). For example, the validation and calibration of maps and models based on an AVHRR classification with a higher spatial detail classification derived by Landsat Thematic Mapper has been successful in boreal forests (Fazakas and Nilsson, 1996) and in tropical areas (Mayaux and Lambin, 1997). At a different scale, Marsh et al. (1994) used airborne video data to validate the classification of Amazon forest types in a Landsat TM classification exercise. By far the most frequent method of validating Level I classifications has been through aerial photointerpretation.

LEVEL II CLASSES

For forest management tactical and operational planning, Level I classes discussed in the previous section are much too general; for these purposes, Level II and Level III maps are usually required. Note that even Level II maps are still fairly general, and a primary use is as a starting point in generating still more detailed maps, such as forest productivity maps (Clerke et al., 1983; Franklin et al., 1997a), or perhaps simply as a way of organizing or stratifying classification projects of smaller land areas (He et al., 1998). Another use of Level II maps might include ways of depicting the environmental context within which the more detailed maps — such as maps of forest stands, wildlife habitat, and harvesting blocks — can be considered. The typical forest inventory approach has been to go directly to the most detailed level of mapping required (the stands), and generalize the map categories by moving backward through the levels. This assumes the forest inventory database can serve the purpose intended (i.e., timely data, attributes of interest, and so on). The approach considered here is to define the mapping categories at each level, and use a different set of data (remote sensing) to classify and map those categories in a nested fashion.

At successively larger mapping scales, application of the principles of the climatic or physiographic approach to any spatially explicit data produces smaller and smaller mapping units with higher spatial and categorical detail, but there is a limit to the amount of detail that can be provided without recourse to more precise differences in class attributes. Resource management maps must use new criteria that are more narrowly defined; terminology can be confusing, but these generally conform to either the vegetative or ecosystematic approaches (Kimmins, 1997), the vegetation structure or environmental factors approaches (Franklin and Woodcock, 1997), or, using the older and even more general terminology, the parametric or landscape approaches (Mabbut, 1968; Robinove, 1981).

These are the more detailed mapping systems in which individual components of vegetation or a combination of landscape attributes are used to map or label classes. The scale of mapping might be anything between about 1:5,000 and 1:25,000 and the minimum mapping unit might be as small as a few hectares or less. The approach is more or less compatible and consistent with the use of photomorphic units in standard, metric aerial photointerpretation; typically, the classes are devised using traditional forestry concepts such as forest species composition and structure as the divisive criteria. Because it is integrative, by using traditional forestry attributes together with soils and topography the ecosystematic approach may contain the greater potential (Bailey, 1996; Spies, 1997).

Typically, the process is to divide the single physiographic forest cover class into several covertypes according to either the vegetative or ecosystematic approaches depending on the purpose of the mapping and the available data. By progressively narrower definitions of classes, the forest covertype is deconstructed systematically into smaller and smaller units, finally yielding the forest stand. The final product might show classes consisting of lifeform classes (e.g., conifers and deciduous), species and structural differences (e.g., pine and aspen classes, open or closed canopy), or ecological communities. For example, in a vegetative approach, the forest is separated from other landscape features because it is comprised of land with trees present. The forest area can be divided, perhaps using an approximation of the number of trees (e.g., forest >500 trees per hectare, woodland <500 trees per hectare). Each of these areas may be divided still further using more detailed forest attributes: dbh, height, crown closure, and age. In an ecosystematic approach, reference to soils and topographic features might be used to refine the classification to the forest community level of detail. That approach is described in a later section.

Structural Vegetation Types

In classifying different Level II classes, such as forest covertypes, there is no substitute for an examination of the spectral response pattern in the available image bands and transforms. From this study, a judgment can be made as to whether these data are likely to provide the necessary discrimination. This judgment can flow from a basic understanding of the behavior of biophysical variables, some of which are used in the description of classes such as vegetation amount and cover, and their influence on remotely sensed data (e.g., Jensen, 1983; Curran, 1980; Leckie, 1990a,b). For relatively simple classification purposes, it is often not necessary to acquire a detailed physical understanding of the spectral response; such understanding is more critical in still more detailed classifications (later sections in this chapter) and in continuous variable estimation (Chapter 7), but not necessarily in a limited generalization procedure. The single largest impediment to more widespread use of classification procedures may well be the lack of understanding and familiarity with the necessary software, rather than a limited appreciation of the physics involved. Silva (1978: p. 22) suggested that "The user of a remote sensing system frequently is able to process data for relatively simple applications without serious concern for radiation and instrumentation."

Some might argue that a Level II classification is not really a simple application; but in essence, at this level of a classification hierarchy the goal is simply to reduce the variance in the remote sensing image data set to a number of broad classes. Usually, something on the order of 15 to 25 classes are required in forested environments. Occasionally, many more classes have been separated using subsequent overlays of forest cover, crown closure, stand development, topographic data, or other attributes (Congalton et al., 1993). Initially, however, the idea in a Level II classification is to classify the features in the image into fewer mappable landscape units, perhaps based on a single criterion such as dominant species or canopy cover. These features of interest are usually readily observable on aerial photographs, and usually also in simple image products such as image enhancements generated from digital remote sensing data. A dominant-species forest covertype classification is an example of such a Level II classification that is useful and required in operational forestry settings.

To drive a Level II classification using standard classifers, a general statistical understanding of the different class reflectance patterns is needed. For example, the influence of an increase in relative vegetation amounts in two classes — readily visible as a different color on the imagery or tone on an aerial photograph — is a predictable decrease in mean red spectral response and an increase in mean near-infrared spectral response. Simple Level II classes, such as an open and closed conifer forest, would vary in their respective amounts of vegetation, and therefore, in their mean red and near-infrared spectral response. A closed forest canopy would typically appear darker in the red band (more absorption) and brighter in the near-infrared (more scattering) than an open forest canopy. Therefore, mean red and near-infrared spectral response should be useful in discriminating these two forest covertype classes. The opposite relationship has occasionally been found in situations usually attributed to the contribution of the understory in the open stand (Ahern et al., 1991). This finding has reemphasized the critical role that local knowledge of forest conditions can play in understanding the spectral response pattern and the correct use of the image data.

Considering different species in layers is a relatively simple though effective way of considering forest covertype structure that lends itself well to an interpretation of multispectral reflectance differences. One such system based on cover was used in the Botswana Kalahari in mapping two classes of woody vegetation (Ringrose and Matheson, 1991): (1) multistory vegetation cover (representing dense browse) and (2) single-story vegetation cover (representing relatively less-dense browse). Another, based on covertype differences and hierarchical ecological principles of classification, was developed for the Kananaskis Valley within the Subalpine Forest Region in Alberta (Legge et al., 1974). Following the development of that early classification scheme, a map showing three forest types and eight landcover classes was required for a portion of the Kananaskis Valley (Franklin et al., 1994). A vegetation classification based on dominant species, conforming to the Anderson et al. (1976) Level II system was devised based on limited field work at 197 field sites and extensive photointerpretation using a 1:40,000 scale, black and white aerial photographs. The separation of the forest covertypes of interest was accomplished using spectral data extracted from a 1984 August Landsat TM image; the elevation,

TABLE 6.2
A Summary of the Classification Accuracy Based on Different Combinations of Spectral and Topographic Data in Forest Covertype Classification in a Montane Forest Region of Alberta

Forest Covertype	Classification Accuracy[a]		
	Landsat TM	SPOT	Spectral/DEM
Lodgepole Pine	81.5	59.7	100.0
White Spruce	53.4	58.7	71.8
Mixed Conifer	27.8	48.2	34.7
Mixed Conifer/Deciduous	77.8	83.3	100.0

[a] Compared to field identification at more than 200 sample sites visited on the ground and described according to percent cover in layers (structural).

Source: Modified from Franklin et al. (1994).

slope, and aspect data were extracted from gridded DEM data (originally produced from an interpolated 1:50,000 contour map). Table 6.2 contains a summary of the classification accuracy obtained by applying the discriminant analysis decision-rule (built using the training areas) to the 100 independent test sites. Overall accuracy was 66% with spectral data alone, 79% with spectral and DEM data.

Another example of forest covertyping by remote sensing was provided by Gonzalez-Rebeles et al. (1998) in the Rio Bravo/Rio Grande Region, following methods developed in the Gap Analysis Project (Scott et al., 1996):

1. Map vegetation or land covertypes.
2. Model vertebrate distributions (geographic locations data and/or habitat association models).
3. Delineate land management categories (ratings of protection).
4. Overlay 1, 2, and 3 to determine if gaps exist in the correspondence between vegetation covertype, species distributions, and management/protection categories.

Each step is critical to the Gap analysis, but the first step, that of mapping land covertypes, can be definitive. Typically, a combination of Landsat TM imagery, aerial videography, aerial photography, field reconnaissance, and other ancillary information are employed (Lillesand, 1996; Murtha et al., 1996). A good example of the approach was implemented in Utah, where Homer et al. (1997) described the classification of 24 Landsat scenes. The process was to perform unsupervised clustering, and then develop the relationship between the clusters and field classes by photointerpretation and field visits. Subsequently, each class was modeled using ecological rules that included topographic information from a DEM. A key feature of this process was the maintenance of data lineage, such that there was the ability to both step up and step down the classification hierarchy to less detailed or finer classes, respectively. The final map of the state yielded 36 covertypes and was found to be approximately 75% correct.

Similar results have been reported when using satellite optical data, with or without a DEM, and this type of Level II classification in a variety of different boreal (Franklin, 1987), temperate (Bolstad and Lillesand, 1992), and tropical forests (Skidmore, 1989). Level II classifications improved significantly when the increased spatial and spectral resolution of the TM sensor (over the MSS) was used (Chavez and Bowell, 1988); the TM had much higher information content for use in forestry (Horler and Ahern, 1986). Typically, Level II classification accuracy increased significantly when using Landsat TM compared to SPOT (fewer bands) (Franklin et al., 1994). Unfortunately, despite the higher TM information content, it was found that the data could not be immediately used to create practical forest classifications, e.g., at the Anderson Level III or below (Wolter et al., 1995). The level of the classification was, for the most part, still restricted to the equivalent of Level II classes. At best, a single Landsat TM image could be used to discriminate classes that were intermediate between Level III and Level II (Franklin, 1992, 1994).

The highest possible levels of detail that can be reliably mapped using Landsat TM satellite data, depending on the forest heterogeneity and the methods of analysis, appears to be at the dominant tree species level. For additional accuracy at this level, the user must:

1. Acquire multitemporal image data (Jakubauskas et al., 1998) or another type of remote sensing imagery (e.g., SAR data),
2. Use the DEM data in different ways (Bolstad and Lillesand, 1992), or
3. Use different image processing techniques in the classification process (Skidmore, 1989).

By incorporating knowledge of species phenology with multitemporal TM satellite imagery, it is possible to develop a forest classification with dominant tree species-level precision. Seasonal satellite sensor digital data can capture specific phenology of forest covertypes in Wisconsin, for example (Wolter et al., 1995). Image ratioing and ratio differencing can be used to create new variables which require fewer training areas, reducing (or normalizing) some of the variability caused by atmosphere, soils, climate, and aspect over a forest region. In this same Wisconsin forest, Bolstad and Lillesand (1992) reported that a rule-based approach using Landsat TM data, soil texture information, terrain position, and soil-plant relationships could separate 13 landcover classes corresponding to Anderson Levels II and III with 89% accuracy — a 16% improvement over a standard spectral classification method. As discussed in earlier chapters, digital elevation models are the most obvious ancillary data source for use in this type of classification project; when combined with spectral data DEMs can be used to provide more accurate Level II classifications than the use of spectral data alone (Hutchinson, 1982; Bolstad and Lillesand, 1994; White et al., 1995).

In early studies of mountainous areas with Landsat MSS data, Strahler et al. (1978) and Fleming and Hoffer (1979) showed that if an *a priori* rationale existed to include topography in a classification process, the accuracy could be increased significantly and the level of detail in the mapping could be greatly improved. In one of the earliest combined spectral/topographic classification studies of forests, McLeod and Logan (1980) noted that initial classifications of forest cover could use DEM data in two ways: (1) as additional variables in the classification and (2) to modify prior probabilities of occurrence of forest classes. They decided to do both. The approach was to consider the topographic information in a predictive vegetation model (Figure 6.1) and then combine the results with Landsat-based forest cover and volume maps. First, the spectral and textural data were clustered into a large number of pure clusters, then the clusters were clustered with reference to the likely relationship between topographic data and various species. In a final step, the clusters were labeled as classes to be mapped according to field-calibrated volume estimates and modeled species proportions. By interactively editing the classes using the digital image display, they compared Landsat MSS/DEM classes to air-photo-mapped forest stands and created polygons made to resemble those on forest stand maps. A good fit was obtained because of the strong dependence of forest species and volume on topography, although no final accuracy statistics were generated.

Improved results in the classification and mapping of forest and other vegetation types from remote sensing data using topography have been reported in many areas of the world using the more recent Landsat TM and SPOT image data (Franklin, 1992, 1994; Pickup and Chewings, 1996). Generally, the DEM can provide anywhere from 10 to 30% increase in mapping accuracy.

Another strategy has been to use existing map data or other data to stratify the imagery prior to or during classification procedures. Boresjö Bronge (1999) provided a recent example of an integrated TM/topographic map approach in Sweden; almost the same classification that was derived from manually interpreting a 1:30,000 color infrared photograph could be obtained at close to 90% classification accuracy using Landsat TM and a series of topographic map masks. These masks were organized such that spectral confusion was eliminated in successive stages. Four conifer, three deciduous, eight mire, and eight other classes were mapped. Most of the error was contained in the various mire classes which were spectrally and topographically similar.

Beaubien (1994) introduced the enhancement-classification method (ECM) used subsequently by Cihlar et al. (1997) in the production of the land cover map of Canada. The procedure uses the information visible in an image that has been contrast enhanced and filtered using standardized methods. Groups or clusters of distinct color patterns are labeled (using field knowledge) and classified with a minimum Euclidean distance decision-rule. Depending on the complexity of the area, the quality of the imagery, and the knowledge and experience of the interpreter, Landsat TM imagery can be used to identify the following forest types (Beaubien, 1994):

- Three softwood density classes,
- Two softwood age classes,
- Two hardwood age classes,
- Softwood, mixed wood, and hardwood regeneration,
- Five mortality levels in softwood-mixed wood stands, and
- Two openland classes, totally open and partially vegetated.

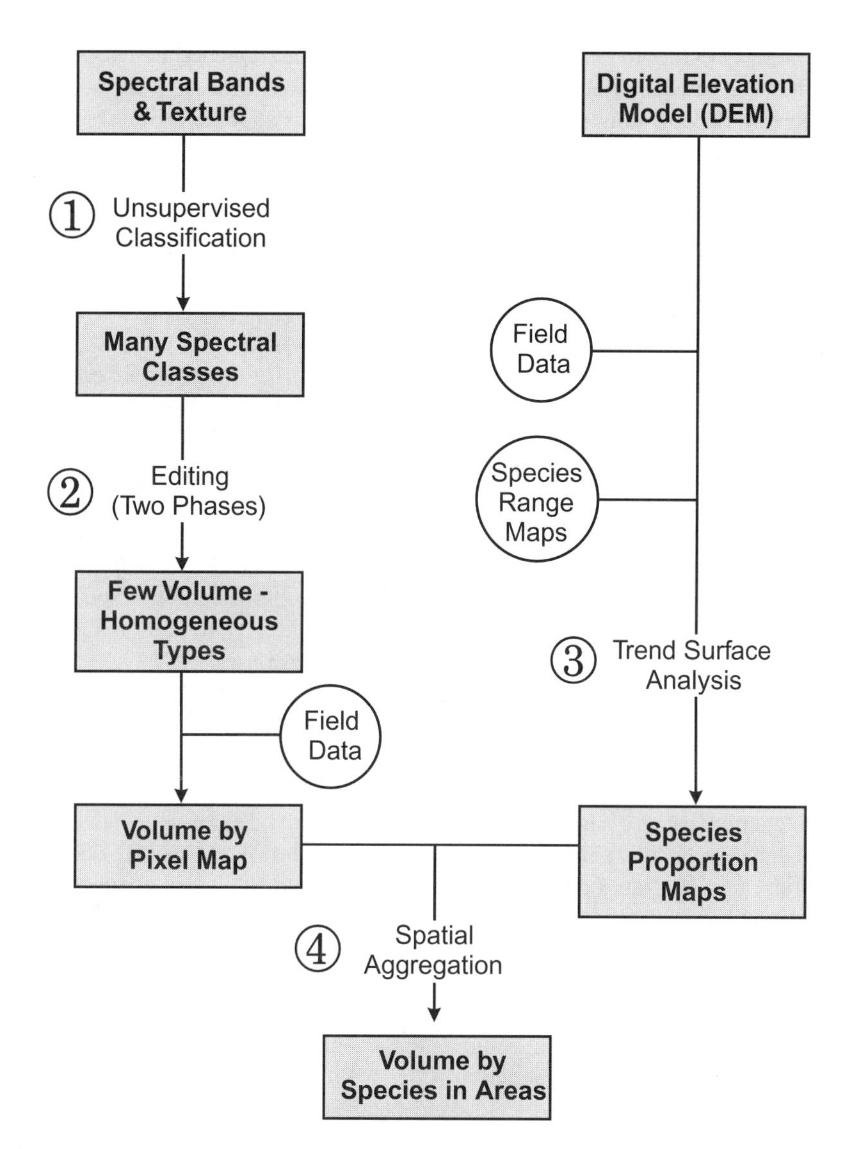

FIGURE 6.1 Development of a predictive vegetation model using topographic information combined with Landsat-based forest cover and volume maps. Here, two different streams of analysis, one based on the image spectral response and DEM data, one based on the use of forest inventory products such as species range maps, come together to allow estimation of volume in pixels and areas. The process uses an unsupervised classification premise initially (step 1), then requires analyst expertise to manually edit the training areas and clusters (step 2). A complete forest inventory with spatially interpolated maps (step 3) and area estimates (step 4) was obtained from relatively coarse spatial resolution satellite remote sensing data, but with less detail and lower accuracy than can be obtained using a photomorphic method and appropriate aerial photographs. (Modified from McLeod, R. G., and T. L. Logan (1980).)

In the northern Quebec boreal forest, users included the Quebec Forest Inventory Section, the Canadian Forest Service Forest Insect and Disease Survey, Hydro-Quebec, several operating pulp and paper companies, and some private companies and individuals engaged in environmental assessments. The main applications were

1. Vegetation cover maps of large northern territories for environmental studies, forest regeneration, and damage caused by eastern spruce budworm;
2. A planning tool for climate change sample plot locations; and
3. Aerial mission support and planning.

Fewer studies have been reported using SAR data for Level II classifications. In the active microwave portion of the spectrum a small, perhaps even an insignificant, difference in spectral response might be found in open and closed forest canopy classes; radar backscatter will better separate completely different features such as wetlands vs. forest vs. urban classes, since woody vegetation and concrete can have very different responses in the longer wavelengths (urban areas will appear much brighter because of the many corner reflectors). The effect of increasing moisture content on soil spectral response in the microwave portion of the spectrum is predictable (i.e., decreased response), as are the perturbations induced by increasing amounts of dead vegetative material (i.e., increased response).

The similar pattern in mean microwave spectral response across forest classes dependent on the various image resolutions has meant that SAR imagery has been selected only infrequently to classify Level II vegetation types of interest. The most obvious reason to select SAR imagery has been that no suitable optical data were available; but because of the relatively widespread availability of optical data, many fewer studies have been conducted separating Level II forest classes using SAR data than with optical/infrared data. SAR classifications tend to be less operational and more research oriented, as familiarity with these data and methods needed to handle SAR data have not yet become common in the user community (Quegan, 1995).

Examples are found in forest speciation or typing in areas with simple forest structures and few species (Rignot et al., 1994). Ahern et al. (1996) examined pure spruce, pine, and aspen stands using multidate, multipolarization data and found nearly 100% accurate discrimination in Alberta. More detailed species composition mapping, for example, in mixedwood stands, has not generally been successful with SAR data and automated classification techniques. Scientists have generally concluded that for Level II (and sometimes even Level I) forest classes, rather than using SAR image tone alone in the image classifications one or more of the following strategies is necessary:

1. Acquisition of multipolarization, multidate, or multifrequency SAR data,
2. Use of image texture analysis,
3. Use of SAR data combined with optical data (data fusion).

Using airborne multifrequency, multidate polarimetric SAR data in Maine, Ranson and Sun (1994a,b) were able to classify lifeform (softwood and hardwood) but not dominant conifer species (hemlock and spruce). These efforts suggested that only a Level I type classification was realistic on a routine basis even if multidate and multifrequency SAR data were made available to the classifier. Multipolarization appears to offer at least one significant advantage (Ahern et al., 1996): only a single image acquisition in one band with opposite send and receive polarization is required to obtain similar results to multidate and multifrequency image acquisitions. Unfortunately, neither multifrequency nor cross-polarized SAR data are yet available from satellite platforms. Airborne SAR systems so equipped are not always available. This situation leaves only the multidate strategy when acquiring imagery, and the use of one of the two image processing approaches (texture analysis or data fusion) when processing imagery.

Data fusion has been used in few situations with SAR data in forestry applications in which optical data are available. Typically, the optical data provide the necessary discrimination with SAR imagery adding a small or possibly no increase in classification accuracy (Paris and Kwong, 1988; Leckie, 1990b; Peddle and Franklin, 1991). Usually, image texture has been considered a more valuable aspect of SAR imagery for use in classification (Ulaby et al., 1986; Weishampel et al., 1994; Imhoff, 1995). As indicated in Chapter 4, texture analysis is not without its own set of challenges; the reliable estimation of texture in noisy, speckle-dominated SAR imagery has never been a simple task (Sheen and Johnston, 1992).

In one study, Wilson (1996) found that simple SAR texture derivatives, such as variance and co-occurrence measures in small windows, could be used to increase the separability of spruce and pine in a boreal forest study area in Alberta; subsequent studies using ERS-1 and Radarsat imagery have confirmed that adding SAR texture derivatives to classifier decision-rules is a more powerful way of separating different forest covertypes in boreal forest environments dominated by conifer species than obtaining additional SAR tone data (such as using different wavelengths or multidate acquisitions) (Franklin et al., 1994; Wulder et al., 1995). In these forest conditions, it is the spatial component of the SAR imagery than can lead to better class discrimination. In hardwood stands, little studied so far, a different conclusion may be reached.

The use of texture has been valuable in Level II classifications using both SAR and optical imagery. For example, Franklin and Peddle (1990) separated four different types of forest cover based on dominant species in a boreal forest in Newfoundland, and found that average classification accuracies were increased from 51.1 to 86.7% when SPOT HRV satellite data were classified together with simple texture derivatives. The largest increases in accuracy were found in the deciduous forest class (predominately white birch (*Betula papyrifera* var. *papyrifera*) stands) in this boreal forest environment. A significant increase in accuracy of approximately 10% was reported in two subsequent study areas in Alberta and New Brunswick in which much more complex forest species composition classes were examined in high spatial resolution airborne imagery (Franklin et al., 2000a).

In classifying general landcover types, decision-rules such as the commonly available statistically based algorithms (for example, ISODATA clustering, linear discriminant analysis, and maximum likelihood) are often employed. Under many conditions, the common statistical classifiers are robust and well behaved; they provide optimal decisions on the covertype class based on simple statistics such as the mean, standard deviation, and covariance of spectral response in classes, and a straightforward formulation of the probabilities of class membership. Results can be occasionally improved by using different formulations and algorithms. Statistical classifiers do not perform well when the data display non-normal distributions (common with DEM and texture data). Improvements in Level II classification accuracy have been obtained by using classifiers that are more suited to the particular data at hand. Evidential reasoning (ER) has been demonstrated in the literature to be theoretically superior to other commercially available classifiers (principally, those that are statistically based) (Peddle, 1995a), and proven to provide higher classification accuracies in subarctic and alpine vegetation classification and active layer depth (permafrost) mapping (Peddle and Franklin, 1992, 1993). A layered classification approach (Jensen, 1978; Franklin and Wilson, 1992) or decision tree approach (Hansen et al., 1996) can outperform a classification which uses the same decision-rule for all data and all classes.

Using Forest Successional Classes

Successional forest classes are difficult to understand and almost impossible to map over large areas based on field surveys alone because of the need to characterize patterns rather than detailed plot locations. Spatial patterns ranging from 0.3 to 3000 km^2 were of interest in the Minnesota Superior National Forest for input to a carbon flux model and a forest growth model; satellite imagery from 1973 and 1983 were available (Hall et al., 1991a). What characterizes this study as different is that the classification of these images was accomplished using not only land cover (e.g., Anderson et al., 1976) hierarchical classes, but more ecologically meaningful classifications which could be described as community patterns (controlled by soil, slope, aspect, and successional history). Based on the changes in these classes, Hall et al. (1991a) were able to describe rates of change at the landscape unit level, and to document the successional direction of the transitions. The power of this approach lies in the ability to infer the state of an ecosystem in years following or preceding that in which ground observations were acquired.

A successional stage classifier was used by Fiorella and Ripple (1993a) in the conifer forests of the Pacific Northwest region of the U.S. Distinguishing old-growth forests and mature forests has been difficult because both successional stages tend to have large trees and high basal and leaf areas. Old-growth stands, however, tend to have larger canopy gaps and a greater heterogeneity of tree sizes than do mature forests; spectrally, these differences resulted in old-growth stands appearing slightly darker in the Landsat TM bands which are most sensitive to absorption (bands 1, 2, 3, 5, and 7). Lower old-growth values were most likely due to shadowing from the uneven tree sizes and the high number of large canopy gaps in old-growth forests.

A lodgepole pine successional gradient in Yellowstone National Park was classified using TM data (Jakubauskas, 1996a,b). Improvements in classification accuracy were obtained when spectral and texture variables were used together. Texture values were high for early successional sites (post-fire), decreased in mid-succession, and then increased as stands progressed to uneven-aged, uneven-height, old-growth conditions. When presented with a full range of successional classes, textures com-

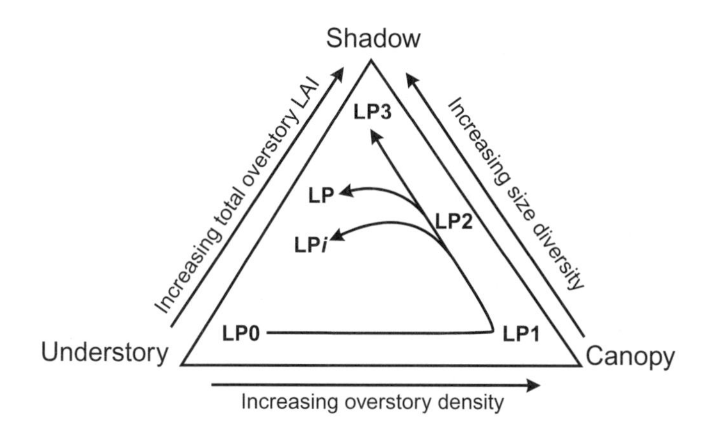

FIGURE 6.2 The interactions between factors influencing lodgepole pine forest spectral reflectance hypothesized as a balance between understory, canopy, and shadow components. A hypothetical vector for the shifting influences of understory, canopy, and shadow upon spectral response measured by a coarse spatial resolution sensor, such as Landsat TM, is shown. LP0 stands are characterized by low density and LAI; LP1 stands have the highest relative overstory density but low size diversity, creating a near-planophile canopy; LP and LP3 stands have the highest LAI and size density, and spectral response is affected primarily by shadow. LPi stands have low LAI values and moderate to high size diversity, reducing light interception by the overstory. (From Jakubauskas, M. E. 1996. *Rem. Sensing Environ.*, 56: 118–132. With permission.)

puted from the relatively coarse resolution TM data were useful, but in full-canopy conditions little discrimination at this scale was possible. The interactions between factors influencing forest spectral reflectance were hypothesized as a balance between understory, canopy, and shadow components (Figure 6.2).

Using image segmentation of texture imagery, followed by classification with a nonparametric decision-rule, four land cover classes (bunch grasses, dense cover of annuals, sparse cover of annuals, and bare soil) were mapped using 13.5 cm spatial resolution digitized aerial color photography from the Jasper Ridge Grassland, California (Lobo, 1997). The procedure was to use the segmentation algorithm to outline areas that were likely to be comprised of similar vegetation types based on their textural similarities, then use a linear discrimination to label those segmented areas as one of the classes. In a second test of this same algorithm in the Bolivian Amazon, four land cover classes (nonflooded alluvial plains forest, lowland seasonally flooded forest, palm forest, and swamp forest) were mapped accurately using Landsat TM imagery of the Chimanes Forest. When the categories remained fairly general (e.g., Table 6.3), the accuracies reported with this method were quite high — 93.91% overall, and all of the error was contained in three nonforest covertypes (not specified). The maps were thought to resemble the maps that would be produced by photointerpretation, principally because of the segmentation by texture. That is, polygonal structures appeared as if they were imposed on the landscape in an interpretation exercise since the segmentation algorithm more closely mimics the photomorphic approach by outlining areas of similar tone and texture.

TABLE 6.3
Four Land Cover Classes (Nonflooded Alluvial Plains Forest, Lowland Seasonally Flooded Forest, Palm Forest, and Swamp Forest) Mapped Using Landsat TM Imagery of the Chimanes Forest in the Bolivian Amazon

Vegetation Type	Classification Accuracy (%)
Swamp forest	100
Palm forest	100
Nonflooded alluvial plains forest	100
Low seasonally flooded forest	100
Three other covertypes	60
Grand Mean	93%

Source: Modified from Lobo (1997).

These results are important because they have illustrated (Lobo, 1997):

1. The successful use of spectral response and texture in a single procedure that worked well with digitized aerial photographs (13.5 cm spatial resolution) and Landsat TM data (30 m spatial resolution) in two different forest environments (tropical and temperate); and
2. The production of map units that more closely resembled those familiar to the map users — the polygonal structures generated by aerial photo-interpretation that capture stand patchiness (He et al., 1998).

In a detailed mapping experiment in a tropical forest using Landsat TM data, Foody et al. (1996) compared a segmented image classification to a conventional per-pixel classification. A significant increase in classification accuracy was obtained in the former procedure; the difference was that segmentation is equivalent to using an object-based classifier. It was much simpler to label objects than to classify individual pixels (Woodcock and Harward, 1992). The segments based on spectral and textural similarities became objects, area polygons, even forest stands. Weighted classification accuracies were generated that considered the likely transition and intermediate conditions that existed in the class continuum; accuracies were high (greater than 80%). The classes were very detailed, including six types of forest separable by age class and other vegetation types such as pasture and fallow ground. Forests of the same age class following different successional pathways were found to have markedly different species composition and spectral reflectance (Figure 6.3). A final classification accuracy of 86.65% was calculated over 11 classes (Foody et al., 1996). One note of caution was provided: the use of a single roll-up accuracy assessment can be misleading, based on the fact that large homogeneous areas were used, often in regions where sample sizes were small, autocorrelated, and geometric confidence might not be high (Adams et al., 1995; Foody, 1999).

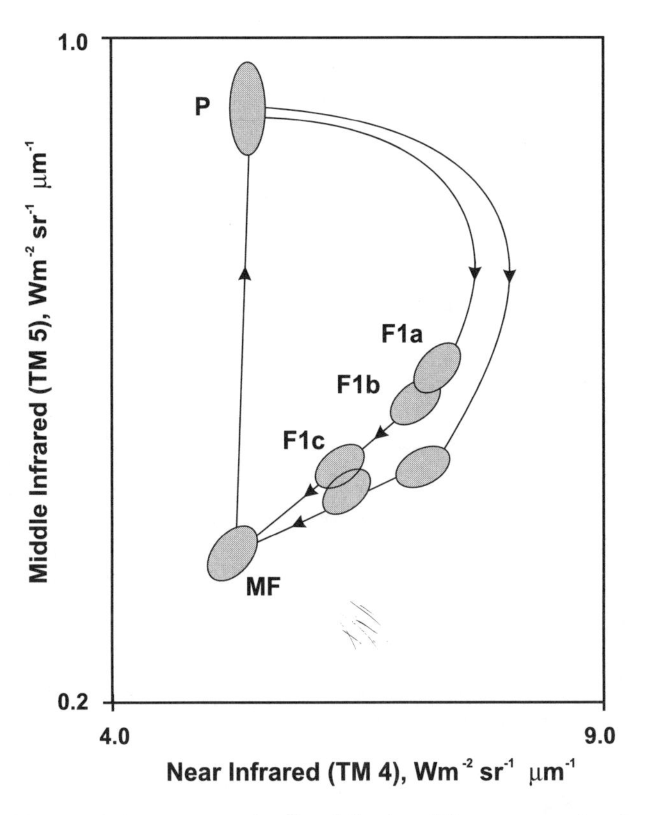

FIGURE 6.3 Forests of the same age class but following different successional pathways are often found to have different species composition and spectral response patterns that are predictable. In this example, the location of tropical forest regenerative classes are shown in a middle-infrared near-infrared (Landsat TM band 5 and band 4) spectral response feature space plot. Two successional pathways may be followed as the forest regenerates, and these classes are spectrally distinguishable for much of the time. The ellipses plotted represent 1 SD from the mean spectral response value. Arrows indicate the potential direction of the successional pathway in spectral response terms, from mature forest (MF at the bottom of the plot) to pasture (P at the top of the plot), along the two different successional pathways. (From Foody, G., G. Palubinskas, R. M. Lucas, et al. 1996. *Rem. Sensing Environ.*, 55: 205–216. With permission.)

For more detailed successional classifications using this type of medium-to-low spatial resolution data, a spectral mixture analysis may be appropriate (Lobo, 1997). Still futher increases in mapping accuracy in a greater range of classes may be obtained. This idea was used to classify multitemporal TM imagery into broad categories of land cover in Brazil, including primary forest and different communities of regrowth vegetation (Adams et al., 1995). Classes were described with reference to the amount of shade that could be expected (e.g., primary forest would have a high shade fraction compared to an open canopy regrowth area which would have a much lower proportion of shade in each pixel). The mixture analysis algorithm created new data or variables for use in a classifier by relating the image data to

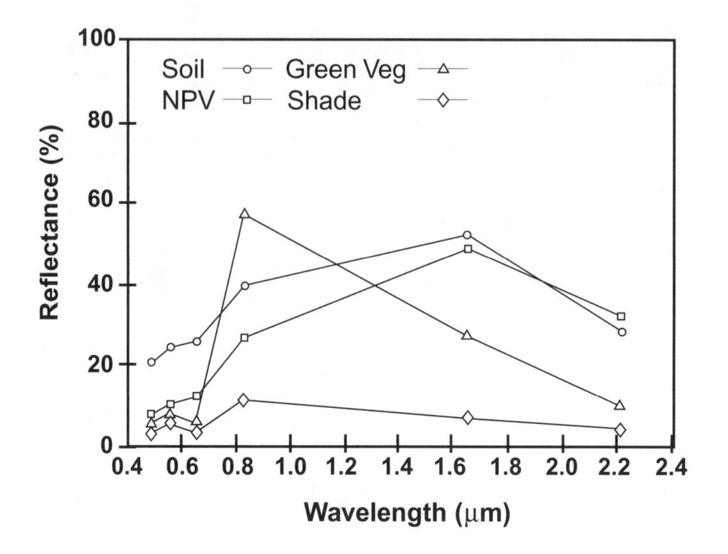

FIGURE 6.4 Spectral mixture analysis (SMA) requires that image data be related to spectral endmembers which are "pure" measurements of the components that contribute spectral response to the pixel (e.g., green vegetation, nonphotosynthetic material such as bark and soil, and shade). Here, Landsat TM pseudospectra of reference endmembers derived from laboratory measurements are shown. (From Adams, J. B., D. E. Sabol, V. Kapos, et al. 1995. *Rem. Sensing Environ.*, 52: 137–152. With permission.)

spectral endmembers, which were pure measurements of the components that contribute spectral reflectance to the pixel (e.g., green vegetation, nonphotosynthetic material such as bark and soil, and shade) (Figure 6.4). In essence, the pixel data were processed to reveal likely proportions of each endmember; then, the classification can proceed with these fractions rather than with the original single (mixed) pixel value for that band.

Overall classification accuracy was estimated at over 90% (Adams et al., 1995). Even so, potential sources of error were traced to similarity of endmember spectra caused by nonspectral attributes in the field (e.g., slash in cutovers resembled bare soils classes) and by spectral differences (e.g., some soils resembled dry grasses). Additional mixing at the subpixel scale was identified as a significant problem; this may be particularly acute in areas with sharp boundaries, but even occurs within supposedly continuous forest classes (Peddle et al., 1999). Another approach was suggested by Sohn et al. (1999):

> *Using a new spectral pattern matching approach with two dates of Landsat data, we were able to map deforestation, secondary regrowth stages of forest, and changes in the intensity of agricultural land use with high accuracy in the maize region of the state of Yucatan.*

The purpose of this mapping by spectral matching techniques was to document changes from Mayan forest management — a process thought to be fundamentally similar in principle to current concepts of sustainable forest ecosystem management

— to modern, mechanized approaches that appear to create landscape changes and patterns that may be unsustainable in the long run.

The key to successful classification of Level II classes has been to move beyond statistical classifiers such as maximum likelihood or mininum distance to means classifiers based on spectral response patterns alone, to include texture, segmentation, topography, new decision-rules, and spectral mixture analysis. Any and all of these image processing steps will provide increases in accuracy that can make the difference between usable map products and maps that are marginal. The methods work on digitized aerial photographs, multispectral scanner and airborne SAR data, and satellite imagery of all types and descriptions. However, even with these improvements, Level II forest classification is often only of passing interest in much of forest management and operational forestry; more detail in the mapping and less methodology to acquire it will be needed before routine use of remote sensing data described here is more likely. An indication of the rich possibilities can be found in considering the various Level III classifications using remote sensing.

LEVEL III CLASSES

SPECIES COMPOSITION

The identification of individual tree species has long been of interest using field spectroradiometric techniques (Gong et al., 1997) and airborne digital imagery (Rohde and Olson, 1972; Hughes et al., 1986; Gerylo et al., 1998). The approach has generally been to sample pixels on the sunlit portion of the crown in several bands and at different times. The different spectral response patterns can typically be related strongly to tree species differences (color, leaf morphology, canopy morphology). The results are classes at Level III; classes at which operational forest management is conducted.

The detail on these maps is impressive; whereas earlier maps stopped at lifeform or covertype, at Level III one of the most important pieces of information is a detailed species composition for the map units. By far the most common method of determining or classifying forest stand species composition is through field work to support the interpretation of aerial photographs (Avery, 1968). First, the area of homogenous species composition — the forest stand — is outlined; then individual tree species are identified on the aerial photography (if the scale and film characteristics permit), or on the ground through a field sampling protocol — a timber cruise or species checklist survey line. Depending on user objectives and project requirements, this process can yield very high accuracies, even with nonphotogrammetric formats. Standard metric aerial photographs are by far the most common type of imagery used in this process, but highly successful identification of loblolly pine (*Pinus taeda*) as the dominant species in plantations has been reported using 35-mm handheld aerial photography (Needham and Smith, 1987). In airborne multispectral video imagery, six different species of bottomland hardwood in Louisiana were successfully identified with over 80% average accuracy by Thomasson et al. (1994) (Table 6.4).

Are photointerpreted species composition stand maps accurate? How to assess the accuracy of a photointerpreted stand map? Field data? Another photointerpreta-

TABLE 6.4
Six Different Species of Bottomland Hardwood and Individual Trees in Louisiana Were Successfully Classified in Multispectral Video Data in Three Different Plots

	Classification Accuracy (%)		
	Run		
Species	1	2	Overall
Cypress	78	76	63
Willow	71	69	68
Ash	92	—	92
Sycamore	75	75	83
Boxelder	67	—	67
Cottonwood	67	—	67
Grand Mean			73%

Source: Modified from Thomasson et al. (1994).

tion exercise? The species may be identified correctly, but in the wrong proportion; or the proportions may be correct, but with a species misidentified or left out. How to compare? Not surprisingly, systematic tests of the repeatability of photointerpretations of species composition for stands are quite rare, but some estimates of accuracy for photointerpretation of species composition have suggested that results are often rather poor. For example, in Canada, Leckie and Gillis (1995: p. 80) reported that in aerial photointerpretation of species composition accuracy using standard metric aerial photographs "A best estimate of species accuracy is that 70 to 85% of the time the species composition is interpreted in the correct order or to within ±25% of the true species proportion for a stand." Higher accuracies are achieved in pure softwood stands, but most likely drop significantly in mixed wood or complex hardwood stands. This does not consider the potentially even larger problem of boundary placement!

Despite this cautionary note, the use of air photos in the task of species identification and general landcover classification is well accepted in the forestry community with very good reason; no other technology has demonstrated consistent improvement (or even equivalent results) over this combination of human image perception and logical analysis of what is perceived (Colwell, 1965; Sayn-Wittgenstein, 1978). What is required in remote sensing is a combination of distinctive spectral response and logical analysis of those data.

Several possible methods exist to tackle the species composition or stand delineation question using digital data. In fairly well-known areas, it is likely that complex analogues could be constructed for use with digital data acquired at the common photo scales. Analogues are typically field and air photo examples that are described and carefully mounted; the analyst's task when interpreting the new area is to find the appropriate analogue that matches most closely the new pattern under study (e.g., Stellingwerf, 1966). Mentally, the ecological setting can be modeled in such a way that new areas can be readily identified based on their similarity to the analogue landscape (Webster and Beckett, 1970; Paijmans, 1970). These procedures rely on the standard differences in the vegetation to be reflected in the air photos; for example, Paijmans (1970) noted that tall *Melaleuca* swamp forest is usually dark-toned, it has a dense canopy and no ground layer, and occurs along lower courses of river on terrain that is inundated for most of the year. *Campnosperma* swamp forest, on the other hand, shows a smooth, even, light gray canopy on air photos; small crowns coupled with possible understory vegetation on peaty soils accounts for the lighter appearance.

This idea, when extended to the digital domain, suggests the construction and maintenance of a digital spectral library, which could be built to include not only spectral response but other aspects such as topography and soil conditions (complete ecological analogues). Computers, rather than human brains, would provide the mechanics of matching the new pattern to the library or catalogue pattern. Analogues are difficult to construct, and because of the almost infinite variety of landscapes and forest conditions, are rare. This will not likely change in the near future for digital spectral libraries, although the ideas underlying the collection of pure end-member spectra may simplify the problem.

Instead, the most common method to identify species has been the development and use of photo selection keys. A manual for Canadian tree recognition on aerial photographs is typical of this approach (Sayn-Wittgenstein, 1978); the manual contains a discussion of the critical importance of scale, film, and filter combinations, and the role that expert knowledge of the ecological setting in which different species occur can play in tree species identification on photography. Many similar selection keys have been produced for tree species recognition on aerial photographs in other regions, including Europe, India, and South America (e.g., Tiwari, 1975; Sayn-Wittgenstein et al., 1978). Local detailed efforts have documented key conditions in smaller areas. Hudson (1991) provided a recent example of the way in which photo selection keys work (Table 6.5). In Dominican Republic West Indian pine

TABLE 6.5
Summary of Distinguishing Airphoto Characteristics for Interpretation of Forest Stands in Montane Areas of Hispaniola

Feature	Pine	Broad-Leaved
Crown shape	Narrowly rounded, open, asymmetrical	Flat, broadly rounded, solid, wide-spreading
Crown margin	Deeply serrate	Smooth, slightly sinnuate
Tone	Light gray	Dark gray
Texture	Rough, broken	Smooth

Source: Modified from Hudson (1991).

TABLE 6.6
Requirements Analysis for a System to Replace Aerial Photographs with Digital Data and Methods

Forest Classification and Sampling Tasks
Segmentation of homogeneous forest types (i.e., pretyping)
Convert boundary pixels of homogeneous forest types to vectors
Link vectors to creat unique polygons
Identify species composition within polygons
Heights of sample trees and stands
Volume estimates of sample trees and stands
Land Use Planning and Associated Tasks
Environmentally sensitive area delineation
Simulation of impacts (such as clearcutting, selective logging, forest renewal, silviculture) and stand development
Identify areas of high risk to natural disasters (landslides, insect devastation, disease)
Change detection, change description, and monitoring
Regeneration surveys
Silvicultural treatment monitoring (thinning, fertilization)
Preventive treatment monitoring (susceptibility and vulnerability ratings)

Source: Adapted from Hegyi et al. (1992).

(*Pinus occidentalis*) forests, trees were identified by their uniquely shaped tree crowns (narrowly rounded, typically asymmetrical, occasionally flat and speading); when viewed in aerial photos, pine crowns had an irregular shape and were often deeply serrated. Their tone was light gray to gray and much lighter than broad-leaved trees, which were darker with characteristically flat or broadly rounded crowns. Stand structure could also be interpreted; pine stands were highly variable, but typically of rougher texture with a less uniform pattern; broad-leaf types were fairly systematic, smooth textured, and only occasionally rough or broken by a bulbous crown. Crown closure estimates were used to stratify the identified stands into stocking classes.

To date, no one has succeeded in duplicating this level of description or the process of identifying forest tree species using digitial data and methods. If one considers only the engineering of sensors, there was once great optimism that digital imagery could completely replace stereoscopic aerial photographs in virtually all such medium- and large-scale forestry applications (Neville and Till, 1991). Practically speaking, much progress was made in identifying the key spectral and spatial characteristics of forest species and building sensors that could capture those characteristics; 10 years ago, it seemed likely that only a few more critical developments were needed to create the right circumstances to allow the automation of the entire Level III classification process. In 1992, Hegyi et al. performed a requirements analysis for a system to replace aerial photography with digital image data and methods for forestry applications (Table 6.6). Airborne digital data appeared more suitable for the automation of the forest classification tasks than digitized aerial photography; it was thought that airborne digital data could meet most forest sampling requirements. A turnkey system with these functional capabilities seemed within reach.

Almost 10 years later, the principle that the digital remote sensing approach will replace aerial photography is virtually unchallenged (Caylor, 2000). But the needed developments in image processing systems and image understanding in order to exploit fully the information contained in high spatial detail imagery has not yet materialized. It appears that generating the digital data to be used in the process was the easy part. Is it only a matter of time before methods are available to completely automate forest stand mapping based on species composition?

Considerable progress has been made in high spatial detail information extraction. With airborne data, the pixels in the image are much smaller than the objects to be classified; individual trees are resolved. The local variance is low because many adjacent pixels have similar values: either the object (the tree crown) or the background. One approach in training the classifier is to select only sunlit tree crown pixels (Hughes et al., 1986); shadows and background pixels can be avoided. In this way, two densities of pine, two densities of pine/spruce mixedwood, and four compositional structures of deciduous/conifer mixedwood were separated in high spatial detail airborne imagery with up to 90% agreement with stands surveyed in the field and separated using species composition, density, and height measurements (Franklin, 1994).

But in mapping the stands, significant problems related to pixel size and variability were noted when using either airborne or satellite imagery; the accuracies dropped considerably outside of training areas and in more variable forest types. Following on from this work, using airborne videography, Gerylo et al. (1998) augmented species composition classifications by including spatial operations, such as crown delineation, prior to the application of maximum likelihood decision-rules. A maxima filter (Wulder et al., 2000) was first used to isolate pixels on crowns from other pixels; rather than simply training the classifier on sunlit tree crowns, the image processing system was used to remove all features that were not likely to be tree crowns. Then, supervised classification using training areas representing each species was applied only to those areas that were deemed to be crowns, thus reducing the error in species classification with understory and shadow pixels (Gerylo et al., 1998). As noted earlier, the final step was to count each crown maxima as a stem in order to derive species composition estimates (Figure 6.5; Chapter 6, Color Figure 1*).

The different species mixtures could thus be discriminated, but mapping such mixtures over large areas in different forest stands must consider all the pixels in the scene, not just those associated with sunlit tree crowns. This is the role of image texture analysis (Franklin et al., 2000a): in some simple forest structures (pure conifer stands in Alberta), classification accuracies approached 75%. In more complex forest stands (hardwood and conifer mixedwood stands in New Brunswick), 65% accuracy was achieved (Franklin et al., 2000a).The key issue appears to be the appropriate use of image characteristics beyond the per-pixel spectral response (Franklin and McDermid, 1993). Most of this discussion has centered on the tradi-

* Color figures follow page 176.

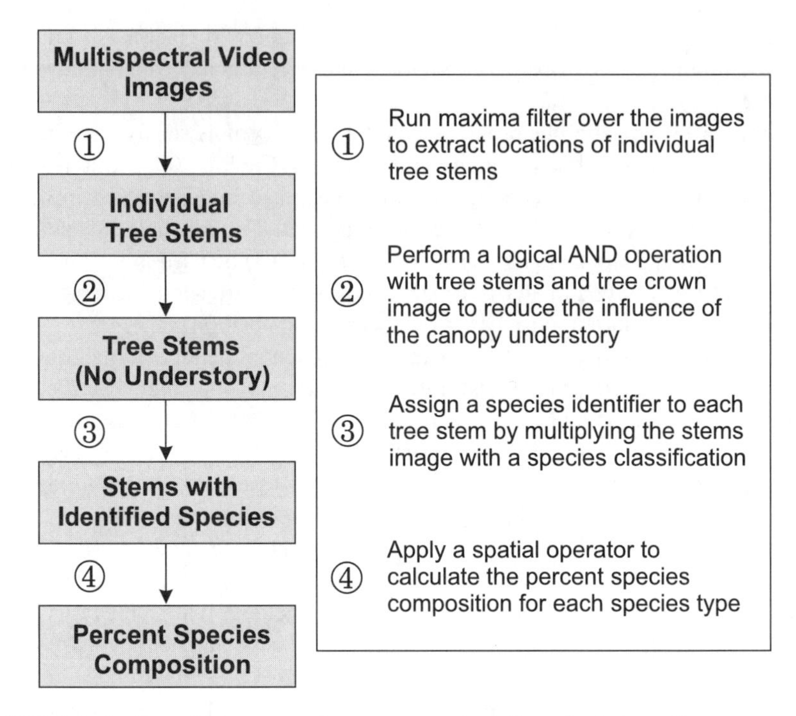

FIGURE 6.5 Outline of the processing steps required for calculation of percent species composition from airborne multispectral videographic imagery with <1 m pixel size. Rather than individual pixels, the emphasis is on "objects" identified in the image — in this case, the objects are individual tree crowns. (From Gerylo, G., R. J. Hall, S. E. Franklin, et al. 1998. *Can. J. Rem. Sensing*, 24: 219–232. With permission.)

tional per-pixel image processing techniques, which are not likely to be very effective or satisfactory in classification of high spatial resolution data. A shift in the analysis methods applied to airborne data has occurred from per-pixel analysis (Treitz et al., 1985; Hughes et al., 1986; Franklin, 1994) to incorporate developments in the use of:

1. Window-based operators (Yuan et al., 1991; Franklin et al., 1996)
2. Spatial feature extraction (Gougeon, 1995; St-Onge and Cavayas, 1995, 1997; Hay et al., 1996)
3. Image segmentation (Lobo, 1997; Green, 2000).

What is needed now is a comprehensive attempt to pull the various algorithms together, such that individual tree crowns are identified and outlined, the species is recognized and classified, the area coverage is assessed, and the stand label is developed and validated. The different approaches should be fairly transparent to the user, but it seems likely that no automated procedure can be constructed without careful input in the form of human interpretation — most obviously, in the form of training data. All the individual processing components have been developed and tested (especially in conifer stands). The most significant aspect has been the study of the spectral differences related to different species — this is where the human/computer interface has been most beneficial.

Multitemporal, multispectral aerial video imagery were evaluated for speciation of cypress, willow, green ash, sycamore, cottonwood, and boxelder in Louisiana (Thomasson et al., 1994). Images were acquired in May and November in three bands (centered at 550, 800, and 1000 nm) from approximately 300 m above ground. Pixel size was 0.329 m, enabling the recognition and classification of individual tree crowns, which were first delineated manually. Overall, 70% classification accuracy was achieved; this included a 10% improvement due to the use of the multitemporal data compared to the use of a single date. In Australia, Herwitz et al. (1998) acquired repeat imagery over 18 years and applied photogrammetric techniques to allow individual tree crown overlays. The study site was a mature tropical lowland rainforest; trees were large with well-defined crowns and branching visible in low-altitude photography. With great care, geometric registration and tree recognition techniques could be used to distinguish tree crown size classes, change in crown size, and mortality of individual trees.

Species vary by leaf angle, crown structure, and color, and hyperspectral data may be more suitable to identify the fine spectral differences that result (Cochrane, 2000). At Harvard Forest in central Massachusetts, hyperspectral imagery was used to classify forest cover based on species composition and foliar chemical characteristics (Martin et al., 1998). For 11 different forest classes mapped as separate polygons (or stands) the airborne hyperspectral imagery were processed to derive estimates of foliar lignin and foliar nitrogen concentrations. The overall appearance of the remote sensing map showed more spatial heterogeneity than the field-classified map; the field map was based on as few as three measurements within each stand polygon delineated by photointerpretation, and the remote sensing map was based on 20 m spatial resolution pixels. Therefore, it was apparent that spatial variation, which was missed by field plot samples, was measured in the remote sensing data (Martin et al., 1998).

This raises a number of interesting questions, not the least of which is (again!) the appropriateness of the air photointerpretation polygonal interpretation for representing the spatial organization and composition of forests. The production of intensive canopy maps showing complete spatial coverage of biophysical variables, not just species and individual tree crowns, illustrates the complexity of this issue yet again; and such maps will undoubtedly increase in value as new evaluative methods for management practices, such as partial harvesting, increase (e.g., Grushecky and Fajvan, 1999).

In a few cases, species composition classification has been reported using low and medium spatial resolution satellite imagery (Franklin, 1994; Wolter et al., 1995). With satellite data, the pixel size is sufficiently large to subsume many individual trees and other features; the pixels are much larger than the objects in the scene (L-resolution). If the density of the objects is more or less uniform, then adjacent pixels in the stand will have more or less similar values. The local variance will be low. The maximum likelihood classifier uses the mean value of spectral response which, when produced using these carefully selected composite pixels, can be shown to be distinct from similarly constituted composite pixels acquired in different inventory stands in the area (Franklin, 1994). This distinction can be made only with very careful selection of stands and training data; this is not an operational procedure, but is useful in some very limited situations in which only standard image analysis techniques can be used.

The most promising approach to species classification with relatively coarse spatial resolution satellite imagery appears to be in the use of spectral mixture analysis. Huguenin et al. (1997) reported a classification of bald cypress and tupelo gum trees in Thematic Mapper imagery in South Carolina. The procedure spectrally matched different candidate backgrounds and possible fractions of the tree species of interest, solving for the combination that most closely matched the observed pixel values. The training data consisted of image pixels that were estimated to contain a relatively consistent amount of the trees and backgrounds where the tree cover was >90%. During training, pixels were examined to determine if the mixture of trees and background was likely to deviate from this estimate; those pixels were modified by an estimate of how much of the background should be removed in order to reduce the potentially distorting influence of the background. The classification provides a soft output; the amount of cypress and tupelo found in each pixel in 10% increments. In an accuracy assessment, 95 of 100 pixels classified as cypress actually did contain cypress, and 93 of 100 pixels classified as tupelo actually did contain tupelo. Results were significantly better than could be obtained with a statistical classifier. Huguenin et al. (1997: p. 724) concluded this novel and promising study by suggesting that:

> *Although the satellite imagery will not replace field work or even the use of aerial photographs, it can potentially reduce the required area of coverage for the field work and photointerpretation efforts ... The ability to classify individual tree species and plant species and report the amount in each pixel has the potential to benefit many other diverse wetland, forestry, agriculture, and ecological applications.*

Ecological Communities

In the hierarchical systems of forest classification, progressively finer detailed levels are typically based on one of three possible criteria:

1. Vegetative (e.g., vegetation structure),
2. Floristic (vegetation species) or
3. Ecological principles (Bailey, 1996).

In the field, the identification of ecological communities (or integrated landscape units) is complex and usually successful only where field studies have established the underlying patterns — the diagnostic environmental variables including climate, parent material, landform, and soil variables (Carter et al., 1999). These patterns may be published in field guides (e.g., Archibald et al., 1996) based largely on accumulated expertise over numerous plots and transects, in which the principles of vegetation ecology have been understood well enough to allow generalizations to other sites. For example, McNab et al. (1999) measured vegetation cover and environmental variables on 79 stratified, randomly located 0.1-ha sample plots in a

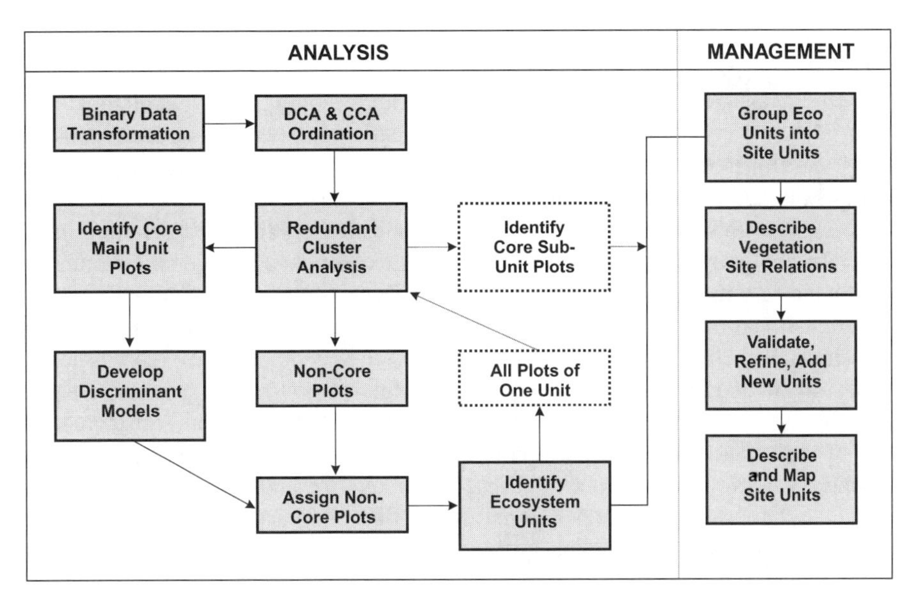

FIGURE 6.6 Field-based ecological unit classification using a combination of multivariate analyses and data transformation techniques. The objective is to identify units that can be distinguished by major differences in physiography, soils, and vegetation — ecological land classification units required in a spatially explicit format in the management of vegetation and large-area monitoring and reporting programs. (From McNab, W. H., S. A. Browning, S. A. Simon, et al. 1999. *For. Ecol. Manage.,* 114: 405–420. With permission.)

4000-ha watershed in North Carolina; 185 inventoried species were associated primarily with the soil A-horizon thickness, soil base saturation, and terrain aspect. Using a combination of multivariate analyses and data transformation techniques, the process of classifying this continuum of vegetation and environment into ecosystem units was made less subjective (Figure 6.6). Note the underlying objective is the creation or recognition of the familiar homogeneous unit; "units that can be distinguished by major differences in physiography, soils and vegetation" (Barnes et al., 1982). In Alabama, Carter et al. (1999) found that nine different land types — unique assemblages of vegetation and environmental factors — could be predicted by various combinations of landform indices, slopes, percent horizon nitrogen, depth, and texture of A and B horizons.

Integrated classifications are comprised of those classification systems in which ecological or community associations become the dominant criteria to separate land units (Zonneveld, 1989). Integrated or ecological classifications are more common and better understood today because of improvements in biogeographical knowledge (Bailey, 1996), but they still suffer from earlier criticisms: lack of precisely defined classification criteria, overemphasis on physical features, potentially useful baseline information lost in producing a single map product, difficulty in incorporating wildlife data, and lack of collection and integration of human-ecological and land-use information (Bastedo and Theberge, 1983). In some ways, ecological land classifications are almost too subjective for use with existing digital techniques; less hard classification decisions and more soft or fuzzy decisions are needed (Rowe, 1996; Foody, 1999).

The major issue is finding ways of generating spatially explicit ecological data from plot or transect observations, and at the same time, classifying the vegetation units and transferring that classification to a base map without obscuring the true complexity and diversity of natural plant assemblages (Cordes et al., 1997). As we have seen, in this classification process as in most others the most common intermediate step is the interpretation of air photos to provide a polygonal structure on which the classification system can be implemented. At the heart of most ecological land classifications for use in operational management are classic photomorphic classifications. But these patterns, while related to vegetation types, are not necessarily dominated by vegetation, and therefore the relationship to spectral response may be even less apparent. Recently, Treitz and Howarth (2000a) considered the use of high spatial detail airborne imagery in the task of classifying such units in northern Ontario. Using 4 m spatial resolution data, the mapping accuracy was not considered appropriate for operational delineation of forest ecosystems (less than 70% correct); one reason is that in these data there are still many objects in each pixel, rather than many pixels per object. The remotely sensed data were problematic (multiple spatial resolution imagery and DEMs were thought necessary), but perhaps more significantly, the authors concluded that more research on forest ecosystem class structure and terrain descriptors was required (Treitz and Howarth, 2000b). The structure of the classes was not particularly well suited to the data available to separate classes.

Essentially, the main problem relates to the classification structure and the breakdown in logic in selection of remotely sensed data to represent class differences. But it is rare to find Anderson Level III forest classifications based solely on the early, relatively coarse resolution satellite data (Huber and Casler, 1990; Wolter et al., 1995); the information content of these images was simply not high enough to support the wide range of fine class distinctions that are needed at this level of mapping detail. Often, specific image variables must be created for this purpose (Franklin, 1994; Franklin and Woodcock, 1997). Again, one of the more powerful options is to acquire and use multitemporal image data.

Multitemporal TM data were used by Mickelson et al. (1998) to map Anderson Level III and IV classes in New England. Of particular interest were deciduous species classes which, when examined at Anderson Level IV, were essentially subcategorical understory classes that could form the basis for more detailed community-level maps. Three seasonal TM images (May, August, and October in different years) were selected: May, because the angiosperm forest types were captured in early bud-break and pre-leaf-out conditions; August, depicting full leaf-on conditions; and October, because of heightened color and senescent leaf condition for maples and oaks. Despite the large number of classes that were mapped (33), the large interval in image dates and field data collection, the high level of compositional and spectral similarity among the classes, and known deficiencies in the training data used to drive the classifier, overall classification accuracies were reported to be good (close to 80%). This study showed that for separating these classes, the least useful imagery was the full leaf-on summer scene.

A second strategy in classification of Level III ecological classes has been to obtain and use DEM data. Gong et al. (1996) mapped ecological land systems using only forest cover data and a DEM as input to a neural network. They used 16 input attributes which included elevation, aspect, slope, eight dominant tree species, and five cover types. The best classification accuracy was just 52%, but the classes were thought to be generally in agreement with an aerial photointerpretation map created earlier, using standard land systems mapping methods. The photo map contained interpreted information on surficial geology, landforms, hydrology, and the relationships to vegetation or ecological communities. At individual sites, the DEM and forest cover maps were a poor substitute for these types of data — but the broad patterns appeared with reasonable agreement. Such predictive vegetation mapping is only a partial solution to the ecological community mapping problem (Franklin, 1995; Florinsky and Kuryakova, 1996; Bolstad et al., 1998). The data and methods lack precision, and are also more suited to the mapping of potential vegetation rather than actual vegetation communities, which are made more complex by disturbances, drainage, and other environmental factors. Two more examples:

1. In a wetland environment, Sader et al. (1995) used National Wetland Inventory (NWI) maps based on high-altitude photography (Tiner, 1990), DEMs, hydric soils interpreted on 1:20,000 orthophotos, and a Landsat TM image to separate softwood wetland, hardwood wetland, and mixed forest wetland from six other forest and land cover types with an overall accuracy in the 80% range using a hybrid classification. A rule-based GIS model, in which the NWI, hydric soils, and slope were the most important variables, provided an additional small increase in classification accuracy.
2. Davis and Dozier (1990) assumed that vegetation cover was a reliable indicator of ecological conditions at a site which they predicted with a combination of DEM and spectral variables. Although the maps were ecologically reasonable, they were weakly related to vegetation pattern, accounting for only one fourth of the variance in the vegetation test data. This weak correlation was attributed to the fact that predictive vegetation mapping did not include obviously important influences, such as management treatments or disturbances.

Typically, classification methods using these data must be supplemented with more complex algorithms; however, more complex algorithms cannot overcome a lack of good input data; "Adequate input data are the most important factor in successful land-systems classification" (Gong et al., 1996: p. 1259). Apparently, based on the work completed so far, satellite remote sensing and DEMs are simply not adequate to generate ecological communities except in the simplest of cases — yet these are the map products with the highest potential to help provide more productive and efficient land use (and land capability) planning and assessment (Rowe, 1996).

Understory Conditions

Two different approaches to remote sensing of understory conditions have been reported:

1. Direct sensing of understory conditions, and
2. Prediction or inference of understory conditions based on overstory characteristics.

In this latter instance, few successful studies have been reported, simply because the forest canopy does not often reliably indicate the ecological dynamics of the hydrologic system, soil type, soil moisture content, ground vegetation, and topographic features below the canopy, which are the key determinants of the understory characteristics. The most obvious exceptions have been where a direct correlation between understory plants and topographic conditions can be determined; but as in the previous section, these are simply maps of potential understory (Davis and Dozier, 1990) rather than actual understory vegetation classifications.

Occasionally, it has been found that the understory does influence spectral response (Spanner et al., 1990) enough that direct mapping can be attempted. Borry et al. (1993) classified poplar (*Populus x euramericana*) stand development stages in Belgium; the understory vegetation significantly influenced stand radiance as measured by the SPOT and Landsat satellite sensor systems. Prior knowledge of the understory did not increase classification accuracy, which was generally good (Table 6.7). This was because the effect of the understory was small compared to the effect of the stand development stage; in other words, understory was predicted by stand development. As others had found, stands in younger stages of development were more readily discriminated than older stands because of structural differences.

In the Mixedwood Section of the Canadian Boreal Forest Region (Rowe, 1972), mixed species stands are among the most challenging types of forests to manage (Smith, 1986). Within these mixed stands, information about the understory component is important for spruce management planning because of its contribution to future timber supply (Morgan, 1991; Lieffers et al., 1996) and habitat diversity (Lieffers and Beck, 1994). Using Landsat TM data acquired during summer (deciduous leaf-on) and spring (deciduous leaf-off) conditions, Hall et al. (2000a) classified

TABLE 6.7
The Most Accurate Landsat Multitemporal Per-Pixel Classification of Poplar Development Stage in 16 Forest Stands in Belgium

Classification Accuracy (%) of Stands in Different Development Stages			
Older	Middle	Young	Average
45	65	87	66
52	70	90	71
51	68	83	67
Grand Mean			68%

Source: Modified from Borry et al. (1993).

16 classes of overstory and understory land cover to 71% accuracy when compared to photointerpretation that had been field-checked at 71 field test sites. The leaf-on image was used to identify the pure deciduous and mixedwood stands, and the leaf-off image was used to detect the presence of understory (if any) within these stands (Ghitter et al., 1995; Hall and Klita, 1997). The land cover classification system was based on descriptors of overstory stand structure and understory distribution, and was derived by a combination of class definition and statistical separability rules.

A conifer understory map produced by interpretation of 1:10,000 scale color infrared aerial photographs acquired during the early spring compared favorably to the leaf-on/leaf-off multitemporal TM satellite image classification (Hall et al., 2000a; Chapter 6, Color Figure 2). Because of the complex overstory/understory combinations in this forest region, one of the most important findings of the study was confirming the importance of devising and validating the class structure itself. Species composition and crown closure are among the significant factors that influence spectral response (Guyot et al., 1989), and these factors were incorporated into the understory land cover classification system. Classification accuracy was affected, in part, by the patchy spatial distribution of conifer understory within polygons defined by overstory species composition and structure. The map product has informational value as a planning aid at the landscape scale and may serve as validation data to national land cover mapping initiatives (Hall et al., 2000a).

These understory classification results compared favorably with the results of a earlier, single TM image understory classification project in the California Sierran mixed-conifer zone. There, Stenback and Congalton (1990) reported 69% classification accuracy in detection of three canopy closure classes, and understory presence or absence. A spectral pattern analysis technique, whereby the individual bands and several transformations of the bands (e.g., principal components, textures, and several ratios) was used to select the appropriate spectral discriminators for the understory conditions at hand. It is important to note, however, that these studies approached the understory problem under very specific conditions (i.e., open canopies or conifer beneath deciduous overstory). Forest understory that is more complex will not generally be amenable to remote sensing techniques with coarse resolution pixel sizes; a high spatial detail sensor designed to acquire data in canopy gaps would be more suitable in many understory mapping conditions.

7 Forest Structure Estimation

> The … proper interpretation of remote sensor data requires a thorough understanding of the temporal and spatial characteristics inherent in the vegetation cover types present and of the related changes in spectral response.
>
> — R. M. Hoffer, 1978

INFORMATION ON FOREST STRUCTURE

The spatial and statistical output from a classification procedure comprises one of the major information products on forest condition available by remote sensing; generally, a second set of forestry information products is obtained by continuous variable estimation procedures. Classification produces information on the features that are contained in the list of classes imposed on the image data; the result is typically a classification map. Continuous variable estimation produces information on features that vary continuously over the landscape depicted in imagery. The result may be a map or an image in which the tones correspond to the level or value of the feature of interest and vary over the extent of the map. The process can become more complex when continuously varying forest conditions are used in the process of classification. This is not usually a problem in conventional vegetation typing or species composition of stands; the map is derived via the usual logic of classification (Zsilinsky, 1964; Avery, 1968). But typing and compiling species composition are only two of the structural attributes of forest stands that are of interest, usually as part of a general forest inventory. Some of the other forest attributes of interest might include:

1. Forest crown closure,
2. Diameter at breast height (dbh),
3. Volume,
4. Height,
5. Stem density
6. Age, and
7. Stage of development.

Some of these attributes can be considered as forest conditions in either discrete classes or as continuously varying attributes to be estimated at some level of precision.

As in species classifications, aerial photography has been instrumental in developing maps of these forest structures almost exclusively through the photomorphic approach followed by field work, but also through direct image interpretations by manual means (e.g., height calculations by parallax or shadows, crown closure estimation using templates, etc.). Species composition has been classified using digital image classification techniques — with high spatial detail imagery — but generally without the level of acceptance accorded the aerial photographic approach, for a wide range of reasons, not the least of which is the difficulty in generating conventional maps with the digital methods (see Chapter 6). Digital classification has been used less frequently when the objective is to map other forest structures, because this type of mapping resembles more the estimation of a continuous variable rather than a discrete categorization. Classification of different density or height classes has been described (Franklin and McDermid, 1993), but applications of remote sensing aimed at these continuous aspects of the forest inventory have been driven largely by empirical or semiempirical model estimation. Unlike classification, which is typically driven by a statistical understanding of what the spectral response patterns mean, such models are based more on the relationships incorporated in a fundamental understanding of the physically based radiative transfer in forests.

A plethora of such studies have been reported attempting to estimate individual forest parameters such as crown closure, basal area, or volume, as independent variables which can be predicted or estimated using a calibrated remote sensing image. The general approach is to:

1. Establish a number of field observation sites in a forest area,
2. Collect forest condition information at those sites,
3. Acquire imagery of the sites,
4. Locate the sites on the image,
5. Extract the remote sensing data from these sites,
6. Develop a model relating the field and spectral data, and finally,
7. Use the model to predict forest parameters for all forest pixels based on the spectral data.

Typically, the objective is to predict the selected field variable through model analysis, with the available remote sensing data as the dependent variables. Then, the model is inverted to predict the independent variable (such as stand volume or density) over large areas of forest. In other words, the spatially explicit remote sensing data are considered the predictors of the locally known field parameters so that the remote sensing image can be used to map that parameter across the image landscape. The remote sensing data are inverted to provide predictions of the desired field variables. Intuitively this seems reasonable; users are aware of the fact that the remote sensing data are dependent on the field data, not the other way around. The common tool is model inversion; models developed through experimental or normative designs are used to describe the relationships contained within a forest/remote sensing data set. The aim is to generate new insights which can guide the field scientists and help new applications become possible.

The physically based models are built mathematically on theoretical models that are typically designed to quantify advances in the ability to predict target and radiation interactions (Jupp and Walker, 1997). The model is driven by the principles of radiation physics to relate spectral properties to biophysical properties (Gerstl, 1990). The model is derived from current experimental understanding of radiation physics, geometry, and energy/chemical interactions. The role of such models in advancing the science of remote sensing cannot be overestimated; but typically, remote sensing data analysts and forestry users have little contact with these models. Their complex and demanding structure have meant that they will likely remain in the domain of the remote sensing instrumentation and radiation specialists (Silva, 1978; Woodham and Lee, 1985; Teillet et al., 1997) rather than the applications specialists (Landgrebe, 1978b; Strahler et al., 1986; Cohen et al., 1996b; Franklin and Woodcock, 1997).

Empirical models might be constructed using the understanding derived from physically based models coupled with laboratory, field, and actual or simulated remote sensing data. Empirical remote sensing studies are plentiful — image classifications, for example, are almost completely empirical. This is the probable way in which most users of remote sensing data will learn and apply their experiences. The empirical approach is a data-driven approach; learning proceeds from understanding the data, data acquisition and the specific conditions under which models derived from those data were inverted. The form of the model can be inferred from physical considerations, while specific model parameters are estimated from empirical data. Unfortunately, purely empirical models have the disadvantage of being highly site specific (Waring and Running, 1998; Friedl et al., 2000). This modeling situation has given rise to an intermediate approach based on a set of semiempirical studies that are hybrids of the purely empirical and theoretical physical models. For example, a statistical (empirical) model of the relationships between reflectance and a canopy characteristics, such as leaf area index (LAI), may be augmented by a physical understanding of the processes involved; the effect of leaf angle, leaf distribution, and leaf shape might be modeled within the larger relationship between reflectance and leaf area well-established through vegetation indices such as the normalized-difference vegetation index (NDVI) (e.g., Chen and Cihlar, 1996; White et al., 1997).

Canopy reflectance models based on geometric-optical modeling approximations of physical processes represent an example of an emerging semiempirical method in remote sensing; these models contain a mix of data-driven relationships and theoretical understanding to provide answers only available in more sophisticated or demanding experimental settings. Li and Strahler (1985) developed one of the first such models — the geometric-optical reflectance model, commonly referred to as the Li-Strahler model. Using the model in California, Woodcock et al. (1997) reported that the model appeared to confirm what had been learned in numerous empirical studies — namely, that canopy reflectance is dominated by canopy cover — and that the advantages of using a canopy reflectance model over an empirically derived relationship were marginal, or at least unclear. The application of forest reflectance modeling and coupling such models to physically based models that incorporate growth and topography is in its infancy (Kimes et al., 1996; Gemmell, 2000). In particular, invertible canopy models are currently scarce and impractical for operational use due to their complexity and our still-evolving understanding; for example, Gemmell (2000) found that multiangle data were useful in improving the accuracy of forest characteristics derived by inversion, but that more extensive testing and validation over larger areas and different forest conditions was essential to better understand the limits of the methods. With a modest investment in training, such models could be used by applications specialists as well as the model developers.

While specific results will vary, empirical methods used in one area to generate a relationship between spectral response and forest conditions generally can be applied, with few modifications, elsewhere. But when using some types of remote sensing data, such as Landsat TM data, and empirical models such as linear regression techniques, other difficulties arise (Salvador and Pons, 1998a,b):

1. Typically low dynamic range of the data; generally, higher correlations can be obtained if log transformations are applied (Ripple et al., 1991; Baulies and Pons, 1995). For example, with respect to leaf biomass, after a certain density is reached, doubling that parameter will not affect the spectral response, but a log transform can help establish linear relationships;
2. Extensive atmospheric and geometric corrections are needed;
3. Difficulty in reducing sensitivity to extraneous factors (a standard feature of the normative remote sensing approach) (Gemmell, 2000);
4. Generally low spatial resolution relative to the objects under scrutiny — trees (Wynne et al., 2000), and;
5. Generally, small sample sizes often resulting in fewer degrees of freedom than required for extensive use.

Perhaps the most important advantage of this approach is its accessibility. There are probably few users in forest management situations who are unable to find the resources to complete the simple normative design that is required to establish a purely empirical relationship between spectral response and, for example, canopy cover. All that is needed are the basic remote sensing infrastructure components, an airborne or satellite remote sensing image, and some field work. The methods are slightly more demanding than classifications, but probably not by much (Franklin, 1986; Walsh, 1987; Franklin and McDermid, 1993).

While the exact form and nature of the empirical relationship will not remain stable as conditions change, it is also true that the relationship will rarely differ dramatically from what has already been reported or observed in an area. For example, the normal relationship between cover and red reflectance is expected to be expressed in a moderate negative correlation between the two variables because increasing cover (larger tree crowns, more leaves) results in more red light absorption (greater photosynthesis activity). Less red light will be detected by a sensor above the canopy. Perhaps the exact relationship is found to be an R value of –0.49. It is possible but not likely that the correlation between red reflectance and cover will be found to be +.49 in another, similar area. More likely, the new area will have a negative relationship of approximately the same magnitude. One interpretation of

TABLE 7.1
Relative Importance of Forest Variables in Explaining Airborne C-Band SAR Backscatter in 93 Alberta Boreal Forest Stands

Covertype	Rank-Order Variables
Hardwood	Volume/ha, biomass/ha, mean age, pine cover
Pine	Hardwood cover, pine cover, crown volume, crown closure
Spruce	No statistically significant relationships were found

Source: Adapted from Ahern et al. (1996).

this relationship might be that remote sensing images of a certain type of young stand are almost always brighter in the visible portion of the spectrum than older stands of that type. This relationship is as likely to be found in one location as in another. If the usual (or normative) relationship between cover and reflectance is one of decreasing reflectance with age for a given species, then this will be more or less likely to be true in New Brunswick as in Finland, Argentina, or Indonesia. The normal relationship must be established, tested, and understood in order for applications of the relationship to be developed.

Similar logic and approaches have been reported using SAR imagery. In particular, Ahern et al. (1996) exhaustively tested for relationships between SAR backscatter and boreal forest stand structure measures, but none of the statistical relationships were strong enough to suggest that C-band backscatter might be capable of providing reliable estimates of stand structural parameters. Different species differed in the strength and significance of the relationships (Table 7.1). Wilson (1996), using a sample of stands from the same data set, took a different approach. First, multiple regression equations were developed that included SAR backscatter and texture measures to predict mean height of spruce and pine stands; standard errors were less than 15%. Then, the stands were grouped by forest inventory classes for height and crown closure. SAR data could provide discrimination of broad height and crown closure classes at reasonable accuracies (Table 7.2). So, despite low correlation between spectral response and a forest variable on a pixel-by-pixel basis, high levels of classification accuracy could still be generated over broader classes and areas. This is one approach to achieving a more successful (i.e., more useful) remote sensing estimation of a continuously varying forest condition; create logical classes and reduce the problem to a classification decision. After all, 42 to 57% classification accuracy of crown closure into four classes is not high; under most circumstances, however, this would be considered much better (more useful!) than nothing.

The success of this empirical inversion idea has generated a vast literature comprised of specific studies and experiments. Many of these studies can be seen as contributing insights to satisfy the growing need to understand the appropriate role of remote sensing in providing information to sustainable forest management goals (Franklin and McDermid, 1993). A number of early empirical studies have served to demarcate the boundary for the use of airborne (Irons et al., 1987, 1991;

TABLE 7.2
Discrimination of Height and Crown Closure Classes Using Airborne SAR Imagery and Texture Variables in 66 Conifer Stands in Alberta

	Classification Accuracy (%)	
Conifer Type	**Height Classes**	**Crown Closure Classes**
Pine Stands	71	42
Spruce Stands	71	57

Source: Modified from Wilson (1966).

Neville and Till, 1991; Miller et al., 1991) and satellite remote sensing (Franklin, 1986; Butera, 1986; Danson, 1987; Walsh, 1987) in forest inventory assessment beyond which correlations are probably too tenuous — or too far from the normative — to support the endeavor. These studies were followed by a number of systematic attempts to integrate satellite remote sensing into forest inventory compilations of large areas (De Wulf et al., 1990; Brockhaus and Khorram, 1992; Bauer et al., 1994) and detailed studies of smaller areas designed to confirm or refine the empirical relationships for certain species or forest types of interest (Ripple et al., 1991; Danson and Curran, 1993; Franklin and Luther, 1995).

Empirical relationships between inventory variables such as canopy closure, stand volume, and species composition and airborne spectral response are typically stronger than those obtained from satellite sensors (Franklin and McDermid, 1993). This is probably because of the higher dynamic range and smaller pixel sizes commonly acquired by airborne sensors. But new satellite sensor data with improved characteristics are increasingly available and will continue to be tested. What is of interest here is a general assessment of remote sensing in estimating the kinds of forest variables that are of interest in compiling a forest inventory. Currently remote sensing is limited to the following generally successful applications (discussed in greater detail in following sections):

- Remote sensing data can be used to stratify forest covertypes at the broad level into classes of density, biomass, or volume; such strata are more pronounced in areas of significant topographic relief, which can be used to enhance the spectral differences and the actual differences in the target variable likely to be more related to topographic (ecosystematic or environmental factors) differences than to forest spectral response conditions;
- Remote sensing data can be used to stratify forest canopy (crown) characteristics; this procedure will be more successful in large (perhaps extensively managed) areas with a simple canopy structure and few species (e.g., plantations) which are relatively flat; this works well because the differences in the reflectance recorded by the satellite sensor will be dominated by changes in crown closure and density rather than by topography;

- Remote sensing data can be used to construct composite structural indices that can be used to differentiate forest stands, and to understand spectral response, in order to better employ the predominately L-resolution satellite imagery in forest inventory assessment (e.g., in classifications).

FOREST INVENTORY VARIABLES

FOREST COVER, CROWN CLOSURE, AND TREE DENSITY

Several early studies established that Landsat and SPOT satellite remote sensing data were related to forest cover, stand age, and crown closure (Walsh, 1980; Poso et al., 1984; Franklin, 1986; Butera, 1986; Horler and Ahern, 1986). The relationships were similar to those understood to be in effect with small-scale (high-altitude) aerial photographs; for example, decreasing visible reflectance (darker image tones) would be associated with increasing crown development. As a stand grows and ages the areas between the crowns are no longer visible, and the shadows cast by the crowns on each other deepen (Figure 7.1). Larger crowns would absorb more light, but reflect more strongly in the infrared (Butera, 1986; Franklin, 1986). The strongest correlations were typically found with the infrared bands (De Wulf et al., 1990) because greater atmospheric penetration would create deeper shadows from larger trees, and because of the large contrast and greater dynamic range.

In 28 stands of Corsican pine (*Pinus nigra* var. *maritima*) in England, a poor relationship between SPOT HRV near-infrared reflectance and forest canopy cover was found (Danson, 1987). The explanation was that, rather than a function of vegetation amount, the variation in the amount of shadow within the canopy influenced the strength of the relationship. Few significant relationships between SPOT HRV measured reflectance and lodgepole pine stands in Alberta were found (Franklin and McDermid, 1993); much stronger relationships with reflectance measured at higher spatial resolution by an airborne sensor were thought to be a result of the higher dynamic range in the data and the smaller pixel size. Again, shadowing effects were thought to be the dominant influence on the spectral response. A stepwise multiple regression predicting cover and density using seven variables of tone and texture extracted from red, green, and near-infrared bands of a 2.5 m spatial resolution airborne image yielded adjusted R^2-values between 0.63 and 0.66 in 14 lodgepole pine stands; this was reduced to 0.45 in the satellite data.

After a fire in lodgepole pines stands in Yellowstone National Park, Jakubauskas (1996a,b) found that TM spectral response was dominated by soil reflectance. As a stand progressed to later successional stages, the spectral response was increasingly dominated by the forest canopy, until maximum canopy closure occurred at approximately 40 years post-fire. As stands developed further, the overstory density declined, but live basal area, height, LAI, and site diversity increased. The larger gaps in the canopy, species composition, and the canopy size of individual trees began to dominate the spectral response. Stands thinned by beetle-induced mortality occupied a middle position in that spectral response, and stands that had been opened up were again influenced largely by understory and soil characteristics. Correlations to height, basal area, and biomass were reasonably strong between lodgepole pine

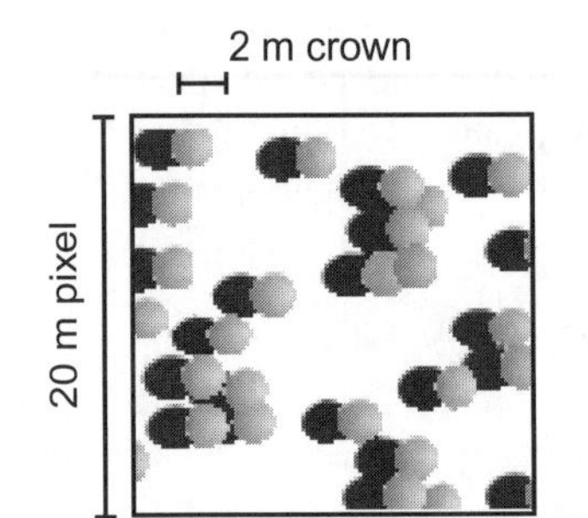

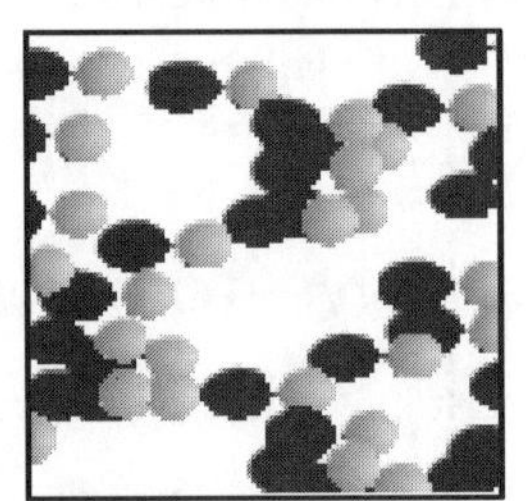

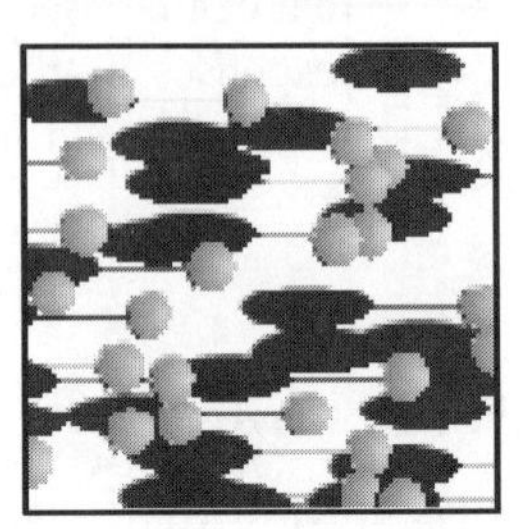

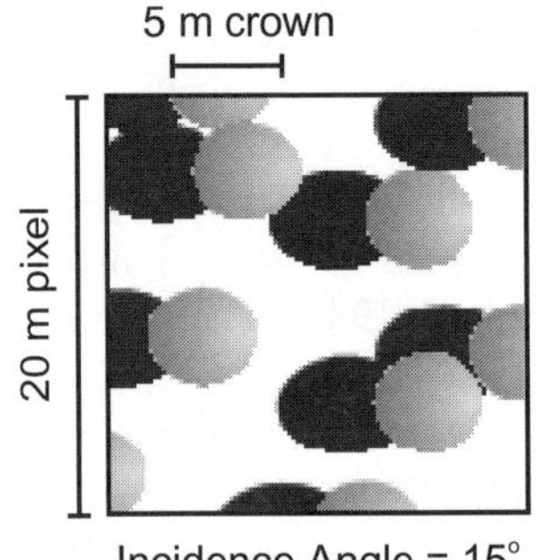

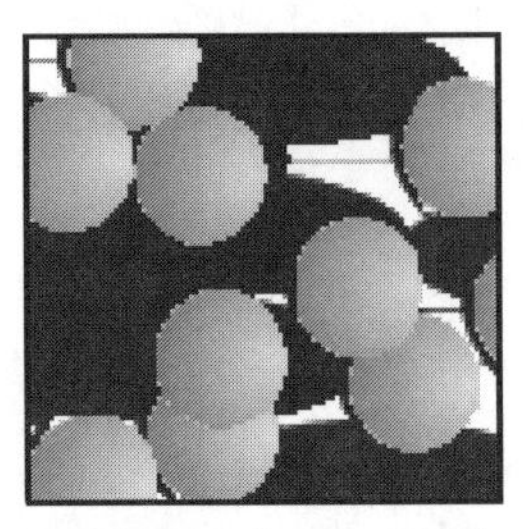

FIGURE 7.1 The geometrical-optical modeling approach considers that spectral response, in areas where the pixel size is larger than the objects (trees), is a combination of shaded and sunlit components. Here, the influence of the relationships is shown with (a) randomly located small trees and different sun angles and again with (b) different tree crown sizes. The amount of shadow and sunlit tree crown and the amount of area visible between the trees varies with the modeled characteristics. The ideal use of the GO model would be to construct a lookup table using all possible variations in the area of interest and then to examine the actual data relative to the modeled data to determine correspondence. If there were marked differences between the predicted and actual spectral response, then perhaps the area had been subjected to an unidentified change (e.g., canopy had been reduced by disturbance). (From Jupp, D. L. B., and J. Walker. 1997. *The Use of Remote Sensing in the Modeling of Forest Productivity*, pages 75–108, Kluwer, Dordrecht. With permission.)

stand conditions and Landsat TM data (Jakubauskas and Price, 1997); correlations to density, size diversity, mean diameter, and number of species were moderate; correlations to understory measures (number of seedlings, understory species, total cover) were weak.

These and other studies have led to the understanding that the effect of increasing or decreasing age, dbh, height, volume, and so on are all second- or third-order effects on remotely sensed image data; the principal influence on the spectral response is the illumination geometry (target-sensor-solar conditions) followed by the amount of vegetation viewed from above. As cover approaches full crown closure, the correlation between reflectance and these biophysical variables approaches zero; "... stand reflectance is primarily dependent on the density, size, and arrangement of crowns and the reflectances of illuminated and shadowed components in the stand, and indirectly on other attributes (site quality, species composition, age) through their effects on these former characteristics" (Gemmell, 1995: p. 296).

The main problem seems to be a fundamental one (Holmgren and Thuresson, 1998): the sensor detects reflectance only from the top of the canopy. If the canopy is open, the reflectance can be correlated with other attributes, such as understory characteristics which may be indirectly related to the target variables; if the canopy is closed the extent to which other parameters can be predicted seems to depend on the extent to which a closed canopy can predict them. In Oregon forests there was "little predictability in the spectral response of conifer forests beyond about 200 years of age, or once old-growth characteristics are attained ... forest stand conditions continue to evolve, but spectral changes appear uncorrelated with that development" (Cohen et al., 1995a: p. 727). In many forests, crown closure will reach a maximum (perhaps reaching 100%) and basal area and structural complexity will continue to increase, but the remotely sensed signal is not significantly affected by these increases (Franklin, 1986).

Canopy Characteristics on High Spatial Detail Imagery

Shadowing related to tree size may be the dominant influence on stand reflectance when high spatial resolution imagery are considered (St-Onge and Cavayas, 1995). A pixel in this type of image will characterize only a small part of a tree crown, shadow, or understory. The texture of the forest stand is generated by the light and dark tones created by individual tree crowns. Texture thus holds the most promise for automated forest cover or density estimation (Eldridge and Edwards, 1993; St-Onge and Cavayas, 1997). Customized texture windows — based on the range derived from image semivariograms calculated over the stand — were useful for estimating canopy coverage in one study in Alberta (Franklin and McDermid, 1993). More frequently, as we have seen, image texture has been used to help classify individual species in a stand (Fournier et al., 1995) and subsequently, to classify species composition (Franklin et al., 2000a).

Image understanding techniques have been developed to delineate tree crowns (Gougeon, 1995; Brandtberg, 1997) and then build a better estimate of crown closure, stem density, and species composition (Gerylo et al., 1998) (Chapter 7, Color Figure 1*). This idea was preceded by attempts to better estimate species proportions, and hence cover, on digitized aerial photographs (Meyer et al., 1996; Magnusson, 1997). A typical process might resemble the three-step procedure applied by Gougeon (1997) to airborne multispectral data acquired with spatial resolutions ranging from 30 to 100 cm:

- Individual tree crown delineation: Using the areas of shade between trees, an algorithm was designed to find local maxima (bright spot assumed to be the crown apex) and local minima (dark spot assumed to be the deepest shadows between crowns). Then, by following the valleys of dark areas

* Color figures follow page 176.

between bright areas, the tree crowns were delineated. A rule-base of tree crown sizes was invoked to separate each crown completely from adjacent bright areas (e.g., impossibly large crowns were separated). Comparisons of the resulting tree crown sizes to field crown estimates were within 7%. Once crowns were separated, estimates of stem density and canopy closure could be generated with fixed or geographic windows.

- Individual tree crown classification: Spectral signatures were used in a standard supervised classification procedure to identify the species associated with each delineated crown. The key here was to treat each isolated tree crown, delineated in the first step, as an object rather than to apply individual per-pixel classification. Options for classification included the use of the brightest pixels, the average of the sunlit portion of the crown, or the mean value within the delineated crown. Classification accuracies with four or five coniferous species were typically in the 72 to 81% range, depending on the original image spatial resolution.
- Forest stand delineation: Using the delineated and classified tree crowns, an algorithm was designed to regroup crowns into stands based on three derived variables in fixed windows: stem density, canopy closure, and species concentrations. An unsupervised classifier was applied to reduce the crown groupings still further, based on a minimum stand area criterion. The results were converted to a vector base to obtain polygons which closely resembled those mapped using aerial photography in a traditional approach to forest inventory.

This image understanding approach based on individual tree crown delineation appeared to work best with images having a spatial resolution <1 m, but conceivably will work well enough with 1 m satellite data to justify more extensive use and development (Wulder, 1999). Most of the current work has been reported in pure or mixed coniferous stands, or stands with only simple combinations of one or two conifer and deciduous tree species (Gerylo et al., 1998). Deciduous tree crowns are much less simple (typically with multiple maxima and less distinct edges), and are thus more difficult to delineate using existing procedures (Warner et al., 1999).

Despite tremendous growth and recent successes in specific areas, such as crown segmentation algorithms (Gougeon, 1995; Hill and Leckie, 1999), this area of image analysis in forestry appears surprisingly poorly developed; a poor relation to the better tested and better understood algorithms for use with L-resolution data. Imagery with 1 m or better spatial resolution have been available for decades, and are now available from satellite platforms. These remote sensing data are ready-made for forestry applications, and are ideally suited to help answer some of the same questions now addressed by extensive field and aerial photographic work, at lower cost and higher accuracy. Yet comparatively few successful projects using these imagery systems for species composition mapping, crown closure estimation, age, and structure mapping have been reported. Perhaps the difficulty in geometric correction and registration has prevented more widespread use of the data. Perhaps the preoccupation with the relatively coarse resolution satellite imagery has prevented a more concerted effort in the airborne arena.

Certainly, the image processing tools for use in high spatial detail applications, with or without high spectral resolution (hyperspectral imagery), are increasing in sophistication and value; after considerable development, the application of high spatial detail multispectral imagery in forestry remains underexploited, but the potential is being recognized.

Forest Age

It would be difficult to argue that forest age can be remotely sensed directly; many forests are hundreds of years old, but there are not many leaves hundreds of years old growing on trees! But forest age is surprisingly difficult to specify directly, even in the field. Typical measurements are age since establishment for young forests, or basal area weighted age in traditional forest inventories; this latter measure may not increase by a simple one-year-per-year. For example, with a thinning treatment the basal area weighted age can actually decrease. What is the age of a shelterwood stand, for example? Site productivity can be a major confounding factor in trying to estimate age directly, even when the species and density are reasonably uniform.

Direct correlation of stand age and remote sensing spectral response approaches a classic "nonsense" correlation (De Wulf et al., 1990). Age is a descriptor or surrogate of forest conditions but not itself an attribute (Cohen et al., 1995a); in essence, a third-order effect on spectral response is what is of interest — the changes in physical structure and composition — such as the size and density of tree species are captured over time in a variable called age. In conifer forests in Oregon, Cohen et al. (1995a) related these changes to satellite spectral response in broad age classes (e.g., <80, 80 to 200, and >200 years), but it is important to remember that these differences in physical structure and composition that are characteristic of aging stands are not directly sensed — rather, it is the differences in illumination, absorption, and shadows which are related to the different sizes and density of trees (Gemmell, 1995). Thus, there is only a small possibility that the relationships between age and spectral response will be strong or invariable across a forested area. It is more likely that remote sensing data will be suggestive of age, or age classes, because of the differences in tree size and density, understory, canopy development, nutrient status, and species among young and old forest stands.

In very young (<35 years) homogeneous Douglas-fir stands in Oregon, a strong relationship between age and TM reflectance was found (Fiorella and Ripple, 1993). Using a neural network model in this same region, Kimes et al. (1996) considered variability within regenerating clearcuts (<50 years). In predicting the year logged, the model network yielded an RMSE of approximately 8 years using the image data alone; adding topograpic data decreased these errors to approximately 5 years. Further decreases in errors in predicting stand age from Landsat TM and topographic data were obtained using site-specific information such as planting year, site preparation, and species planted. These decreases in error were considered evidence that the Landsat imagery contained unexplained variability at the scale of the sites related to the number of replantings, density of plantings, variations in site preparation, and soil conditions. When that variability was accounted for (using texture measures), the neural network model appeared to produce satisfactory inference of forest age

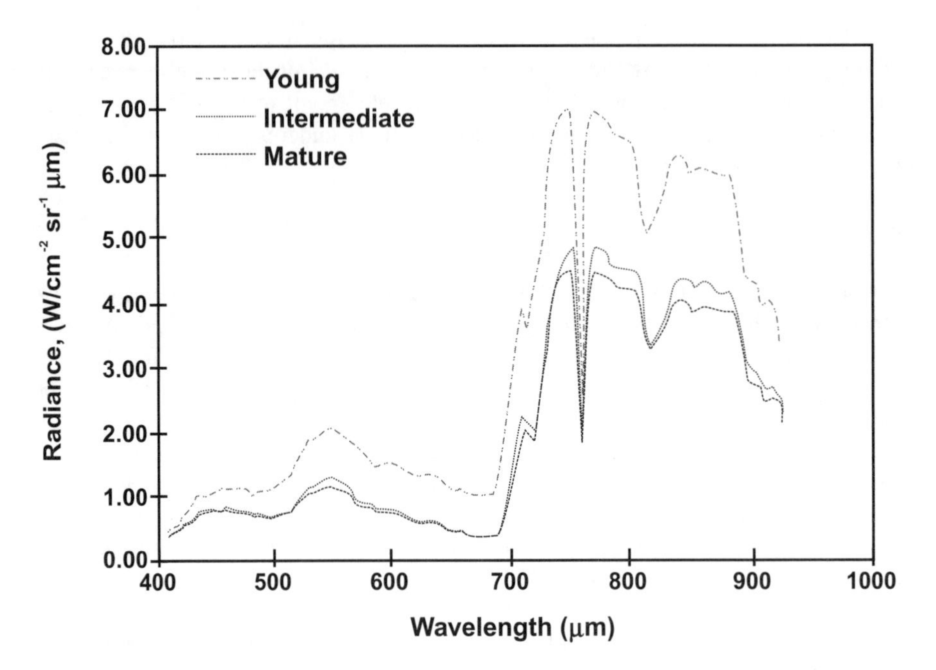

FIGURE 7.2 Mean spectral response curves suggest that as forests age, spectral response changes even if pixel sizes are small. Here, for a few Douglas-fir forest age classes, the differences were large enough that classification of age can be accomplished with a high degree of accuracy. Age differences were largest in the younger age classes. (From Niemann, K. O. 1995. *Photogramm. Eng. Rem. Sensing*, 61: 1119–1127. With permission.)

for young stands in this region. The performance of the neural network model provided a baseline or standard against which the more desirable physically based invertible reflectance and growth models could be developed and compared.

Another approach to infer forest stand age is to define the relationship between age and vegetation structure, development, or type in a classification (Niemann, 1995). Using this approach, stand development was classified as a surrogate of stand age in airborne imagery acquired over a forest area on Vancouver Island, British Columbia. The measured reflectance and age differences among the stands were created as a result of management activities, such as planting and harvesting. Stands greater than 60 years were generally mixed species with a significant understory; stands between 20 and 60 years were typically more uniform, with a closed canopy without the understory; stands less than 20 years were more open (plantations) (Figure 7.2). Reported classification accuracies for the corresponding age differences were on the order of 70% correct. In intensively managed areas, such as plantations, age may be a surrogate for a thinning cycle or other management treatment (De Wulf et al., 1990).

In another example, Hall et al. (1991a) used an understanding of boreal forest succession following a fire to classify different forest stand ages on satellite imagery. The use of successional age classes in forest sites in Puerto Rico, Costa Rica, and Mississippi was reported by Sader et al. (1989). The investigation was designed to explore the possible link between NDVI, canopy foliage biomass, and woody biomass in main bole and tree branches. Confusing the analysis were the topographic effects on image and forest conditions, and the multiple successional pathways that could be interpreted from periodic historical aerial photography (e.g., grass to brush to forest regeneration; row crops to grass to forest regeneration). Across a wide range of forest conditions, weak or no significant correlation was found between successional age class and various transformations of Landsat TM data, and only weak to moderate correlation to similar bands of 10 m spatial resolution airborne multispectral scanner data. As in Oregon (Fiorella and Ripple, 1993; Kimes et al., 1996) and in British Columbia (Niemann, 1995), stand age only in the very young stages of regrowth could be predicted reasonably well. These young areas were almost always brighter (less shadows) and more variable (more openings through which the background could be observed) than older forest areas.

Peterson and Nilson (1993) used successional age trajectories of forest reflectance and calibrated these with field observations of stand age. An age trajectory could be constructed in two ways (Nilson and Peterson, 1994):

1. Collect multiple measurements at the same location over time (the classic Location Through Time or LTT) design;
2. Collect observations at different locations thought to represent the different conditions which exist over time — this approach is sometimes known as the Location Through Condition (LTC) design, and is similar to the classic chronosequence.

The long-lived and highly variable nature of most forests would preclude extensive use of LTT sampling. An appropriate use of LTC sampling, though, requires that airborne or satellite spectral response patterns from numerous sites be available to represent distinct stages of development as the forest ages; in some ways this is similar to the earlier spectral library concept and suffers from the same problems — in an almost infinite variety of conditions, how to acquire such a huge number of observations? One way is to model the reflectance that would be generated if a sensor was there to record it. Fewer observations are needed, but there may be a sacrifice in precision and accuracy. Using airborne radiometer data, Nilson and Peterson (1994) found that the primary controls on reflectance in their forest stands were changes in canopy closure, tree/canopy LAI, species composition, and background reflectance. Because the age dependence of reflectance was strong, the reflectance of a stand at any time during its existence could be predicted by the successional age trajectory. Then, the important tasks were to periodically check different stands with airborne or satellite images in which the effects of sun angle, view angle, and phenology were controlled.

The management implications of this work suggest that imagery be acquired periodically, in the same stands, to determine if there is a difference between the expected reflectance (from the successional age trajectory model coupled with a growth model) and the observed reflectance (Jupp and Walker, 1997). Note that the model would be constructed using LTC sampling, but the ongoing monitoring would be done by LTT sampling. Controlling the differences not due to actual changes would be possible. If no differences were found between predicted and observed reflectance, then the stand would be thought to be experiencing normal forest development. However, if significant differences were found, then perhaps a disturbance agent or large-scale change in growth conditions should be considered likely. This would require detailed investigation. In the Oregon Transect project, a similar logic was used in forest stand LAI estimation (Peterson and Waring, 1994). The remote sensing imagery were used to determine if the ecosystem process model FOREST-BGC predictions of LAI were reasonable (Running, 1994); if the remote sensing imagery suggested that LAI was higher or lower on a given site than the model prediction (given the various assumptions of climate and soils), then (1) further investigation on the ground was warranted or (2) model parameterization and functioning could be subjected to additional testing and potentially improved to bring predictions in line with observations. In essence, once remote sensing observations confirmed that there was a discrepancy in the observations vs. model predictions, a new task for remote sensing and field visits was to try to identify the cause of the differences.

Tree Height

The uses of aerial photography in tree height estimation are well known to forest managers and remote sensing scientists (Titus and Morgan, 1985; Avery and Berlin, 1992; Sader et al., 1989; Kovats, 1997), as are the uses of height in the development of other information of interest to forest management; for example, in using allometric relationships with the crown diameter to predict other forest variables such as volume (Hall et al., 1989b; Hall et al., 1993). Digital airborne and satellite remote sensing has not been very successful in producing reliable estimates of tree or canopy height; in essence, the biophysical relationships between height and spectral response are rarely strong enough to justify model development. The few exceptions can be found in highly site-specific studies designed (1) to relate standard photogrammetric principles to shadows on imagery; for example, Shettigara and Sumerling (1998) and (2) to classify or estimate height as a relative attribute in a few general height classes.

In this latter instance, for example, two height classes of semideciduous forest were mapped from Landsat TM data in an Amazon study area — Class 1: semideciduous forest, and Class 2: tall semideciduous forest (Marsh et al., 1994). However, retaining these two classes did not necessarily represent a particularly logical classification structure for the area, and combining them into a single class improved the overall classification accuracy. Active sensors such as radar and lidar represent a more promising solution to remote tree height estimation (Hyyppä et al., 1997). Airborne SAR image data, under certain specific conditions (such as pure conifer stands with simple structure) have been found to be significantly correlated with tree heights (Weishampel et al., 1994). In Alberta, subsequent predictive models for an independent sample of stands indicated that standard errors were similar to those contained with the existing GIS forest inventory (derived by air photo parallax measurements) (Wilson, 1996).

Of the available remote sensing instruments, it appears that lidar measurements of tree height have the greatest potential. Since the early 1980s, lidars have been used experimentally to improve estimates of stand forest biomass and volume (Maclean and Krabill, 1986; Nelson et al., 1988). Early problems included the fact that the laser profiler obtained heights from the shoulder of the tree crown, as well as the peak; comparisons to field measurements showed that the spot lidar would systematically underestimate tree height. The system worked better in softwood stands where tree crowns were more distinct. Tree height variability was greater in the lidar data of hardwood stands when compared with field measurements. A refinement is to include lidar estimates of canopy density (or porosity); such an estimate can be produced by considering the number of times the laser pulses directly to the ground. The lidar-generated tree heights could be used in estimates of biomass and volume, but tree diameter variation accounted for much of the variation in site-to-site biomass estimates because tree diameter is far and away the most important component in biomass and volume equations. Combining lidar sensors with a spatially explict remote sensing device, such as a digital camera or spectrograph, will provide the ideal solution to the problem of remote height determination (Means et al., 1999; Lefsky et al., 1999a,b).

Structural Indices

Structural indices based on field measurements (Lahde et al., 1999; Latham et al., 1998) and remote sensing measurements (Cohen and Spies, 1992) have been sought for a variety of forest management applications that require information on a composite or summary measure of forest stand structure. In general, structure is an important factor in affecting many ecological responses (Lindenmayer et al., 1999; Spies, 1997). From a remote sensing perspective, the importance of structure lies in the use of linear combinations of field data interpreted as structural indices. It is hoped that these indices can be used to replace individual forest attributes — with their generally low correlation with spectral response — with a composite of field measurements that represents adequately the differences amongst forest stands of interest, but increases the strength of the overall relationships (Gemmell, 1995). Cohen and Spies (1992) first developed a structural complexity index to capture the structural diversity and upper canopy conditions of closed canopy (young to old-growth) stands in the Pacific Northwest. The idea was to compare field-determined structure with data obtained from image semivariograms (Cohen et al., 1990). These semivariograms were hypothesized to capture subtle structural characteristics over the area of the stand, or a transect sample through the stand. Support for their interpretation was obtained when higher spatial detail imagery (SPOT 10 m panchromatic) showed stronger correlations to the structural index than did lower spatial detail imagery (Landsat TM) (Cohen and Spies, 1992).

Danson and Curran (1993) developed a composite variable called canopy volume to describe the surfaces of leaves and branches in three dimensions. Canopy volume is a synthetic variable, constructed from field measures of density, dbh, height, and to a lesser extent, cover, using principal components analysis. In a plantation forest, they hypothesized that canopy volume would be greatest for the older, thinned stands with a few large trees (low density). The composite variable would be directly related to remotely sensed response because a large canopy volume would result in greater interaction with the canopy and a lower spectral response. By reducing the field data to a single composite variable with a stronger hypothetical relationship than the original individual field variables to spectral response, they sought to clarify the causal relationships between forest structure and spectral response (Danson and Curran, 1993: p. 61–62):

> *In the young stands, tree density was high, there were few gaps within the canopy, and the volume of the canopy was low. The LAI and biomass of these young stands had, however, already reached high levels. In this environment there will have been little penetration of radiation into the canopy, and as a result the level of reflected radiation for the canopy as a whole was relatively high.*
>
> *The interaction of near infrared radiation with the young stands will have been in the form of multiple scattering due to the higher reflectance and transmittance of the needles at these wavelengths. However, there will have been little penetration of radiation deep into the canopy because of the absence of gaps, and the level of reflected radiation would therefore again be relatively high.*
>
> *In the older stands the tree density was low because of the removal of trees by thinning. The individual trees and therefore the canopy volume were large, and there were many gaps within the canopy. However, LAI, biomass, and canopy cover were maintained at a high level. In this environment, the initial penetration and subsequent absorption of red radiation will have been great, giving rise to a larger amount of canopy shadow and lower levels of reflected radiation. Similarly, the penetration of near infrared radiation will also have been high with multiple scattering and absorption taking place deep within the canopy. A smaller percentage of the incident near infrared radiation will therefore have emerged from the top of the canopy giving rise to lower recorded near infrared radiance.*
>
> *It is proposed that it was this set of mechanisms that gave rise to the observed dependence of … radiance on the age of the stands ….*

In a British Columbia study area, a structural complexity index obtained by applying principal components analysis to field-measured stand parameters such as basal area, dbh, stem density, and crown diameters was augmented with estimates of variability of some of the parameters (Hansen et al., 2001). A strong loading by field measurements of stand basal area, stem counts, and crown diameter was expected, and was in fact necessary in order to interpret the meaning of the composite index. However, the index was also highly correlated to stand height ($R = 0.75$), and stand age ($R = 0.81$), which were not source variables, but were independently related to variables in the structural complexity index. Strong correlation coefficients between the index and each of the individual structural variable supported the suggestion that the value of the index lies in the ability to capture the variance found in multiple stand parameters in a single attribute (Cohen and Spies, 1992). The strong correlation between structural complexity and stand age, for example, supported the use of a structural complexity index as a surrogate variable for stand age. In this area, as is sometimes common elsewhere, age sampling by increment cores can be unreliable.

The generation of composite field indices, such as canopy volume (Danson and Curran, 1993) and structural complexity (Cohen and Spies, 1992; Hansen et al., 2001) has been accompanied by the search for a similar composite index in multispectral remote sensing imagery. Why relate individual spectral bands to the structural complexity index when the same logic used to create the index can be applied to the image data? The NDVI transformation was examined by several authors (Sader et al., 1989; Cohen et al., 1995a), but in many forests NDVI does not appear to be a good predictor of stand structure variables. One problem with the NDVI is that it uses only the red and near-infrared bands, and the shortwave infrared bands would appear to have important information that would thus not be included in the index.

Earlier work has shown that TM data transformed into the brightness, greeness, and wetness data space with the Tasseled Cap coefficients (Crist and Cicone, 1984) were sensitive to structural characteristics of forests (Horler and Ahern, 1986; Cohen et al., 1992). TM wetness (not necessarily an interpretation of water content) is heavily weighted by the contrast between shortwave-infrared and visible bands. In Figure 7.3 the TM wetness index is plotted against the structural complexity index for the 38 conifer stands sampled in British Columbia by Hansen et al. (2001). The correlation coefficient for this relationship was significant and moderate (R = 0.58).

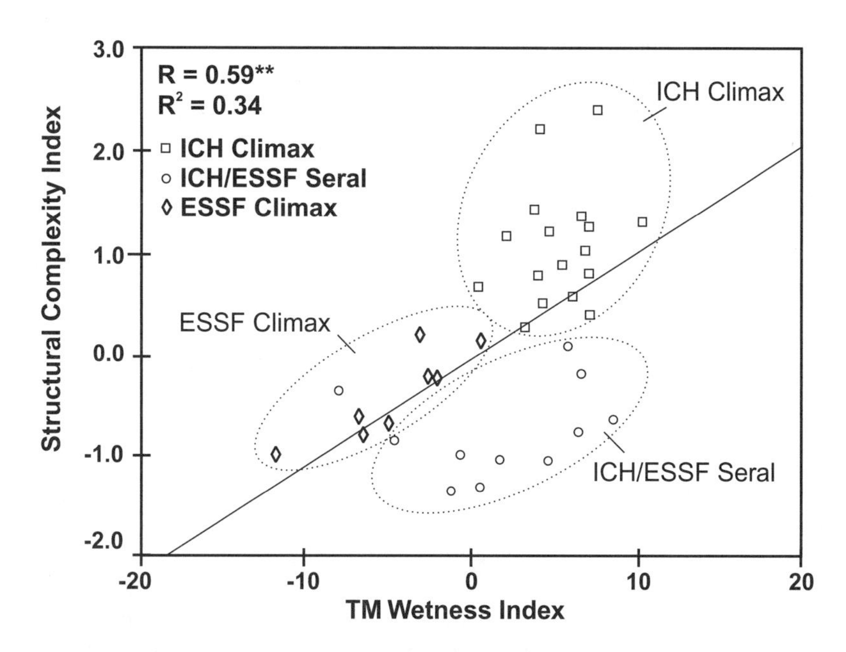

FIGURE 7.3 Landsat TM wetness index values are plotted against a field-based structural complexity index (SCI) comprised of measures of basal area, height, and crown dimensions for 38 conifer stands sampled in British Columbia (Hansen et al., 2001). The position of the stands on the graph indicates stands known to differ by geographically (altitudinal zonation) different "structural" and "wetness" positions. The forest stands differ in their structural complexity measured on the ground and can be distinguished spatially using the TM wetness index. In areas without detailed forest inventory information, such as age classes, the wetness index may be a useful surrogate measure of stand development.

TABLE 7.3
A Moderate to Strong Positive Relationship between Landsat TM Wetness and Stand Age Based on a Sample of 38 Forest Stands in British Columbia

Stand Type	R Value
Cedar Hemlock	+0.76
Spruce Fir	+0.76

Source: Modified from Hansen et al. (2000).

Stands of different types tended to occupy different structural strata, and these strata were strongly related to elevation. For example, mature and old-growth cedar-hemlock stands had high structural complexity values and high TM wetness values; mature and old-growth Engelmann spruce (*Picea engelmannii*) and subalpine fir stands were characterized by lower structural complexity. They had negative TM wetness values. With decreasing elevation, warmer temperatures and lower snow loads allowed the dominance of western red cedar and western hemlock. This species composition was characterized by an increase in crown diameter, crown closure, tree diameter, and stand height compared to Engelmann spruce-subalpine fir stands. The larger crown diameters of the western hemlock and western red cedar trees resulted in higher canopy moisture amounts being captured within a pixel value, and therefore higher wetness values than pixels containing trees with smaller crown diameter and lower crown closure.

Seral succession stands contain mixtures of Douglas-fir, pine, aspen, and other non-zonal or seral species (Braumandl and Curran, 1992). Field measures in these stands exhibited a wider range of species composition (than mature and old-growth stands) and often contained a deciduous vegetation component. Tree diameters, stand height, and crown diameters were generally smaller, but more variable, in these seral stands than in the mature and old-growth stands. The structural complexity index values were relatively low, but TM wetness values were more variable (Hansen et al., 2001).

For mature and old-growth stands, Table 7.3 shows a moderate to strong positive relationship between TM wetness and stand age (R = 0.76). Wetness and several stand structural parameters (crown closure, dbh, stand height, crown diameter, and structural complexity) also increased together, while wetness and stand density (stems per hectare) exhibited a negative relationship. These relationships indicate that as stands matured and structural complexity developed, tree diameter, stand height, and crown diameter increased together while stem density decreased. This resulted in an increase in crown closure, whereby a greater proportion of stand canopy was viewed by the TM sensor. The strong relationship between wetness and the structural complexity index (R = 0.83) appeared to justify the suggestion by Cohen and Spies (1992) that the index captured the structural attributes of a stand in a single index, and could be predicted by the TM wetness value of a pixel. The relationships between wetness and seral/late seral stand structure were less impressive; however, significant correlation coefficients were found between wetness and height to canopy (standard deviation, R = 0.59); crown diameter (standard deviation, R = 0.68); and crown diameter (R = 0.59) (Hansen et al., 2001). This may have been an indication of the sensitivity of the wetness index to canopy roughness or variability.

Cohen and Spies (1992) and Cohen et al. (1995a) studied TM wetness and structural complexity within closed canopy Douglas-fir/western hemlock stands (greater than 80% crown closure) and suggested that the influence of background vegetation was "noise" in the wetness signal (Cohen and Spies, 1992). They also found a negative correlation between wetness and stand age and structure. By including a wider range of crown closure classes and species composition in the analysis, and despite (or because of) the influence of background vegetation, weaker, but still usable significant relationships between TM wetness and structural attributes for seral, late seral, mature, and old-growth conifer stands were obtained in British Columbia (Hansen et al., 2001). The understory reflectance of a stand may contribute to the TM wetness component when discriminating between different biogeoclimatic zones. As the amount of understory reflectance in a pixel decreased and the proportion of canopy reflectance increased, wetness values also increased. When the BC data were divided into two crown closure classes (<85% crown closure and >85% closure), contrasting relationships were found. Figure 7.3 (a,b) shows that Stands with less than 85% crown closure exhibited a positive correlation between SCI and wetness. The structural complexity of stands greater than 85% crown closure was negatively correlated to wetness (similar to Cohen et al., 1995a).

A similar phenomenon has been recognized in both conifer forest (Cohen and Spies, 1992) and agricultural scenes (Crist et al., 1986). Generally, wetness values increase as a forest stand or crop matures until maximum crown closure or maximum canopy cover is achieved, then decrease as shadows and components other than canopy foliage begin to dominate the spectral response. While individual bands are less sensitive to the increased shadowing after this point, an index comprised of a linear combination of those bands appears more related to subsequent, more subtle, crown development.

BIOMASS

Biomass, an estimate of the total living/dead organic material expressed as a weight per area (e.g., kilograms per hectare), has been of greatest interest when aggregated over regional conditions (Penner et al., 1997; Schroeder et al., 1997; Fang et al., 1998). For example, at the county scale of resolution, Brown et al. (1999) produced a map of biomass density and pools of all forests in the eastern U.S. by converting inventoried wood volume estimates of aboveground and belowground biomass. Combining these estimates with AVHRR satellite data produced maps with 4- × 4-km grid cells; these could be aggregated to show the spatial distribution of biomass within each county and state. These products are useful, but are too coarse for many forest management purposes except the larger, strategic ones — for example, involving Kyoto reporting requirements. Instead, spatially explicit estimates of stand or ecosystem biomass are now sought by managers as one component of the carbon cycling budget for a given forest, and as an input to important criteria and indicators of sustainable forestry such as percentage of biomass volume by general forest type (indicator 4.1.4 in the Canadian Council of Forest Ministers, 1997; see Table 2.1). Increasingly, biomass estimation is required at the stand level.

Traditionally, stand biomass estimates are derived by the same process as regional estimation of biomass, by conversion of stem volume estimates from the forest inventory database (Aldred and Alemdag, 1988). In less-well-inventoried areas of the world, biomass estimates may be developed through forest covertype volume tables (Brown and Lugo, 1984). The estimate begins with single tree estimates by species and site types. The appropriate local allometric equations are developed to partition the estimate into foliar, branch, stem, and root biomass estimates, or perhaps into two components: aboveground and belowground woody biomass components (e.g., Lavigne et al., 1996). A recent strategy is to develop a large-scale system for biomass estimation. Such an approach assumes that better biomass estimates can be generated by referencing all available information in a multistage approach — the forest inventory, the available satellite and airborne imagery, and data collected in the field in permanent sample plots (Czaplewski, 1999; Fournier et al., 1999).

Airborne and satellite remote sensing in the optical portion of the spectrum has not often been used to generate biomass estimates — this application can be recognized as very difficult to achieve based solely on the spectral response pattern. In one review, Waring et al. (1995b) indicated that no available satellite remote sensing technique was sensitive to differences in standing and dead forest biomass above approximately 100 Mg dry mass per hectare. This situation has not yet changed, thus removing large areas of the forest from the current remote sensing biomass application. But below this level can be found a significant number of potential targets — for instance, tropical regenerating areas in which biomass accumulation is rapid (Sader et al., 1989; Shimabukuro and Smith, 1995; Malcolm et al., 1998) and areas with naturally occurring low woody biomass levels and slow accumulation rates (e.g., black spruce bogs of Alaska) (Kasischke et al., 1994).

A few studies have attempted to predict forest biomass using the relationships between reflectance and crown closure, crown size, and species. In Japan, Lee and Nakane (1997) estimated biomass with a variety of vegetation indices obtained from Landsat TM imagery in predominantly deciduous stands (comprised of *Quercus serrata*, *Castanea crenata* and *Carpinus laxiflora*), pine (*Pinus densiflora*) forests, and a Japanese cedar (*Cryptomeria japonica*) plantation. The NDVI was best in predicting biomass in the pine stands (R value was 0.85). In the deciduous and cedar plantations, the difference between band 5 and 7 (called the DVI) was the best predictor (R value = –0.83 for cedar, and +0.80 for deciduous stands) (Table 7.4). The sign of the relationship between biomass and DVI changed in the cedar plantation compared to the deciduous stands. The cedar plantation was comprised of stands with relatively similar age classes. Trees were of a uniform diameter and canopy height. The spectral response was more scattered due to the sharp, triangular cedar crown. The DVI was thought to be more sensitive to vegetation density than to leaf moisture content and color. The changes in the relationships among the species were attributed to the influences of different shadowing and leaf biomass.

Using reflectance data acquired from a helicopter over 31 stands of black spruce in Minnesota, Peddle et al. (1999) found that a linear regression analysis of biomass

TABLE 7.4
The Relationship between Landsat TM-Based DVI (Difference between Bands 5 and 7) and Biomass in 33 Stands of Deciduous and Cedar Plantations in Japan

Stand Type	R Value
Cedar	–0.83
Deciduous	+0.80

Source: Adapted from Lee and Nakane (1997).

(kg/m^2) provided an R^2 of 0.83 and an SE of 1.73 kg/m^2. The shadow fraction obtained by spectral mixture analysis was optimal for predicting biomass, NPP, and LAI in that northern conifer forest environment. In a very different forest environment in Brazil, Shimabukuro and Smith (1995: p. 68) interpreted spectral mixture analysis images such that

> *The young eucalyptus presents a higher proportion of vegetation and a lower proportion of shade compared to old eucalyptus. The pine forest presents a higher proportion of shade compared to the eucalyptus forest .... The difference images derived from the vegetation fraction images detect the green biomass variation in the study area.*

Biomass estimation with these subtle spectral differences was thought to be comparable in accuracy, and the effort may be only a fraction of that required in tropical forest field biomass estimation. The fact that similar results have been obtained in these two widely different forest environments is encouraging for the general robustness of the mixture modeling approach.

The canopy penetration potential of long-wavelength microwave energy has long been of interest in biomass estimation. In forest biomass studies using ERS-1 satellite data, Wang et al. (1994) and Kasischke et al. (1994) recommended a series of steps in evaluating the potential of radar data to measure differences in aboveground biomass:

- Study the relationship between the radar signature and the biophysical parameter of interest;
- Develop and evaluate hypotheses designed to determine the cause/effect relationships;
- Develop and validate techniques to derive the desired biophysical parameter from the SAR data.

With very small measured differences in backscatter (3 to 4 dB) derived from the ERS-1 SAR, linear correlation coefficients ranging up to 0.93 were observed for estimates of various components of biomass (bole, branches, foliage) and field measurements in 15 young loblolly pine stands in North Carolina. Over a larger range of stands, including mature trees, a lower correlation was observed (Table

TABLE 7.5
Summary of Linear Correlation Coefficients between ERS-1 SAR Derived Backscatter Values and Components of Dry Weight Biomass in Canopy Trees in 15 Mature and Young Loblolly Pine Stands

	Combined Mature and Young Stands		Young Stands Only	
Loblolly Pine Biomass	**R**	***p***	**R**	***p***
Bole biomass	0.47	0.077	0.90	0.0005
Stem biomass	0.50	0.057	0.90	0.0005
Needle biomass	0.50	0.025	0.93	0.0005
Canopy biomass	0.56	0.029	0.92	0.0005
Total biomass	0.49	0.077	0.91	0.0005

Source: Modified from Kasischke et al. (1994).

7.5). The problem of low dynamic range was traced to the short wavelength and steep incidence angle of the ERS-1 SAR system.

Incidence angle effects have been found to be more important than forest composition in determining forest backscatter, except at small incidence angles (Warner et al., 1996); for example, large incidence angles were necessary to increase boreal forest clearcut visibility in airborne C-band SAR data (Banner and Ahern, 1995). Under a closed or dense canopy, large incidence angles (shallow depression angle) allow the scattering of microwaves to be dominated by volume-scattering components, such as twigs and foliage in tree crowns and branches (Dobson et al., 1992) rather than direct double-bounce and topographic effects (van Zyl, 1993). A combination of empirical studies with theoretical scattering models (Chauhan et al., 1991; Sun et al., 1991) may be needed to understand the cause and effect of microwave scattering in forests (Kasischke et al., 1994). As a research tool, such models can provide powerful insights into the different scattering mechanisms occurring during remote sensing data collection, illuminating the dynamics of canopy orientation, composition, moisture conditions, density, and topographic effects (Hinse et al., 1988; Rauste, 1990; Yatabe and Leckie, 1995). But it is probable that such models, like the radiative transfer and geometrical optical models in the visible and infrared portions of the spectrum, will continue for some time to be far too complex and demanding for use in operational forest management settings (Ahern et al., 1991; Skelly, 1990). The MIMICS model, described briefly in Chapter 4, for example, requires more than 20 input parameters to characterize forest canopy and soil properties (Ulaby et al., 1990; Beauchemin et al., 1995).

Multipolarization appears to contain additional biomass information that can be extracted for forest stands. Le Toan et al. (1992) reported that backscatter relationships with X-, C-, L-, and P-band multipolarized airborne radar observations differed by forest components in 33 pine stands in southern France. The strongest relationships between radar returns and forest components differed by band (Table 7.6); the

TABLE 7.6
The Forest Components Providing the Strongest Explanation of Variance in Four Different Airborne SAR Bands

SAR Band	Component
X-Band data	Twigs and needles
C-Band data	Foliage and branches
L-Band data	Branches and tree trunk
P-Band data	Tree trunk and ground

Source: Modified from Le Toan et al. (1992).

results of these relationships, they felt, demonstrated experimentally the use of SAR data in retrieving forest biomass with "a precision suitable for operational forest management and ecosystem studies." The best relationships were observed using P-band (Figure 7.4), whereas most satellite SAR sensors operate at shorter wavelengths (C- and L-band). The strength of the relationship above 100 tons/ha was not tested since the stands in their study were all relatively young (less than 50 years old).

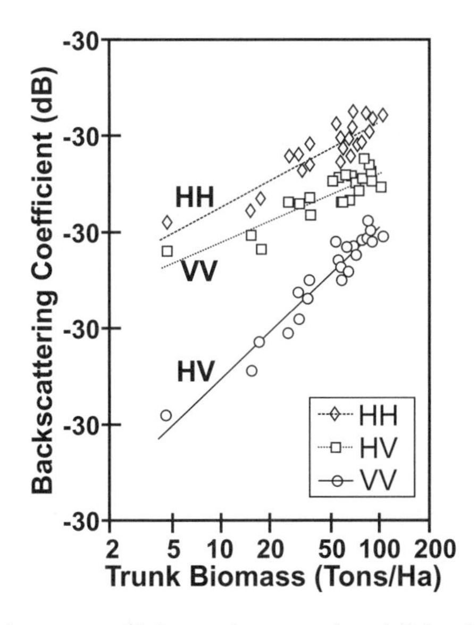

FIGURE 7.4 SAR backscatter coefficients at long-wavelength P-band in three different polarizations (HH, HV, and VV) as a function of trunk biomass (log scale) for forests from 8 to 46 years old.The longer-wavelength SAR data is dominated by the trunk-ground interaction; increasing backscatter is associated with increasing biomass and stand development in these young conifer plantations. Cross-polarization appears to increase the biomass-related differences in the measurements. (From Le Toan, T., A. Beaudoin, J. Riom, et al. 1992. *IEEE Trans. Geosci. Rem. Sensing,* 30: 403–411. With permission.)

Ranson and Saatchi (1992), Lang et al. (1994), and Sun and Ranson (1998) reported a series of studies designed to understand the links between radar images and disturbance regime, forest dynamics, and biomass. This understanding is constructed by using four working components together:

1. Models of radar backscattering,
2. Airborne SAR image acquisitions,
3. Detailed field studies, and
4. Image simulations.

Working at the microscale, small branches were the dominant influence on C-band SAR signals received from conifer canopies, but at longer wavelengths (P-band), depending on the branching characteristics (hemlock vs. pine), the dominant signal effect could be from the direct reflection off the boles or the ground (Ranson and Saatchi, 1992; Lang et al., 1994). Tall, straight, red pine boles in a plantation with a smooth forest floor had a much stronger correlation to radar signals than did natural hemlock stands above a rough forest floor. This interpretation of model results was extended to an image-based classification analysis that incorporated texture in mapping gaps in natural canopies (Sun and Ranson, 1998). Forest strip-cutting in natural stands only slightly reduced radar backscattering, but the spatial patterns (and resulting differences in biomass) could be correctly estimated.

Multifrequency P- and C-band SAR data were used to map forest biomass with a ratio of P- to C-band (Ranson and Sun, 1994a,b); the ratio may enhance the correlation to biomass by compensating for image variations due to radar incidence angle. Earlier, Hussin et al. (1991) attempted to build a system of simultaneous equations, similar to the traditional system of allometrics, in Florida slash pine (*Pinus elliottii*) plantation biomass estimation. Normally, in Florida as elsewhere, biomass equations are driven with field estimates of height and basal area. L-band SAR 11 m spatial resolution data were used as substitutes for estimates of height and basal area in equations of biomass. The following biomass equation, with a $R^2 = 0.977$ MSE <0.001, $n = 35$, was produced:

$$\text{BIOMASS} = -0.02539 + 1.51070\ \text{HT}^{-1} + 0.45944\ \text{BA}^{-1/2}$$

where: $\text{HT}^{-1} = 0.15272 - 0.00210\ (\text{Radar}) + 7.321\text{E-}6\ (\text{radar})^2$ and,

$$\text{BA}^{-1/2} = 1.05813 - 0.00918\ (\text{radar}) + 2.36\text{E-}5\ (\text{radar})^2$$

Height and basal area were predicted using the radar backscattering coefficients, and then biomass was calculated using those estimates (Hussin et al., 1991). There was less than 1% bias associated with the biomass estimates in a validation data set when the radar height and basal area estimates were compared to the field estimates in estimating biomass. Overall, however, field estimates of height and basal area provided superior biomass estimates. Most problematic were the height estimates, which were found to cause up to 4.17% bias in the biomass estimates from the L-band SAR data. The equations were suitable for stands up to about 31 years of age — the time during which the tree height, basal area, and biomass all increased rapidly. Like optical/infrared reflectance, microwave backscatter coefficients were dominated by interactions in the canopy and appear uncorrelated with additional stand development once the canopy had closed. The saturation of the backscatter was believed to be due primarily to crown closure (Hussin et al., 1991: p. 430):

> *As the stands mature and crown closure approaches 100% there is [more] backscatter from the top of the canopy and less penetration into the canopy layer with a lower scattering of the radar signal within the canopy, which reduces the interaction between the radar signal and the individual trees.*

The hope of robust forest biomass prediction using available SAR sensor design appears less likely based on this emerging understand of the physics involved. One suggestion is that SAR data can be used to predict some forest conditions after stratifying the study area by crown closure or LAI; additional research has shown the value of additional stratification of those areas with a likelihood of producing stronger relationships, i.e., away from steep slopes and sparse vegetation cover (Franklin et al., 1994). By overlaying the image on a DEM, it is possible to avoid steep slopes and areas of SAR layover/foreshortening; but areas that have maintained a normal incidence angle can be retained and studied further. In areas with a higher NDVI measured by the TM sensor, there may be a greater likelihood of the stand having full canopy closure; such areas can be avoided in the SAR image, because they are dominated by volume scattering rather than containing a strong double-bounce. One of the problems with this approach is that successively stratifying areas to restrict the analysis typically results in a reduction of the area. Then, the relationships could be applied to only a small fraction of the forest stands of interest. Also, more than one image — a multitemporal approach (Kuntz and Siegert, 1999) — would be required.

The recent deployment of even longer-wavelength radar sensors, such as the CARABAS VHF SAR developed in Sweden, illustrates a more promising measurement design. This system, at long (very high radiofrequency) wavelengths that essentially pass unimpeded through vegetation canopies, provides a strong signal from the main stem of trees. In early trials, CARABAS could accurately estimate forest standing biomass in large, mature forests (up to 750 m^3/ha) in Sweden (Israelsson et al., 1997) and over 1000 m^3/ha in France (Fransson et al., 2000).

Two recent studies have demonstrated that scanning lidar devices can make accurate measurements of stand height, aboveground biomass, and basal area in deciduous forests of the eastern U.S. (Lefsky et al., 1999a) and Douglas-fir/western hemlock forests in the U.S. Pacific Northwest (Means et al., 1999). A comprehensive analysis of the L-resolution lidar waveform, which was transformed to estimate the bulk canopy transmittance and the vertical distribution of reflective canopy surfaces (Lefsky et al., 1999b), produced a three-dimensional canopy volume variable that could predict total aboveground biomass with an R^2 value of 0.91 in 22 plots in Oregon stands of young, mature and old-growth conifer forests. The lidar quantified the volume of filled and empty space in the canopy and could distinguish euphotic and oligophotic canopy zones, enabling the quantification of canopy structure, including multiple canopy layers (Lefsky et al., 1999b). The potential for accurate lidar remote sensing estimates in areas of dense canopy cover and high LAI is of increasing interest as lidar devices become more widely available (Baltsavias, 1999) and the required processing methods better understood (Dubayah and Drake, 2000).

VOLUME AND GROWTH ASSESSMENT

VOLUME AND GROWTH

Volume prediction using remote sensing is strongly related to biomass prediction and, in fact, uses the same basic principles. In essence, what is sought is a remote sensing relationship with crown size and closure, and the desired volume (or dbh) estimate. This same approach has been used in estimating volume and dbh from aerial photography (Smith, 1986; Hall et al., 1989b) and digitized aerial photography (Bolduc et al., 1999). In light of earlier discussions, it should come as no surprise to find that volume estimates from satellite imagery are based on the weak relationships between spectral response and stand structure, as captured by imagery, in the differences in illumination and shadowing of forest canopies. The earliest idea behind volume assessment with remote sensing was that enough correlation would exist between what is detected by the sensor and the stand volume, such that the volume could be predicted from the sensor data without necessarily separating out or even understanding the various influences on the signal (Poso et al., 1984).

A few studies have attempted direct standing volume assessment from Landsat or SPOT satellite sensor data; almost all have commented that the very low dynamic range in the Landsat or SPOT satellite data is a significant factor limiting use for volume inventory at the level of detail required for forest management (Ripple et al., 1991; Danson, 1987; Salvador and Pons, 1998a,b). In future sensors, it is hoped that the quantization noise can be reduced while the radiometric resolution and spatial resolution are increased (Trotter et al., 1997). Even so, in areas without the resources to conduct forest volume inventory, rough (or class-based) volume estimates from satellite imagery can be very useful (Hall et al., 1998; Gerylo et al., 2000).

A common approach has been to use the satellite imagery in stratification prior to volume assessment (Borry et al., 1993). In one case, Ripple et al. (1991) studied the spectral response of Landsat TM and SPOT data in 46 homogeneous Douglas-fir stands in Oregon, and found that inverse curvilinear relationships existed between softwood volume and many of the spectral bands (Figure 7.5). Since the understory had a highly reflective shrub and herb layer, the young conifer stands with low softwood volume had a higher radiance than the older stands with higher volume and having more shadows. In another study of heterogeneous stands in a Mediterranean site, Salvador and Pons (1998a) correlated Landsat data with stand volume and basal area and found, as have many others, that while the small dynamic range in spectral response limited the explanatory power of the image data, there is another problem: how many plots are needed to establish the relationship between spectral response and volume? If too few plots were used to develop the relationship, a tight fit could be found, but the results applied only locally; the relationship will weaken as distance from the calibration sites increases (Iverson et al., 1990). If too many plots were used, there was a greater possibility of introducing large-scale uncon-

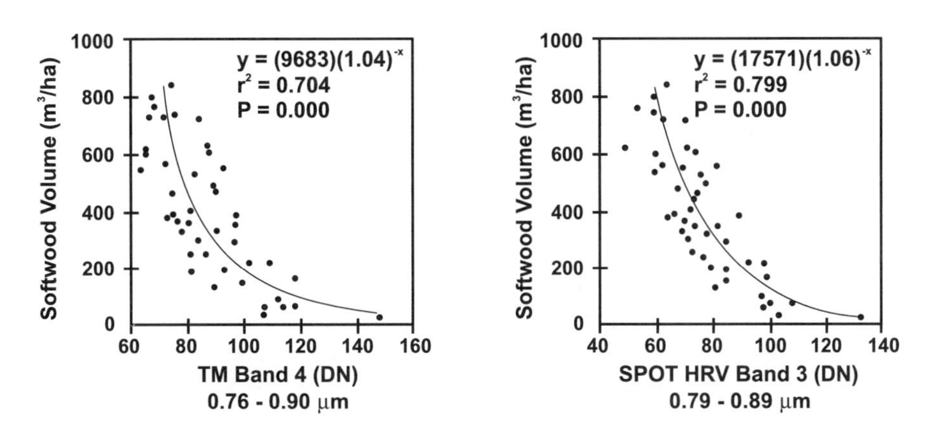

FIGURE 7.5 The spectral response of Landsat TM and SPOT data in 46 homogeneous Douglas-fir stands in Oregon; shown are inverse curvilinear relationships between the softwood volume and many of the spectral bands. As the stands mature and increase in volume, the view from above the canopy is increasingly dominated by shadow effects and increased scattering of near-infrared light. (From Ripple, W. J., S. Wang, D. L. Isaacson, et al. 1991. *Int. J. Rem. Sensing,* 12: 1971–1977. With permission.)

trolled variation, such as changes in climate, soils, geology, forest management, structure, or natural understory. These could easily increase the variability of spectral response. Since there is a reasonably large investment in establishing field plots in support of a remote sensing volume study, fewer plots rather than too many plots have often been used. The result is that there appears to be a general overestimate of the power of the few predictive equations that have been developed.

Great care must be exercised in collecting field data to develop spectral models to predict forestry variables; any comparison of one data source to the other is a potential source of uncertainty. An area that is very heterogeneous or different from the calibration sites will create a large potential source of error in the estimates. Working in a British Columbia forest, Gemmell (1995) noted that the utility of the TM data to predict volume was dependent on:

1. The homogeneity of the forest stand,
2. The spatial scale, and
3. The range of volume being investigated.

Shadowing within the stand was the dominant control on satellite spectral response because the greatest contribution to stand reflectance was the illuminated background (Gemmell, 1998). As trees aged, the crowns remained fairly constant. The importance of the illuminated foliage within the pixel was therefore small, but progressively more and more of the background was obscured. This effect was particularly pronounced in TM band 5 (shortwave infrared), but not as noticeable in band 4 (near-infrared). The effect of the increased scattering by more leaves was hypothesized to be larger in the near-infrared band. There was no sensitivity of spectral response in any band to volumes greater than 400 m^3 ha^{-1}. The dependence of spectral response on topography can reinforce the apparent dependence of spectral response on volume (Gemmell, 1998). This will occur if the higher stand volumes are found on favorable sites which coincide with the sun-sensor geometry. In other words, south-facing slopes had the highest volumes and were also brighter because of the topographic effect (Gu and Gillespie, 1998). Practical interpretation of the scale problem meant that the model had to be developed in larger areas, perhaps by averaging several smaller stands in order to reduce the variance in spectral response and field conditions that were unrelated to volume.

Trotter et al. (1997) studied the high yield, short rotation coniferous forest plantations of New Zealand. At the pixel and stand scale (i.e., averages 100 ha and by wood volume increment), correlations between Landsat spectral response in the near-IR and red bands and volume were very low but significant (Trotter et al., 1997: p. 2220):

> *The usual interpretation of models which provide a weak but significant relation between wood volume and Landsat TM data is that they should provide acceptable estimates over large areas, provided that the residuals are unbiased.*

In 8 of the 10-ha blocks, the model yielded 80 m^3 ha^{-1} RMSE — too high to be used as a basis for optimizing harvest strategies. A forest stand size of about 10 ha would be the maximum size at which accurate forecasts of wood volume would begin to be useful for very detailed harvest planning and harvest optimization. So, there was no real prospect to average even larger areas and hence sharpen the estimates that way. However, even at this level of accuracy the remote sensing estimates of volume provided a useful additional source of information on the overall wood volume trends within larger stands, and could be used to find areas of anomalously high and low wood volume. In the plantation environment, the canopy was not as reflective as the forest floor (grasses and tree ferns) and the correlation between Landsat spectral response and wood volume was correspondingly negative. Unlike natural stands (e.g., boreal or montane conifers), the plantation trees were similar in height without the tall and uneven structures, providing little contrast. The correlation was further weakened as shadowing effects did not change much as density increased.

It is common to deal with forest heterogeneity by using either stand- or plot-level averages of volume and remote sensing data (Franklin, 1986). To find stronger relationships, simply replace continuous wood volume measurements with classes of volume. The problem then becomes one of predicting the mean volume for a range of forest stand conditions (Franklin and McDermid, 1993). It has often been found that grouping the data set according to spatial boundaries derived from homogeneous TM data enabled a stronger relationship between forest characteristics and the TM imagery to be determined (Gemmell, 1995). In homogeneous sites there are fewer problems with registration, and stand spectral response would be less dependent on background effects.

In Finland, even homogeneous stands tend to be quite patchy. This situation led Tokola and Kilpelainen (1999) to quantify errors in remote sensing volume prediction that might be a result of the natural variability in volume near the stand edge and in the interior of the stand. By separating edge pixels from interior stand pixels with a spatial filter (a simple edge detector), the correlation to stand volume was significantly different (higher) than if all the pixels in the stand were used to predict volume. Still, the best R^2 was 0.51, a relationship that had too little predictive power to be of much use. To deal with the spatial heterogeneity issue, the probability of including the edge areas (perhaps based on the size of the stands) could be used to weight the sampled field observations. If errors in satellite image analysis increased the number of pixels on the forest stand border, then a heuristic error model would be able to handle the increased error in volume estimates.

Increasingly, combinations of remote sensing and traditional inventory procedures are being undertaken; the impetus for these activities is the increased demand for timely and accurate forest cover, volume, biomass, and health reports over large areas, such as biomes or countries. It is recognized that remote sensing cannot by itself satisfy the demand for such information — yet. Early efforts focused on demonstrating that nested scales of imagery in conjunction with ground based data and a GIS could successfully generate landscape and regional estimates of forest cover (Iverson et al., 1989a, 1990). The idea was to select a number of calibration sites on the ground, in medium scale, medium spatial detail imagery (e.g., Landsat TM or scanned aerial photography), and in small-scale, low-detail imagery (e.g., AVHRR, SPOT VEGETATION, MODIS), and develop regression functions that scale (Moody and Woodcock, 1995) or otherwise aggregate pixels and patterns (Bensen and MacKenzie, 1995) between each successive level.

An example of this approach was reported in Sweden. Fazakas and Nilsson (1996) integrated TM, AVHRR, and digital forest stand maps with Sweden's National Forest Inventory (NFI) to generate regional and national statistics on forest volume. The Swedish NFI is based on a mix of permanent and temporary sample plots; permanent sample plots are revisited every five years. No spatial estimate of volume can be generated because the sampling network is too sparse. The TM imagery were used to create a regression model, based on a large number (2750) of 10 m NFI field plots. The model could predict stand volume by species, in rough categories, per hectare. Then, within the AVHRR pixels in a calibration area, the TM data were used to predict mean values for total wood volume, Scots pine (*Pinus silvestris*) volume, Norway spruce (*Picea abies*) volume, and birch (*Betula verrucosa* and *B. Pubescens*) volume. New regression models, now employing the AVHRR spectral response (NDVI values), were generated and applied to the corresponding AVHRR data set covering the area of southern Sweden. Regional estimates of forest volumes and biomass were then generated (Fazakas et al., 1999).

Nonparametric k-nn estimates of forest parameters are used in areas of reasonably simple forest structure (Tokola et al., 1996; Tomppo, 1988, 1990). Plot data are used together with the corresponding pixel values as a massive lookup table to estimate a range of parameters at all pixel locations. The underlying premise is that plots with similar stand characteristics will have similar spectral values, and somewhere in the reference data there is something close to the desired relationship to be estimated. The advantage is that all field parameters are estimated simultaneously, and no global form of the model relating spectral variables and field variables is assumed. The process works in this way (Holmgren et al., 1999):

1. Given a pixel value at any location, calculate the Euclidean (or other distance) to all spectral values in the reference data set;
2. Select the "k" closest field sample plots (in terms of spectral distance, not physical distance);
3. Estimate the unknown parameters at the target location as a weighted average of the values of the k sample plots (usually a decaying exponential weight is used).

Volume (and biomass) are intricately linked to forest growth, and much of the same logic applies in developing remote sensing estimates of these parameters. But growth, in and of itself, is not at present a direct remotely sensible attribute of the forest (Franklin and Luther, 1995). Rather, some of the physiological conditions that lead to, or are highly diagnostic of, forest growth can be described by remote sensing measurements. In the next section, the key growth variable LAI is described; LAI may be a critical variable that can be used in forest growth assessment when coupled with ecosystem process models. The key to improved growth estimates from such models is accurate estimation of LAI. Modeled forest growth, with remote sensing driving the variables, such as LAI, may soon develop into a powerful source of information on forest productivity that can outperform standard growth and yield methods. The improved performance is expected because the process models are more closely related to the actual physiological functioning in forests (Waring and Running, 1998).

Occasionally, remote sensing data have been directly correlated with forest growth estimates (Franklin and Luther, 1995). In nine stands aggregated from 72 permanent sample plots in New Brunswick, Ahern et al. (1991) found that net annual volume change (cubic meters per hectare per year) was correlated with some of the Landsat Thematic Mapper bands at levels high enough to provide significant predictive equations; these equations were all improved with the addition of independently derived estimates of stand age. Explaining these relationships required a local knowledge of the stand development in New Brunswick; for example, the previous 10-year history of spruce budworm defoliation and the site quality of the plots which results in an understanding of the role of the understory as the stand is opened up (by cumulative defoliation). This approach is unlikely to be used as a basis for forest management, but can potentially reveal anomalous growth or areas of decline that could be subjected to additional inspection on the ground.

Leaf Area Index (LAI)

Leaf area index (LAI) is the leaf area per unit ground area, a dimensionless index usually defined in units of m^2/m^{-2}. In some remote sensing studies, LAI is expressed as a one-sided ratio of leaf area to projected ground area; in others, all-sided LAI is measured. LAI is an important structural attribute of forest ecosystems because of its potential to be a measure of energy, gas, and water exchanges. For example, physiological processes such as photosynthesis, transpiration, and evapotranspiration are a function of LAI (Pierce and Running, 1988). Accordingly, LAI and forest covertype are the two critical inputs available from remote sensing that are required to run ecological process models to estimate growth and productivity across the landscape (Running et al., 1986; Bonan, 1993; Peterson and Waring, 1994). LAI may be estimated at a variety of scales, and with a plethora of different instruments and techniques (Chen and Cihlar, 1995). In the field, LAI can be estimated using litterfall sampling, sapwood allometrics, and light interception observations, all of which are labor-intensive and impractical for larger stands and landscapes (Smith et al., 1991; Fassnacht et al., 1994).

Remote sensing estimation of LAI is based on the knowledge that green leaves interact selectively with solar radiation (Tucker, 1979; Sellers, 1985; Jasinski, 1996). Much of the near-infrared energy is reflected by foliage (Knipling, 1970; Gausman, 1977); much of the visible energy (dominated in the red portion of the spectrum) is absorbed by photosynthetic pigments (Waring et al., 1995a). Vegetation indices such as NDVI or the simple ratio (SR) can be used to capture the way in which red and near-IR reflectance differ in a single measure. The common approach to LAI estimation is empirical or semiempirical modeling; the same approach discussed above, involving correlation of spectral indices with field estimates and the extension of such estimates over larger areas with regression (Curran et al., 1992; Peddle et al., 1999; Wulder et al., 1996) or a canopy reflectance model (Huemmrich and Goward, 1997). A series of problems have been noted in this approach, including the fact that vegetation indices and LAI relations are known to differ:

1. Across ecoregions (Turner et al., 1999),
2. Among broad-leaf and needle-leaf species (Nemani et al., 1993), and
3. Possibly among conifer species (Kozlowski et al., 1991).

Two simple methods have been reported to increase the sensitivity of NDVI to increasing LAI, particularly when observations are acquired with high spatial detail or medium spatial resolution imagery:

1. Use of stand structure information from the GIS (e.g., stem density, stocking or crown closure estimates) (Franklin et al., 1997b), and
2. Use of image texture to capture the spatial variability in NDVI over small areas (Wulder et al., 1996, 1998).

Remotely sensed reflectance data are actually related to the fraction of incident photosynthetically active radiation (FPAR) absorbed by the canopy (Gower et al., 1999) and are only related to LAI to the extent that FPAR is related to LAI (Bonan, 1993; Chen, 1996). One simple LAI model assumes that light absorbed by the canopy can be approximated by the Beer-Lambert Law (Waring and Running, 1998). Light is attenuated by the canopy with an extinction coefficient (k) (after Jarvis and Leverenz, 1983):

$$LAI = (\ln I_z/I_o)/-k \tag{7.1}$$

where I_z is the PAR measurement under the canopy and I_o is the incoming solar radiation in open areas. This relationship flattens after reaching a threshold, which

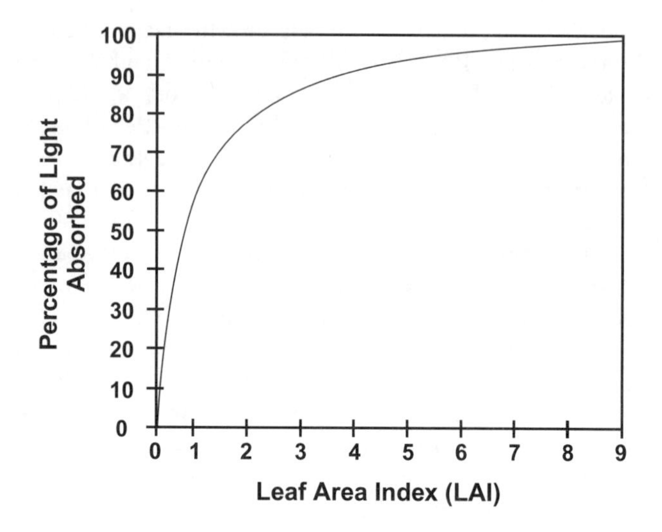

FIGURE 7.6 The relationship between light absorption and LAI (Beer-Lambert law); at higher LAI values the differences in visible light absorption are relatively small, resulting in a saturation of the remotely sensed signal obtained at the top of the canopy when LAI exceeds about 4.

can be seen in Figure 7.6. Based on this model, the relation between LAI and any particular vegetation index based on reflectance will be curvilinear. This results in a predictive equation that is at best imprecise (e.g., by using an exponential equation) and at worst useless when stand LAI exceeds approximately 3 to 4 (Spanner et al., 1990, 1994; Baret and Guyot, 1991).

Basically, as more and more leaves are added they are obscured by the first few layers of leaves; current remote sensing data and methods are very powerful when estimating the difference between no leaves and one leaf, between one leaf and two, between two leaves and three, but not much more. Chen and Cihlar (1995) reviewed the issues in developing empirical LAI relationships to satellite vegetation indices from the perspective of the field data required; in addition to saturation of the signals, clumping of foliage in the stand could lead to widely variable LAI estimates to which remote sensing data will not compare favorably. A one-time application of a single clumping coefficient may be needed (White et al., 1997). This can also account for site errors in allometric LAI. In a comparison of three different field techniques and a two different satellite remote sensing relationships, "Accurate LAI estimates using field-correlated satellite data was possible if all sources of error were identified from ground up" (White et al., 1997: p. 1725).

One of the more difficult issues is the translation of allometric, projected, and optically effective LAI to total LAI estimates. Using airborne imagery over the Boreas sites, the following steps were followed in processing the imagery to create an LAI map (Chen et al., 1999a):

1. DN converted to radiance using sensor calibration coefficients;
2. Radiance converted to reflectance using estimated downwelling irradiance after atmospheric corrections;

3. Geographical registration using onboard GPS;
4. Resampling of all pixels to standard size and north-south orientation;
5. BRDF correction to remove differences across the scene due to view angles;
6. Red and near-infrared reflectances normalized to nadir view and standard solar zenith angle;
7. LAI calculated from SR-based algorithm derived from transect measurements within the site.

Near-infrared reflectance decreased with increased LAI (Chen et al., 1999a). Several earlier studies had suggested that near-infrared reflectance should increase with increased LAI. The subtle reflectance differences were thought to be caused by differences in the forest variables that reduce reflectance by increasing shadows (e.g., crown size) and the forest variables that increase reflectance by increasing scattering (e.g., leaf orientation). As Chen et al. (1999a: p. 27957), state for conifer stands

> *The successful use of vegetation indices, formulated using red and near infrared bands, greatly depends on the delicate balance between reflectance-reducing shadow effects and the reflectance-enhancing multiple-scattering effects in the NIR band, while in the red band, the shadow effects always dominate the signals*

In aspen stands a different relationship was observed because the shadow effects were less pronounced.

The approach has been used successfully in a variety of forest conditions, but usually in pure conifer forests and with a variety of sensor systems. In one early study with aerial data, Gong et al. (1992) reported reasonable success using 5 m spatial resolution airborne data in several western U.S. conifer stands; R^2 values were reported between 0.81 and 0.92 for predicted LAI in the range of 0 to 3 at 8 ponderosa pine stands in Oregon using a sample of different spectral data or derivative spectra. Subsequent tests by Spanner et al. (1994) compared different airborne and satellite spectral data in their ability to predict LAI at these sites. The higher spatial resolution data appeared to provide better LAI estimates. Those studies were aimed at estimating LAI at the time of airborne image acquisition; Curran et al. (1992) estimated seasonal LAI using multitemporal satellite imagery. The main difference was that a seasonal estimate of LAI on the ground (from litter traps) was required to calibrate the remotely sensed estimate (Table 7.7).

Fassnacht et al. (1997) reported an empirical study of the use of Landsat TM imagery in estimating forest LAI as the first step towards the remote estimation of LAI in the Great Lakes states. Because this study lacked a complete sample of hardwood and softwood LAIs upon which to base a relationship between spectral indices and field measurements, major differences were found between those results and several earlier LAI studies in Oregon with different canopy closure conditions in similar stands and stand structures (Peterson et al., 1987; Spanner et al., 1990). The same relationships between spectral indices and field measurements (this time based on allometry) were tested, but with a different range of LAIs subjected to different image processing techniques. In the red band, the Wisconsin results differed significantly from the Oregon results. The difference was partially attributed to the

TABLE 7.7
Predictive Regression Relationship of the Form LAI = a + b*TMNDVI for 16 Plots on Three Dates in Florida Slash Pine Stands

Date	n	a	b	SE	R^2
Feb. 1988	8	–14.31	32.25	0.33	0.86
Sept. 1988	8	–20.02	43.62	0.83	0.82
March 1989	8	–10.80	26.29	0.52	0.83

Source: Modified from Curran et al. (1992).

differences in understory (ferns beneath jack pine) and a relatively crude atmospheric correction which probably overcorrected the Oregon red band data (Peterson, 1997). This situation highlights the critical need for atmospheric corrections in this type of continuous variable estimation procedure; with only a simple added dark-object correction, the data provided seriously misleading results, particularly when the red band was used in the denominator of a vegetation index (such as NDVI).

In Wisconsin, Fassnacht et al. (1997) found that the near-infrared band had the strongest relationships to LAI, consistent with radiation theory which predicts increasing near-infrared scattering with increasing amounts of vegetation, but not consistent with the earlier Oregon studies or the study by Chen et al. (1999a) at the Boreas site; flat or decreasing near-infrared responses were found there. Apart from the different specific relationships, the power of this empirical — normative — design in the remote estimation of LAI can be appreciated when considering that what was learned can be applied with confidence elsewhere (Fassnacht et al., 1997):

- Estimates of LAI should be based on samples across the full range that can be expected within a geographical region for forests of each leaf habit;
- The range of LAIs obtained for forests of different leaf habits should overlap as much as is ecologically possible;
- Mixedwood stands should be treated separately because of their different spectral characteristics from pure deciduous and conifer stands (Chen et al., 1997; Wulder et al., 1996);
- Separate measurements of LAI may be needed on each stratum of the canopy depending on the extent and variability of understory and other canopy layers (Chen, 1996);
- Larger data sets may be needed to smooth out variability unrelated to LAI (note the similarity to the problem with remote volume estimation — in essence, estimates are only valid over large areas);
- Seasonal dependence of the relationships must be understood and measured (Curran et al., 1992; Chen, 1996).

Overcoming the difficulties of remote sensing LAI estimation for a regional forest or even an individual forest stand can be a formidable undertaking. Some of the reasons for pursuing remote LAI estimates, apart from the obvious advantage of spatial coverage, are the difficulty and site-specific nature of most available field methods (e.g., allometric sapwood equations). Now, looking at the remote sensing procedures that have emerged as necessary, it may be reasonable to ask, Where can it lead?

Obviously, for those interested in using ecosystem process models to improve growth estimates or to estimate NPP, or for those empirically monitoring the annual variability in stand productivity, LAI can be a potent stand discriminator and model input. LAI is a useful structural variable in defoliation and health assessment; as noted in the following chapter, LAI is a damage assessment and diagnostic tool. LAI is a critical component in estimates of growth efficiency, a diagnostic measure that can indicate stand health and vulnerability (Waring et al., 1980; Waring, 1987). To some, spatially explicit LAI is an example of an unsurpassed classificatory tool that can remove much of the subjectivity in class thresholds — perhaps moving practitioners to consider more functional classification systems (Graetz, 1990; Running et al., 1994). Of the many sustainable forest management criteria and indicators, it seems likely that knowledge of LAI can be used in developing better estimates of biomass, forest cover, and some individual stand conditions (e.g., canopy transparency).

But for many traditional forest management applications, remote sensing LAI represents a new type of information not readily incorporated into planning. There is a need to determine the relationships between LAI and traditional forestry variables, such as basal area or stem density (Buckley et al., 1999). How else to set up and apply treatments with the aim of achieving recommended LAI conditions? On the other hand, is there a practical need for LAI estimates? Does having an LAI estimate of a stand alter any management decisions on the ground — harvesting or silviculture, for example? Just because LAI can be remotely sensed with some confidence does not automatically mean that LAI must be made available to forest managers. On the other hand, because LAI can be remotely sensed with some confidence does suggest that of all the possible remote sensing variables, this one should be given some attention. Perhaps some creativity and experimentation is required to determine the full range of applications that remotely sensed LAI can fulfill in sustainable forest management.

8 Forest Change Detection

The principal advantage of Landsat, or any satellite data, is their repetitive nature.

— M. Price, 1986

INFORMATION ON FOREST CHANGE

Change to a forest may be apparent only after long periods of time, a result of many almost imperceptible and yet powerful forces. Many forests are slow-growing and relatively long-lived. Forests can give the impression of stasis, climax, an almost unchanging timeless character. But change is a defining characteristic of forests, in landscape pattern and function, occurring at virtually all spatial and temporal scales. An example might be the creation of a soil horizon layer in a conifer forest, predictable by considering the climate conditions, litterfall, and microbial activity. Successional changes, growth changes, changes as a result of structural and age processes, all accrue slowly and with generally small daily, weekly, monthly, even annual variability. Change can also be rapid and transformative; for example, leaves can change color and cell structure overnight. Powerful, even cataclysmic, forces can arrive with little or no warning. Examples might include a wildfire, an insect outbreak, a windthrow, a harvesting operation, or a prescribed burn.

In managing forests, change frequently follows deliberate human decision making and is welcome and predictable — management is often thought of as a way of regulating changes on the landscape. Change following operations in a local forest company may be unknown or unavailable at another level (e.g., regional forest authority or national inventory). Change is sometimes undesirable, often seemingly random. Detection and monitoring many such forest changes across large areas are two of the most important tasks that remote sensing can accomplish in support of sustainable forest management. An important question has emerged that must be addressed by remote sensing (Coppin and Bauer, 1996): Which changes need to be detected and how often? There may be requirements to map changes that are not detectable in the imagery, and there are changes that can be detected, but are not of interest. There needs to be a balance between changes that are statistically identifiable by remote sensing and are of significance for forest applications.

Any approach to forest change detection requires a well-prepared data set, and a specific set of ground observations to calibrate the changes from one type of forest condition to another. The imagery must be in near-perfect registration, with interband

TABLE 8.1
Example Change Detection Contingency Matrix for Three Classes at Two Different Times

	Reference Data[a]								
Class Data[a]	F1	F2	F3	F1F2	F1F3	F2F1	F2F3	F3F1	F3F2
F1	x								
F2		x							
F3			x						
F1F2				x					
F1F3					x				
F2F1						x			
F2F3							x		
F3F1								x	
F3F2									x

[a] Hypothetical three-class example. F1, F2, and F3 represent three different forest types; the class confusion matrix must consider the possible misclassification of pixels in two dates.

Source: Modified from Congalton and Brennan (1998).

and intraband noise reduction (Gong et al., 1992) to reduce misidentification of changes that result from differing image geometry. The imagery should be converted to a standard quantitative measurement such as reflectance (Saint, 1980), or at least converted to a normalized data set or index (Lyon et al., 1998) referenced to a single master observation (Mas, 1999). Among the multitude of possible change detection approaches, an optimal technique must be selected that can provide the best detection of changes and the least error (Cohen and Fiorella, 1999). Issues of change accuracy assessment (Congalton and Brennan, 1998; Biging et al., 1999) must be addressed, over and above the accuracy assessment considerations in single date image applications such as classification. For example, in a classification change detection project, the usual contingency table expands to much greater size (Table 8.1), with consequences for sampling and field work when possible changes between two image classifications are considered likely.

Early change detection work focused on the use of aerial photographs in the interpretation of vegetation change. The need for total coverage in a short period of time (for example during insect detection surveys) resulted in very high costs (Beaubien, 1977). Aerial photographic methods can make sense over historical time periods and in two main types of change detection applications:

1. Detailed vegetation assessments of change at large spatial scales (i.e., fine resolution but small extent or area coverage), or
2. Broad landcover assessments of change at small spatial scales (i.e., coarse resolution [less detail] over a large spatial extent; large area coverage with Level I or II changes).

In the first instance, over small areas, "studies based on aerial photography may be used for very detailed assessment of rates and patterns of change, and to test hypotheses regarding these factors" (Price, 1986: p. 486). As always in interpreting aerial photographs, there is the difficulty related to the boundaries of vegetation types (Abeyta and Franklin, 1998) which when "recognized from the interpretation of photographs will not always coincide with those derived from ground-level studies using classical methods for the description of vegetation; depending on film type, filtering, image scale, time of acquisition and mode of analysis" (Price, 1986: p. 486). The homogeneity assumption can create difficulties that cascade throughout the use of the data. Usually, though, the original photos are stored and can be accessed easily, and are readily interpreted without specialized training or equipment.

The ease of acquisition and interpretation of photography guarantees that, in many change detection applications, this type of data is an appropriate choice (Pitt et al., 1997). In fact, at the operational level most change detection is probably conducted by people looking at the newest photography and comparing it to the GIS database, or even their personal knowledge of the management unit. Aerial photography is under continual improvement (Caylor, 2000), and in its many forms (e.g., film size can be metric, supplemental, oblique, and high-altitude, and film emulsions can be natural color, color infrared, reversal (or positive), and panchromatic) continues to be an indispensible management tool in change applications (Lowell et al., 1996).

At large spatial extents, coarse changes in landcover or vegetation type can be considered using aerial photographs. Principally, historic vegetation patterns would be of interest. The difficulties in using uncontrolled photomosaics and variable radiometry aerial photographs over large areas are reasonably well known and relatively easily accommodated by experienced air photointerpreters. For example, Burns (1985) used aerial photographs covering three large test areas in Lousiana, Kansas, and Arizona, and compared the results of change detection to those obtained from Landsat imagery. Only Level I landcover changes (from forest to agriculture, agriculture to urban, range to agriculture, and range to urban) could be reliably detected by Landsat and confirmed by aerial photographic work over a five-year period. Accuracies were estimated to exceed 75% in all categories for large areas.

Aerial photographic techniques will certainly be required when considering landscape changes in the era before routine satellite observations were collected. For example, Turner (1990) used the manual interpretation of air photos dating from the 1930s through the 1980s to monitor changes in eight Level I and II landcover categories in Georgia (urban, agricultural, transitional, improved pasture, coniferous forest, upper deciduous forest, lower deciduous forest, and water). The forest classes were defined by a canopy cover of at least 50%, and an estimated average tree height of 3 m. Photographs at three sites using three aerial photographic scales (1:20,000, 1:40,000, and 1:60,000) were examined over a 50-year interval. Each photo pair was viewed under a mirror stereoscope and the interpreted land cover polygons transferred to an acetate overlay. A grid with cells representing 1 ha was then placed over the acetate, and the land cover representing the greatest proportion of each cell was digitized to create a raster database. Differences at each time period could be summarized by area and location, and the raster database subjected to landscape pattern analysis.

Digital methods of change detection were largely developed for, and applied to, satellite imagery to take advantage of the new repetitive, synoptic digital data (Saint, 1980; Howarth and Wickware, 1981). At first, such methods were not widely used — possibly because of the relatively coarse spatial resolution of the early satellite data obtained by the Landsat MSS sensor — but more likely, as in other remote sensing applications, users experienced difficulty in interpreting the data (Wickware and Howarth, 1981; Singh, 1989; Coppin and Bauer, 1996). As new satellite and airborne images became available, it appears more likely that remote sensing data acquired repetitively at short intervals and with consistent image quality will be a necessary database for forest change detection (Mas, 1999).

The simplicity of the basic idea of digital remote sensing change detection is deceptive (Donoghue, 1999): consider a pixel or group of pixels over time and determine the likelihood of change. The basic changes in spectral response caused by forest harvesting, silviculture, and natural disturbance are similar; typically, following the removal or significant decrease of forest canopy cover there is an increase in visible reflectance and a decrease in near-infrared reflectance. The greater the amount of forest removed, the greater the changes in reflectance that are observed. Similar patterns have been observed in SAR imagery; cleared areas are brighter in SAR spectral response than are forested surfaces.

Several early problems in digital change detection have been overcome with time and experience. For example, it was felt that small changes of local interest could not be detected reliably by satellite remote sensing (the spatial resolution problem). This issue has largely disappeared as the types of changes that can be detected have become better understood and the data options have increased. Experience has enabled greater confidence in the application; in Brazil, for example, separating acacia and eucalyptus plantations from natural forest was more readily accomplished with TM compared to MSS data because of the improved spectral, spatial, and radiometric resolutions (Deppe, 1998). Another reason was that the field data were collected at a time closer to the acquisition date of the TM data — this will often be the case. In any event, in digital change detection it has often been found that the TM data are actually too fine and need to be generalized to reduce the tremendous data volume to a more manageable level. Principal components analysis is often the data reduction tool of choice (Fung and LeDrew, 1987).

Recently, the use of satellite imagery in change detection applications has flourished; change detection is one of the most powerful reasons for using digital remote sensing data, and certainly satellite remote sensing imagery (Lunetta and Elvidge, 1999). Continual refinement of the methods of change detection by satellite and airborne remote sensing has been provided by numerous reported examples of forest changes caused by natural disturbances, such as floods (Michener and Houhoulis, 1997), winds (Ramsay et al., 1998; Mukai and Hasegawa, 2000), wildfires (Koutsias and Karteris, 2000; Salvador et al., 2000), insects and disease (Leckie et al., 1992; Franklin and Raske, 1994), and other forest decline phenomena (Yuan et al., 1991; Brockhaus et al., 1993).

The digital methods force more precise answers to questions of methodology in change detection than those required by manual aerial photointerpretation of land-cover or vegetation types:

- What is a significant change?
- How does one assess the accuracy of change detection?

The first question is typically addressed by establishing thresholds of change. The method of establishing thresholds depends on the image analysis technique, but likely involves a type of training data collected in known change locations (Malila, 1980; Fung and LeDrew, 1988; Cohen et al., 1998). Identifying the specific nature of change in those areas detected with a high probability of change would no doubt require field or air photo work. Often, the only way to check on the early image data is through interpretation of historical air photographs (Hansen et al., 2000). Assessing the accuracy of change detection typically involves images that were acquired in the past, often under less than ideal conditions. Sampling for accuracy assessment in this situation is problematic. In addition, a wide variety of possible sources of error in assessing accuracy in a change detection project originate in the classification scheme, registration problems, and change detection algorithms (Biging et al., 1999).

Which algorithm will be able to detect change reliably but not misfire? While the techniques are variable, two broad approaches are common, based on Johnson and Kasischke (1998):

1. Data transformation (e.g., image differencing, PCA), and
2. Change labeling (e.g., regression, classification).

The classification approach is generally indicated when the differences between the two images to be compared are large (e.g., very different ground conditions, different seasons, or different sensors). The idea is to provide a classification of each date separately, and then compare the results (Jakubauskas et al., 1990; Franklin and Wilson, 1991b; Mas, 1999). Comparative studies have shown that if the change is large and distinct (e.g., clearcuts, fires, or urban development), then classification techniques can be highly effective. The classification approach can also reduce the influence of other factors, such as differing radiometric properties, by independently placing the spectral responses in the appropriate classes before comparing information from different dates (Pilon et al., 1988; Mas, 1999). A disadvantage of this approach is that, even though many changes that are smaller than individual pixels can occur, only a complete change in class membership will be detected (Foody and Boyd, 1999). Despite having no standardized change detection protocol, digital methods of change detection and identification are increasingly considered for use with all types of airborne imagery, including digitized aerial photography (Price, 1986; Meyer et al., 1996) and SAR data (Cihlar et al., 1992).

Image differencing has been accomplished using many different algorithms, ranging from simple subtraction to complex statistical manipulations such as expressed in a principal components analysis (Fung and LeDrew, 1987). Comparisons of different image differencing, and classification procedures has been the focus of several recent studies aimed at developing an optimal change detection technique (Muchoney and Haack, 1994; Collins and Woodcock, 1996; Mas, 1999; Morisette et al., 1999). Image differencing using the original bands, or a transformation of the original bands, requires greater attention to radiometric issues and may also present information that is more difficult to interpret. Rather than a simple class-by-class summary, image differences must be related to the changing feature on the ground; changes in reflectance, for example. More complex change detection procedures are typically an elaboration of the concept of image differencing and may be still more difficult to interpret. Change vector analysis, for example, provides a magnitude of change and a directional vector for detected changes in imagery, but these outputs appear to be inadequately described in the literature. Their use may be subject to uncertainties not yet fully understood (Cohen and Fiorella, 1999).

Generally, differences are small in the performance of the change detection algorithms tested to date. Most are readily available in commercial image processing systems. A more important factor may be the different types of data that are available. There may be a difficulty in detecting change on recent satellite imagery compared to coarser resolution historical data; this coarser resolution data may be in the form of a satellite image (e.g., Landsat MSS data) or a polygonal database generated by aerial photography and field surveys. The polygonal data represent a special form of the change detection problem; rarely will it be possible to compare polygon to polygon. Even in the traditional task of forest inventory change, it is more typical that the inventory is completely replaced rather than updated in a change detection procedure (Lowell et al., 1996).

Instead, tools such as the Polygon Update Program (PUP) (Wulder, 1998a) have been devised to examine pixels within polygonal structures such as forest stands. Not only can the forest inventory guide the change detection analysis to the areas of highest interest, but the polygons themselves can provide a way of organizing the landscape such that the changes are reported as aggregated within polygons. This process has been termed polygon decomposition (Wulder, 1998a) and refers to the process of analyzing previously delineated polygon areas using ancillary digital information acquired from an independent source. Often, the mix of vegetation is of interest within the polygonal structures or forest stands (Carpenter et al., 1999). The idea is to use those independent data typically acquired through remote sensing to provide insight into the internal characteristics of the polygonal area, typically delineated using aerial photointerpretation. The polygon, or vector, data are used as the context for the analysis of remote sensing, or raster, data. The polygonal data represent areas of generalization, but the remotely sensed data can be used to make measurements or aggregate information in a meaningful way within those generalized areas.

In essence, the polygonal information is a way of structuring or stratifying the remote sensing information for analysis (Varjö, 1996); another way to view this process is to consider that the remote sensing data are a way of explaining the polygonal structure. The fusion of the raster and vector data allows for the augmentation of current information in the previously delineated polygon areas. The current information available from the remotely sensed data may be physical properties such as spectral response values (Chalifoux et al., 1998), or categorical properties such as the result of an image classification or change detection procedure.

HARVESTING AND SILVICULTURE ACTIVITY

CLEARCUT AREAS

Forest harvesting by clearcutting has long been monitored by satellite remote sensing, with accuracies suitable for operational mapping in many different types of forests and with a variety of sensors. Since forest clearings are generally visible in hardcopy aerial and satellite imagery, both analogue or manual interpretation and digital approaches have been used to:

1. Detect forest clearcuts (Drieman, 1994; Banner and Ahern, 1995; Pilon and Wiart, 1990; Yatabe and Leckie, 1995; Murtha and Pollock, 1996);
2. Map clearcut boundaries (Rencz, 1985; Hall et al., 1989a; Hall et al., 1991c; Hall et al., 2000b);
3. Direct field sampling to areas of high likelihood of change (Kux et al., 1995; Varjö, 1996);
4. Provide information on successful legal enforcement of protected areas (Fransson et al., 1999);
5. Provide landscape-level summaries of area changes (Hansen et al., 2000).

The principal reason to consider satellite imagery in the task of clearcut mapping is the reduced cost compared to aerial photographic surveys and field mapping of cutblocks. Before cost savings can be realized, it is necessary to show that the same levels of accuracy that are obtainable using traditional methods are possible with satellite remote sensing techniques. For example, in Alberta, the two major physical criteria for accepting an alternative image source for cutover update were (1) cutover area accuracy and (2) boundary placement accuracy. Using standard manual photomorphic techniques, Hall et al. (1989a) showed that overall cutover area accuracies were 86.7, 89.5, and 86.9% on medium-scale airphotos, Landsat TM, and MSS imagery, respectively. Overall, cutover boundary placement errors for air photo techniques, Landsat TM, and MSS imagery were 30.1, 24.9, and 38.3 m, respectively (Figure 8.1).

In a cost analysis, Landsat TM images offered a 12:1 cost savings in data acquisition over aerial photography (Hall et al., 1989a). The MSS imagery were not recommended for operational mapping of clearcuts, but the TM data were considered an appropriate alternative to the use of air photos, at least in the type of forest studied (predominately conifer stands). This study was recently updated using IRS 5.8 m panchromatic satellite data with a similar conclusion; under certain circumstances satellite remote sensing imagery can provide cutblock updates comparable to those acquired with aerial photographic methods (Hall et al., 2000b). Errors were even

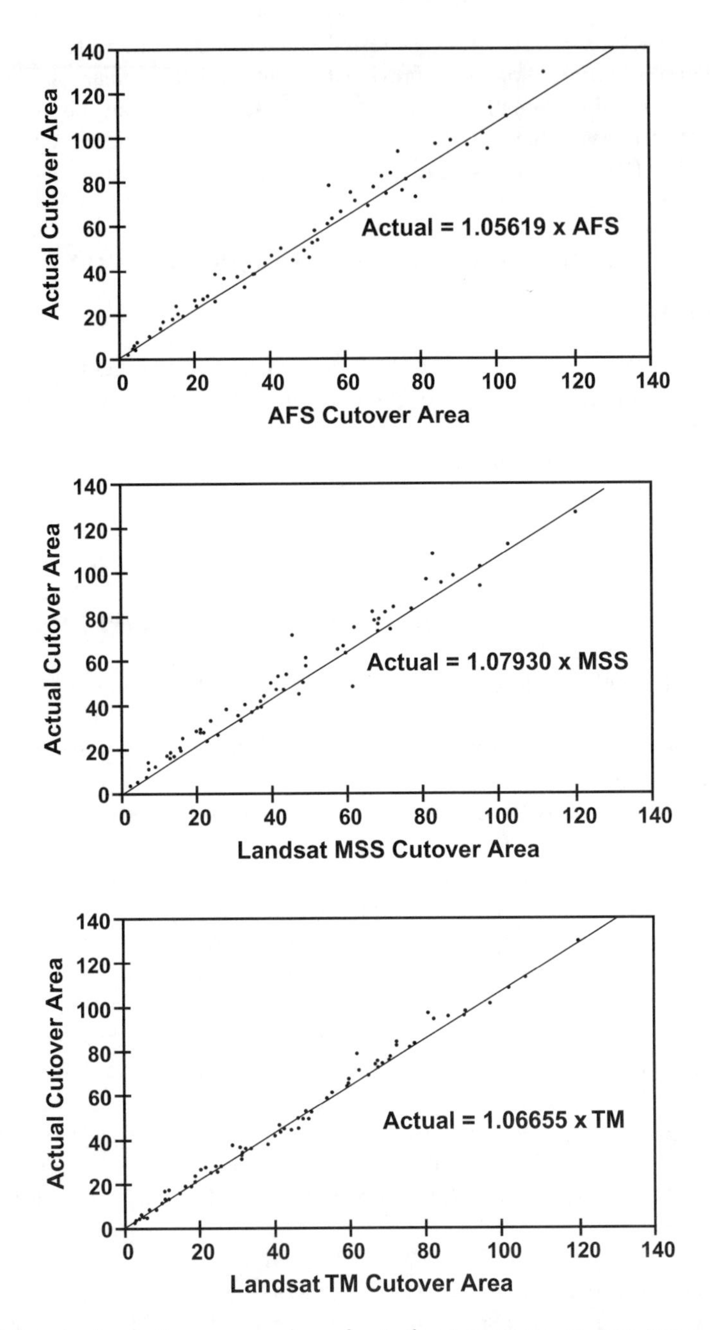

FIGURE 8.1 A simple linear regression of actual cutover area vs. two types of image interpretations based on Landsat MSS and TM data. Using enlarged color composites, TM-predicted cutover area was within guidelines suggested for area and boundary placement of cutovers in Alberta. (From Hall, R. J., A. R. Kruger, J. Scheffer, et al. 1989. *For. Chron.,* 65: 441–449. With permission.)

lower than with the TM imagery; boundary placements ranged from 16 to 20 m of 1:20,000-scale photogrammetric measurements. However, visual interpretation is a time-consuming and labor-intensive method for large-area mapping (Sader, 1995).

Using six Landsat satellite images of a 1.2-million-hectare area in the central Oregon Cascade Range, Cohen et al. (1998) mapped cutovers between 1972 and 1993. All images were resampled to 25 m, masked using a DEM to eliminate lower elevation agricultural areas, transformed to Tasseled Cap vegetation indices, subtracted from previous images to create image differences, and classified using an unsupervised clustering algorithm. Comparison of the resulting harvest map with an independent reference database (using three different methods) indicated that an overall accuracy of greater than 90% was achieved. This is an important study not only for the demonstration of mapping cutovers with high accuracy from satellite data; the area covered in the application was so large, and covered such a long time period, that to attempt this mapping in any other way is almost inconceivable.

In Canada, several studies have been reported that confirm the utility and accuracy of clearcut mapping from digital satellite data. Using Landsat TM band 5 difference images in Nova Scotia, cutover area estimates differed by a maximum of 10% when compared to traditional aerial photograph mapping (Rencz, 1985); this difference was almost entirely attributed to other environmental changes such as gravel pits, flooded areas, and blowdown, and to the prevalence of small cutovers less than 1.5 ha in size in mixedwood stands. Using SPOT panchromatic images and simulated Radarsat imagery, clearcuts were mapped in Alberta (Banner and Ahern, 1995); very high levels of agreement were obtained, with errors decreasing with greater spatial resolution and when using nadir imagery (Figure 8.2).

Using multiseason airborne C-band SAR imagery for clearcut detection in Newfoundland, total clearcut areal error was less than 4% when compared to a control sample of clearcuts mapped using 1:12,500-scale color aerial photographs (Drieman, 1994). With SAR data, image interpretation concerns exist because of the strong dependence on topography and the typically low inclination angles (Edwards and Rioux, 1995). Great care must be employed in selecting image dates for comparison because of the large range of backscatter response that can be obtained from vegetation targets seasonally (Cihlar et al., 1992). Single date, single polarization, single incidence angle SAR data are typically presented as black and white gray-scale imagery, which can be difficult to interpret because of their highly textured and speckled appearance.

In tropical areas, the opportunities for field observations and the ancillary data (e.g., air photos and topographic maps) necessary for investigating forest changes may be lacking, making satellite imagery and digital methods an ideal information approach (Sader, 1995). Lowry et al. (1986: p. 904) suggested that "the accurate and ready delineation of cleared areas and plantations indicates that SAR is a reliable remote sensor for estimating and monitoring tropical deforestation and to some extent reforestation." In comparing airborne and simulated satellite C-band SAR data and Landsat TM data, a very high level of agreement was obtained in providing annual estimates of large (1000 to 10,000 ha) and medium (100 to 1000 ha) clearings in Brazil (Kux et al., 1995).

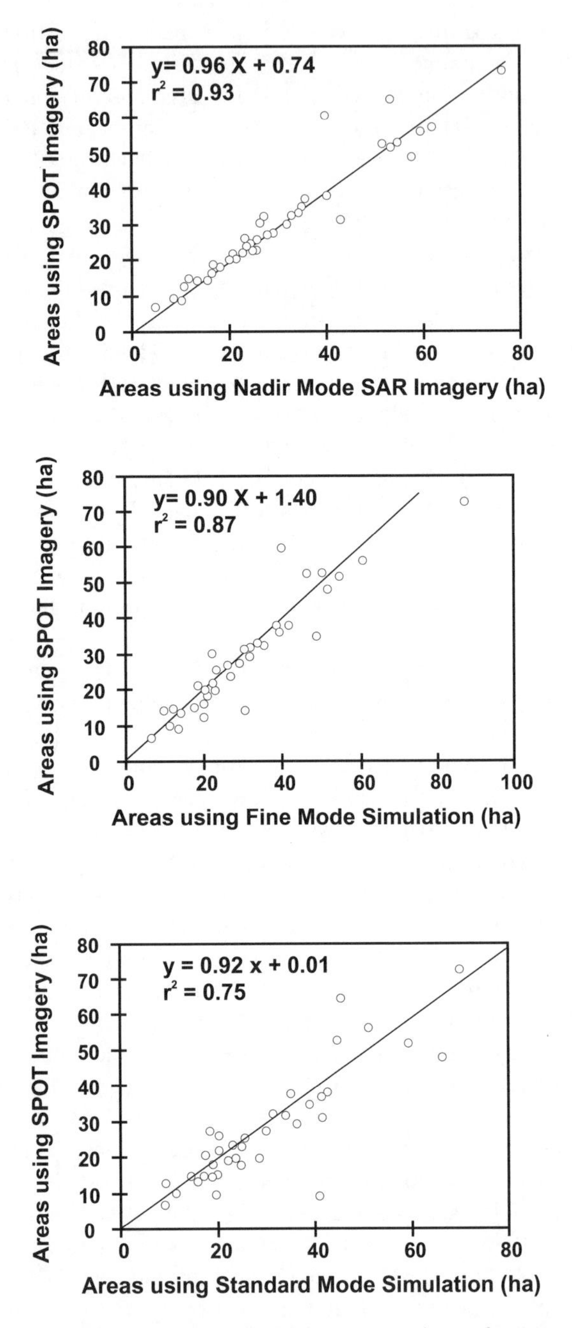

FIGURE 8.2 Strong relationships were found in a comparison of cutover areas measured manually using SPOT panchromatic imagery and three types of simulated Radarsat data in Alberta. Areas with steeper topography, variable forest types, and more variable cutting practices would likely be more difficult to interpret. (From Banner, A. V., and F. J. Ahern. 1995. *Can. J. Rem. Sensing*, 21: 124–137. With permission.)

Deforestation in Rhondonia due to human occupation was estimated to have reached at least 52 million hectares by 1996 (INPE, 1998). In order to better understand the deforestation process and assess some of the effects of the long-term occupation, 1977, 1985, and 1995 Landsat imagery were classified separately and compared (Alves et al., 1999). Deforested areas included pastures, annual and perennial crops, burned areas, and secondary vegetation. More than 90% of the total deforestation in the 1985 Landsat image, and 81% of the total deforestation in the 1995 Landsat image, occurred within 12.5 km of the areas deforested in 1977. High rates of forest depletion were linked to new settlements and roads into previously remote areas.

More automated methods of change detection have been developed and preliminary results are encouraging for their use in boreal forests. In Finland, a method was developed to analyze changes that deviate from normal vegetative succession. Such changes are usually rather rapid and of small areal extent when compared to the area changes related to natural vegetative succession (Häme et al., 1998). A typical example of such a change is a clearcut, but even damage caused by insects can be profound over a short period of time when compared to a successional change. The system used two images acquired on different dates as input:

- To reduce the mixed pixel effect, find homogeneous areas that can be used as seeds for clustering image data;
- Based on these training data, apply a clustering procedure separately to each image, and then list and name the cluster pairs by referring to a common index (in this case the NDVI);
- Transfer the clusters from the first image to the second image in the series, and note statistical differences in clusters in this second image (e.g., a high standard deviation);
- In those clusters with statistical differences between the first and second image, scale the differences and note the direction of the change in the vector;
- Indicate using output channels (such as the NDVI) the direction (positive or negative value) and magnitude of change.

The method was tested in southern Finland and was found to reliably detect and identify clearcuts (Häme et al., 1998). In addition, the method provided information on forest damage even though the actual magnitude of the change was small compared to the magnitude of change in clearcut areas.

An extensive system of change detection was implemented by Varjö (1996) in Finland; the aim was to find a method that could be used to check existing updates done by field or aerial photointerpretation, and subsequently, to direct field efforts to areas where the updates were not in accord with the remote sensing method. Clearcut areas and thinned and holdover removal stands were separated using multitemporal Landsat TM data after radiometric and geometric corrections were applied. The classifier worked within stands delineated the traditional way (by aerial photointerpretation); the mean and variance of reflectance was considered in each stand in each of the two images. When comparing the image classification results to recorded treatments, almost 7% of the stands were recommended for field inspection because of discrepencies between the observations by satellite and the existing records over the two-year period of the study. The suggestion was made that for a 10-year inventory cycle, fewer than one third of the stands now visited on the ground would need to be surveyed.

A variation of this approach has been used in the Hinton, Alberta region to map clearcuts over a two-year period with Landsat TM data. Figure 8.3 shows the binary image (clearcuts shown in black) that existed in 1996, and the additional harvesting that took place in 1997 and 1998. The accumulation of clearcutting is shown as an input to the application of landscape metrics to determine the spatial structure of the area, discussed later in this chapter.

PARTIAL HARVESTING AND SILVICULTURE

Compared to clearcut and harvest block detection by remote sensing, fewer studies have examined the effect of silvicultural activities or partial harvesting on the spectral response of forests (Gerard and North, 1997). Typically, the disturbance to the forest canopy resulting from these activities is much less than that which occurs during clearcutting (Chapter 8, Color Figure 1*). Subsequently, it is more difficult to use satellite spectral response, particularly in their detection. In manual interpretation of satellite imagery, partial harvesting in mixedwood stands known to occur in one Alberta study area was not consistently mapped (Hall et al., 1989a). The tonal differences were simply too small to be noticed by the image analysts when mapping at 1:100,000 scales or smaller.

In Scandinavia, most forests stands are subject to between one and three thinnings before the final clearcutting (Olsson, 1994). Normally, between 20 and 50% of the basal area is removed in one thinning; commercial logs are usually removed but the cutting waste remains. The material left on the ground, and the gaps created, present a possible spectral response that can be detected if the spectral, spatial, and radiometric resolutions of the sensors are adequate (Figure 8.4). After thinning, Landsat TM image reflectance increased in all bands except the near-infrared (band 4). Because the visible bands were sensitive to the amount of canopy that was sunlit and the amount of shading of the ground that occurred with a dense forest cover, "it can be assumed that the change in shadow patterns is an important factor behind the reflectance increase" (Olsson, 1994: p. 229). In the near-infrared portion of the spectrum, a small decrease in reflectance could be generally attributed to a reduction in the proportion of photosynthetically active canopy, the covering of the ground with cutting debris, and the changes in tree species proportions. In the middle infrared portion of the spectrum (e.g., TM bands 5 and 7), there was less diffuse scattering of light. Measurements in these areas can be as sensitive to shadow patterns as the visible bands (Horler and Ahern, 1986; Chen et al., 1999a).

A similar partial harvesting/silvicultural situation exists in New Brunswick. In one recent study, 424 balsam fir stands were found with changes detected by a

---

* Color figures follow page 176.

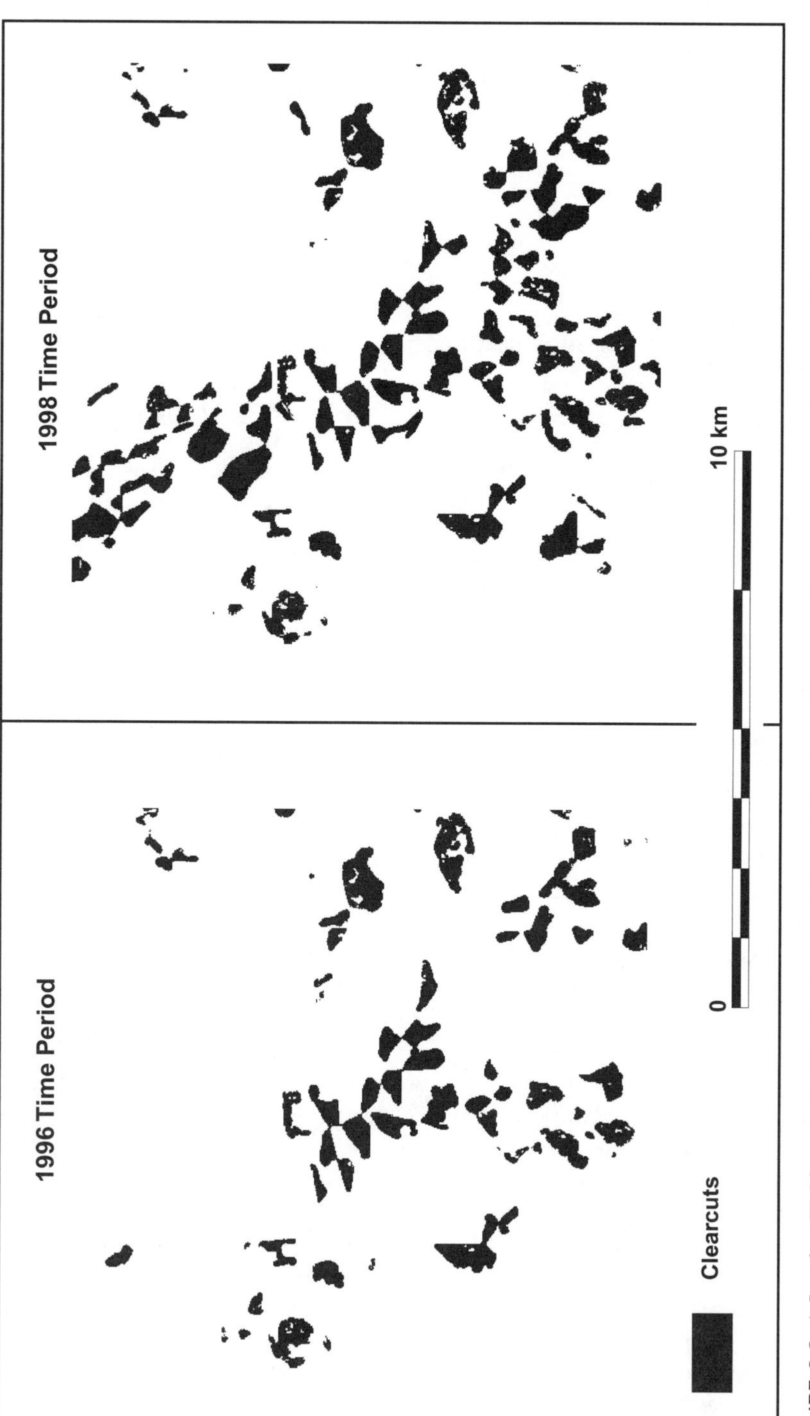

FIGURE 8.3 A Landsat TM image classification to reveal clearcuts in a boreal forest environment suggests the power of the change detection approach for this forestry application. In 1996 the clearcuts (shown as black patches), many of which were more than 10 years old, could be accurately delineated and separated from the surrounding forest mosaic (white background). By overlaying a 1998 Landsat TM image, new clearcut areas could be readily distinguished from the older cuts and the mature or young forests of the area. (Example provided by L. M. Moskal, University of Kansas.)

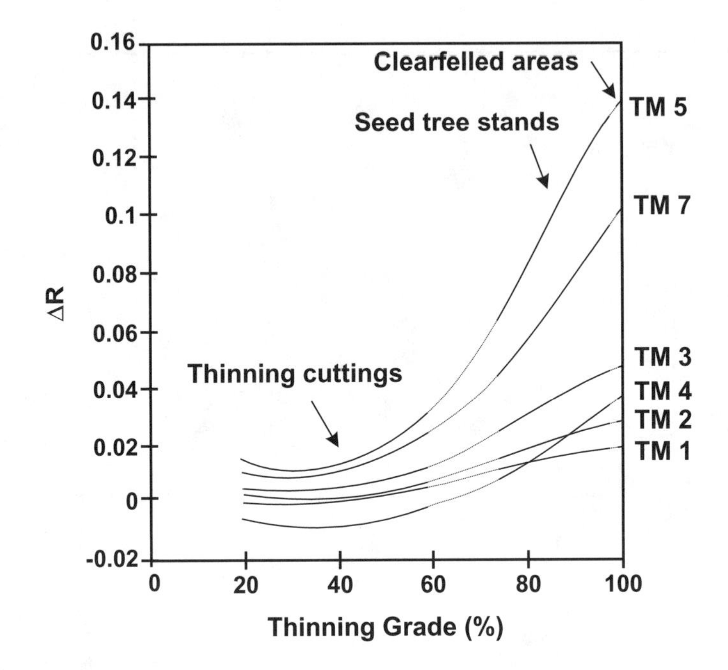

FIGURE 8.4 Annual reflectance change after cutting as a function of thinning as observed by the Landsat TM sensor in a boreal forest environment. Thinned areas had a much smaller difference in annual spectral response compared to seed tree areas which, in turn, had a smaller difference than was observed in clearcut areas. (From Olsson, H. 1994. *Rem. Sensing Environ,* 50: 221–230. With permission.)

1992–1997 Landsat Thematic Mapper remote sensing classification procedure (Franklin et al., 2000b). First, the image data were corrected for atmospheric effects, then transformed to brightness/greenness/wetness indices. The resulting six indices (three from each of the two image dates) were selected for classification using a maximum likelihood classifier. Areas that had been cutover in the intervening years were distinct in that they showed increased brightness, decreased greenness, and decreased wetness. In these stands, some 76,882 pixels were found to have changed in roughly equal proportions in three classes of change: light, moderate, and severe. Light changes were attributed to partial harvesting and precommercial thinning, and moderate changes were attributed to clearcutting with legacy patches and some hardwood selection cutting. Severe changes were clearcuts. The classification accuracy was estimated to be approximately 70%.

The effect of these physical changes to forests has been difficult to predict; typically, reflectance in all bands would increase with a reduction in basal area. However, in some areas a reduction in basal area has been followed closely by an increase in leaf area as the understory responds to the opening of the canopy (Franklin et al., 2000b) (Chapter 8, Color Figure 2). This increase in leaf area can decrease reflectance in visible bands and increase near-infrared reflectance; the opposite effect to that observed by Olsson (1994) in areas with little or no vegetative understory. In areas of spruce budworm mortality, stands with a significant deciduous component

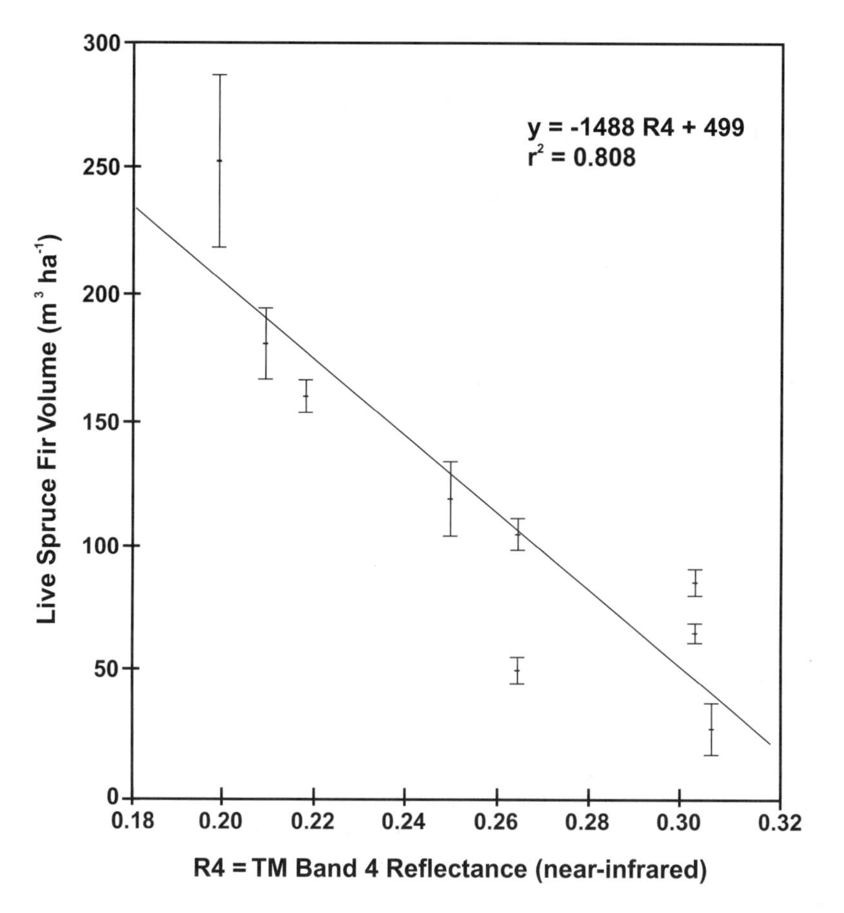

FIGURE 8.5 A higher near-infrared reflectance in spruce-fir stands thinned by spruce budworm tree mortality in New Brunswick. As the crown was opened up, near-infrared reflectance increased as a result of the exuberant understory beneath the conifer canopy. Total LAI increased despite the reduction in canopy LAI. In another area, the opposite effect may be observed; a reduction in canopy leaf area could cause a reduction in near-infrared reflectance viewed from above the canopy. (From Ahern, F. J., T. Erdle, D. A. MacLean, et al. 1991. *Int. J. Rem. Sensing*, 12: 367–400. With permission.)

showed a negative relationship between conifer volume and near-infrared reflectance (Ahern et al., 1991). That is, stands with lower spruce-fir volumes — caused by tree mortality rather than thinning or harvesting — had increased near-infrared reflectance because of the exuberant understory growth (Figure 8.5).

Regeneration

Regeneration surveys by aerial and field methods are a standard practice in many forest jurisdictions. What is needed is an assessment of stocking levels and planting success. This information can be obtained by plot-based or strip cruising, often coupled with air photography. Remote sensing — through supplemental aerial photography

(Zsilinszki, 1970; Hall and Aldred, 1992) — has long played a significant role in such regeneration surveys through direct estimation of cover, seedling, and stem counts.

Satellite remote sensing methods of forest regeneration assessment are much less common (Fiorella and Ripple, 1993b; Lawrence and Ripple, 1999). One approach is to consider the reflectance characteristics of the development of forest stands over time, from establishment or initiation (usually by disturbance) through the thinning phase, into stand maturity, and the various end points for forest ecosystems. Initially, as the new plants begin to grow the regenerating area would appear bright in all bands, gradually decreasing as decreasing amounts of the soil surface and understory were visible to the sensor. Increased absorption by greater concentrations of pigments in the canopy leaves, and increased shadowing, may reduce reflectance still further. Along these lines, Peterson and Nilson (1993) and Nilson and Peterson (1994) introduced the concept of the stand reflectance trajectory as discussed in Chapter 7. By this, it was meant that each of the stages in the development of the stand — for example, the successional stages — could be considered predictable in terms of reflectance.

In Tanzania, Prins and Kikula (1996) reported that detection of strong coppicing from roots and stumps in miomba woodland (*Brachystegia* and *Julbernadia*) was possible using Landsat MSS data acquired in the dry season after the first year of fallow. In northern California, 30 Landsat images acquired over almost 30 years were used to track the reflectance changes in clearcuts and regenerating areas (Kiedman, 1999). Images were calibrated and normalized so that differences in reflectance could be observed and quantified over time. A spectral mixture analysis approach was used; each pixel was modeled to determine the vegetation, soil, and shade fraction based on an extensive endmember library. As a single stand was observed by plotting the reflectance measurements over time, the endmember fractions changed in a predictable way according to the physical changes in the proportion of vegetation, soil, and shade induced by the clearing, regeneration, and maturing of the vegetation in the stand. The year a stand was cut was obvious by the significant reduction in the vegetation fraction; as the stand regenerated, the vegetation fraction gradually increased and the soil fraction gradually decreased. Eventually, as the forest canopy closed and the stand reached maturity, the vegetation fraction decreased and the shade fraction increased.

These studies provide good examples of the type of data and forest models that remote sensing can provide forest managers for regeneration assessment. First, depending on vegetation phenology and image characteristics, it should be possible to detect regeneration soon after disturbance has occurred. Then, based on observation of spectral response over time, and the calibration of those patterns with field data, a monitoring tool can be designed that is relatively inexpensive, covers large areas, and is quantitative.

NATURAL DISTURBANCES

FOREST DAMAGE AND DEFOLIATION

Forest impact is defined as the "net effect of an organism, after all beneficial and detrimental influences have been balanced, on the quantity and quality of the multiple resources expected from the land" (Alfaro, 1988: p. 281). Forest damage is a negative impact, and is generally considered to have occurred when there is (1) a reduction in growth or (2) actual mortality of trees. Forest damage can arise from a wide range of environmental and artificial causes originating from biological, hydrological, and atmospheric sources. Damage may be caused by forest insects, various diseases, fungi, mechanical or physical forces (such as machines, flooding, and winds) (Ramsey et al., 1998; Mukai and Hasegawa, 2000). Damage may be caused by forest decline phenomena linked to air pollution (Tømmervik et al., 1998), climatic stress, or changing stand dynamics.

The concept of forest damage is intrinsically related to the general concept of forest ecosystem health — one of the principal indicators underlying a sustainable forest management strategy. Core indicators of health usually include plant and site characteristics; dendrology, mensuration, crown assessments (density, transparency, diameter, ratio, and dieback); crown and bole damage; altered wood quality; soil chemistry; root disease; and presence, condition, or absence of bioindicator plants. Remote sensing inputs to these broad areas have been relatively few (Riley, 1989); forest health monitoring will continue to be principally a field-based activity — "it is only through careful field observations that any statements will be possible about the status of individual tree species and forest ecosystems today and in the future" (Innes, 1992: p. 52). Repeated measurement of crown density, discoloration and dieback, needle retention, premature leaf loss, and shoot death are known to be subjective. Such measurements are demonstrably useful in forest management, as careful training showed that between-stand variability was greater than it was in assessment of plantations — in other words, even-aged and predominantly single-species stands (Innes and Boswell, 1990). Most forest health assessments occur in conditions that are much less ideal.

Others have stressed the unecessary subjectivity and high cost of such detailed field observations, coupled with a strong desire to make indicator measurements not currently feasible; for example, "to generate forest damage maps in real time, providing a versatile and powerful tool for forest managers" (Reid, 1987: p. 429). There is a clear need for continued development of forest health monitoring by remote sensing. Two approaches appear viable (Dendron Resources Inc., 1997):

1. Detecting indicators or markers of physiological response to stress (derived from leaf reflectance, canopy chemistry, and bioindicator plants), and
2. Capturing long-term changes in health and vigor by classifying and measuring characteristics of stand development.

Managers need to know where, when, and why certain biotic agents cause changes in structure, composition, growth, and development of the forest (Stoszek, 1988). While no single inventory and monitoring method is likely to be found for all types of forest damage and aspects of forest ecosystem health, there would appear to be a clear role for remote sensing based on remote (spectral response) detection of differences in color (Murtha, 1976; Rock et al., 1986) and detection of a loss of plant chlorophyll, turgidity, foliage, or other growing organs (Hoque et al., 1992).

The approach is to try and relate differences in remotely sensed spectral response to chlorosis (yellowing), dehydration, foliage reddening, or foliage reduction over time; the idea is that these differences can be classified, correlated, interpreted, or otherwise related to known damage or stress conditions such as defoliation caused by insect activity (Hall et al., 1983) or differences in internal leaf structure caused by ozone and pollutants (Essery and Morse, 1992). This approach has led to successful use of remote sensing in at least three forest damage applications:

1. Early detection of stress and damage,
2. Mapping of damage extent, and
3. Quantification of damaged forest conditions.

Spectral observations of healthy and stressed vegetation have resulted in better understanding of the possible image characteristics that must be interpreted. The general reflectance characteristics of healthy and stressed leaves are well known (Reid, 1987: Figure 8.6); a blue shift and a reduced infrared reflectance appear to be the dominant effects. As the plant becomes less healthy, the increased reflectance differences between the red and infrared portions of the spectrum, the red-edge of plant reflectance (Horler et al., 1983), is shifted toward the blue end of the spectrum (shorter wavelengths). This blue shift has been noted in many different settings (Rock et al., 1986; Miller et al., 1991), and may even be universally applicable to green plants under stress (Essery and Morse, 1992).

However, Essery and Morse (1992: Figure 8.7) illustrated another possible reflectance curve that can result from stress on vegetation. The blue shift occurred, but the near-infrared reflectance was increased rather than decreased. Their inter-

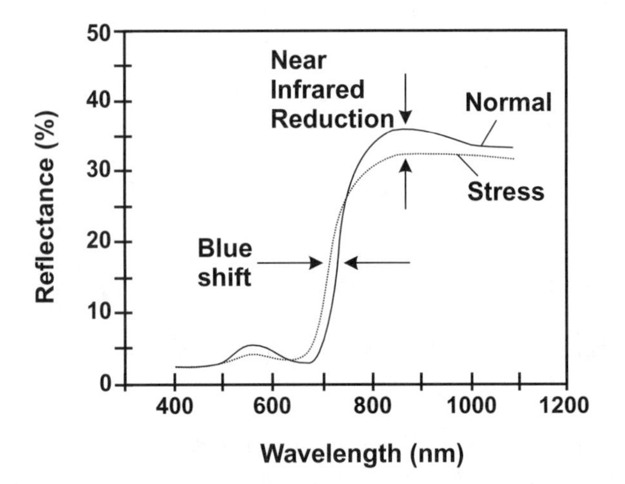

FIGURE 8.6 Spectral response curve shows the expected changes in green leaves under stress. A small reduction in green light reflectance, an increase in red light reflectance, and a reduction in near-infrared light reflectance has been observed. The shift of the red-edge of leaf reflectance to shorter wavelengths, called the blue shift, is a universal property of leaves under stress. (From Reid, N. J. 1987. *Environ. Sci. Technol.*, 21: 428–429. With permission.)

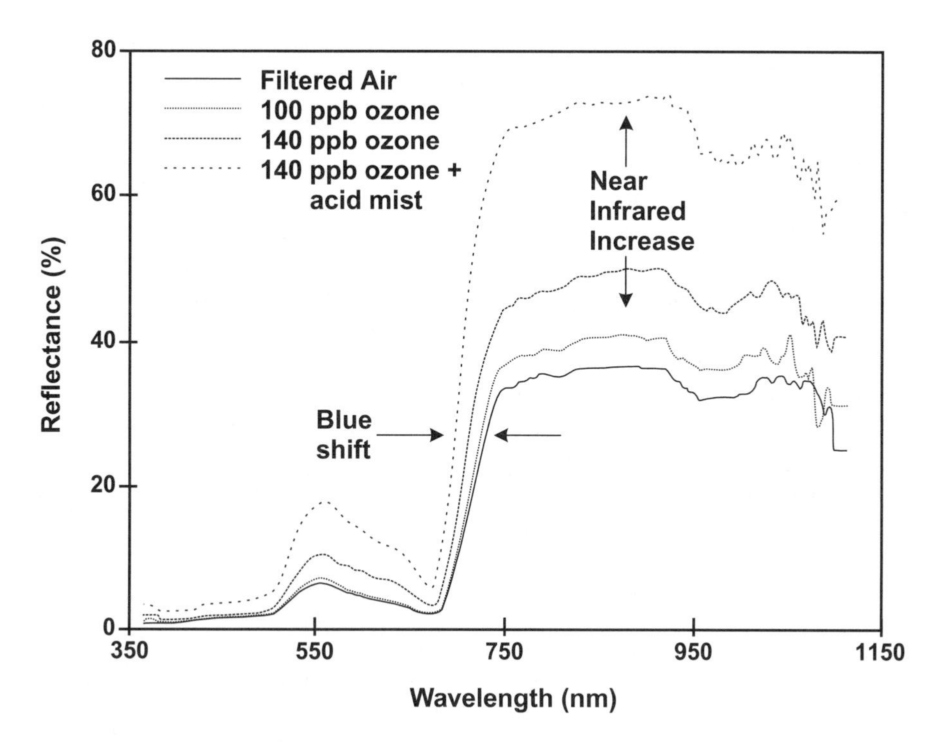

FIGURE 8.7 Spectral reflectance curves for control, ozone treatments, and acid mist treatments. Here, the characteristic blue shift and increased red light reflectance are observed, but green light reflectance and near-infrared reflectance also increased with increasing stress. These effects could be attributed to increased leaf dessication. (From Essery, C. I., and A. P. Morse. 1992. *Int. J. Rem. Sensing*, 13: 3045–3054. With permission.)

pretation was that ozone and acid mist exposure caused dehydration and the initial stages of senescence, which resulted in higher near-infrared reflectance (Guyot et al., 1989). Several different stressors can create the same physiological response in leaves, or a different response depending on a number of factors including the original status of the vegetation, and duration of the stress. Thus, a profile of the healthy condition might be necessary in order to detect and understand a deterioration in health.

Several early studies have helped establish the value of remote sensing in early detection and forest damage assessment and have led the way for methodological improvements meant to create an operational application in some forest areas. In British Columbia, manual interpretation of color infrared photography was just as effective (at approximately 20% of the cost) as traditional field surveys in identifying trees in the early stages of a spruce bark beetle (*Dendroctonus rufipennis* Kby) attack — while foliage was still visually green (Murtha and Cozens, 1985). Many of those trees later died. It was not possible to determine beforehand in the photography which of the attacked trees would be killed and which would recover (by pitching-out the beetles). For the attacked spruce trees "loss of greeness, the dryness, and the lack of spectral variation among the foliage age classes produces light-toned, highly reflective, mono-hued tree crowns which seem to 'glow like a halo' on CIR photos" (Murtha, 1985: p. 99). Spruce needles on trees attacked by the spruce beetle passed from the green attack stage to the gray (mortality) stage with a gradual fading of leaf reflectance.

Mountain pine beetles (*Dendroctonus ponderosae* Hopk.), on the other hand, typically turn foliage on lodgepole pine trees bright red for a short period of time following infestation (Murtha and Wiart, 1987). Successful mountain pine beetle infestation monitoring programs were designed to detect this red-attack stage of the infestation using remote sensing. But even this large color change was only partially visible when manual interpretation techniques were used with satellite imagery. When interpreting SPOT HRV color composite imagery (Sirois and Ahern, 1988), the minimum red-attack damage detectable was approximately 1 to 2 ha in size, wherein 80 to 100% of the crowns were red. This threshold of detection was too great to be practical for mountain pine beetle control programs — there, the requirement is to detect infestations of five or more trees. Subsequent studies with digitized photos and to a lesser extent, satellite imagery, indicated promise that a green-attack (current) model of mountain pine beetle infestations could be developed in mature forest communities in British Columbia (Murtha and Wiart, 1989a,b).

One of the leading causes and indicators of forest damage is defoliation, which in turn can arise from a number of causes. Damage and defoliation are not equivalent measures. For example, damage, which is measured as tree mortality or growth reduction, may be suspected after defoliation, which is measured as a reduction in leaf area, becomes noticeable, but many forests can experience some degree of defoliation without noticeable effects on growth or accumulated reserves. Forest managers and scientists have developed the concepts of forest susceptibility and forest vulnerability to help differentiate between levels of defoliation and consequential forest damage. These ideas have led to an intermediate type of remote sensing application, between mapping damage after it has already occurred and predicting the future occurrence of forest damage. This latter application might be of great interest to managers requiring as much lead time as possible in prescribing treatments and modeling effects. Ideas for remote sensing of forest susceptibility and vulnerability are described in the next section.

Attempts to map defoliation (and subsequently, forest damage) have been reported with digital aerial sensors (Yuan et al., 1991; Ahern et al., 1992), satellite sensors (Dottavio and Williams, 1983), and with methods ranging from image classification (Franklin, 1989; see Chapter 8, Color Figure 3), to spectral color shifts (Rock et al., 1988), to stand spectral retrogression techniques (Price and Jakubauskas, 1998). With remote sensing imagery, it is often possible to relate observed differences in color to differences in leaf area (Leckie et al., 1992). Forest defoliation has been mapped on this principle by aerial sketch mappers and by photointerpreters (Murtha, 1972) in forestry for many years — large-scale forest insect infestations have been monitored by observers in aircraft since the 1920s, and annual sketch mapping of forest damage is now routine in North America. When delineating infestations of forest defoliators, observers in aircraft mentally average the level of defoliation for a reasonably sized (but still small) area, and record the average infestation of this area and of adjacent areas to produce an aerial sketch map. The size of the area mentally averaged by the observer depends on the variability of the defoliation, the speed and height of the aircraft, and the scale of the map used to record the information. Such sketch maps have met a variety of needs in the forest community; however, with the rapid increase in the use of forest inventory data in digital format, the traditional methods of recording infestation damage may lack the precision required.

Beaubien and Jobin (1974b: p. 450), in their study of color infrared photography and early satellite images, noted that "remote sensing techniques can potentially provide the forest manager with a more rapid and accurate damage assessment, and permanent records of information useful in the study of ecological factors affecting forest insect pests." In recent years, the infestations of several species of forest insects have been successfully delineated with satellite remote sensing imagery, although operational procedures have not yet been developed and the accuracy of these studies have rarely been compared to the routine sketch-mapping products (Bucheim et al., 1985; Ciesla et al., 1989; Joria et al., 1991; Brockhaus et al., 1992, 1993; Franklin, 1989; Franklin and Raske, 1994; Franklin et al., 1995b). Only a few of the available studies are summarized here.

Light, moderate, and severe blackheaded budworm defoliation classes were mapped with Landsat TM data with 82% accuracy in the balsam fir forests of Newfoundland (Luther et al., 1991). Classification of hemlock looper defoliation, typically more damaging since more foliage is removed by this insect during feeding, was 93% correct. The spectral differences were consistent with expectations: a strong inverse relationship between near-infrared reflectance and increased damage, for example. Using Landsat TM images before and after defoliation by western spruce budworm (*Choristoneura occidentalis* Freeman) occurred in a subalpine forest in Oregon, 21 plots of damage in two classes were separated with 86% accuracy (Franklin et al., 1995b). Based on aerial videography data acquired in September and October after the final year of the infestation, more than 70% accuracy was obtained in these same plots. Achieving this level of accuracy depended on understanding the existing stand structure through use of a reference image (Franklin et al., 1995b; Cohen and Fiorella, 1999). Aspen defoliation classification, also using before and after images, even without extensive field data, provided very high accuracies in identifying defoliated and healthy aspen stands (Hall et al., 1984). In one review and case study, Michener and Houhoulis (1996: p. 13) concluded that when change detection analysis was based on data acquired immediately prior to and following a discrete disturbance event, spectral change could be related to ecological changes with a reasonably high degree of certainty "… otherwise, spectral changes associated with a specific disturbance may be confounded with land use change, annual phenological differences, climate, and other factors that differ between pre- and post-disturbance imagery."

Eastern spruce budworm damage in balsam fir stands in Newfoundland was classified with up to 100% accuracy using a single date SPOT HRV imagery by stratifying the stands prior to classification using the available forest inventory data (Franklin and Raske, 1994). In a three-way comparison between the remote sensing defoliation classification, ocular field estimates in discrete plots within stands, and sketch-mapping products, the sketch maps were the least accurate. In some ways this is not surprising — sketch maps are designed to yield large-area depictions of defoliation rather than impact (MacLean, 1990) and are not usually thought to be accurate at the point or even stand level. Or are they? It has always been the case that "people responsible for control operations need more accurate surveys and more detailed information (such as tree species attacked, stand and site conditions) not always provided by sketch mapping" (Beaubien and Jobin, 1974b: p. 450). The level of precision in the satellite and airborne remote sensing classifications, however, suggests the possible form of an operational remote sensing defoliation mapping procedure (Franklin and Raske, 1994):

1. Stratify satelite imagery by inventory (or classification if inventory not available);
2. Provide seed estimates of defoliation by field or aerial surveys;
3. Generate equations that relate the change in reflectance to the amount of defoliation;
4. Apply and reiterate the procedures to classify the entire map area.

Some areas have experienced significant forest defoliation as a result of (hypothesized) anthropogenic activities leading to leaf chlorosis (Khorram et al., 1990; Brockhaus et al., 1993; Ekstrand, 1994a) sometimes referred to as forest decline (Ardö, 1998). For example, acid mine tailings and associated forest damage were mapped by airborne sensors in eastern Ontario (Levesque and King, 1999; Walsworth and King, 1999), as part of a study to develop 'a forest health index' from multispectral airborne digital camera imagery (Olthof and King, 2000). A soil metal concentration gradient was observed with distance from the tailings, and there were significant leaf reflectance properties correlated with this gradient. In another example, regression models were developed to predict the percent defoliation in forests in the Black Hills of North Carolina from digital Landsat and SPOT data (Brockhaus et al., 1992); the cause of the defoliation was thought to be related to acid rain and ozone deposition (Table 8.2). In southern Sweden, Ekstrand (1990) examined the relationship between Landsat satellite sensor data and spruce needle loss in 25 forest stands. Natural stand variations such as species composition and percent hardwood

TABLE 8.2
Models Predicting Defoliation Based on Landsat and SPOT Satellite Data and Elevation in 21 Plots in the Black Mountains, North Carolina

Model	SE	R^2
%Defoliation = 112.75 – 2.46*(TMband4)	10.46	0.65
%Defoliation = 132.64 – 2.99*(HRVband3)	14.29	0.55
%Defoliation = 39.42 – 2.09*(TMband4) + 0.28 * (elevation)	7.77	0.87
%Defoliation = 39.04 – 2.63*(HRVband3) + 0.38 * (elevation)	9.90	0.80

Source: Adapted from Brockhaus et al. (1993).

in the understory, seasonal changes, and additional varibility caused by atmospheric effects, solar angles, and topography were controlled in the analysis. Spectral shifts in stands where needle loss ranged from 10 to 40% included significantly increased visible spectral response and reduced near-infrared spectral response.

These applications indicate that remote sensing data, even at fairly coarse satellite pixel resolutions, can provide unique data on change caused by damage and defoliation agents. If the structure of the stand is known beforehand, and the agent of change is at least suspected, there are few impediments to the routine detection and mapping of the changes that result at the stand level. This application is one in which remote sensing data are not actually competing directly with an older, established technology, such as aerial photographs; sketch mapping cannot be considered a serious competitive approach when the large areas, stand-level geometric precision, and quantitative data requirements (Gillis and Leckie, 1996) are examined. There is a role for such data, but remote sensing is clearly part of the answer to future forest defoliation and damage surveys. It is not that difficult to envision a forest health network that relies on satellite remote sensing, field observations, and other monitoring and sampling measurements.

Mapping Stand Susceptibility and Vulnerability

One way in which forest damage and forest defoliation have been considered separately has been in the development of stand susceptibility and vulnerability models. Such models are reasonably well-developed to evaluate some of the more common insects or forest pests in North America; these management tools are necessary in forecasting the degree of insect defoliation, and the associated impact on forest ecosystems.

Susceptibility is the probability of defoliation. Stand susceptibility is sometimes called a hazard rating, used to help plan insect population control strategies in the short term, and planning silvicultural control to reduce the amount of defoliation in the long term. The response of forest stands to insect attack is generally referred to as stand vulnerability, often summarized in risk ratings which might be used to reflect the difference between the ability of a stand to withstand defoliation and continue growing vs. one unable to recover due to insufficient resources (Waring, 1987). For example, Coyea and Margolis (1994) used historical reconstructions of forest growth efficiency to predict tree mortality following budworm attack, suggesting that such measures are sensitive, physiologically based predictors of health. Patterns of forest growth may indicate vulnerability indirectly by measuring the accumulated reserves or the ability of the stand to produce defensive compounds.

In general, forest susceptibility and vulnerability to insects may be a consequence of a large number of factors, including the intensity, duration, and size of the outbreak, the proximity of the outbreak foci, the abundance of habitat, insect nutritional requirements, populations of predators and parasites, spraying activities and other management actions, and environmental factors associated with climate (Waring and Schlesinger, 1985). Risk and hazard rating systems try to summarize these factors for specific regional settings with a reduced set of predictors that are easy to acquire and understand (Speight and Wainhouse, 1989). The resulting systems are considered essential for developing effective pest management strategies (Hudak, 1991), and have been traditionally developed using a combination of field, climate, and forest inventory variables. For example, in eastern Canada, one vulnerability rating system for the eastern spruce budworm is based on three measures (Raske, 1986): (1) the abundance of host species in the stand, (2) the degree of stand maturity, and (3) a measure of mean climatic conditions.

More complex hazard and risk rating systems exist for spruce budworm that include estimates of growth and stocking density, since observations have shown that insect abundance and distribution may be related to the growth pattern of forests (MacLean, 1980; MacLean and Porter, 1994). In the eastern U.S., hazard ratings for gypsy moth, *Lymantria dispar* (L.), rely on stand structure (basal area, species composition) and insect population dynamics (Houston and Valentine, 1977; Liebhold et al., 1993). One susceptibility rating system for bark beetles in western Canada is based entirely on forest structure and uses only basal area, age, density, and location in the ratings (Shore and Safranyik, 1992). One criterion used in developing this system was that most of the data should be obtainable from the existing forest inventory.

In only a few cases have remote sensing data been studied to determine their possible role in mapping stand susceptibility and vulnerability (Luther et al., 1997). But from the earliest days of forest defoliation mapping by satellite (Dottavio and Williams, 1983), the use of imagery to predict the occurrence of damage, rather than simply map the results of the disturbance, was thought promising. For example, a predictive forest decline model was developed using Landsat TM and digital elevation data for an area in North Carolina (Khorram et al., 1990); the objective was to determine areas that were likely to decline in future based on the conditions in areas that had declined in the past. The model fit was reasonably good ($R^2 = 0.85$), although a lack of field data and changing environmental conditions prevented the reliable extension of the model over time and space.

Remote sensing may be useful in stand susceptibility and vulnerability ratings systems by providing:

1. More complex structural information than can be obtained from the forest inventory, and
2. Variables not currently provided in the forest inventory or not available with enough precision.

These two possibilities were explored by correlating Landsat TM data acquired before and after a blackheaded budworm infestation in a conifer forest in Newfoundland (Luther et al., 1997). Since forest structure and forest growth rates appeared to be closely related to stand susceptibility in this second-growth balsam fir ecosystem, the idea developed that perhaps remote sensing data could be used to implement these relationships in a predictive model through the links between spectral response and forest growth and structure (Franklin and Luther, 1995). The probability of attack (susceptibility) was generally well predicted using a combination of remote sensing and forest inventory variables (Table 8.3); stand vulnerability was also well

TABLE 8.3
Results of Predictions from Optimal Logistic Regression Models for Balsam Fir Susceptibility and Vulnerability to Blackheaded Budworm Defoliation in Newfoundland Using Predictors Obtained by Sketch Mapping, Landsat TM Image Classification, and the Existing Forest Inventory[a]

Model	Percent Correct[b]
Susceptibility	80.6
Pre-outbreak vulnerability	66.7
Post-outbreak vulnerability	77.8

[a] The susceptibility model predicts the probability that a plot will be defoliated. The vulnerability models predict the probablility that a plot will experience reduced growth following defoliation. The difference between the pre-outbreak and post-outbreak vulnerability models is that defoliation variables can be included in the post-outbreak models. Pre-outbreak models would be useful for developing insect control options, whereas the post-outbreak models would be useful for developing salvaging strategies.
[b] Average percent classification accuracy, checked in 45 field plots, using various combinations of modeling variables including the percent defoliation observed in sketch mapping, percent defoliation predicted by Landsat TM spectral response, age, height, and cover classes from the forest inventory database.

Source: Modified from Luther et al. (1997).

predicted in areas that were susceptible (Table 8.3). The overall conclusions were that (Luther et al., 1997: p. 88–89):

- The analysis of forest susceptibility indicated that younger stands with relatively lower basal area and tree density were preferentially defoliated by the blackheaded budworm; this could be predicted with spectral, field, or forest inventory data with decreasing levels of accuracy, respectively.
- Vulnerability or damage expressed as reduced growth could be predicted using spectral values measured before the outbreak, because the spectral values were strongly related to forest structure and moderately related to the growth efficiency and vigor of the vulnerable stands.
- Susceptibility and vulnerabilty forecasts based on Landsat TM data acquired prior to defoliation by the blackheaded budworm resulted in higher classification accuracy than forecasts based on forest inventory data.
- Vulnerability forecasts improved when estimates of defoliation derived from aerial surveys were included in post-outbreak vulnerability models; further improvements were possible if remote sensing data were acquired during peak defoliation in each year and used to classify defoliation.
- The best predictions of susceptibility and vulnerability combined selected satellite spectral measurements with forest inventory data. These models

produced classification accuracies of 81, 67, and 78% for predicting susceptibility and pre- and post-outbreak vulnerability, respectively.

FIRE DAMAGE

Detection of active fires and fire scars has been a major application of remote sensing in forestry and will continue to be so; few observational platforms can match the accuracy and consistency of remote sensing in mapping and monitoring the dynamics of burning and burnt vegetation over large areas (Landry et al., 1995; Koutsias and Kareteris, 2000). Global observation of fire occurrence on a daily basis is available using the AVHRR, SPOT VEGETATION, or EOS MODIS sensors (Running et al., 2000).

Fire scars can be mapped to create fire history maps (Salvador et al., 2000) and to quantify carbon exchanges with the atmosphere (Kasischke et al., 1993), but operational examples of fire mapping and input to forest management are typically found at larger spatial scales. For example, the observed differences in burn severity were linked to the degree of vegetation change in a pine forest ecosystem in northern Michigan (Jakubauskas et al., 1990). Two dates of Landsat imagery approximately 10 years apart were classified, compared in a GIS overlay procedure, and then sorted into classes according to a burn severity index developed from a near-infrared to red ratio using a satellite image acquired immediately after the fire. The largest differences in spectral values for different types of burn were recorded in this index (Figure 8.8). The amount of change experienced by the pine forest decreased with decreasing burn severity.

A similar analysis in a burnt area in Oregon quantified the effect of the wildfire on the landscape patterns for this area (Kushla and Ripple, 1998). Pre- and post-burn imagery and image transformations were examined to create estimates of canopy cover, which then were used in a classification procedure. The differences in classes and canopy cover in the two image dates were observed to change in successional patterns; for example, early seral stages increased in area while closed forest decreased. Clearings after the burn were much more variable in size and more complex in shape than the clearcutting pattern in areas adjacent to the burn. The wildfire enhanced landscape diversity and increased edge, which was interpreted to mean the fire probably had a negative effect on northern spotted owl habitat. In a Mediterranean environment, Chuvieco (1999) mapped changes in spatial structure resulting from a large fire using both classification data and continuous reflectance data (in the form of the NDVI statistic). Interpretation was consistent with theoretical homogenization of the landscape following a large, stand-replacing fire.

A spectral unmixing approach was used by Vine and Puech (1999) with SPOT imagery in their Mediterranean study area. Spectral response data from six images acquired after the burn were related to shadow, vegetation, and soil fractions. For pixels that had been burned, these fractions changed predictably with different vegetation regrowth patterns. By clustering the image data, regrowth classes were created that represented areas of different burn severity (Jakubauskas et al., 1990). Also revealed were areas of different original vegetation structure and topography.

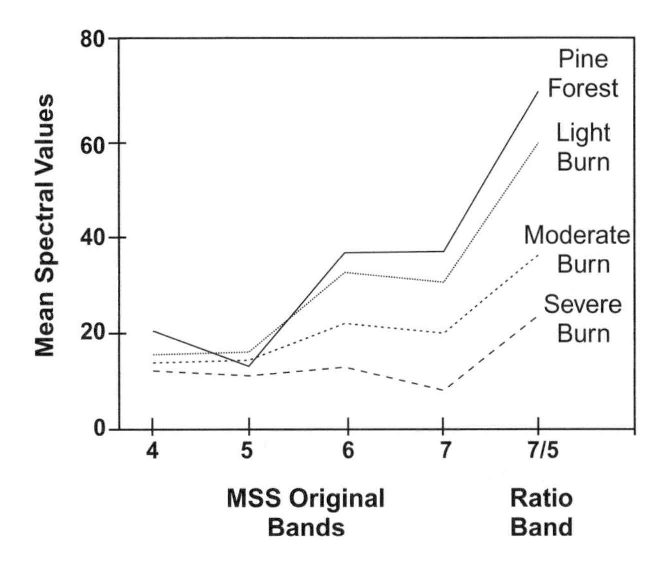

FIGURE 8.8 The change in pine forest mean spectral response with increasing burn severity caused by wildfire. After burning, the severely burned areas were much darker in all bands of the Landsat MSS; a ratio of red and near-infrared light (7/5) shows the high level of spectral distinctiveness between the unburned and burned forest areas. These areas of different burn severity subsequently followed different successional trajectories which could be predicted from the Landsat image data. (From Jakubauskas, M. E., K. P. Lulla, and P. W. Mausel. 1990. *Photogramm. Eng. Rem. Sensing*, 56: 371–377. With permission.)

CHANGE IN SPATIAL STRUCTURE

Natural and human disturbances, ecological succession, and recovery from previous disturbances are all forces that modify ecosystem pattern within the landscape. These forces alter ecosystem heterogeneity with various effects on species diversity. A small disturbance may increase the heterogeneity of a landscape and increase habitat niches. For example, a fire or windthrow opens the forest floor to sunlight, promoting growth of herbaceous plants, grasses, and bushes, thus improving habitat quality for a variety of organisms such as pollinating insects and ungulates. A severe disturbance, such as a major landslide or flood, either may decrease landscape heterogeneity by removing the elemental configuration, or increase heterogeneity by changing only a part of the structure (Forman and Godron, 1986). Remote sensing of disturbances for the express purpose of quantifying the structure of the landscape (and change in structure) is a relatively new application (Sachs et al., 1998).

In addition to horizontal patterns, disturbance affects the vertical heterogeneity of ecosystems and landscapes. In an undisturbed landscape, horizontal landscape structure tends to become increasingly homogeneous with time (with maturity), while vertical structure becomes higher and more heterogeneous (Forman and Gordon, 1986). With disturbance, landscape structure tends toward the inverse pattern; heterogeneous horizontal structure and more homogeneous vertical structure. However, in some cases both vertical and horizontal diversity are affected similarly. For example, a catastrophic fire or large clearcut reduces the structural heterogeneity of a forest by eliminating both the vertical layers and horizontal patches of vegetation. Forest habitat diversity has been linked to vertical structural complexity, that is, to canopy density, variation in tree size, coarse woody debris, understory density, litter depth, presence of snags, and fallen trees (Hansen et al., 1991; Rickers et al., 1995).

Fragmentation of a landscape occurs when land cover patches are dissected by disturbance (Forman, 1995). Fragmentation leads to smaller patches, more distant patches, and increases in edge/area ratios, and has been associated with the spatial density of roads, pipelines, and other dissection factors that divide patches and sever corridors; for example, in the Bighorn National Forest of Wyoming, 30 landscape pattern metrics were used to illustrate that fragmentation had increased over time in relation to the existence of roads and clearcuts (Tinker et al., 1998). By frequently monitoring landscape composition and pattern, signs of landscape disturbance would warn of approaching changes with sufficient time to respond against irreversible damage. Tolerance thresholds for these signs would include developing lower limits for patch size decreases, upper limits for the density of linear features, and so on (Silbaugh and Betters, 1997).

Extending the measurement of change in remote sensing images to incorporate change in landscape structure has become an important aspect of many forest management questions at the same time that it has emerged as an important question in its own right. Haines-Young and Chopping (1996), for example, interpret the way that changes in different landscape indices can be used to assess the spatial implications of the various design guidelines that have been proposed to promote sustainable forms of forestry in the U.K. Forest management must take into account the larger landscape in which forest stands and ecosystems occur; it is the concern with aggregated effects that is of prime importance (Swanson et al., 1997). An emerging view is that over large areas and long time periods foresters should attempt to manage within the spatial and temporal variability of the natural disturbance regime. This requires that the spatial pattern imposed on the natural landscape by disturbances be discerned and quantified in order for management prescriptions, such as harvesting plans, to mimic them as closely as possible (Spies and Turner, 1999). Remote sensing may be one of the few tools available that is well designed to help understand the patterns and to monitor how well management mimics the disturbance regime.

Fragmentation Analysis

Forest fragmentation can be defined as "the division of large, comparatively homogeneous tracts of forest into a heterogeneous mixture of much smaller patches" (Reed et al., 1996: p. 267). No single landscape metric captures all aspects of fragmentation (Davidson, 1998; Jaeger, 2000); instead, a suite of selected metrics may be useful in interpretation of landscape change, and must be carefully considered relative to the type of change (the patches) and the background matrix (the forest mosaic). Using dense/sparse forest and various agriculture units in Honduras, Kammerbauer and Ardon (1999) interpreted a shape complexity index based on perimeter/area ratios in an area that experienced a reduction in forest cover of 25% between

1955 and 1975, and a further 11% between 1975 and 1995. The changing patterns were thought to represent a fragmentation trend related to physical, ecological, demographic, and social policy processes independently documented for this local (9.5 km^2 watershed) area.

In another example, the extent of landscape fragmentation in the New Jersey Pine Barrens region was documented by illustrating significant changes in selected landscape metrics between a 1972 and 1988 landscape (Luque et al., 1994). The two landscapes were classified and compared in their landscape metrics. As a result of human disturbances such as housing development and logging activities over the 16-year period, a range of landscape metrics including fractal dimension, diversity, and contagion decreased while dominance, disturbance, and edge indices increased at the landscape level. At the patch level of analysis, the mean size of forested patches decreased significantly. These results indicated a trend to a more dissected or fragmented landscape over time as a result of human disturbances.

The history of the fragmentation process in Wyoming was studied using maps based on aerial photography extending from 1950 to 1993 (Reed et al., 1996). A comparison of covertypes with habitat classes and clearcut/undisturbed maps was conducted for each year of interest in the sequence. Decreased mean patch size, mean patch perimeter, increased fractal dimension and total edge, and decreased dominance were found. The changes in these metrics were interpreted to be consistent with increased fragmentation. The area studied was barely manageable using aerial photographs; Larson et al. (1997) have suggested that even larger study areas must be considered since many important processes are only observed at the landscape scale (Cohen et al., 1995b). Such large-area studies must be based on data that are consistent (no spatially autocorrelated error) and complete (not sampled) (Sachs et al., 1998). Only satellite remote sensing data have these characteristics.

In their work in British Columbia, Sachs et al. (1998) reported a large region of the interior forests to be in the early stages of fragmentation. This analysis was based on the classification of Landsat imagery acquired in 1975 and 1992. A number of landscape metrics and the change in those metrics over time were shown to be related to such issues as the viability of populations of old-growth forest interior dwelling species (in particular, the northern spotted owl). The main concern was to document the hypothesized increased fragmentation of the study area due to harvesting. Human disturbance was shown to have affected 8.4% of the study area outside protected areas between 1975 and 1992. Mature and older conifer forest area decreased more than 10%, accompanied by decreases in mean conifer patch size and the percentage of interior forest area. The patches created by harvesting were smaller and simpler in shape than the conifer forest matrix in which they occurred. The rate of change — 0.49% per year — was estimated to be at the low end of the range of disturbance rates for a variety of temperate forests throughout the world.

In a large forested area on the border of China and North Korea, 1972 and 1988 Landsat imagery were classified into forest and nonforest classes — basically a Level I analysis of covertype (Zheng et al., 1997). Much of the change detected was a result of clearcutting; partial harvesting had increased in this area after 1980 as a result of government policies encouraging selective harvest over clearcutting, but

TABLE 8.4
A Summary of Land Cover Class Transitions for Each of the Classes Mapped in the Landsat MSS and TM Change Analysis between 1975 and 1997 in British Columbia near Revelstoke National Park

Class	Area (ha)		Change in Area
	1975	1997	(%)
Recent Cuts	1,263	3,168	150
Recent Burns	2,411	316	–86
Immature Forest	13,270	16,617	25
Cedar Hemlock	23,177	22,488	–3

Source: Modified from Hansen et al. (2000).

the Level I classification scheme did not show those areas as changes. The method was unable to distinguish between natural and human disturbances — not enough training data for use in the classification procedure. Increased forest fragmentation, interpreted in landscape metrics, resulting from human disturbance activities was supported by considering areas inside and adjacent to forest reserves where known (very low) levels of activity took place in the time period of interest. The annual rate of forest cover loss was 0.73% over the 16-year period for the study area, a comparable rate to that reported by Spies et al. (1994) in the U.S. Pacific Northwest for a similar period.

In many of these studies, the goal was to determine the overall or average rate of change across some management area — usually a large area. But within large areas there may be significant differences in the spatial distribution of changes (Miller et al., 1998). Remote sensing is an ideal tool to examine this possibility. For example, in New Jersey (Luque et al., 1994) and in Wyoming (Tinker et al., 1998), some watersheds were more severely impacted than the relatively low average rates of change might indicate — the spatial distribution of change was not constant. Hansen et al. (2000) reported large differences in the impacts found in different watersheds in a study of the fragmentation of forest classes in an area adjacent to the Sachs et al. (1998) study in the North Columbia Mountains. Some watersheds contained virtually no forest changes (apart from age class differences) between 1975 and 1997, while the overall study area experienced an increased area of harvesting of over 150%. The overall patterns are isolated in Table 8.4, which contains the summary of land cover class transitions for each of the classes mapped in the analysis.

Overall, the most obvious change expressed in two map products based on 1975 and 1997 Landsat imagery was contained in the Recent Cuts class: 150.75% change between 1975 and 1997 (Hansen et al., 2000). The area of recent cuts in 1975 was approximately 1263 ha; this increased to approximately 3168 ha in 1997. The area in Recent Burns decreased substantially from 1975 to 1997, perhaps as a direct result of fire suppression activities. Burns accounted for approximately 14% of the change, primarily in the 140- to 250-year age class; harvesting accounted for more than 75% of the total area changed in these two age classes of the forest. Clearly, the major differences in the forest classes were directly attributable to the increased area of forest cutblocks in the forest stands, which appeared to have been accurately depicted in the remote sensing classification and mapping exercise.

Habitat Pattern and Biodiversity

Wildlife habitat mapping from satellite remote sensing data has long been of interest because of the potential that highly accurate and effective maps can be generated from the synoptic, repetitive, and consistent reflectance data over large areas of wilderness (Avery and Haines-Young, 1990; Aspinall and Veitch, 1993; Lauver and Whistler, 1993). A common approach is to combine environmental factors in a GIS to produce ecological land classifications useful for habitat delineation and biodiversity assessment (Davis and Dozier, 1990; Stoms, 1992; Stoms and Estes, 1993). As noted earlier in Chapter 6, such an approach typically falls short of predicting ecological communities as they are currently understood and mapped in the field, except in the simplest of environments.

Habitat classes are, unfortunately, almost always similarly unattainable from direct predictive mapping procedures. Several early studies of elk (Huber and Casler, 1990) and grizzly bear habitat (Craighead et al., 1985; Butterfield and Key, 1985) emphasized the need to standardize remote sensing and field methods, and anticipated many of the problems that would be encountered in a transition to an operational satellite remote sensing habitat mapping procedure, including:

1. The difference between spectrally distinct landcover classes and *interpreted* habitat classes,
2. The use of imagery together with ancillary data such as DEMs and biophysical land classifications,
3. The difficulty in field verification of habitat suitability and use, and
4. The need to ensure consistency in mapping across ever larger management units (ultimately encompassing whole mountain ranges and ecophysiographic regions).

Despite significant progress in key areas and numerous illustrative case studies (e.g., Debinski et al., 1999), the idea of habitat mapping by remote sensing remains largely undeveloped. Because satellite remote sensing imagery is often only one of several different sources of information that can be accessed by wildlife managers, one of the most significant questions to emerge is: How do we best use all of the available data in the habitat mapping task? This question, and many of the larger procedural issues, may apply equally in the development of satellite remote sensing habitat mapping protocols for several wide-ranging, wilderness species. Analysis of habitat patterns typically involves two steps: (1) generate a model of habitat for a given species, and (2) generate a model of habitat attributes across the landscape.

Many species do not have well-defined habitat requirements that are understood and accessible to forest managers (Griffiths et al., 2000), although a few well-known examples have been reported and these could serve as a remote sensing model for other species. For example, Oregon spotted owls appeared to select nest sites in areas with a high percentage of old-growth and mature forests adjacent to the nests (Ripple et al., 1991); such low fragmentation sites may be preferred because of structurally suitable trees for nests, ameliorated microclimates, suitable forage substrates, refuges from predators, or sufficient prey. In addition, patch size was found to be important, as was patch density (Ripple et al., 1997). Each of these was highly correlated with the percentage of old-growth and mature forests (Ribe et al., 1998). The Mexican spotted owl (Peery et al., 1999) was studied for similar habitat requirements; nests were found to be located in the midst of a particular mix of forest patches comprised of different conifer species and ages. Landsat imagery was used to determine the appropriate mixture of patch and matrix units, and the likely optimal configuration to provide recommended "protected activity centers" surrounding existing nests.

Habitat and forest cover are not necessarily equivalent but have sometimes been considered so (Jorge and Garcia, 1997). For example, in studying the American pine marten (*Martes americana*) in Utah, Hargis et al. (1999) considered forest with a fairly closed canopy and abundant coarse woody debris to be prime marten habitat. They then quantified habitat fragmentation from input maps of classified Landsat TM imagery recoded into forest and nonforest categories using five landscape metrics: the percentage of landscape in openings, edge density, mean proximity index, mean nearest neighbor distance between open areas, and mass fractal dimension. Marten captures were negatively correlated with loss of forest habitat and martens showed significant responses to landscape pattern, particularly the size and proximity of open areas. Suggested management implications included a comparison of progressively outward cutting rather than clustered or single patch clearcutting, so that maximum areas of undisturbed forest remained for use by martens.

A method of mapping forest-breeding songbird microhabitat by modeling climate, DEM, and remote sensing data in Ohio was recently developed (Dettmers and Bart, 1999). Spatially explicit variables included slope, land surface morphology, land surface curvature, water flow accumulation downhill, an integrated moisture index, and a Landsat TM land cover classification of forest covertypes. At key data locations with observations of presence of the target organisms (flycatchers and warblers), the variables were collected and used to develop a regression model which subsequently was applied to generate maps delineating good habitat for each species. The correlations were generally low because many variables not included in the models were known to influence avian density (e.g., food availability, predator abundance), but were too difficult to collect and use. Obviously, this is a classic normative methodology in which the normal relationship is barely known, let alone understood. However, the models performed better than did random models, and the maps provided new insights into how environmental variables affected habitat quality. An implication of this study is that detailed data on plant species composition or structure may not always be necessary to construct useful predictive habitat models.

Similarly, Venier and Mackey (1997) used TM-derived land cover and a topographic wetness index (TWI) to map songbird habitat and then assessed the spatial configurations of the predicted habitat patches in developing population estimates.

Including edge and area sensitivity resulted in a 50% reduction in the perceived available habitat. The use of habitat suitability classes (or index values grouped into classes) as input to a caribou fragmentation analysis in interior British Columbia has also been described (Hansen et al., 2000). The principal causes of the forest changes were timber harvesting and wildfires. Observed directional changes in the spatial metrics calculated for each time period appeared to confirm the observed spatial effects of forest fragmentation recognized in the literature:

1. A loss of total habitat area,
2. Increased patch abundance and density,
3. Decreased patch size,
4. More edges,
5. Less core area,
6. Simplified patch shape, and
7. A wider, more dispersed patch configuration over time.

Similar maps of pseudo-habitat have been based on the Landsat TM greenness component (Mace et al., 1999); the idea was to establish the relationship between greenness variability and habitat suitability for grizzly bears, and then monitor changes in greenness and spatial patterns using landscape metrics. The goal of habitat managers should be the "conversion of satellite imagery to a validated map with vegetal and physiognomic descriptions" (Mace et al., 1999: p. 376). The problems relate to the fundamental difficulty in finding a way to match desired (habitat) classes with distinct spectral classes that can actually be mapped with acceptable precision across the wide and variable landscapes of interest (Dechka et al., 2000).

Clearly this application has tremendous potential and daunting challenges. Understanding and implementing a spatial analysis of the habitat of a single species is difficult to grasp; when considering habitat and biodiversity, the ecological issues alone threaten to overwhelm even the most tenacious spatial data analyst. Simply put, remote sensing can contribute to the needs in this area, but until those needs are clarified, stated in terms of classification or field variables with some potential to be derived by remote sensing, the application will almost certainly continue to remain underdeveloped. This application of remote sensing is wide open. At the moment, the most useful information products that remote sensing can provide in biodiversity and habitat assessment are thought to be

1. Classification information on forest covertypes, forest ecosystems, and forest stands,
2. Change information, and
3. Direct observations of spectral response which can be converted into forest structural information of interest, such as crown closure, stand volume, and tree species compostion estimates.

Direct remote sensing contributions to biodiversity and habitat studies, then, has taken the form of converting these information products into source information for models of habitat (resource quality and quantity, perhaps linked to indicator species)

(Debinski and Brussard, 1992). This approach is most powerful when the indicator species hypothesis has been established. A second approach has been to develop correlations with multiscale maps of species richness (Jorgensen and Nohr, 1996; Walker et al., 1992; Stoms and Estes, 1993). The role of remote sensing has been to map habitat and to help build models of habitat that can be used in biodiversity assessment. In only a few cases have remotely sensed data been used alone; the complexity of habitat and biodiversity issues in real landscape terms suggest that only multisource spatial data can lead to a successful application.

9 Conclusion

"… the Hercynian forest was unimaginably ancient, literally pre-historic … *intacta aevis et congenita mundo prope immortali sorte miracula excedit* (coeval with the world, which surpasses all marvels by its almost immortal destiny)"

— Pliny, 23–79 A.D. (Quoted in Schama, 1995)

THE TECHNOLOGICAL APPROACH — REVISITED

A wave of biological awareness, and a growing recognition among the people of the world that environmental limits exist, and may be rapidly approaching, has transformed forest management in recent years. A sustainable forest management approach is urgently sought as one of the critical mechanisms to ensure a balance in human resource use and natural world coexistence. The development of remote sensing in forestry can be considered as one essential component of the required management system — one of several technologies in the expanding continuum of technological resources deployed during forest management planning and operations. The developers of remote sensing and, increasingly, many users, consider remote sensing to be a decisive tool, a flexible method, one critical part of a technological approach designed to satisfy the large and growing need for forest information and knowledge.

Forest ecosystems must be understood and human impacts quantified and managed well enough, despite an inherent lack of precision and knowledge in virtually all forest management decisions, to address growing concerns over forest biodiversity, long-term productivity and health, and public involvement and responsibility. Adaptive strategies and adherence to scientific principles of design, observation, and change, are needed continuously to invent, reinvent, and improve forest practices. Foresters and resource management professionals increasingly are faced with the need to answer questions about forests that were not considered part of earlier forest management strategies, and for which few data or information sources are available. Now, and increasingly in the future, forest management will be tied to monitoring criteria and indicators to determine forest ecosystem sustainability and the sustainability of management practices, such as harvesting, planting, silviculture, and suppression or enhancement of natural processes. Perhaps a minimum set of indicators would include measures of (Whyte, 1996):

- Yield of all forest products harvested,
- Growth rates, regeneration, and condition of the forest,
- Composition and observed changes in flora and fauna,
- Environmental and social impacts of harvesting and other activities,
- Cost, productivity, and efficiency of forest management.

All forest management is dependent on reliable sources of information of forest conditions, forest values, and forest practices summarized in these indicators. Forest certification based on indicators is increasingly considered by producers and users of forest products as a viable component of a sustainable forest management strategy for the world forests. There may never be a universal set of agreed-upon indicators, based on the various motivations for their derivation (National Research Council, 1998). But whether an indicator approach or some other approach to monitoring is selected or evolves, there is little doubt that remote sensing is one of the sources of information and insight that forest management must possess and use wisely.

A description of the forest resource — a forest inventory — is the reservoir of critical information in forest management. Forestry today is increasingly defined by values, cultures, communities, and politics (Shepard, 1993), but the inventory continues to represent a critical component (Carter et al., 1997). The construction and maintenance of the forest inventory is dependent on remote sensing — principally in the form of aerial photography. The modern forest inventory has evolved into an enormous, complex, and multifaceted digital database and associated analytical tasks contained within a geographical information system (GIS). A forest inventory GIS contains information on forest structure, composition, and development, and is based on the concepts of first delineating then treating homogeneous, or acceptably heterogeneous, forest stands. For most of the past century, these forest stands and the larger forest ecosystems and landcover patterns of which they are a part have been recognized, mapped, and described by skillful human interpretation of aerial photographs; producing a forest inventory from scratch without access to aerial photography is almost inconceivable today. In some ways, the modern digital forest GIS inventory exists in the form that it does because it contains exactly the type of information that can be acquired through careful application of the first, most commonly available form of remote sensing, namely, aerial photography.

Practically speaking, remote sensing has not yet reached the high levels of availability, understanding, and ease of use of aerial photographs. In only a few years, this situation will be completely changed! Foresters are now engaged in creative thinking about what the forest inventory should contain; one part of this thinking is focused on efforts to satisfy the information needs as required in a forest certification and sustainable forest management criteria and indicators approach. The inventory will be constructed around GIS technology, which itself is undergoing rapid development and change (Aronoff, 1989; Longley et al., 1999). Remote sensing is emerging as a critical component of the forestry information system of the future — remote sensing of forest conditions and forest changes. The forest inventory will contain information related at different scales to ecosystem processes (productivity, water balance, biogeochemical cycling), forest structure, species richness and biodiversity, and landscape variability and structure. Supplementing traditional stand descriptions (such as crown closure) with dynamic variables such as LAI, will help create linkages to ecosystem models that can simulate fundamental processes underlying forest growth and can integrate climatic variability and disturbance (Waring and Running, 1999). It will be just as inconceivable to build and maintain this future inventory without access to digital remote sensing as it would be to build today's inventories without access to aerial photography.

Remote sensing was a term originally coined in the 1960s to represent the new possibilities suggested by the analysis of digital images acquired in many different parts of the electromagnetic spectrum. The past provides an interesting perspective, but the fact is that remote sensing data and methods have experienced an almost unbelievable rate of improvement. Today, remote sensing is a sometimes bewildering array of technology, data, and methods — a full technological package that includes the collection and analysis of data from instruments in ground-based, atmospheric, and Earth-orbiting positions, evolving linkages with GPS data and GIS data layers and functions, and an emerging modeling component. Spatial resolution, spectral resolution, radiometric resolution, temporal resolution — the fundamental characteristics of remote sensing imagery — all are improving rapidly. The cost of remote sensing data acquisition, computer support, and image processing software has plummeted. The availability and quality of ancillary data, GIS data, has improved dramatically. The importance of remote sensing as a component in education of practicing resource management professionals has long been recognized and is increasingly relevant. There has never been a time in which remote sensing data and image processing functionality were more available, more inexpensive, more widely understood. All indications are that these trends will stabilize or continue — faster, less expensive, better, more widely available.

And yet remote sensing as a field is still perhaps too "research" oriented and adheres too strongly to pure scientific approaches. Remote sensing has not yet developed an intensive focus on practical forestry applications, despite early efforts to identify the useful remote sensing data characteristics in forest classification, inventory, and monitoring (Sohlberg and Sokolov, 1986; Eden and Parry, 1986; Howard, 1991). Today, very few of these applications are operational in forest management anywhere in the world. Remote sensing has displayed many if not all the classic signs of a new discipline or field of science — a rapid rush to develop the field has meant that perhaps too little attention has been paid to the end users, and the end uses, of the technology. In turn, the end users have perhaps ignored or neglected to consider fully the emerging remote sensing perspective. There are clear signs that these problems are fading as the real challenge of remote sensing in forest management is recognized: the conversion of remote sensing data to information that can fulfill a need expressed in forest management terms. At least two weak points exist and must be addressed:

1. The ability to acquire and process remote sensing data into information products, and
2. The ability to understand and act on the implications and the knowledge derived from the available and created information.

Many forest management problems today are a result of management decisions applied over small areas, which have undesirable, aggregated consequences as larger areas are considered. There is a need to consider multiple scales, temporally and spatially — only remote sensing methods can provide these data. An example application of increasing relevance is the need to monitor increasing fragmentation and stand simplification caused by forest harvesting activities, across entire watersheds and ecoregions over several rotations. It has been noted often that the problem of estimating the net productive area is common to many tropical timber producers (Vanday, 1996); apparently, this problem cannot be overcome with sporadic, single-scale aerial photographic coverage, and only field-based (e.g., plot or cruise) inventory methods. But this problem can be reconciled through multiscale and multitemporal remote sensing, and the resulting efficient, spatially explicit, and accurate GIS database. Remote sensing is a way of increasing and extending confidence in the forest inventory database and field sampling. Such issues can now be addressed by virtue of the capability to observe phenomena at multiple scales with accurate and reliable digital remote sensing technology.

Remote sensing has begun to synthesize into a logical framework or method of study that is far superior, in aggregate, to analogue interpretations and isolated observations in single-field or modeling studies. A well-designed remote sensing activity can yield data which are synoptic and repetitive, and with appropriate methods can often generate consistent information and results over large areas and long periods of time that are simply unavailable in any other way. Models of forest ecosystem dynamics, for instance, require input of quantitative information on the biophysical and biochemical properties of vegetation at seasonal or annual time steps, with details within individual forest stands, and extending spatially across the landscape or region (Coops et al., 1998). This information can only feasibly be obtained by remote sensing (Lucas et al., 2000). Remote sensing methods and data must be considered within the continuum of information needs that exist or will be generated to support sustainable forest management plans in local, regional, and global settings. The unique, synoptic, and repetitive perspective offered by remote sensing — from above — in digital formats compatible with geographical information systems (GIS) and at multiple scales, will be increasingly valuable if the goal of achieving sustainability in decision making concerning Earth's remaining forest resources is to be realized.

Remote sensing is a young, dynamic, new science — assertive, confident, and visionary. The full potential has yet to be realized or even properly understood. Earlier breakthroughs in the use of remote sensing were created by new understanding of the data (e.g., as contained in the interpretation of NDVI) (Dale, 1998) and wider availability of the infrastructure to support remote sensing applications (e.g., the proliferation of GIS image processing tools on the desktop) (Longley et al., 1999). What new themes appear likely to help propel remote sensing to new breakthroughs in understanding and use in forest management? The window of discussion is best restricted to the relatively short term — perhaps to consider what can be accomplished in the next five, maybe ten, years. Writing in the year 2000, can anyone confidently predict remote sensing and forest management breakthrough applications and infrastructure conditions in 2050, or even 2020? About twenty years ago, the

Landsat TM, SPOT, and Radarsat sensors were on the drawing boards of three different countries, or in various stages of construction and testing. Their launches were between 3 and 15 years away. Airborne hyperspectral and lidar sensors had yet to be deployed; 50 years ago the first artificial satellite had not yet been built, let alone launched; 100 years ago the Wright brothers were building the world's first successful airplane, but still dreaming of the first flight.

Understanding Pixels — Multiscale and Multiresolution

The multi concept has a long history in remote sensing — the natural environment, the Earth system, and individual forest ecosystems are multifaceted, and from the earliest days it was understood that in order to acquire information on that environment, remote sensing must be multispectral, multiresolution, multitemporal (Colwell, 1983). The idea was usually to try and match remote sensing observations to the feature of interest; if possible, acquire data from several different altitudes and with different sensor packages. But often, few remote sensing opportunities were available and people had to make do with data that captured only a portion of the multifaceted environmental phenomena of interest — theoretically, it was understood that this approach was seriously deficient, but practically, little could be done about it.

With satellite sensors now collecting data in several different portions of the spectrum at spatial resolutions ranging from 1 m to 1 km, and the dramatically increasing availability worldwide of many different kinds of airborne sensors, multi-remote-sensing has now truly arrived. It is now almost always possible to collect imagery known to be much more closely related to the optimal spatial, spectral, and temporal resolutions at which the environmental feature of interest exists. The sensors themselves have improved; for example, data are now downloaded routinely in 16 bits. The situation is not ideal, but much improved: it is now much more possible to generate or acquire data from which the maximum amount of information on the forest attributes of interest can be derived.

At the heart of the remote sensing enterprise is the pixel; the fundamental unit of analysis that must be understood if remote sensing observations are to be of increasing use in forestry. With increasing availability of multiresolution data, the pixel can exist as an organizing principle, a nested or hierarchical structure within which different levels of information can reside. This information will scale between levels, and may also be directional. Apart from capturing forest information at different scales, there are now several ways in which pixels can be interpreted.

The pixel can be considered as a spectral response pattern — even a hyperspectral response pattern — that integrates all the environmental features (and adjacency effects) contributing spectral response within the instrument's instantaneous field of view. Thus, interpreting pixels by their integrated nature results in applications consistent with the way in which the integrated terrain unit concept evolved (Robinove, 1981). The pixel can be considered to be comprised of a weighted function of features contributing spectral response — the spectral mixture of sunlit and shaded tree crowns, background, and understory (Li and Strahler, 1985). The pixel can be considered within the spatial context in which it occurs, and can be decomposed within forest stand polygons, segmented, or placed within some areal or geographic context to understand better the actual information contained therein. The pixel is an instantaneous observation, to be considered in concert with the increasing number of data layers in a georeferenced GIS. With multiple remote sensing images, acquired over time, nested and spatially coherent, the pixel becomes a sampling tool.

Many are aware that for every complex question there is a simple, and usually wrong, answer. The complex question in remote sensing relates to the very nature of the pixel and the methods available to understand it (Cracknell, 1998); What environmental factors are responsible for the signal detected and recorded in a pixel in a remotely sensed image? Finding a simple and right answer to this question motivates much activity in remote sensing — a driving force in remote sensing forestry applications. Extracting information from the pixel, and groups of pixels, remains the highest priority in remote sensing. In answering the question, remote sensing must be of value, must provide information unavailable in other ways, must go beyond provision of mere information, do so cost-effectively, and with the potential to create new insights and knowledge in forest management and forest science.

Driven by the increasing need for forest information, forestry remote sensing applications continue to emerge; converting pixels to maps, pixels to models, pixels to monitoring observations. The methods of moving from data to information products continue to increase in complexity. They are sometimes astonishingly complex, yet often fall short of accomplishing simple goals. Much promise is extended for automated classifications relying on simple image pixel spectral response patterns or context in simple thresholding or identification problems, but for most tasks, a complex integration of human interpretation skills, GUI-software, and raw computer power is the only foreseeable way in which quality information products of real value in satisfying information needs will be generated. The philosopher's stone — the promised land — the El Dorado of remote sensing — completely automated methods of remote sensing, a seamless flow from data to information products — continues to inspire, but is a long, long, way from realization. A more modest trend to standardized protocols in certain key areas such as supervised classification, for example, may make possible the smooth integration of larger and larger areas in greater spatial detail than was previously possible.

Aerial Photography and Complementary Information

Aerial photography has long been the remote sensing method of choice in forestry. In light of the trends of the past few years in digital image analysis, sensor development, and sensor deployment, forestry use of aerial photographs — itself a reasonably young science — will continue to decline in relation to digital remote sensing (Caylor, 2000). It seems possible, even likely, in the not so distant future, that film will no longer be used as a remote sensing medium, replaced by digital sensors of higher sensitivity, greater range, and more reliable storage. Modern (digital) remote sensing originated in, but will shortly completely overshadow, the acquisition and manual interpretation of aerial photographs. It is abundantly clear that aerial photographs alone cannot provide the necessary amounts and quality of information required for sustainable forest management. Even the information required *now* cannot be provided by aerial photography; but can remote sensing provide the information required without aerial photography? Is the type of information that is required helping to make digital remote sensing the observational platform of choice in forestry?

Visual analysis, including aerial photointerpretation and, increasingly, combined visual/digital analysis of imagery, will likely continue in a crucial role in forestry. A key condition is that the imagery, photographic or digital, remain simple to acquire and use. Visual analysis requires relatively little training or equipment (compared to digital remote sensing), and human capability in pattern recognition and reasoning will not soon be duplicated in any (usable) computer environment. No simple substitute for the delineation and compilation of forest stand maps is yet available; candidate procedures based on spatial overlays (Congalton et al., 1993) and image segmentation (Woodcock and Harwood, 1992; Ryherd and Woodcock, 1997) remain poorly developed, and have been tested in only a few forests (Hagner, 1990). It is expected that the current prominent role of aerial photography in providing forestry information will be gradually reduced in the initial stages of forest inventory and map database development. The information required to understand stand development and fundamental ecosystem processes, such as photosynthesis and hydrological cycling, are only available through remote sensing.

The visually based imagery such as aerial photography seem best suited as a stratification tool that can then be improved with access to more precise descriptions of the dynamic, changing forest conditions. In time, forest stratification will shift from photomorphic methods to spectral methods (Wynne and Oderwald, 1998), a remote sensing, satellite-assisted forest inventory. This perspective is necessary in order to understand carbon dynamics, an issue that will be of increasing urgency as the requirement to place forest management within a regional and global context approaches. Only through digital remote sensing and access to comprehensive spatial databases can required reporting under the Kyoto Protocol be accomplished, for example.

This scenario presupposes the continuation of the forest stand as the basis for management and forest operations. Indications are that this may not be a logical way in which to organize the forest under the still-evolving ecosystem management paradigm. Here, the requirement is for forest stands to be aggregated, or otherwise connected, to comprise forest ecosystems that are the fundamental management unit more readily employed in considerations of biodiversity, physiological processing, carbon dynamics, and so on. Are forest stands, recognized and delineated on the basis of their appearance on aerial photographs, to continue to shape and direct forest management activities such as prescriptions and treatments? It seems more likely that the more comprehensive data layers in the GIS and the more scientifically based digital remote sensing observations will be used to create new stand-like units that represent quantifiable, repeatable, consistently identifiable units of management. The role of soils, for example, in forest growth and productivity will be more readily integrated in management by explict recognition, rather than simply assumed through relations with forest structure in units interpreted from aerial photography. As individual tree detection and monitoring improves, it appears likely that a dynamic forest inventory of the true phenomenological unit of the forest — the trees — will become desirable.

Actual Measurement vs. Prediction — The Role of Models

Now this is a safe (fearless!) prediction — five or ten years out there will be increased modeling in all aspects of geospatial data analysis and forest management. Models are essential in modern remote sensing, and the increased availability of remote sensing imagery will guarantee a virtual avalanche of empirical and semiempirical models to be built, tested, and applied — usually in the normative scientific design. Most remote sensing image analysis can be considered a modeling exercise; predicting the occurrence of spectral response patterns relative to a forest condition of interest, but without rigid control of confounding variables. Even now, keeping track of the many remote sensing models — classificatory, predictive, explanatory, and exploratory — that have been developed and deployed in forestry applications is a Sisyphean task.

Physically based radiative transfer models will continue to improve as radiation physics are better understood in forest canopies. These improvements are essential to greater understanding of the potential forestry uses of remote sensing data; the topographic effect under geotropic forest conditions (Gu and Gillespie, 1998), for example, must be better understood and adequately modeled before robust corrections can be developed for wider application. But physically based models are unlikely to make an immediate impression in remote sensing forestry applications because so few remote sensing analysts can access and apply them. The role of such models will continue to be restricted to improvements in understanding the physical process of radiative transfer in the various different wavelength intervals of interest. It is hoped that this understanding will continue to trickle down to commercial software developers and applications specialists, in much the same way that more complex, physically based atmospheric corrections have slowly made their way into mainstream applications in recent years (Richter, 1997).

On the other hand, statistical and semiempirical models such as the geometrical-optical or GO canopy reflectance models used to understand remote sensing imagery, are becoming understandable themselves. The GO model and its microwave twin, the backscattering model, have created a potential breakthrough in the way remote sensing data are used, since they provide the opportunity for image analysts to invert the complex physical interactions through mechanistic simulation (Strahler et al., 1986). There are many fewer terms to consider. The models can be used to explain image data in greater detail than previously thought possible; even to unmix relatively coarse resolution satellite image pixels such that the delicate interplay between radiation in the sunlit and shadowed portions of the forest canopy can be understood and used to interpret the image spectral response. Ultimately, there may be less reason to acquire remote sensing data indiscriminately. Instead, remote sensing may be used as a way to check on routine model predictions — the model predicts that the spectral response of this forest should be X, Y, and Z; now, deploy a sensor and determine if the model predictions are correct.

Modeling in forest management represents a potentially revolutionary innovation; and the number and quality of models have grown considerably in recent years. Spatial models for landscape structure, landscape models for ecosystem productivity and disturbance, stand models for dynamics and productivity, tree and crown models for competition and growth, leaf, root, carbon, nutrient, and hydrological models; there are models and more models for questions of interest in management, ecosystem analysis, and forest science. Linking these models together and providing for hierarchical predictions tied to forest management concerns such as spatially distributed logging plans have sometimes been considered a secondary issue. Now, the models are emerging from the development stage with a new or renewed focus on practical applications (Shugart, 1998).

Typically, forest models require a constant stream of initialization and validation data. For many, only remote sensing seems capable of generating the data streams necessary to both parameterize and confirm model predictions at the many different scales involved. Over time, models of productivity, successional dynamics, and forest fragmentation seem particularly likely to both: (1) make an impact in forest management planning, and (2) to require remote sensing data for continued use. Ecosystem process models now rely on remotely sensed LAI and forest covertype (Franklin et al., 1997b), and have been shown to provide improved performance with remotely sensed estimates of biochemical conditions (e.g., nitrogen) (Lucas et al., 2000). Research into biochemical conditions of forest canopies has increased as the key driving variables of forest dynamics have been clarified (Wessman, 1990; Curran, 1992; Zagolski et al., 1996). Can fully functional forest growth and yield models based on ecophysiology, climate, and remote sensing biophysical and biochemical status be far behind? It seems likely that routine remote sensing assessment of landscape structure and dynamics will soon be possible.

Remote Sensing Research

Both theoretical and applied research drive remote sensing applications and, no doubt by their very nature, will provide for surprising new insights and developments. The research issues that would enable remote sensing to provide more effective information for forest management can be easily listed:

- Develop a more confident understanding of the forest information content of remotely sensed spectral response, for example, and provide more powerful ways of interpreting and quantifying that information;
- Increase the synergy between remotely sensed data and GIS data layers developed from remote sensing and other sources;
- Make better use of field data and the results of aerial photointerpretation when analyzing spectral information;
- Find ways to maximize the efficiency of algorithms that combine human and computer image understanding, and traditional image processing functionality; and,
- Simply build information systems that respond to the information needs of the user.

Assertion of issues is one way to indicate the likely directions of research and the specific developments that can be expected in the near term.

Remote sensing research is eclectic and fragmented, designed and executed according to disciplinary goals and new opportunities. This is part of the attraction and fascination of remote sensing! The literature of remote sensing research and applications is largely concentrated in a few key journals devoted to the field, but important remote sensing papers can be found across the library map — in engineering, natural sciences, physical science, computer science, and practical (trade) journal venues. There is a vast literature developed annually from the numerous remote sensing symposia and conferences. It is necessary to read those journals, attend those meetings, be connected, in order to gain an appreciation of the breadth and depth of remote sensing research. As the field has expanded, perhaps there is more time for contemplation (Curran, 1985), but the field is still moving fast, and is quick to cross-fertilize and create new challenges.

The key drivers in remote sensing research can be used to indicate potential trends and can be divided into three main themes: (1) research related to remote sensing data, (2) research related to remote sensing methods, and (3) research applications. This latter theme might appear obvious, but as has been noted in this book and elsewhere (e.g., Anger, 1999; Olson and Weber, 2000), remote sensing has not always been attentive to applications, sometimes behaving as if remote sensing research on data and methods were a sufficient end. The issue is not how to conduct remote sensing research, but how to ensure that remote sensing research leads to greater effectiveness in forestry applications.

Obviously, increased understanding of remote sensing data and finding ways to improve data — in quality, quantity, and newness — has provided a sustained series of stimuli in remote sensing research and applications. This can be forecast to continue steadily, perhaps even increase in importance. For example, quality in sensor design and data flow will continue to improve; unlike some consumer products (e.g., toasters, freezers, automobiles), the engineering design criteria in remote sensing have by no means stabilized or been exhausted. A new sensor design or configuration can overturn the field overnight. Relatively simple extensions of existing sensor designs, for example, will provide bountiful new data sets in currently underutilized portions of the spectrum, such as the shortwave infrared (Babey et al., 1999) for use in forestry applications. Data quality is related closely to image processing and the correct or appropriate specification of information products. The ready availability of historical remote sensing data, the multiple platform options now in place, and shifting political and pricing policies have dramatically increased the quantity of remote sensing data — a variation on the multiremote-sensing concept.

In remote sensing, at least, there really is something new under the sun. The new data options are truly exciting. Increased spatial, spectral, radiometric, temporal, and angular resolutions, all within a few years; this is an incredibly challenging time in which to consider remote sensing applications. Some of the highlights:

- The Earth Observing System (EOS) has started operation with the Terra satellite carrying the MODIS sensor, among others (Running et al., 2000). MODIS represents a multiresolution (250, 500, 1000 m) design, and will acquire global coverage daily.

- Airborne lidar systems have recently improved to the point that routine remote sensing determination of canopy height, aboveground biomass, volume, canopy structure, and basal area calculation are feasible (Lefsky et al., 1999b; Baltsavias, 1999).
- The quality and resolution of videography, frame cameras, and digital imaging sensors continue to sharpen, such that softcopy may soon rival "hardcopy" in equivalent information content (Caylor, 2000) — ultimately, at a fraction of the cost, time, and effort, and with considerable improvements in downstream data processing potential.
- Hyperspectral imaging from airborne platforms has literally overwhelmed the earlier multispectral "dimensional" limits to analysis, opening up opportunities to design applications to detect individual chemical bonds, water absorption bands, and pigment concentration of foliage (Treitz and Howarth, 1999).
- Three new SAR satellites are planned for launch in 2001 and 2002 (van der Sanden et al., 2000), adding polarization diversity and polarimetry to a range of resolutions and swath widths (ENVISAT, ALOS PALSAR, Radarsat-2).
- Ikonos-2 and other high spatial detail satellites have opened a new era of satellite remote sensing with 1 m or better spatial resolution in multispectral packages.

All that remains of the new data questions are the computer hardware (Um and Wright, 1999) and software issues; "not yet powerful enough to make large raster data files the medium of choice for most photo users" (Caylor, 2000: p. 18). Not yet, perhaps, but soon.

Methods of remote sensing — methods of analyzing remote sensing imagery, extracting data, and evaluating remote sensing information — are the final frontiers. This is the litmus test against which remote sensing technology will succeed or fail; can remote sensing deliver the information products that forest managers want and need? In forestry, methods of extracting information from digital images are not well developed or purpose-designed. Existing procedures are overwhelmingly statistical, rigid, often without clear decision points, based more on counting, simple distributional criteria, and brute force pixel pushing than intelligence and design. Barriers include relatively poor integration with GIS, poor integration with field methods, poor communication between systems, even between different remote sensing datasets. The information extracted, and the way information extraction has occurred, have rarely been consistent with the needs (and even more rarely with the preferences) of forest management.

This is not to criticize the work that has been accomplished; necessary steps have been taken in the evolution of remote sensing in the service of forestry decision making. More rule-based, soft or fuzzy image analysis systems have been developed because the information that is contained in remote sensing imagery about forests is often soft or fuzzy. A greater emphasis on spatial operators and integration of human interpretation with computer processing has expanded the use of image data, simulating better the processes that skilled human image interpreters use. Texture, context, image segmentation, more complete integration with GIS data layers, spatial reasoning, and knowledge-based image analysis, all are needed if maximum value is to be extracted from remote sensing in forest management.

Remote sensing research based on new data and methods of image analysis can create new opportunities for insight and understanding of the practices of sustainable forest management. The applications are straightforward; better classifications, estimation of biochemical and biophysical variables, better change detection and validation of measurements. The objective of greater ability to extract forest information from remote sensing is not a goal in and of itself, but rather a signpost along the way. The goal of remote sensing research in forestry must continue to be increased knowledge about forests — and human and natural impacts — and the wisdom to use that knowledge in the search for ways to ensure human and natural world coexistence.

References

Abeyta, A., and J. Franklin. 1998. The accuracy of vegetation stand boundaries derived from image segmentation in a desert environment. *Photogramm. Eng. Rem. Sensing,* 64: 59–66.

Adams, J. B., D. E. Sabol, V. Kapos, R. A. Filho, D. Roberts, M. O. Smith, and A. R. Gillespie. 1995. Classification of multispectral images based on fractions of endmembers: applications to land-cover change in the Brazilian Amazon. *Rem. Sensing Environ.,* 52: 137–152.

Ahearn, S. C. 1988. Combining Laplacian images of different spatial frequencies (scales): implications for remote sensing analysis. *IEEE Trans. Geosci. Rem. Sensing,* 26: 826–831.

Ahern, F. J. 1992. Satellite data applied to forest management. *Rem. Sensing Can.,* 20: 45.

Ahern, F. J., T. Erdle, D. A. MacLean, and I. D. Kneppeck. 1991. A quantitative relationship between forest growth rates and Thematic Mapper reflectance measurements. *Int. J. Rem. Sensing,* 12: 387–400.

Ahern, F. J., A. C. Janetos, and E. Langham. 1998. Global Observation of Forest Cover: a CEOS' Integrated Observing Strategy. Pages 103–105 in *Proc. 27th Int. Symp. Rem. Sensing Environ.,* ERIM, Ann Arbor, MI.

Ahern, F. J., D. G. Leckie, and D. Werle. 1993. Applications of Radarsat SAR data in forested environments. *Can. J. Rem. Sensing,* 19: 330–337.

Ahern, F. J., I. McKirdy, and J. Brown. 1996. Boreal forest information content of multi-season, multi-polarization C-band SAR data. *Can. J. Rem. Sensing,* 22: 456–472.

Ahern, F. J., and J. Sirois. 1989. Reflectance enhancements for the Thematic Mapper: an efficient way to produce images of consistently high quality. *Photogramm. Eng. Rem. Sensing,* 55: 61–67.

Ahern, F. J., W. C. Skelly, K. McDonald, and C. M. Pearce. 1991. Studies of microwave sensitivity to black spruce forest biomass with the MIMICS model. Pages 234–240 in *Proc. 14th Can. Symp. Rem. Sensing,* Canadian Aeronautics and Space Institute, Ottawa.

Albrecht, J. 1999. Universal analytical GIS operations — a task oriented systematisation of data structure-independent GIS functionality. Pages 577–591 in M. Craglia and H. Onsrud, Eds., *Geographic Information Research: A Trans-Atlantic Perspective.* Taylor and Francis, London.

Aldred, A. H., and I. S. Alemdag. 1988. *Guidelines for Forest Biomass Inventory.* Petawawa National Forestry Inst. Rep. PI-X-77. Canadian Forest Service, Natural Resources Canada, Ottawa.

Aldred, A. H., and J. J. Lowe. 1978. *Application of Large-scale Photos to a Forest Inventory in Alberta.* FMI Inf. Rep. FMR-X-107. Canadian Forest Service, Natural Resources Canada, Ottawa.

Alfaro, R. I. 1988. Pest damage in forestry and its assessment. *Northwest Environ. J.,* 4: 279–300.

Allen, T. F. H. 1998. The landscape level is dead: persuading the family to take it off the respirator. Pages 35–54 in D. L. Peterson and V. T. Parker, Eds. *Ecological Scale: Theory and Applications.* Columbia University Press, New York.

Allen, T. F. H., and T. B. Starr. 1982. *Hierarchy: Perspectives for Ecological Diversity.* University of Chicago Press, Chicago.

Allen, T. R. 2000. Topographic normalization of Landsat Thematic Mapper data in three mountain environments. *Geocarto Int.,* 15: 13–19.

Alves, D. S., J. L. G. Pereira, C. L. De Sousa, J. V. Soares, and F. Yamaguchi. 1999. Characterizing landscape changes in central Rondonia using Landsat TM imagery. *Int. J. Rem. Sensing,* 20: 2877–2882.

Anderson, G. S., and B. J. Danielson. 1997. The effects of landscape composition and physiognomy on metapopulation size: the role of corridors. *Landscape Ecol.,* 12: 261–271.

Anderson, J. R., E. E. Hardy, J. T. Roach, and R. E. Witmer. 1976. *A Land Use and Land Cover Classification System for Use with Remote Sensor Data.* U.S. Geological Survey Professional Paper 964. Washington, DC.

Anger, C. D. 1999. Airborne hyperspectral remote sensing in the future? Pages 1–5, Vol. I, in *Proc. 4th Int. Symp. Airborne Rem. Sensing 21st Can. Symp. Rem. Sensing*. Canadian Aeronatics and Space Institute, Ottawa, Canada.

Anonymous. 1995. *Criteria and Indicators for the Conservation and Sustainable Management of Temperate and Boreal Forests. The Montreal Process.* Canadian Forest Service, Natural Resources Canada, Ottawa.

Anonymous. 1999. *Canada's Model Forest Program. Achieving Sustainable Forest Management through Partnership.* Canadian Forest Service, Natural Resources Canada, Ottawa.

Anuta, P. E. 1977. Computer assisted analysis techniques for remote sensing data interpretation. *Geophysics,* 42: 468–481.

Aplet, G. H., N. Johnson, J. T. Olson, V. A. Sample, Eds., 1994. *Defining Sustainable Forestry.* Island Press, Washington, D.C.

Archibald, J. H., G. D. Klappstein, and I. G. W. Corns. 1996. *Field Guide to Ecosites of Southwestern Alberta.* Canadian Forest Service, Spec. Rep. 8. Northern Forestry Centre, Edmonton, Alberta.

Ardö, J. 1998. *Remote Sensing of Forest Decline in the Czech Republic.* Lund University Press, Lund, Sweden.

Aronoff, S. 1989. *Geographic Information Systems: A Management Perspective.* WDL Publications, Ottawa, Canada.

Aspinall, R., and N. Veitch. 1993. Habitat mapping from satellite imagery and wildlife survey data using a Bayesian modeling procedure in a GIS. *Photogramm. Eng. Rem. Sensing,* 59: 537–543.

Atkinson, P. M., and N. J. Tate, Eds. 1999. *Advances in Remote Sensing and GIS Analysis.* Wiley, Chichester, U.K.

Avery, T. E. 1968. *Interpretation of Aerial Photographs.* 2nd ed. Burgess, Minneapolis, MN.

Avery, T. E. 1978. *Forester's Guide to Aerial Photo Interpretation.* USDA Forest Service, Agriculture Handbook No. 308. U.S. Department of Agriculture, Washington, DC.

Avery, T. E., and G. L. Berlin. 1992. *Fundamentals of Remote Sensing and Airphoto Interpretation.* 5th ed. Macmillan, New York.

Avery, T. E., and H. E. Burkhard. 1994. *Forest Measurements.* 4th ed. McGraw-Hill, Boston.

Avery, M. I., and R. H. Haines-Young. 1990. Population estimates for the dunlin Calidris alpina derived from remotely sensed satellite imagery of the Flow Country of northern Scotland. *Nature,* 344: 860–862.

Babey, S. K., and C. D. Anger. 1989. A Compact Airborne Spectrographic Imager (CASI). Pages 1028–1031 in *Proc. 12th Can. Symp. Rem. Sensing Int. Geosci. Rem. Sensing Symp.* Canadian Aeronautics and Space Institute, Ottawa.

Babey, S. K., C. D. Anger, S. B. Achal, T. Ivanco, A. Moise, P. R. Costella, and J. DeBliek. 1999. Development of a next generation Compact Airborne Spectrographic Imager: *casi-2*. Pages 229–239, Vol. I, in *Proc. 4th Int. Symp. Airborne Rem. Sensing 21st Can. Symp. Rem. Sensing.* Canadian Aeronautics and Space Institute, Ottawa.

Bachelord, E. P., and D. M. Griffith. 1994. Continuing professional forestry education: a summary of issues. Pages 107–122 in *Forestry Education: New Trends and Prospects.* FAO Forestry Paper #123. FAO, Rome.

Bailey, R. G. 1996. *Ecosystem Geography.* Springer-Verlag, New York.

Bailey, R. G., R. D. Pfister, and J. A. Henderson. 1978. Nature of land and resource classification — a review. *J. For.,* 76: 650–655.

Baker, W. L., J. J. Honaker, and P. J. Weisberg. 1995. Using aerial photography and GIS to map the forest-tundra ecotone in Rocky Mountain National Park, Colorado, for global change research. *Photogramm. Eng. Rem. Sensing,* 61: 313–320.

Baltsavias, E. P. 1999. Airborne laser scanning: existing systems, firms and other sources. *ISPRS J. Photogramm. Rem. Sensing,* 54: 164–198.

Band, L. E., D. L. Peterson, S. W. Running, J. Coughlan, R. Lammers, J. Dungan, and R. Nemani. 1991. Forest ecosystem processes at the watershed scale: basis for distributed simulation. *Ecol. Model.,* 56: 171–196.

Banner, A. V., and F. J. Ahern. 1995. Incidence angle effects on the interpretability of forest clearcuts using airborne C-HH SAR imagery. *Can. J. Rem. Sensing,* 21: 64–66.

Banner, A. V., and F. J. Ahern. 1995. Forest clearcut mapping using airborne C-band SAR and simulated Radarsat imagery. *Can. J. Rem. Sensing,* 21: 124–137.

Baret, F., and G. Guyot. 1991. Potentials and limits of vegetation indices for LAI and APAR assessment. *Rem. Sensing Environ.,* 35: 161–173.

Barnes, B. V., K. S. Pretziger, T. A. Spies, and V. H. Spooner. 1982. Ecological forest site classification. *J. For.,* 80: 493–498.

Barnsley, M. 1984. Effects of off-nadir view angles on the detected spectral response of vegetation canopies. *Int. J. Rem. Sensing,* 5: 715–728.

Barnsley, M., and S. A. W. Kay. 1990. The relationship between sensor geometry, vegetation-canopy geometry and image variance. *Int. J. Rem. Sensing,* 11: 1075–1083.

Barnsley, M. 1999. Digital remotely-sensed data and their characteristics. Pages 451–466 in Longley, P. A., M. F. Goodchild, D. J. Maguire, and D. W. Rhind, Eds. *Geographical Information Systems.* 2nd ed. Wiley, New York.

Baskent, E. Z. 1997. Assessment of structural dynamics in forest landscape management. *Can. J. For. Res.,* 27: 1675–1684.

Baskent, E. Z. 1999. Controlling spatial structure of forested landscapes: a case study towards landscape management. *Landscape Ecol.,* 14: 83–97.

Baskent, E. Z., and G. Jordan. 1995. Characterizing spatial structure of forested landscapes: a hierarchical perspective. *Can. J. For. Res.,* 25: 1830–1849.

Bass, B., R. Hansell, and J. Choi. 1998. Towards a simple indicator of biodiversity. *Environ. Monit. Assess.,* 49: 337–347.

Bastedo, J. D., and J. B. Theberge. 1983. An appraisal of interdisciplinary resource surveys (ecological land classification). *Landscape Planning,* 10: 317–334.

Bateman, I., and A. Lorett. 1998. Using a GIS and large area databases to predict yield class: a study of Sitka spruce in Wales. *Forestry,* 71: 147–168.

Battaglia, M., and P. J. Sands. 1998. Process-based forest productivity models and their application in forest management. *For. Ecol. Manage.,* 102: 13–32.

Bauer, M. E., T. E. Burk, A. R. Ek, P. R. Coppin, S. D. Lime, T. A. Walsh, D. K. Walters, W. Befort, and D. F. Heinzen. 1994. Satellite inventory of Minnesota forest resources. *Photogramm. Eng. Rem. Sensing,* 60: 287–298.

Baulies, X., and X. Pons. 1995. Approach to forest inventory and mapping by means of multispectral airborne data. *Int. J. Rem. Sensing,* 16: 61–80.

Bayer, T., R. Winter, and G. Schreier. 1991. Terrain influences in SAR backscatter and attempts to their correction. *IEEE Trans. Geosci. Rem. Sensing,* 29: 451–462.

Beaubien, J. 1977. Remote sensing for evaluating forest insect damage. *Proc. 10th Annu. Northeastern For. Dist. For. Insect Work Conf.*, Quebec, Canada.

Beaubien, J. 1979. Forest type mapping from Landsat digital data. *Photogramm. Eng. Rem. Sensing,* 45: 1135–1144.

Beaubien, J. 1994. Landsat TM satellite images of forests: from enhancement to classification. *Can. J. Rem. Sensing,* 20: 17–26.

Beaubien, J., J. Cihlar, G. Simard, and R. Latifovic. 1999. Land cover from multiple Thematic Mapper scenes using a new enhancement-classification methodology. *J. Geophys. Res.,* 104 D22: 27909–27920.

Beaubien, J., and L. Jobin. 1974a. ERTS-1 imagery for broad mapping of forest damage and covertypes of Anticosti Island. *Canadian Surveyor,* 28: 164–166.

Beaubien, J., and L. Jobin. 1974b. Forest insect damage and cover types from high-altitude color-IR photographs and ERTS-1 imagery. *Proc. ISP Comm. VII Symp. Rem. Sensing Photointerpret.,* Banff, Alberta, Canada.

Beauchemin, M., K. P. B. Thompson, and G. Edwards. 1995. Modelling forest stands with MIMICS: implications for calibration. *Can. J. Rem. Sensing,* 21: 518–526.

Beauchesne, P., J. P. Ducruc, and V. Geradin. 1996. Ecological mapping: a framework for delineating forest management units. *Environ. Monit. Assess.,* 39: 173–186.

Becker, M. 1999. How to characterize the diversity of ecosystems? *Sustainable Forest Management: Contribution of Research.* IUFRO Occas. Pap. No. 9. International Union of Forest Research Organizations, Vienna.

Behan, R. W. 1997. Scarcity, simplicity, separatism, science — and systems. Pages 411–417 in K. A. Kohm and J. F. Franklin, Eds. *Creating a Forestry for the 21st Century. The Science of Ecosystem Management.* Island Press, Washington, DC.

Bellehumeur, C., and P. Legendre. 1998. Multiscale sources of variation in ecological variables: modeling spatial dispersion, elaborating sampling designs. *Landscape Ecol.,* 13: 15–25.

Benediktsson, J. A., P. H. Swain, and O. K. Ersoy. 1990. Neural network approaches versus statistical methods in classification of multisource remote sensing data. *IEEE Trans. Geosci. Rem. Sensing,* 28: 540–551.

Benkelman, C. A., W. B. Cohen, D. Stow, and A. Hope. 1992. High resolution digital imagery applied to vegetation studies. Pages 126–131 *in Proc. ISPRS Comm.,* VII Vol. XXIX Part B7. XVII Int. Congr. Photogramm. Rem. Sensing, Washington, DC.

Benson, B. J., and M. D. MacKenzie. 1995. Effects of sensor spatial resolution on landscape structure parameters. *Landscape Ecol.,* 10: 113–120.

Bergen, K., J. Colwell, and F. Sapio. 2000. Remote sensing and forestry: collaborative implementation for a new century of forest information solutions. *J. For.,* 98: 5–9.

Berlyn, G. P., and P. M. S. Ashton. 1996. Sustainability of forests. *J. Sustainable For.,* 3: 77–89.

Berry, J.K., and W. J. Ripple. 1996. Emergence of Geographic Information Systems in Forestry. Pages 107–128 in McDonald, P. and J. Lassoie, Eds. *The Literature of Forestry and Agroforestry.* Cornell University Press, Ithaca, New York.

Beven, K. J., and E. F. Wood. 1983. Catchment geomorphology and the dynamics of runoff contributing areas. *J. Hydrology,* 65: 139–158.

Bezdek, J. C., R. Ehrlich, and W. Full. 1984. FCM: fuzzy c-means clustering algorithm. *Comput. Geosci.,* 10: 191–203.

Biging, G. S., D. R. Colby, and R. G. Congalton. 1999. Sampling systems for change detection accuracy assessment. Pages 281–308 in R. Lunetta and C. D. Elvidge, Eds. *Remote Sensing Change Detection: Environmental Monitoring Methods and Applications.* Taylor & Francis, London.

Biging, G. S., M. Dobbertin, and E. C. Murphy. 1995. A test of airborne multispectral videography for assessing the accuracy of wildlife habitat maps. *Can. J. Rem. Sensing,* 21: 357–367.

Binaghi, E., P. Madella, M. Grazia Montesano, and A. Rampini. 1997. Fuzzy contextual classification of multisource remote sensing images. *IEEE Trans. Geosci. Rem. Sensing,* 35: 326–339.

Birot, Y. 1999. Sustainable forest management: a strong demand by society to the scientific community. *Sustainable Forest Management: Contribution of Research.* IUFRO Occas. Pap. No 9. International Union of Forest Research Organizations, Vienna.

Blackburn, G. A. 2000. Relationships between spectral reflectance and pigment concentrations in stacks of deciduous broadleaves. *Rem. Sensing Environ.,* 70: 224–237.

Blackburn, G. A., and E. J. Milton. 1997. An ecological survey of deciduous woodlands using airborne remote sensing and geographical information systems (GIS). *Int. J. Rem. Sensing,* 18: 1919–1935.

Blair, J. B., D. L. Rabine, and M. H. Hofton. 1999. The Laser Vegetation Imaging Sensor: a medium-altitude digitisation-only airborne laser altimeter for mapping vegetation and topography. *ISPRS J. Photogramm. Rem. Sensing,* 54: 115–122.

Blaszcynski, J. S. 1997. Landform characterization with geographic information systems. *Photogramm. Eng. Rem. Sensing,* 63: 183–191.

Bobbe, T., D. Reed, and J. Schramek. 1993. Georeferenced airborne video imagery. *J. For.,* 91: 34–37.

Bolduc, P., K. Lowell, and G. Edwards. 1999. Automated estimation of localized forest volume from large-scale aerial photographs and ancillary cartographic information in a boreal forest. *Int. J. Rem. Sensing,* 20: 3611–3624.

Bolstad, P. V., and T. M. Lillesand. 1991. Semi-automated training approaches for spectral class definition. *Int. J. Rem. Sensing,* 13: 3157–3166.

Bolstad, P. V., and T. M. Lillesand. 1992. Improved classification of forest vegetation in northern Wisconsin through a rule-based combination of soils, terrain, and Landsat Thematic Mapper data. *For. Sci.,* 38: 5–20.

Bolstad, P. V., W. Swank, and J. Vose. 1998. Predicting southern Applachian overstory vegetation with digital terrain data. *Landscape Ecol.,* 13: 271–283.

Bonan, G. 1993. Importance of leaf area index and forest type when estimating photosynthesis in boreal forests. *Rem. Sensing Environ.,* 43: 303–314.

Bordelon, M. A., D. C. McAllister, and R. Holloway. 2000. Sustainable forestry Oregon Style. *J. For.,* 98: 26–34.

Boresjö Bronge, L. 1999. Mapping boreal vegetation using Landsat TM and topographic map data in a stratified approach. *Can. J. Rem. Sensing,* 25: 460–474.

Bormann, B. T., M. H. Brookes, E. D. Ford, A. R. Kiester, C. D. Oliver, and J. F. Weigand. 1994. *A Framework for Sustainable-Ecosystem Management.* Gen. Tech. Rep. PNW-GTR-331. USDA Forest Service, Pacific Northwest Research Station, Portland, OR.

Borry, F. C., P. B. DeRoover, M. M. Leysen, R. R. DeWulf, and R. E. Goossens. 1993. Evaluation of SPOT and TM data for forest stratification: a case study for small-size poplar stands. *IEEE Trans. Geosci. Rem. Sensing,* 31: 483–490.

Boyce, M. S., and A. Haney, Eds. 1997. *Ecosystem Management.* Yale University Press, New Haven, CT.

Boyle, T. J. B. 1991. Biodiversity of Canadian forests: current status and future challenges. *For. Chron.,* 68: 444–453.

Brace, L. G., and I. E. Bella. 1988. Understanding the understory: dilemma and opportunity. Pages 69–86 in J. K. Samoil, Ed. *Management and Utilization of Northern Hardwoods.* Canadian Forest Service Inf. Rep. NOR-X-296. Northern Forestry Centre, Edmonton, Alberta.

Bracher, G., and P. Murtha. 1994. Estimation of foliar macro-nutrients and chlorophyll in Douglas-fir seedlings by leaf reflectance. *Can. J. Rem. Sensing,* 20: 102–115.

Bradley, G. A., and F. Ulaby. 1981. Aircraft radar response to soil moisture. *Rem. Sensing Environ.,* 11: 419–438.

Brandtberg, T. 1997. Towards structure-based classification of tree crowns in high spatial resolution aerial images. *Scand. J. For. Res.,* 12: 89–96.

Braumandl, T. F., and M. P. Curran, Eds. 1992. *A Field Guide for Site Identification and Interpretation for the Nelson Forest Region.* B.C. Ministry of Forests, Nelson, British Columbia.

Bricker, O. P., and M. A. Ruggiero. 1998. Toward a national program for monitoring environmental resources. *Ecol. Appl.* 8: 326–329.

Brockhaus, J. A., and S. Khorram. 1992. A comparison of SPOT and Landsat TM data for use in conducting inventories of forest resources. *Int. J. Rem. Sensing,* 13: 3035–3043.

Brockhaus, J. A., S. Khorram, R. Bruck, and M. V. Campbell. 1993. Characterization of defoliation conditions within a boreal montane forest ecosystem. *Geocarto Int.,* 1: 35–42.

Brockhaus, J. A., S. Khorram, R. Bruck, M. V. Campbell, and C. Stallings. 1992. A comparison of Landsat TM and SPOT HRV data for use in the development of forest defoliation models. *Int. J. Rem. Sensing,* 13: 3235–3240.

Brown, D. G., J. D. Duh, and S. A. Drzyzga. 2000. Estimating error in an analysis of forest fragmentation change using North American Landscape Characterization (NALC) data. *Rem. Sensing Environ.,* 71: 106–117.

Brown, S. L., and A. E. Lugo. 1984. Biomass of tropical forests: a new estimate based on forest volumes. *Science,* 223: 1290–1293.

Brown, S. L., P. Schroeder, and J. S. Kern. 1999. Spatial distribution of biomass in forests of the eastern USA. *For. Ecol. Manage.,* 123: 81–90.

Bruzzone, L., and D. Fernandez Prieto. 2000. An adaptive parcel-based technique for unsupervised change detection. *Int. J. Rem. Sensing,* 21: 817–822.

Bryant, D. 1997. *The Last Frontier Forests: Ecosystems and Economies on the Edge.* World Resources Institute, Washington, DC.

Bucheim, M. P. and T. M. Lillesand. 1989. Semi-automated training field extraction and analysis for efficient digital image classification. *Photogramm. Eng. Rem. Sensing,* 55: 1347–1355.

Bucheim, M. P., A. L. Maclean, and T. M. Lillesand. 1985. Forest cover type mapping and spruce budworm defoliation detection using simulated SPOT imagery. *Photogramm. Eng. Rem. Sensing,* 51: 1115–1122.

Buckley, D. S., J. G. Isebrands, and T. L. Sharik. 1999. Practical field methods of estimating canopy cover, PAR, and LAI in Michigan oak and pine stands. *North. J. Appl. For.,* 16: 25–32.

Budge, A. 1999. Appendix A of the Report on UCGIS meeting: a summary. *Photogramm. Eng. Rem. Sensing,* 65: 1004.

Buiten, H. J. 1993. Image interpetation: visual or digital? Pages 507–513 in Buiten, H. J., J. G. P. W. Clevers, Eds. *Land Observation by Remote Sensing: Theory and Applications.* Gordon Breach, Amsterdam.

Burgess, D. W., P. Lewis, and J.-P. A. L. Muller. 1995. Topographic effects in AVHRR NDVI data. *Rem. Sensing Environ.,* 54: 223–232.

Burke, I. C., D. S. Schimel, C. M. Yonker, W. J. Parton, L. A. Joyce, and W. K. Lauenroth. 1990. Regional modeling of grassland biogeochemistry using GIS. *Landscape Ecol.,* 4: 45–54.

Burkholder, E. F. 1999. Spatial data accuracy as defined by the Global Spatial Data Model. *Surv. Land Inf. Syst.,* 59: 26–30.

Burnett, M. R., P. V. August, J. H. Brown, Jr., and K. T. Killingbeck. 1998. The influence of geomorphological heterogeneity on biodiversity: I. A patch-scale perspective. *Conserv. Biol.,* 12: 363–370.

Burns, G. S. 1985. Comparative evaluation of digital change detection methods in forestland and rangeland environments using Landsat multispectral scanner data. Pages 200–207 in *Proc. Pecora 10 Symp.*, American Society for Photogrammetry and Remote Sensing, Bethesda, MD.

Burroughs, P. A. 1986. *Principles of Geographical Information Systems for Land Resources Assessment.* Clarendon Press, Oxford.

Burroughs, P. A., and R. A. McDonnell. 1998. *Principles of Geographical Information Systems.* Oxford University Press, Oxford.

Bush, T. F., and F. T. Ulaby. 1978. Crop inventories with radar. *Can. J. Rem. Sensing,* 4: 81–87.

Butera, M. K. 1986. A correlation and regression analysis of percent canopy closure versus TMS spectral response for selected sites in the San Juan National Forest, Colorado. *IEEE Trans. Geosci. Rem. Sensing,* 24: 122–128.

Butterfield, B. A., and C. W. Key. 1985. Mapping grizzly bear habitat in Glacier National Park using a stratified Landsat classification: a pilot study. Pages 58–66 in *Proc. Grizzly Bear Habitat Symp.,* USDA Forest Service Intermountain Research Station, Ogden, Utah.

Cain, D. H., K. Riitters, and K. Orvis. 1997. A multi-scale analysis of landscape statistics. *Landscape Ecol.,* 12: 199–212.

Campbell, J. B., and L. Ran. 1993. CHROM: a C program to evaluate the application of the dark object subtraction technique to digital remote sensing data. *Comput. Geosci.,* 19: 1475–1499.

Canadian Council of Forest Ministers. 1995. *Defining Sustainable Forest Management. A Canadian Approach to Criteria and Indicators.* Canadian Forest Service, Natural Resources Canada, Ottawa.

Canadian Council of Forest Ministers. 1997. *Criteria and Indicators of Sustainable Forest Management in Canada.* Tech. Rep. Canadian Forest Service, Natural Resources Canada, Ottawa.

Canadian Institute of Forestry. 2000. *Forests and Climate Change.* Canadian Institute of Forestry, Position Paper, Ottawa.

Cannon, R. L., J. V. Dave, J. C. Bezdek, and M. M. Trivedi. 1986. Segmentation of a Thematic Mapper image using the fuzzy c-means clustering algorithm. *IEEE Trans. Geosci. Rem. Sensing,* 24: 400–408.

Carlotto, M. J. 1998. Spectral shape classification of Landsat Thematic Mapper imagery. *Photogramm. Eng. Rem. Sensing,* 64: 905–913.

Carpenter, G. A., M. N. Gjaja, S. Gopal, and C. E. Woodcock. 1997. ART neural networks for remote sensing: vegetation classification from Landsat TM and terrain data. *IEEE Trans. Geosci. Rem. Sensing,* 35: 308–325.

Carpenter, G. A., S. Gopal, S. Macomber, S. Martens, and C. E. Woodcock. 1999. A neural network method for mixture estimation for vegetation mapping. *Rem. Sensing Environ.,* 70: 138–152.

Carr, J. R., and F. Pellon de Miranda. 1998. The semivariogram in comparison to the co-occurrence matrix for classification of image texture. *IEEE Trans. Geosci. Rem. Sensing,* 36:1945–52.

Carter, D. R., L. G. Arvantes, D. Brackett, V. Boychera, and S. Sager. 1997. A decision support system for timber harvest scheduling. *J. For.,* 97: 12–18.

Carter, R. E., M. D. MacKenzie, and D. H. Gjerstad. 1999. Ecological land classification in the Southern Loam Hills of south Alabama. *For. Ecol. Manage.,* 114: 395–404.

Caylor, J. 2000. Aerial photography in the next decade. *J. For.,* 98: 17–19.

Chalifoux, S., F. Cavayas, and J. T. Gray. 1998. Map-guided approach for the automatic detection on Landsat TM images of forest stands damaged by the spruce budworm. *Photogramm. Eng. Rem. Sensing,* 64: 629–635.

Chauhan, N. S., R. H. Lang, and K. J. Ranson. 1991. Radar modeling of a boreal forest. *IEEE Trans. Geosci. Rem. Sensing,* 29: 627–638.

Chavez, P. S., Jr. 1986. Digital merging of Landsat TM and digitized NHAP data for 1:24,000 scale image mapping. *Photogramm. Eng. Rem. Sensing,* 52: 1637–1646.

Chavez, P. S., Jr. 1988. An improved dark-object subtraction technique for atmospheric scattering correction of multispectral data. *Rem. Sensing Environ.,* 24: 459–479.

Chavez, P. S., Jr., and J. A. Bowell. 1988. Comparison of the spectral information content of Landsat Thematic Mapper and SPOT for three different sites in the Phoenix, Arizona region. *Photogramm. Eng. Rem. Sensing,* 54: 1699–1708.

Chen, J. M. 1996. Canopy architecture and remote sensing of the fraction of photosynthetically active radiation absorbed by boreal conifer forests. *IEEE Trans. Geosci. Rem. Sensing,* 34: 1353–1368.

Chen, J. M. 1999. Spatial scaling of a remotely sensed surface parameter by contexture. *Rem. Sensing Environ., 69*: 30–42.

Chen, J. M., T. A. Black, and R. S. Adams. 1991. Evaluation of hemispherical photography for determining plant area index and geometry of a forest stand. *Agric. For. Meteorol.,* 32: 1656–1665.

Chen, J. M., and J. Cihlar. 1995. Quantifying the effect of canopy architecture on optical measurements of leaf area index using two gap size analysis methods. *IEEE Trans. Geosci. Rem. Sensing,* 33: 777–787.

Chen, J. M., and J. Cihlar. 1996. Retrieving leaf area index of boreal conifer forests using Landsat TM images. *Rem. Sensing Environ.,* 55: 153–162.

Chen, J. M., P. D. Blanken, T. A. Black, M. Guilbeault, and S. Chen. 1997. Radiation regime and canopy architecture in a boreal aspen forest. *Agric. For. Meteorol.,* 86: 107–125.

Chen, J. M., S. G. LeBlanc, J. R. Miller, J. Freemantle, S. E. Loechel, C. L. Walthall, K. A. Innanen, and H. P. White. 1999a. Compact Airborne Spectrographic Imager (CASI) used for mapping biophysical parameters of boreal forests. *J. Geophys. Res.,* 104 D22: 27945–27958.

Chen, J. M., J. Liu, J. Cihlar, and M. L. Goulden. 1999b. Daily canopy photosynthesis model through temporal and spatial scaling for remote-sensing applications. *Ecol. Model.,* 124: 99–119.

Chen, L. C., and J. Y. Rau. 1993. A unified solution for digital terrain model and orthoimage generation from SPOT stereopairs. *IEEE Trans. Geosci. Rem. Sensing,* 31: 143–1252.

Chen, R. C. 1998. An annotated guide to Earth remote sensing data and information resources for social science applications. Pages 209–228 in Liverman, D., E. F. Moran, R. R. Rindfuss, and P. C. Stern, Eds., *People and Pixels: Linking Remote Sensing and Social Science*. National Academy Press, Washington, DC.

Cheng, K. S., H. C. Yeh, and C. H. Tsai. 2000. An anisotropic spatial modeling approach for remote sensing image rectification. *Rem. Sensing Environ.,* 73: 46–54.

Chiou, W. C., Sr. 1985. NASA image-based geological expert system development project for hyperspectral image analysis. *Appl. Opt.,* 24: 2085–2091.

Christian, C. S. 1958. The concept of land units and land systems. Pages 74–81 in *Vol. 20 Proc. 9th Pacific Sci. Congr.*

Christian, C. S., and G. A. Stewart. 1968. Methodology of integrated surveys. Pages 233–280 in *Proc. Conf. Aerial Surv. Integrated Stud. (Toulouse 1964).* UNESCO, Paris.

Chuvieco, E. 1999. Measuring changes in landscape pattern from satellite images: short-term effects of fire on spatial diversity. *Int. J. Rem. Sensing,* 20: 2331–2346.

Chuvieco, E., and R. G. Congalton. 1988. Using cluster analysis to improve the selection of training statistics in classifying remotely sensed data. *Photogramm. Eng. Rem. Sensing,* 54: 1275–1281.

Ciesla, W. M., C. W. Dull, and R. E. Acciavatti. 1989. Interpretation of SPOT-1 color composites for mapping defoliation of hardwood forests by gypsy moth. *Photogramm. Eng. Rem. Sensing,* 55: 1465–1470.

Cihlar, J., J. Beaubien, R. Latifovic, and G. Simard. 1999. *Land Cover of Canada 1995 Ver. 1.1.* Digital data set documentation, Natural Resources Canada, Ottawa, Ontario.

Cihlar, J., J. Beaubien, Q. Xiao, J. Chen, and Z. Li. 1997a. Land cover of the Boreas region from AVHRR and Landsat data. *Can. J. Rem. Sensing,* 23: 163–175.

Cihlar, J., J. Chen, and Z. Li. 1997b. On the validation of satellite-derived products for land applications. *Can. J. Rem. Sensing,* 23: 381–389.

Cihlar, J., R. Latifovic, J. Chen, J. Beaubien, and Z. Li. 2000. Selecting representative high resolution sample images for land cover studies. Part 1. Methodology. *Rem. Sensing Environ.,* 71: 26–42.

Cihlar, J., D. Manak, and N. Voisin. 1994. AVHRR bidirectional reflectance effects and compositing. *Rem. Sensing Environ.,* 48: 77–88.

Cihlar, J., T. J. Pultz, and A. L. Gray. 1992. Change detection with synthetic aperture radar. *Int. J. Rem. Sensing,* 13: 401–414.

Cihlar, J., Q. Xiao, J. Chen, J. Beaubien, K. Fung, and R. Latifovic. 1998. Classification by progressive generalization: a new automated methodology for remote sensing multichannel data. *Int. J. Rem. Sensing,* 19: 2648–2704.

Civco, D. L. 1989. Topographic normalization of Landsat Thematic Mapper digital imagery. *Photogramm. Eng. Rem. Sensing,* 55: 1303–1309.

Clark, M. J., A. M. Gurnell, E. J. Milton, M. Seppala, and M. Kyostila. 1985. Remotely sensed vegetation classification as a snow depth indicator for hydrological analysis in sub-arctic Finland. *Fennia,* 163: 195–216.

Clerke, W. H., J. L. Christensen, and J. K. Dooley. 1983. Mapping prime timberland using Landsat and gridded soil databases. Pages 112–118 in *Proc. 9th Int. Symp. Machine Processing Rem. Sensed Data.* Laboratory for Applications of Remote Sensing, Purdue University, West Lafayette, Indiana.

Clough, D. J. 1972. Preliminary benefit/cost analysis of Canadian satellite/aircraft remote sensing applications. Pages 3–28 in *Proc. 1st Can. Symp. Rem. Sensing.* Canadian Aeronautics and Space Institute, Ottawa.

Cochrane, M. A. 2000. Using vegetation reflectance variability for species level classification of hyperspectral data. *Int. J. Rem. Sensing,* 21: 2075–2087.

Cohen, W. B., and M. Fiorella. 1999. Comparison of methods of conifer forest change detection with Thematic Mapper imagery. Pages 89–102 in R. Lunetta and C. D. Elvidge, Eds. *Remote Sensing Change Detection: Environmental Monitoring Methods and Applications.* Taylor & Francis, London.

Cohen, W. B., M. Fiorella, J. Gray, E. Helmer, and K. Anderson. 1998. An efficient and accurate method for mapping forest clearcuts in the Pacific Northwest using Landsat imagery. *Photogramm. Eng. Rem. Sensing,* 64: 293–300.

Cohen, W. B., M. E. Harmon, D. O. Wallin, and M. Fiorella. 1996a. Two decades of carbon flux from forests of the Pacific northwest. *BioScience,* 46: 836–844.

Cohen, W. B., J. D. Kushla, W. J. Ripple, and S. L. Garman. 1996b. An introduction to digital methods in remote sensing of forested ecosystems: focus on the Pacific Northwest, USA. *Environ. Manage.,* 20: 421–435.

Cohen, W. B., and T. A. Spies. 1992. Estimating structural attributes of Douglas-fir/western hemlock forest stands from Landsat and SPOT imagery. *Rem. Sensing Environ.,* 41: 1–17.

Cohen, W. B., T. A. Spies, and G. A. Bradshaw. 1990. Semivariograms of digital imagery for analysis of conifer canopy structure. *Rem. Sensing Environ.,* 29: 669–672.

Cohen, W. B., T. A. Spies, and M. Fiorella. 1995a. Estimating the age and structure of forests in a multiownership landscape of western Oregon, USA. *Int. J. Rem. Sensing,* 16: 721–746.

Cohen, W. B., T. A. Spies, F. J. Swanson, and D. O. Wallin. 1995b. Land cover on the western slopes of the Central Oregon Cascade Range. *Int. J. Rem. Sensing,* 16: 595–596.

Colby, J. D. 1991. Topographic normalization in rugged terrain. *Photogramm. Eng. Rem. Sensing,* 57: 531–537.

Collins, J. B., and C. E. Woodcock. 1996. An assessment of several linear change detection techniques for mapping forest mortality using multitemporal Landsat TM data. *Rem. Sensing Environ.,* 56: 66–77.

Colwell, R. N. 1965. The extraction of data from aerial photographs by human and mechanical means. *Photogrammetria,* 20: 211–228.

Colwell, R. N. 1968. The field of photointerpretation. Pages 1–15 in *Proc. 2nd Semin. Aerial Photointerpret. Dev. Can.* Queen's Printer and Controller of Stationery, Ottawa.

Colwell, R. N. 1983. *Manual of Remote Sensing.* 2nd ed. American Society for Photogrammetry and Remote Sensing, Bethesda, MD.

Congalton, R. G. 1997. Exploring and evaluating the consequences of vector-to-raster and raster-to-vector conversion. *Photogramm. Eng. Rem. Sensing,* 63: 425–434.

Congalton, R. G., and M. Brennan. 1998. Change detection accuracy assessment: pitfalls and considerations. Pages 919–932 in *Proc. ASPRS Resour. Technol. Inst. Annu. Meet.* American Society for Photogrammetry and Remote Sensing, Bethesda, MD.

Congalton, R. G., and K. Green. 1999. *Assessing the Accuracy of Remotely Sensed Data: Principles and Practice.* CRC Press, Boca Raton, FL.

Congalton, R. G., K. Green, and J. Teply. 1993. Mapping old-growth forests on National Forest and Park Lands in the Pacific Northwest from remotely sensed data. *Photogramm. Eng. Rem. Sensing,* 59: 529–535.

Connors, R., and C. Harlow. 1980. A theoretical comparison of texture algorithms. *IEEE Trans. Pattern Anal. Mach. Intelligence,* 2: 204–222.

Cooke, F. M. 1999. *The Challenge of Sustainable Forests: Forest Resource Policy in Malaysia, 1970–1995.* University of Hawaii Press, Honolulu.

Cooper, P. R., D. E. Friedman, and S. A. Wood. 1987. The automatic generation of digital terrain models by stereo. *Acta Astron.,* 15: 171–180.

Coops, N. C. 1999. Linking multiresolution satellite derived estimates of canopy photosynthetic capacity and meteorological data to assess forest productivity in a *Pinus radiata* (D. Don) stand, *Photogramm. Eng. Rem. Sensing,* 65: 1149–1165.

Coops, N. C., and D. Culvenor. 2000. Utilizing local variance of simulated high spatial detail imagery to predict spatial pattern of forests. *Rem. Sensing Environ.,* 71: 208–260.

Coops, N. C., and R. H. Waring. 2001. The use of multi-scale remote sensing imagery to derive regional estimates of forest growth capacity using 3–PGS. *Rem. Sensing Environ.,* in press.

Coops, N. C., R. H. Waring, and J. J. Landsberg. 1998. Assessing forest productivity in Australia and New Zealand using a physiologically based model driven with averaged monthly weather data and satellite-derived estimates of canopy photosynthetic capacity. *For. Ecol. Manage.,* 104: 113–127.

Coppin, P. R., and M. E. Bauer. 1996. Digital change detection in forest ecosystems with remote sensing imagery. *Rem. Sensing Rev.,* 13: 207–234.

Cordes, L. D., F. M. R. Hughes, and M. Getty. 1997. Factors affecting the regeneration and distribution of riparian woodlands along a northern prairie river: the Red Deer River, Alberta, Canada. *J. Biogeogr.,* 24: 675–695.

Coufal, J., and D. Webster. 1996. The emergence of sustainable forestry. Pages 147–167 in P. MacDonald and J. Lassoie, Eds. *The Literature of Forestry and Agroforestry.* Cornell University Press, Ithaca, NY.

Coughlan, J. C., and J. L. Dungan. 1997. Combining remote sensing and forest ecosystem modeling: an example using the Regional HydroEcological Simulation System (RHES-Sys). Pages 135–158 in Gholz, H. L., K. Nakane, and H. Shimoda, Eds. *The Use of Remote Sensing in the Modeling of Forest Productivity.* Kluwer, Dordrecht.

Coulombe, M., and M. Brown. 1999. *Forest Management Certification Programs.* The Society of American Foresters 1999 Task Force Report. Bethesda, MD.

Coyea, M. R., and H. A. Margolis. 1994. The historical reconstruction of growth efficiency and its relationship to tree mortality in balsam fir ecosystems affected by spruce budworm defoliation. *Can. J. For. Res.,* 24: 2208–2221.

Cracknell, A. P. 1998. Synergy in remote sensing — what's in a pixel? *Int. J. Rem. Sensing,* 19: 2025–2047.

Craighead, J. J., F. L. Craighead, and D. J. Craighead. 1985. Using satellites to evaluate ecosystems as grizzly bear habitat. Pages 101–112 in *Proc. Grizzly Bear Habitat Symp.,* USDA Forest Service Intermountain Research Station, Ogden, Utah.

Crist, E. P. 1985. A TM Tasseled Cap equivalent transformation for reflectance factor data. *Rem. Sensing Environ.,* 17: 301–306.

Crist, E. P., and R. C. Cicone. 1984. A physically-based transformation of Thematic Mapper data — the TM Tasseled Cap. *IEEE Trans. Geosci. Rem. Sensing,* 22: 256–263.

Cross, A. M., D. C. Mason, and S. J. Dury. 1988. Segmentation of remotely sensed images by a split-and-merge process. *Int. J. Rem. Sensing,* 9: 1329–1345.

Curran, P. J., 1980. Multispectral remote sensing of vegetation amount. *Prog. Phys. Geogr.,* 4: 315–341.

Curran, P. J. 1985. *Principles of Remote Sensing*. Longman, London.

Curran, P. J. 1987. Remote sensing methodologies and geography. *Int. J. Rem. Sensing,* 8: 1255–1275.

Curran, P. J. 1992. Estimating foliar chemical concentration with the Airborne Visible/Infrared Imaging Spectrometer (AVIRIS). Pages 705–708 in *Proc. ISPRS Comm.* VII, Vol. XXIX B7. ISPRS XVIIth Congr., Washington, DC.

Curran, P. J. 1994. Imaging spectroscopy. *Prog. Phys. Geogr.,* 18: 247–266.

Curran, P. J., J. Dungan, and H. L. Gholz. 1992. Seasonal LAI measurements in slash pine using Landsat TM. *Rem. Sensing Environ.,* 39: 3–13.

Curran, P. J., and J. A. Kupiec. 1995. Imaging spectroscopy: a new tool for ecologists. Pages 71–88 in F. M. Danson and S. E. Plummer, Eds. *Advances in Environmental Remote Sensing*. John Wiley & Sons, London.

Curran, P. J., and H. D. Williamson. 1985. The accuracy of ground data used in remote-sensing investigations. *Int. J. Rem. Sensing,* 6: 1637–1651.

Czaplewski, R. L. 1999. Multistage remote sensing: toward an annual National Inventory. *J. For.,* 97: 44–48.

Dai, X., and S. Khorram. 1999. A feature-based image registration algorithm using improved chain-code representation combined with invariant moments. *IEEE Trans. Geosci. Rem. Sensing,* 37: 2351–2362.

Daily, M. 1983. Hue-saturation-intensity split-spectrum processing of Seasat radar imagery. *Photogramm. Eng. Rem. Sensing,* 49: 349–355.

Dale, V. H. 1998. Management of forests as ecosystems: a success story or a challenge ahead? Pages 50–68 in M. L. Pace and P. M. Graffman, Eds. *Success, Limitations, and Frontiers in Ecosystem Science.* Springer-Verlag, New York.

Danson, F. M. 1987. Preliminary evaluation of the relationships between SPOT-1 HRV data and forest stand parameters. *Int. J. Rem. Sensing,* 8: 1571–1575.

Danson, F. M., and P. J. Curran. 1993. Factors affecting the remotely sensed response of coniferous forest plantations. *Rem. Sensing Environ.,* 43: 55–65.

Davidson, C. 1998. Issues in measuring landscape fragmentation. *Wild. Soc. Bull.,* 26: 32–37.

Davis, B. A., and J. R. Jensen. 1998. Remote sensing of mangrove biophysical characteristics. *Geocarto Int.,* 13: 55–64.

Davis, F. W., and J. Dozier. 1990. Information analysis of a spatial database for ecological land classification. *Photogramm. Eng. Rem. Sensing,* 56: 605–613.

Davis, F. W., and S. Goetz. 1990. Modeling vegetation pattern using digital terrain data. *Landscape Ecol.,* 4: 69–80.

Davis, T. J., and C. P. Keller. 1997. Modelling uncertainty in natural resource analysis using fuzzy sets and Monte Carlo simulation: slope stability analysis. *Int. J. Geogr. Inf. Sci.,* 11: 409–434.

Debinski, D. M., and P. F. Brussard. 1992. Biological diversity assessment in Glacier National Park, Montana: I. Sampling design, in D. H. Mackenzie, D. E. Hyatt, and V. J. McDonald, Eds. *Proc. Int. Symp. Ecological Indicators, Elsevier, Essex, UK.* Pages 393–407.

Debinski, D. M., K. Kindscher, and M. E. Jakubauskas. 1999. A remote sensing and GIS-based model of habitats and biodiversity in the Greater Yellowstone ecosystem. *Int. J. Rem. Sensing,* 20: 3281–3291.

Dechka, J. A., D. R. Peddle, S. E. Franklin, and G. Stenhouse. 2000. Grizzly bear habitat mapping using evidential reasoning and maximum likelihood classifers: a comparison. Pages 393–402 in *Proc. 22nd Can. Symp. Rem. Sensing.* Canadian Aeronautics and Space Institute, Ottawa.

De Laubenfels, D. J. 1975. *Mapping the World's Vegetation: Regionalizations of Formations and Flora.* Syracuse University Press, Syracuse, NY.

Delorme, R. 1998. GIS protects our natural resources. *GIS World,* (4) 11: 42–44.

Dempster, A. P. 1967. Upper and lower probabilities induced by a multivalued mapping. *Ann. Math. Stat.,* 38: 325–339.

Dendron Resources Inc. 1997. *Bioindicators of Forest Health and Sustainability.* Forest Res. Inf. Pap. No. 138. Ontario Forest Research Institute, Sault Ste. Marie, Ontario.

Deppe, F. 1998. Forest area estimation using sample surveys and Landsat MSS and TM data. *Photogramm. Eng. Rem. Sensing,* 64: 285–292.

Desachy, J., L. Toux, and E. H. Zahzah. 1996. Numeric and symbolic data fusion: a soft computing approach to remote sensing image analysis. *Pattern Recognition Lett.,* 17: 1361–1378.

Dettmers, R., and J. Bart. 1999. A GIS modeling method applied to predicting forest songbird habitat. *Ecol. Appl.,* 9: 152–163.

De Wulf, R. R., R. E. Gossens, B. P. DeRoover, and F. C. Borry. 1990. Extraction of forest stand parameters from panchromatic and multispectral SPOT-1 data. *Int. J. Rem. Sensing,* 11: 1571–1588.

Dikau, R. 1989. The application of digital relief model to landform analysis in geomorphology. Pages 51–78 in J. Raper, Ed. *Three-Dimensional Applications in Geographic Information Systems*. Taylor & Francis, London.

Diner, D. J., G. P. Anser, R. Davies, Y. Knyazikhin, J. P. Muller, A. W. Nolin, B. Pinty, C. B. Schaaf, and J. Stroeve. 1999. New directions in Earth observing: scientific applications of multiangle remote sensing. *Bull. Am. Meteorol. Soc.,* 80: 2209–2228.

Dobson, M. C. 2000. Forest information from synthetic aperture radar. *J. For.,* 98: 41–43.

Dobson, M. C., L. Pierce, K. Sarabandi, F. T. Ulaby, and T. Sharik. 1992. Preliminary analysis of ERS-1 SAR for forest ecosystem studies. *IEEE Trans. Geosci. Rem. Sensing,* 30: 203–211.

Domik, G., F. Lerberl, and J. Cimino. 1988. Dependence of image grey values on topography in SIR-B images. *Int. J. Rem. Sensing,* 9: 1013–1022.

Donaghue, D. N. M. 1999. Remote sensing. *Prog. Phys. Geogr.,* 23: 271–281.

Dottavio, C. L., and D. L. Williams. 1983. Satellite technology: an improved means for monitoring forest insect defoliation. *J. For.,* 81: 30–34.

Dowman, I. J. 1999. Encoding and validating data from maps and images. Pages 437–450 in Longley, P. A., M. F. Goodchild, D. J. Maguire, and D. W. Rhind, Eds. *Geographical Information Systems*. 2nd ed. John Wiley & Sons, New York.

Dralle, K., and M. Rudemo. 1997. Stem number estimation by kernal smoothing of aerial photos. *Can. J. For. Res.,* 26: 1228–1236.

Drieman, J. A. 1994. Forest cover typing and clearcut mapping in Newfoundland with C-band SAR. *Can. J. Rem. Sensing,* 20: 11–16.

Driscoll, R. S., D. R. Betters, and H. D. Parker. 1978. Land classification through remote sensing — techniques and tools. *J. For.,* 76: 656–661.

Dubayah, R. O., and J. B. Drake. 2000. Lidar remote sensing for forestry. *J. For.,* 98: 44–46.

Dubayah, R. O., and P. M. Rich. 1995. Topographic solar radiation models for GIS. *Int. J. Geogr. Inf. Syst.,* 9: 405–419.

Duda, R. O., and P. E. Hart. 1973. *Pattern Classification and Scene Analysis*. John Wiley & Sons, New York.

Duggin, M. J. 1985. Factors limiting the discrimination and quantification of terrestrial features using remotely sensed radiance. *Int. J. Rem. Sensing,* 6: 3–27.

Duggin, M. J., and C. J. Robinove. 1990. Assumptions implicit in remote sensing data acquisition and analysis. *Int. J. Rem. Sensing,* 11: 1669–1694.

Duguay, C. R. 1993. Radiation modelling in mountainous terrain: review and status. *M. Res. Dev.,* 13: 339–357.

Duhaime, R. J., P. V. August, and W. R. Wright. 1997. Automated vegetation mapping using digital orthophotography. *Photogramm. Eng. Rem. Sensing,* 63: 1295–1302.

Dungan, J., L. Johnson, C. Billow, P. Matson, J. Mazzurco, J. Moen, and V. Vanderbilt. 1996. High spectral resolution reflectance of Douglas-fir under different fertilization treatments: experiment design and treatment effects. *Rem. Sensing Environ.,* 55: 217–228.

Durden, S. L., J. D. Klein, and H. A. Zebker. 1991. Polarimetric radar measurements of a forested area near Mt. Shasta. *IEEE Trans. Geosci. Rem. Sensing,* 29: 444–450.

Dutchak, K. 2000. Alberta access update program using IRS 5.8 m panchromatic data. Page 183 in *Proc. 22nd Can. Symp. Rem. Sensing.* Canadian Aeronautics and Space Institute, Ottawa.

Dymond, J. R. 1992. Nonparametric modelling of radiance in hill country for digital classification of aerial photographs. *Rem. Sensing Environ.,* 39: 95–102.

Dymond, J. R., R. C. DeRose, and G. R. Harmsworth. 1995. Automated mapping of land components from digital elevation models. *Earth Surf. Processes Landforms,* 20: 131–137.

Dymond, J. R., and J. D. Shepherd. 1999. Correction of the topographic effect in remote sensing. *IEEE Trans. Geosci. Rem. Sensing,* 37: 2618–2620.

Ecological Stratification Working Group. 1996. *A National Ecological Framework for Canada.* Centre for Land and Biological Resources Research and State of the Environment Directorate. Ottawa.

Eden, M. J., and J. T. Parry, Eds. 1986. *Remote Sensing and Tropical Land Management.* John Wiley & Sons, New York.

Edirisinghe, A., J. P. Louis, and G. E. Chapman. 1999. Radiometric calibration of multispectral airborne video systems. *Int. J. Rem. Sensing,* 2855–2870.

Edwards, G. 1993. The integration of remote sensing and GIS: fundamental questions and new approaches. Pages 873–878 in *Proc. 8 Congr. Assoc. Quebec. Teledetection, 16th Can. Symp. Rem. Sensing.* Canadian Aeronautics and Space Institute, Ottawa, Canada.

Edwards, G., and S. Rioux. 1995. A detailed assessment of relative displacement error in cutover boundaries derived from airborne C-band SAR. *Can. J. Rem. Sensing,* 21: 185–197.

Egenhofer, M. J., and R. G. Golledge, Eds. 1998. *Spatial and Temporal Reasoning in Geographical Information Systems.* Oxford University Press, Oxford.

Ehlers, M. 1997. Rectification and registration. Pages 13–36 in Star, J. L., J. E. Estes, and K. C. McGwire, Eds. *Integration of Geographic Information Systems and Remote Sensing*. Cambridge University Press, Cambridge, U.K.

Ehlers, M., G. Edwards, and Y. Bedard. 1993. Integration of remote sensing with geographic information systems: a necessary evolution. *Photogramm. Eng. Rem. Sensing,* 55: 1619–1627.

Ekstrand, S. 1990. Detection of moderate damage on Norway spruce using Landsat TM and digital stand data. *IEEE Trans. Geosci. Rem. Sensing,* 28: 685–692.

Ekstrand, S. 1994a. Landsat TM based forest damage assessment: correction for topographic effects. *Photogramm. Eng. Rem. Sensing,* 62: 151–161.

Ekstrand, S. 1994b. Assessment of forest damage with Landsat TM: correction for varying forest stand characteristics. *Rem. Sensing Environ.,* 47: 291–302.

Eldridge, N. R., and G. Edwards. 1993. Continuous tree class density surfaces derived from high resolution digital image analysis. Pages 947–952 in *Proc. GIS/LIS'93*. Vancouver, British Columbia.

Eliason, P., L. A. Soderblom, and P. S. Chavez. 1981. Extraction of topographic and spectral albedo function for multispectral images. *Photogramm. Eng. Rem. Sensing,* 48: 1571–1579.

Elkie, P. C., R. S. Rempel, and A. P. Carr. 1999. *Patch Analyst User's Manual: A Tool for Quantifying Landscape Structure.* NWST Tech. Manual TM-002. Northwest Science and Technology, Ministry of Natural Resources, Thunder Bay, Ontario.

Eng, M. 1998. Spatial patterns in forested landscapes: implications for biology and forestry. Pages 42–75 in J. Voller and S. Harrison, Eds. *Conservation Biology: Principles for Forested Landscapes.* UBC Press, Vancouver, British Columbia.

English, M. R., and V. H. Dale. 1999. Next steps for tools to aid environmental decision making. Pages 317–329 in V. H. Dale and M. R. English, Eds. *Tools to Aid Environmental Decision Makers*. Springer-Verlag, New York.

Erdle, T. 1998. Progress toward sustainable forest management: insight from the New Brunswick experience. *For. Chron.,* 74: 378–384.

Erdle, T., and M. Sullivan. 1998. Forest management design for contemporary forestry. *For. Chron.,* 74: 83–90.

Essery, C. I., and A. P. Morse. 1992. The impact of ozone and acid mist on the spectral reflectance of young Norway spruce trees. *Int. J. Rem. Sensing,* 13: 3045–3054.

Estes, J. E. 1985. The need for improved information systems. *Can. J. Rem. Sensing,* 11: 124–131.

Estes, J. E., and T. R. Loveland. 1999. Characteristics, sources and management of remotely-sensed data. Pages 667–675 in Longley, P. A., M. F. Goodchild, D. J. Maguire, and D. W. Rhind, Eds. *Geographical Information Systems*. 2nd ed. John Wiley & Sons, New York.

Estes, J. E., C. Sailor, and L. R. Tinney. 1986. Applications of AI techniques to remote sensing. *Prof. Geogr.,* 38: 133–141.

Estes, J. E., and J. L. Star. 1997. Research needed to improve remote sensing and GIS integration: conclusions and a look toward the future. Pages 176–203 in Star, J. L., J. E. Estes, and K. C. McGwire, Eds. *Integration of Geographic Information Systems and Remote Sensing*. Cambridge University Press, Cambridge, U.K.

Evans, I. S. 1972. General geomorphometry, derivatives of elevation, and descriptive statistics. Pages 17–90 in R. J. Chorley, Ed. *Spatial Analysis in Geomorphology.* Methuen, London.

Evans, I. S. 1980. An integrated system of terrain analysis and slope mapping. *Z. Geomorphol.,* 36: 274–295.

Everitt, J. H., D. E. Escobar, and J. Noriega. 1991. A high resolution multispectral video system. *Geocarto Int.,* 6: 45–51.

Expert Panel on Forest Management in Alberta. 1990. *Forest Management in Alberta: Report of the Expert Review Panel.* Alberta Energy, Forests, Lands and Wildlife, Edmonton.

Fabbri, A. G., K. Fung, and S. M. Yatabe. 1986. *The use of artificial intelligence in remote sensing: a review of applications and current research.* Presented Paper NATO Advanced Study Institute on Statistical Treatments for Estimation of Mineral and Energy Resources. Lucca, Italy, June 22–July 4, 1986.

Fang, J., G. Wang, G. Liu, and S. Xu. 1998. Forest biomass of China: an estimate based on the biomass-volume relationship. *Ecol. Appl.,* 8: 1084–1091.

Farquhar, G. D, S. von Caemmerer, and J. A. Berry. 1980. A biochemical model of photosynthetic CO_2 assimilation in leaves of C_3 species. *Planta,* 149: 78–79.

Fassnacht, K. S., S. T. Gower, M. D. MacKenzie, E. V. Nordheim, and T. M. Lillesand. 1997. Estimating the leaf area index of north central Wisconsin forests using Landsat Thematic Mapper. *Rem. Sensing Environ.,* 61: 229–245.

Fassnacht, K. S., S. T. Gower, J. M. Norman, and R. E. McMurtrie. 1994. A comparison of optical and direct methods for estimating foliage surface area index in forests. *Agric. For. Meteorol.,* 71: 183–207.

Fazakas, Z, and M. Nilsson. 1996. Volume and forest cover estimation over southern Sweden using AVHRR data calibrated with TM data. *Int. J. Rem. Sensing,* 17: 1701–1709.

Fazakas, Z., M. Nilsson, and H. Olsson. 1999. Regional forest biomass estimation by use of satellite data and ancillary data. *Agric. For. Meteorol.,* 98/99: 417–425.

Fedkiw, J., and J. H. Cayford. 1999. Forest management: a dynamic evolving profession. *For. Chron.,* 74: 213–217.

Fent, L., R. J. Hall, and R. K. Nesby. 1995. Aerial films for forest inventory: optimizing film parameters. *Photogramm. Eng. Rem. Sensing,* 61: 281–289.

Ferguson, I. S. 1996. *Sustainable Forest Management.* Oxford University Press, Melbourne, Australia.

Fiorella, M., and W. J. Ripple. 1993a. Determining the successional stage of temperate coniferous forest with Landsat satellite data. *Photogramm. Eng. Rem. Sensing,* 59: 239–246.

Fiorella, M., and W. J. Ripple. 1993b. Analysis of conifer forest regeneration using Landsat Thematic Mapper data. *Photogramm. Eng. Rem. Sensing,* 59: 1383–1388.

Fisher, P. 1997. The pixel — a snare and a delusion. *Int. J. Rem. Sensing,* 18: 679–685.

Fisher, P. F., and R. E. Lindenberg. 1989. On distinctions among cartography, remote sensing, and geographic information systems. *Photogramm. Eng. Rem. Sensing,* 55: 1431–1434.

Fjeldsa, J., D. Ehrlich, E. Lambin, and E. Prins. 1997. Are biodiversity "hotspots" correlated with current ecoclimatic stability? *Biodiversity and Conserv.,* 6: 401–422.

Fleming, M. D., and R. M. Hoffer. 1975. *Computer-Aided Analysis of Landsat-1 MSS Data: A Comparison of Three Approaches, Including a 'Modified Clustering' Approach.* Laboratory for Applications of Remote Sensing Information Note 072475. Purdue University, West Lafayette, Indiana.

Fleming, M. D., and R. M. Hoffer. 1979. Machine processing of Landsat MSS and DMA topographic data for forest cover type mapping. Pages 377–390 in *Proc. 1979 Symp. Machine Process. Rem. Sensed Data.* Laboratory for Applications of Remote Sensing, Purdue University, West Lafayette, Indiana.

Fletcher, R. A., J. McAlexander, and E. Hansen. 1998. STORA: the road to certification. Pages 14–1 to 14–18 in *Case Studies for the Business of Sustainable Forestry: A Project of the Sustainable Forestry Working Group.* The John D. and Catherine T. MacArthur Foundation, Chicago.

Florinsky, I. V., and G. A. Kuryakova. 1996. Influence of topography on some vegetation cover properties. *Catena,* 27: 123–141.

Fogel, D. N., and L. R. Tinney. 1996. *Image registration using multiquadratic functions, the finite element method, bivariate mapping polynomials and the thin plate spline.* NCGIA Tech. Rep. 96–1. University of California, Santa Barbara.

Foglein, J., and J. Kittler. 1983. The effect of pixel correlations on class separability. *Pattern Recognition Lett.,* 1: 401–407.

Food and Agriculture Organization. 1994a. *Readings in Sustainable Forest Management.* For. Pap. 122. FAO, Rome.

Food and Agriculture Organization. 1994b. *Forestry Education: New Trends and Prospects.* For. Pap. 123. FAO, Rome.

Food and Agriculture Organization. 1998. *Guidelines for the Management of Tropical Forests. 1. The Production of Wood.* For. Pap. 135. FAO, Rome.

Foody, G. M. 1986. An assessment of the topographic effect on SAR image tone. *Can. J. Rem. Sensing,* 12: 124–131.

Foody, G. M. 1988. The effects of viewing geometry on image classification. *Int. J. Rem. Sensing,* 9: 1909–1915.

Foody, G. M. 1996. Approaches for the production and evaluation of fuzzy land cover classifications from remotely sensed data. *Int. J. Rem. Sensing,* 17: 1317–1340.

Foody, G. M. 1999. The continuum of classification fuzziness in thematic mapping. *Photogramm. Eng. Rem. Sensing,* 65: 443–451.

Foody, G. M., and D. S. Boyd. 1999. Detection of partial land cover change associated with migration of inter-class transitional zones. *Int. J. Rem. Sensing,* 20: 2723–2740.

Foody, G., G. Palubinskas, R. M. Lucas, P. J. Curran, and M. Honzak. 1996. Identifying terrestrial Carbon sinks: classification of successional stages in regenerating tropical forests from Landsat Thematic Mapper data. *Rem. Sensing Environ.,* 55: 205–216.

Forman, R. T. T., and M. Godron. 1986. *Landscape Ecology,* John Wiley & Sons, New York.

Forman, R. T. T. 1987. The ethics of isolation, the spread of disturbance, and landscape ecology. Pages 213–229 in M. Turner, Ed. *Landscape Heterogeneity and Disturbance.* Springer-Verlag, New York.

Forman, R. T. T. 1995. *Land Mosaics: The Ecology of Landscapes and Regions.* Cambridge University Press, Cambridge, U.K.

Fournier, R. A., G. E. Edwards, and N. R. Eldridge. 1995. A catalogue of potential spatial discriminators for high spatial resolution digital images of individual tree crowns. *Can. J. Rem. Sensing,* 21: 285– 298.

Fournier, R. A., J. Luther, M. Wulder, C.H. Ung, S. Magnussen, L. Guindon, M.C. Lambert, and J. Beaubien. 1999. Mapping forest biomass from inventory and remotely sensed data. Pages 171–178 in Vol. II, *Proc. 21st Can. Symp. Rem. Sensing.* Canadian Aeronautics and Space Institute, Ottawa.

Frank, T. D. 1988. Mapping dominant vegetation communities in the Colorado Rocky Mountain Front Range with Landsat Thematic Mapper and digital terrain data. *Photogramm. Eng. Rem. Sensing,* 54: 1727–1734.

Franklin, J. 1986. Thematic Mapper analysis of conifer forest structure and composition. *Int. J. Rem. Sensing,* 7: 1287–1301.

Franklin, J. 1995. Predictive vegetation mapping: geographic modeling of biospatial patterns in relation to environmental gradients. *Prog. Phys. Geogr.,* 19: 474–499.

Franklin, J., and C. E. Woodcock. 1997. Multiscale vegetation data for the mountains of southern California: spatial and categorical resolution. Pages 141–168 in D. A. Quattrochi and M. F. Goodchild, Eds. *Scale in Remote Sensing and GIS.* CRC Press, Boca Raton.

Franklin, J. F. 1997. Ecosystem management: an overview. Pages 21–53 in Boyce, M. S., and A. Haney, Eds. *Ecosystem Management.* Yale University Press, New Haven, CT.

Franklin, J. F., and R. T. T. Forman. 1987. Creating landscape patterns by forest cutting: ecological consequences and principles. *Landscape Ecol.,* 1: 5–18.

Franklin, S. E. 1987a. Geomorphometric processing of digital elevation models. *Comput. Geosci.,* 13: 603–609.

Franklin, S. E. 1987b. Terrain analysis from digital patterns in geomorphometry and Landsat MSS spectral response. *Photogramm. Eng. Rem. Sensing,* 53: 59–65.

Franklin, S. E. 1991. Image transformations in mountainous terrain and the relationship to surface patterns. *Comput. Geosci.,* 17: 1137–1149.

Franklin, S. E. 1992. Satellite remote sensing of forest type and landcover classes in the Subalpine Region, Kananaskis Valley, Alberta. *Geocarto Int.,* 7: 25–35.

Franklin, S. E. 1994. Discrimination of subalpine forest species and canopy density using digital CASI, SPOT PLA and Landsat TM data. *Photogramm. Eng. Rem. Sensing,* 60: 1233–1241.

Franklin, S. E., and C.F. Blodgett. 1993. An example of satellite multisensor data fusion. *Comput. Geosci.,* 19: 577–583.

Franklin, S. E., D. R. Connery, and J. A. Williams. 1994. Classification of alpine vegetation using Landsat Thematic Mapper, SPOT HRV and DEM data. *Can. J. Rem. Sensing,* 20: 49–56.

Franklin, S. E., and P. T. Giles. 1995. Radiometric processing of aerial and satellite remote sensing imagery. *Comput. Geosci.,* 21: 413–425.

Franklin, S. E., R. T. Gillespie, B. D. Titus, and B. D. Pike. 1994. Aerial and satellite sensor detection of *Kalmia angustifolia* at forest regeneration sites in central Newfoundland. *Int. J. Rem. Sensing,* 15: 2553–2557.

Franklin, S. E., R. J. Hall, L. M. Moskal, A. J. Maudie, and M. B. Lavigne. 2000a. Incorporating texture into classification of forest species composition from airborne multispectral images. *Int. J. Rem. Sensing,* 21: 61–79.

Franklin, S. E., M. B. Lavigne, M. J. Deuling, M. A. Wulder, E. R. Hunt, Jr. 1997a. Landsat TM-derived forest covertypes for use in ecosystem models of net primary production. *Can. J. Rem. Sensing,* 23: 91–99.

Franklin, S. E., M. B. Lavigne, M. J. Deuling, M. A. Wulder, E. R. Hunt, Jr. 1997b. Estimation of forest leaf area index using remote sensing and GIS data for modelling net primary production. *Int. J. Rem. Sensing,* 18: 3459–3471.

Franklin, S. E., M. B. Lavigne, B. A. Wilson, and E. R. Hunt, Jr. 1994. Empirical relations between balsam fir (*Abies balsamea*) forest stand conditions and ERS-1 data in western Newfoundland. *Can. J. Rem. Sensing,* 20: 124–130.

Franklin, S. E., M. B. Lavigne, B. A. Wilson, E. R. Hunt, Jr., D. R. Peddle, G. J. McDermid, and P. T. Giles. 1995a. Topographic dependence of synthetic aperture radar imagery. *Comput. Geosci.,* 21: 521–532.

Franklin, S. E., and J. E. Luther. 1995. Satellite remote sensing of balsam fir forest structure, growth, and cumulative defoliation. *Can. J. Rem. Sensing,* 21: 400–411.

Franklin, S. E., and G. McDermid. 1993. Empirical relations between digital SPOT HRV and casi imagery and lodgepole pine (*Pinus contorta*) forest stand parameters. *Int. J. Rem. Sensing,* 14: 2331–2348.

Franklin, S. E., L. M. Moskal, M. B. Lavigne, and K. Pugh. 2000b. Interpretation and classification of partially harvested forest stands in the Fundy Model Forest using multitemporal Landsat TM data. *Can. J. Rem. Sensing,* 26: 318–333.

Franklin, S. E., and D. R. Peddle. 1987. Texture analysis of digital image data using spatial co-occurrence. *Comput. Geosci.,* 13: 293–311.

Franklin, S. E., and D. R. Peddle. 1990. Classification of SPOT HRV imagery and texture features. *Int. J. Rem. Sensing,* 11: 551–556.

Franklin, S. E., D. R. Peddle, B. A. Wilson, and C. F. Blodgett. 1991. Pixel sampling of remotely sensed digital imagery. *Comput. Geosci.,* 17: 759–775.

Franklin, S. E., and A. Raske. 1994. Satellite remote sensing of spruce budworm defoliation in western Newfoundland. *Can. J. Rem. Sensing,* 20: 37–48.

Franklin, S. E., R. H. Waring, R. W. McCreight, W. B. Cohen, and M. Fiorella. 1995b. Aerial and satellite sensor detection and classification of western spruce budworm defoliation in a subalpine forest. *Can. J. Rem. Sensing,* 21: 299–308.

Franklin, S. E., and B. A. Wilson. 1991a. Spatial and spectral classification methods in remote sensing. *Comput. Geosci.,* 17: 1151–1172.

Franklin, S. E., and B. A. Wilson. 1991b. Vegetation mapping and change detection using SPOT MLA and Landsat TM imagery in Kluane National Park. *Can. J. Rem. Sensing,* 17: 2–17.

Franklin, S. E., and B. A. Wilson. 1992. A three-stage classifier for remote sensing of mountain environments. *Photogramm. Eng. Rem. Sensing,* 58: 449–454.

Franklin, S. E., M. A. Wulder, and M. B. Lavigne. 1996. Automated derivation of geographic window sizes for remote sensing digital image texture analysis. *Comput. Geosci.,* 22: 665–673.

Fransson, J. E. S., F. Walter, and H. Olsson. 1999. Identification of clear-felled areas using SPOT P and Almaz-1 SAR data. *Int. J. Rem. Sensing,* 20: 3583–3593.

Fransson, J. E. S., F. Walter, and L. M. H. Ulander. 2000. Estimation of forest parameters using CARABAS-II VHF SAR data. *IEEE Trans. Geosci. Rem. Sensing,* 38: 720–727.

Frazer, G. W., J. A. Trofymow, and K. P. Lertzman. 1997. *A method for estimating canopy openness, effective leaf area index, and photosynthetically active photon flux density using hemispherical photography and computerized image analysis techniques.* Natural Resources Canada, Inf. Rep. BC-X-373. Pacific Forestry Centre, Victoria, British Columbia.

Friedl, M. A., K. C. McGwire, and D. K. McIver. 2000. An overview of uncertainty in optical remotely sensed data for ecological applications. in C. Hunsaker, M. Goodchild, M. A. Friedl, and T. Case, Eds. *Spatial Data Uncertainty in Ecology.* Springer-Verlag, New York.

Friend, A. D., A. K. Stevens, R. G. Knox, and M. G. R. Cannell. 1997. A process-based, terrestrial biosphere model of ecosystem dynamics (Hybrid v3.0). *Ecol. Model.,* 95: 249–287.

Fritz, L. W. 1996. The era of commercial Earth observation satellites. *Photogramm. Eng. Rem. Sensing,* 62: 39–45.

Frohn, E. 1998. *Remote Sensing for Landscape Ecology.* CRC Press, Boca Raton, Florida.

Fu, K. S. 1976. Pattern recognition in remote sensing of the Earth's resources. *IEEE Trans. Geosci. Electron.,* 14: 10–18.

Fung, T., and E. LeDrew. 1987. Application of principal components analysis to change detection. *Photogramm. Eng. Rem. Sensing,* 53: 1649–1658.

Fung, T., and E. LeDrew. 1988. The determination of optimal threshold levels for change detecting using various accuracy indices. *Photogramm. Eng. Rem. Sensing,* 54: 1449–1454.

Gahegan, M., and J. Flack. 1996. A model to support the integration of image understanding techniques within a GIS. *Photogramm. Eng. Rem. Sensing,* 62: 483–490.

Gao, J. 1999. A comparative study on spatial and spectral resolutions of satellite data in mapping mangrove forests. *Int. J. Rem. Sensing,* 20: 2823–2833.

Gaston, G. G., P. M. Bradley, T. S. Vinson, and T. P. Kolchugina. 1997. Forest ecosystem modeling in the Russian Far East using vegetation and landcover identified by classification of GVI. *Photogramm. Eng. Rem. Sensing,* 63: 51–58.

Gausman, H. 1977. Reflectance of leaf components. *Rem. Sensing Environ.,* 6: 1–9.

Gemmell, F. M. 1995. Effects of forest cover, terrain, and scale on timber volume estimation with Thematic Mapper data in a Rocky Mountain site. *Rem. Sensing Environ.,* 51: 291–305.

Gemmell, F. M. 1998. An investigation of terrain effects on the inversion of a reflectance model. *Rem. Sensing Environ.,* 65: 155–169.

Gemmell, F. M. 2000. Testing the utility of multi-angle spectral data for reducing the effects of background spectral variations in forest reflectance model inversion. *Rem. Sensing Environ.,* 72: 46–63.

Gemmell, F. M., and A. J. McDonald. 2000. View zenith angle effects on the forest information content of three spectral indices. *Rem. Sensing Environ.,* 72: 139–158.

Gerard, F. F., and P. R. North. 1997. Analyzing the effect of structual variability and canopy gaps on forest BRDF using a geometric optical model. *Rem. Sensing Environ.,* 62: 46–62.

Gerstl, S. A. W. 1990. Physics concepts of optical and radar reflectance signatures. *Int. J. Rem. Sensing,* 11: 1109–1117.

Gerstl, S. A. W., and C. C. Borel-Donohue. 1992. Principles of the radiosity method vs. radiative transfer for canopy reflectance modelling. *IEEE Trans. Geosci. Rem. Sensing,* 30: 271–275.

Gerstl, S. A. W., and C. Simmer. 1986. Radiation physics and modelling for off-nadir satellite-sensing of non-Lambertian surfaces. *Rem. Sensing Environ.,* 20: 1–129.

Gerylo, G., S. E. Franklin, R. J. Hall, S. Gooderham, and L. Gallagher. 2000. Modelling forest stand parameters from Landsat Thematic Mapper data. Pages 405–414 in *Proc. 22nd Can. Symp. Rem. Sensing.* Canadian Aeronautics and Space Institute, Ottawa.

Gerylo, G., R. J. Hall, S. E. Franklin, A. Roberts, and E. J. Milton. 1998. Hierarchical image classification and extraction of forest species composition and crown closure from airborne multispectral images. *Can. J. Rem. Sensing,* 24: 219–232.

Ghitter, G. S., R. J. Hall, and S. E. Franklin. 1995a. Variability of Landsat Thematic Mapper data in boreal deciduous and mixed-wood stands with conifer understory. *Int. J. Rem. Sensing,* 16: 2989–3002.

Ghitter, G. S., R. J. Hall and S. E. Franklin. 1995b. Identifying class structure of white spruce understory beneath deciduous or mixedwood stands for improved classification results. Pages 643–649 in *Proc. XVII Can. Symp. Rem. Sensing*. Canadian Aeronautics and Space Institute, Ottawa.

Gholz, H. L., K. Nakane, and H. Shimoda, Eds. 1997. *The Use of Remote Sensing in the Modeling of Forest Productivity.* Kluwer, Dordrecht.

Gibson, D. W. 1999. Conversion is out, measurement is in — are we beginning the surveying and mapping era of GIS? *Surv. Land Inf. Syst.,* 59: 69–72.

Giles, P.T., and S. E. Franklin. 1998. An automated approach to the classification of slope units using digital data. *Geomorphology,* 21: 251–264.

Gillis, M. D., and D. G. Leckie. 1993. *Forest Inventory Mapping Procedures Across Canada.* Petawawa National Forestry Institute, Inf. Rep. PI-X-114. Canadian Forest Service, Ottawa.

Gillis, M. D., and D. G. Leckie. 1996. Forest inventory update in Canada. *For. Chron.,* 72: 138–156.

Glackin, D. L. 1998. International space-based remote sensing overview: 1980–2007. *Can. J. Rem. Sensing,* 24: 307–314.

GOFC Design Team. 1998. *A Strategy for Global Observation of Forest Cover.* Draft Rep. Ver. 1.2. Committee for Earth Observation Satellites, Washington, DC.

Goldberg, M., M. Alvo, and G. Karam. 1983. The analysis of Landsat imagery using an expert system: forestry applications. Pages 493–503 in *Proc. Autocarto Six*. International Cartographic Association, Ottawa, Canada.

Goldberg, M., D. G. Goodenough, M. Alvo, and G. Karam. 1985. A hierarchical expert system for updating forestry maps with Landsat data. *Proc. IEEE,* 73: 1054–1063.

Gong, P., G. S. Biging, S. M. Lee, X. Mei, Y. Sheng, R. Pu, B. Xu, K. P. Schwarz, and M. Mostafa. 1999. Photo ecometrics for forest inventory. *Geogr. Inf. Sci.,* 5: 9–14.

Gong, P., and P. J. Howarth. 1990. An assessment of some factors influencing multispectral land-cover classification. *Photogramm. Eng. Rem. Sensing,* 56: 597–603.

Gong, P., E. F. LeDrew, and J. R. Miller. 1992. Registration-noise reduction in difference images for change detection. *Int. J. Rem. Sensing,* 13: 773–779.

Gong, P., R. Pu, and J. Chen. 1996. Mapping ecological land systems and classification uncertainties from digital elevation and forest-cover data using neural networks. *Photogramm. Eng. Rem. Sensing,* 62: 1249–1260.

Gong, P., R. Pu, and J. R. Miller. 1992. Correlating leaf area index of Ponderosa Pine with hyperspectral CASI data. *Can. J. Rem. Sensing,* 18: 275–282.

Gong, P., R. Pu, and J. R. Miller. 1995. Coniferous forest leaf area index estimation along the Oregon Transect using Compact Airborne Spectrographic Imager data. *Photogramm. Eng. Rem. Sensing,* 61: 1107–1117.

Gong, P., R. Pu, and B. Yu. 1997. Conifer species recognition: an exploratory analysis of *in situ* hyperspectral data. *Rem. Sensing Environ.,* 62: 189–200.

Gonzalez-Rebeles, C., V. J. Burke, M. D. Jennings, G. Cebanos, and N. C. Parker. 1998. Transnational gap analysis of the Rio Bravo/Rio Grande region. *Photogramm. Eng. Rem. Sensing,* 64: 1115–1118.

Goodchild, M. F. 1992. Geographical information science. *Int. J. Geogr. Inf. Syst.,* 6: 31–45.

Goodchild, M. F. 1999. Multiple roles for GIS in global climate research. Pages 277–295 in M. Craglia and H. Onsrud, Eds. *Geographic Information Research: A Trans-Atlantic Perspective.* Taylor & Francis, London.

Goodchild, M. F., and J. D. Proctor. 1997. Scale in a digital geographic world. *Geogr. Environ. Model.,* 1: 5–23.

Goodenough, D. G. 1988. Thematic Mapper and SPOT integration with geographic information systems. *Photogramm. Eng. Rem. Sensing,* 54: 167–176.

Goodenough, D. G., A. S. Bhogal, R. Fournier, R. J. Hall, J. Iisaka, D. Leckie, J. E. Luther, S. Magnussen, O. Niemann, and W. M. Strome. 1998. Earth observation for sustainable development of forests (EOSD). Pages 57–60 in *Proc. 20th Can. Symp. Rem. Sensing.* Canadian Aeronautics and Space Institute, Ottawa.

Goodenough, D. G., D. Charlebois, and S. Matwin. 1994. Automating reuse of software for expert systems analysis of remote sensing data. *IEEE Trans. Geosci. Rem. Sensing,* 32: 525–533.

Goodenough, D. G., P. M. Narenda, and K. O'Neill. 1978. Feature subset selection in remote sensing. *Can. J. Rem. Sensing,* 4: 143–148.

Gougeon, F. 1995. A crown-following approach to the automatic delineation of individual tree crowns in high spatial resolution aerial images. *Can. J. Rem. Sensing,* 21: 274–284.

Gougeon, F. 1997. Recognizing the forest from the trees: individual tree crown delineation, classification and regrouping for inventory purposes. Pages 807–814 in Volume II, *Proc. 3rd Int. Airborne Rem. Sensing Conf. Exhibition.* Environmental Research Institute of Michigan, Ann Arbor.

Gower, S. T., C. J. Kucharik, and J. M. Norman 1999. Direct and indirect estimation of LAI, FAPAR and NPP for terrestrial ecosystems. *Rem. Sensing Environ.,* 70: 29–51.

Graetz, R. D. 1990. Remote sensing of terrestrial ecosystem structure: an ecologist's pragmatic view. Pages 5–30 in R. J. Hobbs and H. A. Mooney, Eds. *Remote Sensing of Biosphere Functioning.* Springer-Verlag, New York.

Graham, L. A., and C. Gallion. 1996. Image processing under Windows NT: a comparative review. *GIS World,* September 9: 36–44.

Graham, L. N., K. Ellison, and C.S. Riddell. 1997. The architecture of a softcopy photogrammetry system. *Photogramm. Eng. Rem. Sensing,* 63: 1013–1020.

Graham, R., and R. E. Read. 1986. *Manual of Aerial Photography.* Focal Press, London.

Green, K. 1999. Development of the spatial domain in resource management. Pages 1–15 in S. Morain, Ed. *GIS Solutions in Natural Resource Management: Balancing the Technical-Political Equation.* Onword Press, Santa Fe, NM.

Green, K. 2000. Selecting and interpreting high-resolution images. *J. For.,* 98: 37–39.

Greenfield, P. H. 2000. Digital imaging basics for natural resources management. *J. For.,* 98: 21–23.

Gregersen, H., A. Lundgren, and N. Byron. 1998. Forestry for sustainable development: making it happen. *J. For.,* 96: 6–10.

Gregory, A. F. 1971. Earth-observation satellites: a potential impetus for economic and social development. *World Cartography,* XI: 1–15.

Gregory, A. F. 1972. What do we mean by remote sensing? Pages 33–37 in *Proc. 1st Can. Symp. Rem. Sensing.* Canadian Aeronautics and Space Institute, Ottawa.

Gregory, A. F., and H. D. Moore. 1986. Thematic mapping from Landsat and collateral data: a review of one company's experience and a forecast of future potential. *Can. J. Rem. Sensing,* 12: 55–63.

Griffiths, G. H., J. Lee, and B. C. Eversham. 2000. Landscape pattern and species richness; regional scale analysis from remote sensing. *Int. J. Rem. Sensing,* 21: 2685–2704.

Grushecky, S. T., and M. A. Fajvan. 1999. Comparison of hardwood stand structure after partial harvesting using intensive canopy maps and geostatistical techniques. *For. Ecol. Manage.,* 114: 421–432.

Gu, D., and A. Gillespie. 1998. Topographic normalization of Landsat TM images of forest based on subpixel sun-canopy-sensor geometry. *Rem. Sensing Environ.,* 64: 166–175.

Gu, X. F., and G. Guyot. 1993. Effect of diffuse irradiance on the reflectance factor of reference panels under field conditions. *Rem. Sensing Environ.,* 45: 249–260.

Guindon, B. 1997. Computer-based aerial image understanding: a review and assessment of its application to planimetric information extraction from very high resolution satellite images. *Can. J. Rem. Sensing,* 23: 38–47.

Guindon, B. 2000. A framework for the development and assessment of object recognition modules from high-resolution satellite images. *Can. J. Rem. Sensing,* 26: 334–348.

Gurney, C. M. 1981. The use of contextual information to improve land cover classification of digital remotely sensed data. *Int. J. Rem. Sensing,* 2: 379–388.

Guruswamy, L. D., and J. A. McNeely, Eds. 1998. *Protection of Global Biodiversity: Converging Strategies.* Duke University Press, Durham, NC.

Guyot, G., D. Guyon, and J. Riom. 1989. Factors affecting the spectral response of forest canopies: a review. *Geocarto Int.,* 3: 3–18.

Hagner, O. 1990. Computer aided forest stand delineation and inventory based on satellite remote sensing. Pages 94–105 in R. Sylvander, Ed. *SNS/IUFRO Workshop on the Usability of Remote Sensing for Forest Inventory and Planning.* Umeå, Sweden.

Haines-Young, R., and M. Chopping. 1996. Quantifying landscape structure: a review of landscape indices and their application to forested landscapes. *Prog. Phys. Geogr.,* 20: 418–445.

Haines-Young, R., D. R. Green, and S. H. Cousins, Eds. 1993. *Landscape Ecology and GIS.* Taylor and Francis, London.

Hall, F. G., D. B. Botkin, D. E. Strebel, K. D. Woods, and S. J. Goetz. 1991a. Large-scale patterns of forest succession as determined by remote sensing. *Ecology,* 72: 628–640.

Hall, F. G., D. R. Peddle, and E. F. LeDrew. 1996. Remote sensing of biophysical variables in boreal forest stands of *Picea mariana. Int. J. Rem. Sensing,* 17: 3077–3081.

Hall, F. G., D. E. Strebel, J. E. Nickerson, and S. J. Goetz. 1991b. Radiometric rectification: toward a common radiometric response among multidate, multisensor images. *Rem. Sensing Environ.,* 35: 11–27.

Hall, F. G., J. R. Townshend, and E. T. Engman. 1995. Status of remote sensing algorithms for estimation of land surface state parameters. *Rem. Sensing Environ.,* 51: 138–156.

Hall, J. P. 1999. Remote sensing and criteria and indicators of sustainable forest management. Pages 367–374 in *Proc., Int. Forum Automated Interpretation High Spatial Resolution Digital Imagery For.* Canadian Forest Service, Pacific Forestry Centre, Victoria.

Hall, R. J., and A. H. Aldred. 1992. Forest regeneration appraisal with large-scale aerial photographs. *For. Chron.,* 68: 142–150.

Hall, R. J., P. H. Crown, and S. J. Titus. 1984. Change detection methodology for aspen defoliation with Landsat MSS digital data. *Can. J. Rem. Sensing,* 10: 135–142.

Hall, R. J., R. V. Dams, and L. N. Lyseng. 1991c. Forest cutover mapping from SPOT satellite data. *Int. J. Rem. Sensing,* 12: 2193–2204.

Hall, R. J., and L. Fent. 1996. Influence of aerial film spectral sensitivity and texture on interpreting images of forest species composition. *Can. J. Rem. Sensing,* 22: 350–359.

Hall, R. J., S. E. Franklin, and G. R. Gerylo. 1998. Estimation of stand volume from high resolution multispectral imagery. Pages 191–196 in *Proc. 20th Can. Symp. Rem. Sensing.* Canadian Aeronautics and Space Institute, Ottawa.

Hall, R. J., and D. L. Klita. 1997. Remote sensing — GIS integration: progress towards defining a conifer understory classification system for use with Landsat TM data. CD-ROM Pap. No. 63, in *Proc. 19th Can. Symp. Rem. Sensing.* Canadian Aeronautics and Space Institute, Ottawa.

Hall, R. J., A. R. Kruger, J. Scheffer, S. J. Titus, and W. C. Moore. 1989a. A statistical evaluation of Landsat TM and MSS data for mapping forest cutovers. *For. Chron.,* 65: 441–449.

Hall, R. J., R. Morton, and R. Nesby. 1989b. A comparison of existing models for dbh estimation from large-scale photos. *For. Chron.,* 65: 114–120.

Hall, R. J., D. R. Peddle, and D. L. Klita. 2000a. Mapping conifer understory within boreal mixedwoods from Landsat TM satellite imagery and forest inventory information. *For. Chron.,* 76: in press.

Hall, R. J., G. N. Still, and P. H. Crown. 1983. Mapping the distribution of aspen defoliation using Landsat colour composites. *Can. J. Rem. Sensing,* 9: 86–91.

Hall, R. J., S. J. Titus, and W. J. A. Volney. 1993. Estimating top-kill volumes with large-scale photos on trees defoliated by the jack pine budworm. *Can. J. For. Res.,* 23: 1337–1346.

Hall, R. J., N. Walsworth, T. Balce, and K. Dutchak. 2000b. Cutblock update with high resolution satellite images. Pages 673–678 in *Proc. 22nd Can. Symp. Rem. Sensing.* Canadian Aeronautics and Space Institute, Ottawa.

Häme, T., I. Heiler, and J. S. Miquel-Ayanz. 1998. An unsupervised change detection and recognition system for forestry. *Int. J. Rem. Sensing,* 19: 1079–1099.

Hammer, R. D. 1998. Space and time in the soil landscape: the ill-defined ecological universe. Pages 105–140 in D. L. Peterson and V. T. Parker, Eds. *Ecological Scale: Theory and Applications.* Columbia University Press, New York.

Hansen, A. J., T. A. Spies, F. J. Swanson, and J. L. Ohmann. 1991. Conserving biodiversity in managed forests. *BioScience,* 41: 382–392.

Hansen, E., R. Fletcher, and J. McAlexander. 1998. Sustainable forestry, Swedish style for Europe's greening market. *J. For.,* 96: 38–43.

Hansen, M., R. Dubayah, and R. DeFries. 1996. Classification trees: an alternative to traditional land cover classifiers. *Int. J. Rem. Sensing,* 17: 1075–1081.

Hansen, M. J., S. E. Franklin, and C. Woudsma. 2001a. Caribou habitat classification and fragmentation analysis of old-growth cedar/hemlock forests in British Columbia using Landsat TM and GIS data. *Rem. Sensing Environ.,* in press.

Hansen, M. J., S. E. Franklin, C. Woudsma, and M. Peterson. 2001b. Forest structure classification in the North Columbia Mountains using the Landsat TM Tasseled Cap Wetness Component. *Can. J. Rem. Sensing,* 27(1): 20–32.

Haralick, R. M. 1986. Statistical image texture analysis. Pages 247–279 in T. Y. Young and K. S. Fu, Eds. *Handbook of Pattern Recognition and Image Processing*. Academic Press, New York.

Haralick, R. M., and H. Joo. 1986. A context classifier. *IEEE Trans. Geosci. Rem. Sensing,* 24: 997–1007.

Haralick, R. M., K. Shanmugam, and I. Dinstein. 1973. Texture features for image classification. *IEEE Trans. Syst. Man Cybern.,* 3: 610–621.

Haralick, R. M., and L. G. Shapiro. 1992. *Computer and Robot Vision.* Addison-Wesley, Reading, MA.

Haralick, R. M., S. R. Sternberg, and X. Zhuang. 1987. Image analysis using mathematical morphology. *IEEE Trans. Pattern Anal. Machine Intelligence,* 9: 532–550.

Hardy, E. C., and R. E. Burgan. 1999. Evaluation of NDVI for monitoring live moisture in three vegetation types of the western USA. *Photogramm. Eng. Rem. Sensing,* 65: 603–610.

Hargis, C. D., J. A. Bissonette, and J. L. David. 1998. The behavior of landscape metrics commonly used in the study of habitat fragmentation. *Landscape Ecol.,* 13: 167–186.

Hargis, C. D., J. A. Bissonette, and D. L. Turner. 1999. The influence of forest fragmentation and landscape pattern on American martens. *J. Appl. Ecol.,* 36: 157–172.

Haring, L. L., J. F. Lounsbury, and J. W. Frazier. 1992. *Introduction to Scientific Geographic Research.* 4th ed. Wm. C. Brown Publishers, Dubuque, Iowa.

Harris, L. D. 1984. *The Fragmented Forest: Island Biogeography Theory and the Preservation of Biotic Diversity.* University of Chicago Press, Chicago.

Hay, G. J., K. O. Niemann, and G. F. McLean. 1996. An object-specific image texture analysis of H-resolution forest imagery. *Rem. Sensing Environ.,* 55: 108–122.

Hayes, L., and A. P. Cracknell. 1987. Georeferencing and registering satellite data for monitoring vegetation over large areas. *Pattern Recognition Lett.,* 5: 95–105.

He, D. C., and L. Wang. 1992. Unsupervised textural classification of images using the texture spectrum. *Pattern Recognition,* 25: 247–255.

He, H. S., D. J. Mladenoff, V. C. Radeloff, and T. R. Crow. 1998. Integration of GIS data and classified satellite imagery for regional forest assessment. *Ecol. Appl.,* 8: 1072–1083.

Heath, G. R. 1956. A comparison of two basic theories of land classification and their adaptability to regional photointerpretation key techniques. *Photogramm. Eng.,* 22: 144–168.

Heath, G. R. 1974. ERTS data tested for forestry application. *Photogramm. Eng.,* 40: 1087–1091.

Hebda, R. J. 1997. Impact of climate change on biogeoclimatic zones of British Columbia and Yukon. Pages 13.1–13.15 in E. Taylor and B. Taylor, Eds. *Responding to Climate Change in British Columbia and Yukon: Vol. 1.* Environment Canada, Ottawa.

Hebda, R. J. 1998. Atmospheric change, forests and biodiversity. *Environ. Monit. Assess.,* 49: 195–212.

Hegyi, F., P. Pilon, and P. A. Walker. 1992. Replacing aerial photos in resource inventories with airborne digital data and GIS. Pages 510–517 *in Proc. ASPRS/ACSM/RT92 Tech. Pap. Vol. 5.* American Society for Photogrammetry and Remote Sensing, Bethesda, MD.

Hegyi, F., and R. V. Quesnet. 1983. Integration of remote sensing and computer assisted mapping technology in forestry. *Can. J. Rem. Sensing,* 9: 92–98.

Henderson, F. M., and A. J. Lewis, Eds. 1998. *Principles and Applications of Imaging Radar.* Vol. 2, John Wiley & Sons, New York.

Henderson, P., and S. Tanimoto. 1974. Considerations for efficient picture output via lineprinter. *Comput. Graphics Image Process.,* 3: 327–335.

Herwitz, S. R., R. E. Slye, and S. M. Turton. 1998. Co-registered aerial stereopairs from low-flying aircraft for the analysis of long-term tropical rainforest canopy dynamics. *Photogramm. Eng. Rem. Sensing,* 64: 397–405.

Hill, D.A., and D. G. Leckie, Eds. 1999. *International Forum on Automated Interpretation of High Spatial Resolution Digital Imagery for Forestry.* Natural Resources Canada, Canadian Forest Service, Victoria, British Columbia.

Hinse, M., Q. H. J. Gwyn, and F. Bonn. 1988. Radiometric correction of c-band imagery for topographic effects in regions of moderate relief. *IEEE Trans. Geosci. Rem. Sensing,* 26: 122–132.

Hobbs, R. J. 1998. Managing ecological systems and processes. Pages 459–484 in Peterson, D. L. and V. T. Parker, Eds. *Ecological Scale: Theory and Applications.* Columbia University Press, New York.

Hoffer, R. M. 1978. Biological and physical considerations in applying computer-aided analysis techniques to remote sensor data. Pages 227–289 in P. H. Swain and S. M. Davis, Eds. *Remote Sensing: The Quantitative Approach.* McGraw-Hill, New York.

Hohl, P., Ed. 1998. *GIS Data Conversion: Strategies, Techniques and Management.* Onword Press, Sante Fe, NM.

Holben, B. N., and C. O. Justice. 1980. The topographic effect on spectral response from nadir-pointing sensors. *Photogramm. Eng. Rem. Sensing,* 46: 1191–1200.

Holmgren, J., S. Joyce, M. Nilsson, and H. Olsson. 1999. Estimating stem volume and basal area in forest compartments by combining satellite image data with field data. *Scand. J. For. Res.,* 15: 103–111.

Holmgren, P., and T. Thuresson. 1997. Applying objectively estimated and spatially continuous forest parameters in tactical planning to obtain dynamic treatment units. *For. Sci.,* 43: 317–326.

Holmgren, P., and T. Thuresson. 1998. Satellite remote sensing for forestry planning — a review. *Scand. J. For. Res.,* 13: 90–110.

Holopainen, M., and G. Wang. 1998. The calibration of digital aerial photographs for forest stratification. *Int. J. Rem. Sensing,* 19: 677–696.

Homer, C. G., R. D. Ramsey, T. C. Edwards, Jr., and A. Falconer. 1997. Landscape cover-type modeling using a multi-scene Thematic Mapper mosaic. *Photogramm. Eng. Rem. Sensing,* 63: 59–67.

Hoque, E., P. J. S. Hutzler, and H. Hiendl. 1992. Reflectance, color, and histological features as parameters for the early assessment of forest damages. *Can. J. Rem. Sensing,* 18: 105–110.

Hord, R. M., and W. Brooner. 1976. Land-use map accuracy criteria. *Photogramm. Eng. Rem. Sensing,* 42: 671–677.

Horler, D. N. H., and F. J. Ahern. 1986. Forestry information content of Thematic Mapper data. *Int. J. Rem. Sensing,* 7: 405–428.

Horler, D. N. H., M. Dockay, and J. Barber. 1983. The red-edge of plant leaf reflectance. *Int. J. Rem. Sensing,* 4: 273–288.

Host, G. E., K. S. Pregitzer, C. W. Ramm, J. B. Hart, and D. T. Cleland. 1987. Landform-mediated differences in successional pathways among upland forest ecosystems in northwestern lower Michigan. *Forest Science* 33: 445–457.

Houghton, J. T., G. J. Jenkins, and J. J. Ephraums, Eds. 1990. *Climate Change — The IPCC Scientific Assessment.* Cambridge University Press, Cambridge, U.K.

Houston, D. R., and H. T. Valentine. 1977. Comparing and predicting forest stand susceptibility to gypsy moth. *Can. J. For. Res.,* 7: 447–461.

Howard, J. A. 1991. *Remote Sensing of Forest Resources.* Chapman and Hall, London.

Howarth, P. J., and G. M. Wickware. 1981. Procedures for change detection using Landsat digital data. *Int. J. Rem. Sensing,* 2: 277–291.

Huber, T. P., and K. E. Casler. 1990. Initial analysis of Landsat TM data for elk habitat monitoring. *Int. J. Rem. Sensing,* 11: 907–812.

Hudak, J. 1991. Integrated pest management and eastern spruce budworm. *For. Ecol. Manage.,* 39: 313–337.

Hudson, W. D. 1991. Photo interpretation of montane forests in the Dominican Republic. *Photogramm. Eng. Rem. Sensing,* 57: 79–84.

Huemmrich, K., and S. Goward. 1997. Vegetation canopy PAR absorption and NDVI: an assessment for ten tree species with the SAIL model. *Rem. Sensing Environ.,* 61: 254–269.

Huete, A. R. 1988. A soil-adjusted vegetation index (SAVI). *Rem. Sensing Environ.,* 25: 295–309.

Hughes, J. S., D. L. Evans, and P. Y. Burns. 1986. Identification of two southern pine species in high resolution aerial MSS data. *Photogramm. Eng. Rem. Sensing,* 52: 1175–1180.

Hugli, H., and W. Frei. 1983. Understanding anisotropic reflectance in mountainous terrain. *Photogramm. Eng. Rem. Sensing,* 49: 671–683.

Huguenin, R. L., M. A. Karaska, D. Van Blaricom, and J. R. Jensen. 1997. Subpixel classification of bald cypress and tupelo gum trees in Thematic Mapper imagery. *Photogramm. Eng. Rem. Sensing,* 63: 717–725.

Hunt, E. R., Jr. 1991. Airborne remote sensing of canopy water thickness scaled from leaf spectrometer data. *Int. J. Rem. Sensing,* 12: 643–649.

Hunt, E. R., Jr. 1994. Relationship between woody biomass and PAR conversion efficiency for estimating net primary production from NDVI. *Int. J. Rem. Sensing,* 15: 1725–1730.

Hunt, E. R., Jr., M. B. Lavigne, and S. E. Franklin. 1999. Factors controlling the decline of growth efficiency and net primary production for balsam fir in Newfoundland assessed using an ecosystem process model. *Ecol. Model.,* 122: 151–164.

Hunt, E. R., Jr., F. C. Martin, and S. W. Running. 1991. Simulating the effects of climatic variation on stem carbon accumulation of a ponderosa pine stand: comparison with annual growth increment data. *Tree Physiol.,* 9: 161–171.

Hunt, E. R., Jr., S. C. Piper, R. Nemani, C. D. Keeling, R. D. Otto, and S. W. Running. 1996. Global net carbon exchange and intra-annual atmospheric CO_2 concentration predicted by an ecosystem process model and three-dimensional atmospheric transport model. *Global Biogeochem. Cycles,* 10: 431–456.

Hunt, E. R., Jr., and S. W. Running. 1992. Simulated dry matter yields for aspen and spruce stands in the North American boreal forest. *Can. J. Rem. Sensing,* 18: 126–133.

Hunter, M., Jr., 1997. The biological landscape. Pages 57–68 in Kohm, K. and J. Franklin, Eds. *Creating a Forestry for the 21st Century. The Science of Ecosystem Management.* Island Press, Washington, DC.

Hussin, Y. A., R. M. Reich, and R. M. Hoffer. 1991. Estimating slash pine biomass using radar backscatter. *IEEE Trans. Geosci. Rem. Sensing,* 29: 427–431.

Hutchinson, C. F. 1982. Techniques for combining Landsat and ancillary data for digital classification improvement. *Photogramm. Eng. Rem. Sensing,* 48: 123–130.

Hyde, R. F., and N. J. Vesper. 1983. Some effects of resolution cell size on image quality. *Landsat Data Users Notes,* 29: 9–12.

Hyman, A. H. 1996. Information presentation for new sensors: a focus on selected sensors of the Earth Observing System (EOS). *Prog. Phys. Geogr.,* 20: 146–158.

Hyyppä, J., J. Pulliainen, M. Hallikainen, and A. Saatsi. 1997. Radar-derived standwise forest inventory. *IEEE Trans. Geosci. Rem. Sensing,* 35: 392–404.

Imhoff, M. L. 1995. A theoretical analysis of the effect of forest structure on synthetic aperture radar backscatter and the remote sensing of biomass. *IEEE Trans. Geosci. Rem. Sensing,* 33: 341–352.

Innes, J. L. 1992. Forest decline. *Prog. Phys. Geogr.,* 16: 1–64.

Innes, J. L. 1993. *Forest Health: Its Assessment and Status.* CAB International, Wallingford, U.K.

Innes, J. L., and R. C. Boswell. 1990. Reliability, presentation, and relationships among data from inventories of forest condition. *Can. J. For. Res.,* 20: 790–799.

Inoue, Y., S. Morinaga, A. Tomita. 2000. A blimp-based remote sensing system for low-altitude monitoring of plant variables: a preliminary experiment for agricultural and ecological applications. *Int. J. Rem. Sensing,* 21: 279–385.

International Systemap Corp. 1997. *The Fundamentals of Digital Photogrammetry.* I.S.M. International Systemap Corp., Vancouver, B.C. (http://www.ismcorp.com).

Iqbal, M. 1983. *An Introduction to Solar Radiation.* Academic Press, Toronto.

Irons, J. R., B. Johnson, and G. Linebaugh. 1987. Multiple-angle observations of reflectance anisotropy from an airborne linear array sensor. *IEEE Trans. Geosci. Rem. Sensing,* 25: 372–383.

Irons, J. R., K. J. Ranson, D. Williams, R. Irish, and F. Huegel. 1991. An off-nadir pointing imaging spectroradiometer for terrestrial ecosystem studies. *IEEE Trans. Geosci. Rem. Sensing,* 29: 66–74.

Isard, S. A. 1989. Topoclimatic controls in an alpine fellfield and their ecological significance. *Phys. Geogr.,* 10: 13–31.

Israelsson, H., L. M. H. Ulander, J. Askne, J. E. S. Fransson, P.-O. Frölind, A. Gustavsson, and H. Hellsten. 1997. Retrieval of forest stem volume using VHF SAR. *IEEE Trans. Geosci. Rem. Sensing,* 35: 36–40.

Itten, K. I. and P. Meyer. 1993. Geometric and radiometric correction of TM data of mountainous forested areas. *IEEE Trans. Geosci. Rem. Sensing,* 31: 764–770.

Iverson, L. R., E. A. Cook, and R. L. Graham. 1989. A technique for extrapolating and validating forest cover across large regions: calibrating AVHRR data with TM data. *Int. J. Rem. Sensing,* 10: 1805–1812.

Iverson, L. R., E. A. Cook, and R. L. Graham. 1990. Estimating forest cover over southeastern United States using TM-calibrated AVHRR data. Pages 1252–1262 in Volume 3 *Proc. Int. Conf. Workshop Global Natural Resour. Monit. Assess.,* ASPRS, Bethesda, MD.

Iverson, L. R., R. L. Graham, and E. A. Cook. 1989. Applications of satellite remote sensing to forested ecosystems. *Landscape Ecol.,* 3: 131–143.

Jaeger, J. A. G., 2000. Landscape division, splitting index, and effective mesh size: new measures of landscape fragmentation. *Landscape Ecol.,* 15: 115–130.

Jakubauskas, M. E. 1996a. Thematic Mapper characterization of lodgepole pine seral stages in Yellowstone National Park, USA. *Rem. Sensing Environ.,* 56: 118–132.

Jakubauskas, M. E. 1996b. Effects of forest succession on texture in Landsat Thematic Mapper imagery. *Can. J. Rem. Sensing,* 23: 257–263.

Jakubauskas, M. E., K. Kindscher, and D. M. Debinski. 1998. Multitemporal characterization and mapping of montane sagebrush communities using Indian IRS LISS-II imagery. *Geocarto Int.,* 13: 65–74.

Jakubauskas, M. E., K. P. Lulla, and P. W. Mausel. 1990. Assessment of vegetation change in a fire-altered forest landscape. *Photogramm. Eng. Rem. Sensing,* 56: 371–377.

Jakubauskas, M. E., and K. P. Price. 1997. Empirical relationships between structural and spectral factors of Yellowstone lodgepole pine forests. *Photogramm. Eng. Rem. Sensing,* 63: 1375–1381.

Jarvis, P. G. and J. W. Leverenz. 1983. Productivity of temperate, deciduous and evergreen forests. Pages 133–144 in O. L. Lange, C. B. Omond, and H. Zeigler, Eds. Physiological Plant Ecology IV. Springer Verlag, New York.

Jasinski, M. F. 1996. Estimation of subpixel vegetation density of natural regions using satellite multispectral imagery. *IEEE Trans. Geosci. Rem. Sensing,* 34: 804–813.

Jazouli, R., D. L. Verbyla, and D. L. Murphy. 1994. Evaluation of SPOT Panchromatic digital imagery for updating road locations in a harvested forest area. *Photogramm. Eng. Rem. Sensing,* 60: 1449–1452.

Jensen, J. R. 1978. Digital land cover mapping using layered classification logic and physical composition attributes. *Am. Cartographer,* 5: 121–132.

Jensen, J. R. 1983. Biophysical remote sensing. *Ann. Assoc. Am. Geogr.,* 73: 111–132.

Jensen, J. R. 1996. *Introductory Digital Image Processing. A Remote Sensing Perspective.* 2nd ed. Prentice-Hall, Englewood Cliffs, NJ.

Jensen, J. R. 2000. *Remote Sensing of the Environment. An Earth Resource Perspective.* Prentice-Hall, Englewood Cliffs, NJ.

Jensen, J. R., F. Qiu, and M. Ji. 1999. Predictive modeling of coniferous forest age using statistical and artificial neural network approaches applied to remote sensor data. *Int. J. Rem. Sensing,* 20: 2805–2822.

Jet Propulsion Laboratory. 1986. *Shuttle Imaging Radar-C Science Plan.* JPL Publ. 86-29. Jet Propulsion Laboratory, Pasadena, CA.

Jobin, L., and J. Beaubien. 1974. Capability of ERS-1 imagery for mapping forest cover types of Anticosti Island. *For. Chron.,* 50: 1–6.

Johnson, R. D. and E. Kasischke. 1998. Change vector analysis: a technique for multispectral monitoring of land cover and condition. *Int. J. Rem. Sensing,* 19: 411–426.

Johnson, L. F., and C. R. Billow. 1996. Spectrometric estimation of total nitrogen concentration in Douglas-fir foliage. *Int. J. Rem. Sensing,* 17: 489–500.

Joria, P., S. C. Ahearn, and M. Conner. 1991. A comparison of SPOT and Landsat Thematic Mapper satellite systems for detecting gypsy moth defoliation in Michigan. *Photogramm. Eng. Rem. Sensing,* 57: 1605–1612.

Jorge, L. A. B., and G. J. Garcia. 1997. A study of habitat fragmentation in southeastern Brazil using remote sensing and geographic information systems (GIS). *For. Ecol. Manage.,* 98: 35–47.

Jorgensen, A. F., and H. Nohr. 1996. The use of satellite images for mapping landscape and biological diversity. *Int. J. Rem. Sensing,* 17: 99–109.

Joy, M. W., B. Klinkenberg, and S. Cumming. 1994. Handling uncertainty in a spatial forest model integrated with GIS. Pages 359–365 in *Proc. GIS'94 Symp.*, Vancouver, British Columbia.

Jupp, D. L. B., and J. Walker. 1997. Detecting structural growth changes in woodlands and forests: the challenge for remote sensing and the role of geometric-optical modeling. Pages 75–108 in Gholz, H. L., K. Nakane, and H. Shimoda, Eds. *The Use of Remote Sensing in the Modeling of Forest Productivity.* Kluwer, Dordrecht.

Justice, C. O., and J. R. G. Townshend. 1981. Integrating ground data with remote sensing. Pages 38–58 in J. R. G. Townshend, Ed. *Terrain Analysis and Remote Sensing.* George Allen and Unwin, London.

Kalliola, R., and K. Syrjanen. 1991. To what extent are vegetation types visible in satellite imagery? *Ann. Bot. Fennici,* 28: 45–57.

Kammerbauer, J., and C. Ardon. 1999. Land-use dynamics and landscape change patterns in a typical watershed in a hillside region of central Honduras. *Agric. Ecosyst. Environ.,* 75: 93–100.

Karimi, H. A., X. Dai, S. Khorram, A. J. Khattak, and J. E. Hammer. 1999. Techniques for automated extraction of roadway inventory features from high resolution satellite imagery. *Geocarto Int.,* 14: 5–16.

Kasischke, E. S., L. L. Bourgeau-Chavez, N. L. Christensen, Jr., and E. Haney. 1994. Observations on the sensitivity of ERS-1 SAR image intensity to changes in aboveground biomass in young loblolly pine forests. *Int. J. Rem. Sensing,* 15: 3–16.

Kasischke, E. S., and N. L. Christensen, Jr. 1990. Connecting forest ecosystem and microwave backscatter models. *Int. J. Rem. Sensing,* 11: 1277–1298.

Kauth, R. J., and G. S. Thomas. 1976. The Tasseled Cap — a graphic description of the spectral-temporal development of agricultural crops as seen by Landsat. Pages 41–51 in *Proc. 1976 Symp. Machine Process. Rem. Sensed Data.* Laboratory for Applications of Remote Sensing, Purdue University, West Lafayette, IN.

Keen, M. 1997. Catalyst for change: the emerging role of participatory research in land management. *The Environmentalist,* 17: 87–96.

Kennedy, R. E., W. B. Cohen, and G. Takao. 1997. Empirical methods to compensate for a view-angle-dependent brightness gradient in AVIRIS imagery. *Rem. Sensing Environ.,* 62: 277–291.

Kershaw, C. D. 1987. Discrimination problems for satellite images. *Int. J. Rem. Sensing,* 8: 1377–1383.

Kettig, R. L., and D. A. Landgrebe. 1976. Classification of multispectral image data by extraction and classification of homogeneous objects. *IEEE Trans. Geosci. Electron.,* 14: 19–26.

Khazenie, N., and M. M. Crawford. 1990. Spatial-temporal autocorrelated model for contextual classification. *IEEE Trans. Geosci. Rem. Sensing,* 28: 529–539.

Khorram, S., J. Brockhaus, R. I. Bruck, and M. V. Campbell. 1990. Modeling and multitemporal evaluation of forest decline with Landsat TM digital data. *IEEE Trans. Geosci. Rem. Sensing,* 28: 746–748.

Kiedman, C. A. 1999. *Assessment of land cover change and regeneration in a northern California forested ecosystem using historical Landsat and GIS.* NCGIA Tech. Rep. 99-1. University of California, Santa Barbara.

Kim, H., and P. H. Swain. 1990. A method for classification of multisource data using interval-valued probabilities and its applications to Hiris data. Pages 75–81 in *Proc. Workshop Multisource Data Integration Rem. Sensing.* NASA Conf. Publ. 3099, Washington, DC.

Kimes, D. S., B. N. Holben, J. E. Nickeson, and W. A. McKee. 1996. Extracting forest age in a Pacific northwest forest from Thematic Mapper and topographic data. *Rem. Sensing Environ.,* 56: 133–140.

Kimes, D. S., and J. A. Kirchner. 1981. Modeling the effects of various radiant transfers in mountainous terrain on sensor response. *IEEE Trans. Geosci. Rem. Sensing,* 19: 100–108.

Kimmins, J. P. 1997. *Forest Ecology: A Foundation for Sustainable Management.* 2nd ed. Prentice-Hall, Upper Saddle River, New Jersey.

King, D. J. 1991. Determination and reduction of cover type brightness variations with view angle in airborne multispectral video imagery. *Photogramm. Eng. Rem. Sensing,* 57: 1571–1577.

King, D. J. 1992. Evaluation of radiometric quality, statistical characteristics, and spatial resolution of multispectral videography. *J. Imaging Sci. Technol.,* 36: 394–404.

King, D. J. 1995. Airborne multispectral digital camera and video sensors: a critical review of system designs and applications. *Can. J. Rem. Sensing,* 21: 245–273.

King, D. J., M. Jollineau, and B. Fraser. 1999. Evaluation of MK-4 multispectral satellite photography in land cover classification in eastern Ontario. *Int. J. Rem. Sensing,* 20: 3311–3331.

Kirman, J. M. 2000. The ability of grades five and six children to use Radarsat satellite images for geography instruction. *Can. J. Rem. Sensing,* 26: 103–110.

Kirman, J. M., and C. Jackson. 1993. Grade 6 children's ability to use a Landsat digital data computer program. *J. Geogr.,* 93: 254–263.

Klijn, F., Ed. 1994. *Ecosystem Classification for Environmental Management.* Kluwer, Dordrecht.

Kloditz, C., A. van Boxtel, E. Carfagna, and W. van Deursen. 1998. Estimating the accuracy of coarse scale classification with high scale information. *Photogramm. Eng. Rem. Sensing,* 64: 127–134.

Knipling, E. B. 1970. Physical and physiological basis for reflectance of visible and near-infrared radiation from vegetation. *Rem. Sensing Environ.,* 1: 155–159.

Kohm, K., and J. Franklin, Eds. 1997. *Creating a Forestry for the 21st Century.* Island Press, Washington, DC.

Koutsias, N., and M. Karteris. 2000. Burned area mapping using logistic regression modeling of a single post-fire Landsat-5 Thematic Mapper image. *Int. J. Rem. Sensing,* 21: 673–687.

Kovats, M. 1997. A large-scale aerial photographic technique for measuring tree heights on long-term forest installations. *Photogramm. Eng. Rem. Sensing,* 63: 741–747.

Kozlowski, T., P. Kramer, and S. Pallardy. 1991. *The Physiological Ecology of Woody Plants.* Academic Press, Toronto.

Kreig, R. A. 1970. Aerial photographic interpretation for land use classification in the New York State land use and natural resources inventory. *Photogrammetria,* 26: 101–111.

Kriebel, K. T. 1978. Measured spectral bidirectional reflectance properties of four vegetated surfaces. *Appl. Op.,* 17: 253–259.

Kummer, D. M. 1992. Remote sensing and tropical deforestation: a cautionary note from the Philippines. *Photogramm. Eng. Rem. Sensing,* 58: 1469–1471.

Kuntz, S., and F. Siegert. 1996. Dipterocarp forest mapping and monitoring by satellite data: a case study from East Kalimantan. Pages 206–227 in A. Schulte and D. Schone, Eds. *Dipterocarp Forest Ecosystems: Towards Sustainable Forest Management.* World Scientific, Singapore.

Kuntz, S., and F. Siegert. 1999. Monitoring of deforestation and land use in Indonesia with multitemporal ERS data. *Int. J. Rem. Sensing,* 20: 2835–2853.

Kushla, J. D., and W. J. Ripple. 1998. Assessing wildfire effects with Landsat Thematic Mapper data. *Int. J. Rem. Sensing,* 19: 2493–2507.

Kuusipalo, J., J. Kangas, and L. Vesa. 1997. Sustainable forest management in tropical rain forests: a planning approach and case study from Indonesian Borneo. *J. Sustainable For.,* 5: 93–118.

Kuusk, A., and T. Nilson. 2000. A directional multispectral forest reflectance model. *Rem. Sensing Environ.,* 72: 244–252.

Kux, H. J. H., F. J. Ahern, and R. W. Pietsch. 1995. Evaluation of radar remote sensing for natural resource management in the tropical rainforests of Acre State, Brazil. *Can. J. Rem. Sensing,* 21: 430–440.

Lacate, D. S. 1969. *Guidelines for Biophysical Land Classification.* Canadian Forest Service Publication No. 1624. Ottawa.

Lachowski, H., P. Maus, and N. Roller. 2000. From pixels to decisions: digital remote sensing technologies for public land managers. *J. For.,* 98: 13–15.

Lahde, E., O. Laiho, Y. Norokorpi, and T. Saksa. 1999. Stand structure as a basis of diversity index. *For. Ecol. Manage.,*115: 213–220.

Landgrebe, D. A. 1978a. The quantitative approach: concept and rationale. Pages 1–20 in P. H. Swain and S. M. Davis, Eds. *Remote Sensing: The Quantitative Approach.* McGraw-Hill, New York.

Landgrebe, D. A. 1978b. Useful information from multispectral image data: another look. Pages 336–374 in P. H. Swain and S. M. Davis, Eds. *Remote Sensing: The Quantitative Approach.* McGraw-Hill, New York.

Landgrebe, D. A. 1983. Land observation sensors in perspective. *Rem. Sensing Environ.,* 13: 391–402.

Landgrebe, D. A. 1997. The evolution of Landsat data analysis. *Photogramm. Eng. Rem. Sensing,* 63: 859–868.

Landry, R., F. J. Ahern, and R. O'Neil. 1995. Forest burn visibility on C-HH radar images. *Can. J. Rem. Sensing,* 21: 204–206.

Landsberg, J., and N.C. Coops. 1999. Modelling forest productivity across large areas and long periods. *Nat. Resour. Model.,* 12: 1–28.

Landsberg, J., and S. T. Gower. 1996. *Applications of Physiological Ecology to Forest Production.* Academic Press, San Diego, CA.

Lang, R. H., N. S. Chauhan, K. J. Ranson, and O. Kilic. 1994. Modeling P-band SAR returns from a red pine stand. *Rem. Sensing Environ.,* 47: 132–141.

Lapointe, G. 1998. Sustainable forest management certification: the Canadian program. *For. Chron.*, 74: 227–230.

Lark, R. M. 1996. Geostatistical description of texture on an aerial photography for discriminating classes of land cover. *Int. J. Rem. Sensing,* 17: 2115–2133.

Larson, D. R., S. R. Shifley, F. R. Thompson, III, B. L. Brookshire, D. C. Dey, E. W. Kurzejeski, and K. England. 1997. Ten guidelines for ecosystem researchers: lessons from Missouri. *J. For.,* 95: 4–9.

Latham, P. A., H. Zuuring, and D. W. Coble. 1998. A method for quantifying vertical forest structure. *For. Ecol. Manage.,* 104: 157–170.

Lauenroth, W.K., C. C. Canham, A. P. Kinzig, K. A. Poiani, M. Kemp, and S. W. Running. 1998. Simulation modeling in ecosystem science. Pages 404–414 in M. L. Pace and P. M. Groffman, Eds. *Successes, Limitations, and Frontiers in Ecosystem Science.* Springer-Verlag, New York.

Lauer, D. T., S. A. Morain, and V. V. Salomonson. 1997. The Landsat Program: its origins, evolution, and impacts. *Photogramm. Eng. Rem. Sensing,* 63: 831–838.

Lauver, C. L. and J. L. Whistler. 1993. A hierarchical classification of Landsat TM imagery to identify natural grassland areas and rare species habitat. *Photogramm. Eng. Rem. Sensing,* 59: 627–634.

Lavigne, M. B., J. E. Luther, S. E. Franklin, and E. R. Hunt, Jr. 1996. Comparing branch biomass prediction equations for *Abies balsamea. Can. J. For. Res.,* 26: 611–616.

Law, B. E., and R. H. Waring. 1994. Combining remote sensing and climatic data to estimate net primary production across Oregon. *Ecolog. Appl.,* 4: 717–728.

Lawrence, R. L., and W. J. Ripple. 1999. Calculating change curves for multitemporal satellite imagery: Mount St. Helens 1980–1995. *Rem. Sensing Environ.,* 67: 309–319.

Leberl, F. 1983. Photogrammetric aspects of imaging radar. *Rem. Sensing Rev.,* 1: 51–58.

Leblon, B., H. Granberg, C. Ansseau, and A. Royer. 1993. A semi-empirical model to estimate the biomass production of forest canopies from spectral variables. Part 1: relationship between spectral variables and light interception efficiency. *Rem. Sensing Rev.,* 7: 109–125.

Leckie, D. G. 1990a. Advances in remote sensing technologies for forest surveys and management. *Can. J. For. Resour.,* 20: 465–483.

Leckie, D. G. 1990b. Synergism of synthetic aperture radar and visible/infrared data for forest type discrimination. *Photogramm. Eng. Rem. Sensing,* 56: 1237–1246.

Leckie, D. G., J. Beaubien, J. R. Gibson, N. T. O'Neill, T. Piekutowski, and S. P. Joyce. 1995. Data processing and analysis for MIFUCAM: a trial of MEIS imagery for forest inventory mapping. *Can. J. Rem. Sensing,* 21: 337–356.

Leckie, D. G., and M. D. Gillis. 1995. Forest inventory in Canada with an emphasis on map production. *For. Chron.,* 71: 74–88.

Leckie, D. G., M. Gillis, F. Gougeon, M. Lodin, J. Wakelin, and X. Yuan. 1999. Computer assisted photointerpretation aids to forest inventory mapping: some possible approaches. Pages 335–344 in *Proc. Int. Forum Automated Interpretation of High Spatial Resolution Imagery for Forestry.* Canadian Forest Service, Pacific Forestry Centre, Victoria, British Columbia.

Leckie, D. G., and D. P. Ostaff. 1988. Classification of airborne multispectral scanner data for mapping current defoliation caused by the spruce budworm. *For. Sci.,* 34: 259–275.

Leckie, D. G., X. Yuan, D. P. Ostaff, H. Piene, and D. A. MacLean. 1992. Analysis of high resolution multispectral MEIS imagery for spruce budworm damage assessment on a single tree basis. *Rem. Sensing Environ.,* 40: 125–136.

Lee, N. J., and K. Nakane. 1997. Forest vegetation classification and biomass estimation based on Landsat TM data in a mountainous region of west Japan. Pages 159–171 in Gholz, H. L., K. Nakane, and H. Shimoda, Eds. *The Use of Remote Sensing in the Modeling of Forest Productivity.* Kluwer, Dordrecht.

Lee, T., J. A. Richards, and P. H. Swain. 1987. Probabilistic and evidential approaches for multisource data analysis. *IEEE Trans. Geosci. Rem. Sensing,* 25: 283–292.

Lefsky, M. A., D. Harding, W. B. Cohen, G. Parker, and H. H. Shugart. 1999a. Surface lidar remote sensing of basal area and biomass in deciduous forests of eastern Maryland, USA. *Rem. Sensing Environ.,* 67: 83–98.

Lefsky, M. A., W. B. Cohen, S. A. Acker, G. G. Parker, T. A. Spies, and D. Harding. 1999b. Lidar remote sensing of the canopy structure and biophysical properties of Douglas-fir western hemlock forests. *Rem. Sensing Environ.,* 70: 339–361.

Legge, A. H., D. R. Jaques, C. E. Poulton, C. L. Kirby, and P. Van Eck. 1974. *Development and application of an ecologically based remote sensing legend for the Kananaskis, Alberta, remote sensing test corridor (Subalpine Forest Region).* University of Calgary, Alberta.

Leighty, R. D. 1987. Organizing the landscape for image understanding purposes. Pages 69–73 in Vol. 758 *Proc. SPIE Image Understanding Man-Machine Interface.* SPIE, Bellingham, WA.

Leopold, A. 1949. *A Sand County Almanac.* Oxford University Press, Oxford.

Leprieur, C. E., J. M. Durand, and J. L. Peyon. 1988. Influence of topography on forest reflectance using Landsat Thematic Mapper and digital terrain data. *Photogramm. Eng. Rem. Sensing,* 54: 491–496.

Le Toan, T., A. Beaudoin, J. Riom, and D. Guyon. 1992. Relating forest biomass to SAR data. *IEEE Trans. Geosci. Rem. Sensing,* 30: 403–411.

Levesque, J., and D. J. King. 1999. Airborne digital camera image semivariance for evaluation of forest structural damage at an acid mine site. *Rem. Sensing Environ.,* 68: 112–124.

Li, H., and J. F. Reynolds. 1993. A new contagion index to quantify spatial patterns of landscapes. *Landscape Ecol.,* 8: 155–162.

Li, X., and A. Strahler. 1985. Geometric-optical modeling of a conifer forest canopy. *IEEE Trans. Geosci. Rem. Sensing,* 23: 705–721.

Liao, J., and H. Guo. 1998. Multifrequency and multipolarization radar data for estimation of forest biomass over the Zhaoqing area of southern China. *Can. J. Rem. Sensing,* 24: 240–245.

Liebhold, A. M., E. E. Simons, A. Sior, and J. D. Unger. 1993. Forecasting defoliation caused by the gypsy moth from field measurements. *Environ. Entomol.,* 22: 26–32.

Lieffers, V. J., and J. A. Beck. 1994. A semi-natural approach to mixedwood management in the prairie provinces. *For. Chron.,* 70: 260–264.

Lieffers, V. J., K. J. Stadt, and S. Navratil. 1996. Age structure and growth of understory white spruce under aspen. *Can. J. For. Res.,* 26: 1002–1007.

Light, D.L. 1996. Film cameras or digital sensors? The challenge ahead for aerial imaging. *Photogramm. Eng. Rem. Sensing,* 62: 285–291.

Likens, G. E. 1998. Limitations to intelligent progress in ecosystem science. Pages 247–271 in M. L. Pace and P. M. Groffman, Eds. *Success, Limitations, and Frontiers in Ecosystem Science.* Springer-Verlag, New York.

Lillesand, T. M. 1996. A protocol for satellite-based land cover classification in the upper Midwest. Pages 103–118 in J. M. Scott, T. H. Tear, and F. W. Davis, Eds. *Gap Analysis: A Landscape Approach to Biodiversity Planning.* American Society for Photogrammetry and Remote Sensing, Bethesda, MD.

Lillesand, T. M., and R. W. Kiefer. 1994. *Remote Sensing and Image Interpretation.* 3rd ed. John Wiley & Sons, New York.

Limp, W. F. 1999. Image processing software: system selection depends on user needs. *GeoWorld,* May, 12: 36–46.

Linden, D. S. 2000. Videography for foresters. *J. For.,* 98: 53–57.

Lindenmayer, D. B., B. G. Mackey, I. C. Mullen, M. A. McCarthy, A. M. Gill, R. B. Cunningham, and C. F. Donnelly. 1999. Factors affecting stand structure in forests — are there climatic and topographic determinants? *For. Ecol. Manage.,* 123: 55–63.

Lippke, B. R. and J. T. Bishop. 1999. The economic perspective. Pages 597–638 in M. L. Hunter, Jr., Ed. *Maintaining Biodiversity in Forest Ecosystems.* Cambridge University Press, Cambridge, U.K.

Liu, J., J. M. Chen, J. Cihlar, and W. M. Park. 1997. A process-based boreal ecosystem productivity simulator using remote sensing inputs. *Rem. Sensing Environ.,* 62: 158–177.

Liverman, D., E. F. Moran, R. R. Rindfuss, and P. C. Stern, Eds. 1998. *People and Pixels: Linking Remote Sensing and the Social Sciences.* National Academy Press, Washington, DC.

Lobo, A. 1997. Image segmentation and discriminant analysis for the identification of land cover units in ecology. *IEEE Trans. Geosci. Rem. Sensing,* 35: 1136–1145.

Longley, P. A., M. F. Goodchild, D. J. Maguire, and D. W. Rhind. 1999. Epilogue: seeking out the future. Pages 1009–1021 in Longley, P. A., M. F. Goodchild, D. J. Maguire, and D. W. Rhind, Eds. *Geographical Information Systems.* 2nd ed. John Wiley & Sons, New York.

Loveland, T. R., J. W. Merchant, D. O. Ohlen, and J. F. Brown. 1991. Development of a land-cover characteristics database for the conterminous U.S. *Photogramm. Eng. Rem. Sensing,* 57: 1453–1463.

Lowell, K. E., G. Edwards, and G. Kucera. 1996. Modelling the heterogeneity and change of natural forests. *Geomatica,* 50: 425–440.

Lowry, R. T., P. Van Eck, and R. V. Dams. 1986. SAR imagery for forest management. Pages 901–906 in Proc., IGARSS'86 Symp. ESA SP-II254, Zurich.

Lucas, N. S. and P. J. Curran. 1999. Forest ecosystem simulation models: the role of remote sensing. *Prog. Phys. Geogr.,* 23: 391–423.

Lucas, N. S., P. J. Curran, S. E. Plummer, and F. M. Danson. 2000. Estimating the stem carbon production of a coniferous forest using an ecosystem simulation model driven by the remotely sensed red-edge. *Int. J. Rem. Sensing,* 21: 619–631.

Ludwig, D., B. Walker, and C. S. Holling. 1997. Sustainability, stability, and resilience. *Conserv. Ecol.,* 1: http://www.consecol.org/Journal/vol1/iss1/art7/index.html.

Lueder, D. R. 1959. *Aerial Photographic Interpretation.* McGraw-Hill, New York.

Lugo, A. E. 1998. Biodiversity and public policy: the middle of the road. Pages 33–45 in L. D. Guruswamy, and J. A. McNeely, Eds. *Protection of Global Biodiversity: Converging Strategies.* Duke University Press, Durham, NC.

Lunetta, R. S. 1999. Applications, project formulation, and analytical approach. Pages 1–20 in Lunetta, R. S., and C. D. Elvidge, Eds. *Remote Sensing Change Detection: Environmental Monitoring Methods and Applications.* Taylor & Francis, London.

Lunetta, R. S., and C. D. Elvidge, Eds. 1999. *Remote Sensing Change Detection: Environmental Monitoring Methods and Applications.* Taylor & Francis, London.

Lunetta, R. S., J. G. Lyon, B. Guindon, and C. D. Elvidge. 1998. North American landscape characterization dataset development and data fusion issues. *Photogramm. Eng. Rem. Sensing,* 64: 821–829.

Luque, S. S., R. G. Lathrop, and J. A. Bognar. 1994. Temporal and spatial change in an area of the New Jersey Pine Barrens. *Landscape Ecol.,* 9: 287–300.

Luscombe, A. P., I. Ferguson, N. Shepperd, D.G. Zimcik, and P. Naraine. 1993. The Radarsat synthetic aperture radar development. *Can. J. Rem. Sensing,* 19: 298–310.

Luther, J., S. E. Franklin, and J. Hudak. 1991. Satellite remote sensing of current year defoliation by forest pests in western Newfoundland. Pages 192–198 in *Proc. 14th Can. Symp. Rem. Sensing.* Canadian Aeronautics and Space Institute, Ottawa.

Luther, J., S. E. Franklin, J. Hudak, and J. P. Meades. 1997. Forecasting the susceptibility and vulnerability of balsam fir forests to insect defoliation with satellite remote sensing. *Rem. Sensing Environ.,* 59: 77–91.

Lyon, J. G., D. Yuan, R. S. Lunetta, and C. D. Elvide. 1998. A change detection experiment using vegetation indices. *Photogrammetric Engineering Remote Sensing* 64: 143–150.

Mabbutt, J. A. 1968. Review of concepts of land classification. Pages 11–28 in G. A. Stewart, Ed. *Land Evaluation: Papers of a CSIRO Symposium.* Macmillan, Melbourne, Australia.

MacArthur, R. H., and E. O. Wilson. 1967. *The Theory of Island Biogeography,* Princeton University Press, Princeton, NJ.

Macdonald, D. S. 1972. Minister's address. Pages iii–vi in *Proc. 1st Can. Symp. Rem. Sensing.* Canadian Aeronautics and Space Institute, Ottawa.

Mace, R. D., J. S. Waller, T. L. Manley, K. Ake, and W. T. Wittinger. 1999. Landscape evaluation of grizzly bear habitat in western Montana. *Conserv. Biol.,* 13: 366–376.

Mackinnon, J., and R. de Wulf. 1994. Designing protected areas for giant pandas in China. Pages 127–142 in R. I. Miller, Ed. *Mapping the Diversity of Nature.* Chapman & Hall, London.

MacLean, D. A. 1980. Vulnerability of fir-spruce stands during uncontrolled spruce budworm outbreaks: a review and discussion. *For. Chron.,* 56:213–221.

MacLean, D. A. 1990. Impact of forest pests and fire on stand growth and timber yield: implications for forest management planning. *Can. J. For. Res.,* 20: 391–404.

MacLean, D. A., and K. B. Porter. 1994. Development of a decision support system for spruce budworm and forest management planning in Canada. Pages 863–872 in *Proc. GIS'94 Symp.* Vancouver, British Columbia.

Maclean, G. A., and W. B. Krabill. 1986. Gross-merchantable timber estimation using an airborne LIDAR system. *Can. J. Rem. Sensing,* 12: 5–18.

Maclean, G. A., and G. L. Martin. 1984. Merchantable timber volume estimation using cross-sectional photogrammetric and densitometric methods. *Can. J. For. Res.,* 14: 803–810.

Madden, M., D. Jones, and L. Vilchek. 1999. Photointerpretation key for the Everglades vegetation classification system. *Photogramm. Eng. Rem. Sensing,* 65: 171–177.

Magnusson, S. 1997. A method for enhancing tree species proportions from aerial photos. *For. Chron.,* 73: 479–487.

Malcolm, J. R., B. L. Zimmerman, R. B. Cavalcanti, F. J. Ahern, and R. W. Pietsch. 1998. Use of Radarsat SAR data for sustainable management of natural resources: a test case in the Kayopo indigeneous area, Para, Brazil. *Can. J. Rem. Sensing,* 24: 360–375.

Malila, W. A. 1980. Change vector analysis: an approach for detecting forest changes with Landsat. Pages 326–335 in *Proc. 6th Annu. Symp. Machine Process. Rem. Sensed Data.* Purdue University Laboratory of Applied Remote Sensing, West Lafayette, Indiana.

Manguel, A., Ed. 1998. *By the Light of the Glow-Worm Lamp: Three Centuries of Reflections on Nature.* Plenum Press, New York.

Marceau, D. J., P. J. Howarth, J. M. Dubois, and D. J. Gratton. 1990. Evaluation of the gray-level co-occurrence matrix method for land cover classification using SPOT imagery. *IEEE Trans. Geosci. Rem. Sensing,* 28: 513–517.

Markon, C. 1992. Land cover mapping of the upper Kuskokwim resource management area, Alaska, using Landsat and a digital database approach. *Can. J. Rem. Sensing,* 18: 62–71.

Marsh, S. E., J. L. Walsh, and C. Sobrevila. 1994. Evaluation of airborne video data for land-cover classification assessment in an isolated Brazilian forest. *Rem. Sensing Environ.,* 48: 61–69.

Martin, M. E., S. D. Newman, J. D. Aber, and R. G. Congalton. 1998. Determining forest species composition using high spectral resolution remote sensing data. *Rem. Sensing Environ.,* 65: 249–254.

Martinez, L. J. M., D. E. R. Vanegas, W. van Wijngaarden, M. F. Quinones, W. Bijker, and D. H. Hoekman. 1996. *GIS for the Colombian Amazonia: the Guaviare Case.* Tropenbos, Colombia.

Mas, J.-F. 1999. Monitoring land-cover changes: a comparison of change detection techniques. *Int. J. Rem. Sensing,* 20: 139–152.

Maser, C. 1994. *Sustainable Forestry: Philosophy, Science and Economics.* St. Lucie Press, Delray Beach, Florida.

Mater, C. M. 1998. Emerging technologies for sustainable forestry. Pages 4–1 to 4–27 in *Case Studies for the Business of Sustainable Forestry: A Project of the Sustainable Forestry Working Group.* The John D. and Catherine T. MacArthur Foundation, Chicago.

Mather, P. 1987. *Comuter Processing of Remotely Sensed Images.* John Wiley & Sons, London.

Matsuyama, T. 1987. Knowledge-based aerial image understanding systems and expert systems for image processing. *IEEE Trans. Geosci. Rem. Sensing,* 25: 305–316.

Mausel, P. W., G. Y. Kalaluka, J. K. Lee, D. E. Mudderman, D. E. Escobar, and M. R. Davis. 1990. Multispectral video analysis of micro-relief dominated poorly drained soils in western Indiana. *J. Imaging Sci. Technol.,* 16: 113–119.

Mausel, P. W., W. J. Kramber, and J. K. Lee. 1990. Optimum band selection for supervised classification of multispectral data. *Photogramm. Eng. Rem. Sensing,* 56: 55–60.

Mayaux, P., and E. Lambin. 1997. Tropical forest area measured from global land cover classifications: inverse calibration models based on spatial textures. *Rem. Sensing Environ.,* 59: 29–43.

McCaffrey, T. M., and S. E. Franklin. 1993. Automated training site selection for large area remote sensing image analysis. *Comput. Geosci.,* 19: 1413–1428.

McCay, D. 1998. Patterns of sand pine invasion into longleaf pine forests in the Florida Panhandle. CD-ROM *Proc. ASPRS Spring Annu. Meet.,* ASPRS, Bethesda, MD.

McCreight, R, C. F. Chen, and R. H. Waring. 1994. Airborne environmental analysis using an ultralight aircraft. Pages 384–392 in Vol. I *Proc. 1st Int. Airborne Rem. Sensing Conf.,* ERIM, Ann Arbor, MI.

McDermid, G. J., and S. E. Franklin. 1995. Remote sensing and geomorphometric discrimination of slope processes. *Z. Geomorphol. Suppl. Band,* 101: 165–185.

McDermid, G. J., J. Hudak, J. Meades, and S. E. Franklin. 1993. Modelling forest growth efficiency using multitemporal Landsat TM data. Pages 621–627 in *Proc. 16th Can. Symp. Rem. Sensing.* Canadian Aeronautics and Space Institute, Ottawa.

McDonnell, R. A. 1996. Including the spatial dimension: using geographic information systems in hydrology. *Prog. Phys. Geogr.,* 20: 159–177.

McGarigal, K., and B. J. Marks. 1995. FRAGSTATS: Spatial pattern analysis program for quantifying landscape structure. *USDA Forest Service General Techn. Rep. PNW-GTR-351.* Corvallis, Oregon.

McGarigal, K., and W. C. McComb. 1995. Relationship between landscape pattern and breeding birds in the Oregon Coast Range. *Ecol. Monogr.,* 65: 235–260.

McKeown, D. M., Jr. 1984. Knowledge-based aerial photointerpretation. *Photogrammetria,* 39: 91–123.

McKeown, D. M., Jr. 1987. The role of artificial intelligence in the integration of remotely sensed data with geographic information systems. *IEEE Trans. Geosci. Rem. Sensing,* 25: 330–348.

McKeown, D. M., Jr., S. D. Cochran, S. J. Ford, J. C. McGlone, J. A. Shufelt, and D. A. Yocum. 1999. Fusion of HYDICE hyperspectral data with panchromatic imagery for cartographic feature extraction. *IEEE Trans. Geosci. Rem. Sensing,* 37: 1261–1271.

McLeod, R. G., and T. L. Logan. 1980. The use of Landsat, digital terrain, and ground sample data as inductively derived information input to a multispectral classifier for wildlands mapping and inventory. Pages 81–93 in *Vol. 10, Harvard Library of Computer Graphics 1980 Mapping Collection.* Harvard University, Cambridge, MA.

McNab, W. H. 1989. Terrain shape index: quantifying the effect of minor landforms on tree height. *For. Sci.,* 35: 91–104.

McNab, W. H. 1993. A topographic index to quantify the effect of mesoscale landform on site productivity. *Can. J. For. Res.,* 23: 1100–1107.

McNab, W. H., S. A. Browning, S. A. Simon, and P. E. Fouts. 1999. An unconventional approach to ecosystem unit classification in western North Carolina, USA. *For. Ecol. Manage.,* 114: 405–420.

Means, J. E., S. A. Acker, D. J. Harding, J. B. Blair, M. A. Lefsky, W. B. Cohen, M. E. Harmon, and W. A. McKee. 1999. Use of large-footprint scanning airborne lidar to estimate forest stand characteristics in the western Cascades of Oregon. *Rem. Sensing Environ.,* 67: 298–308.

Merchant, J. W. 1981. Employing Landsat MSS data in land use mapping: observations and considerations. *Proc. Pecora 7 Symp. Am. Soc. Photogrammetry Rem. Sensing, Bethesda, MD,* 71–91.

Merideth, R. W., Jr. 1981. Doctoral dissertations pertaining to remote sensing and photogrammetry: a selected bibliography. *Photogramm. Eng. Rem. Sensing,* 47: 617–629.

Metzger, J. P., and E. Muller. 1996. Characterizing the complexity of landscape boundaries by remote sensing. *Landscape Ecol.,* 11: 65–77.

Meyer, P., K. Staenz, and K. I. Itten. 1996. Semi-automated procedures for tree species identification in high spatial resolution data from digitized color infrared aerial photography. *ISPRS J. Photogramm. Rem. Sensing,* 51: 5–16.

Meyers, N. 1997. Our forestry prospect: the past recycled or a surprise rich future? *The Environmentalist,* 17: 233–247.

Michelson, D. B., B. M. Liljeberg, and P. Pilesjö. 2000. Comparison of algorithms for classifying Swedish landcover using Landsat TM and ERS-1 SAR data. *Rem. Sensing Environ.,* 71: 1–15.

Michener, W. K., and P. F. Houhoulis. 1996. *Identification and Assessment of Natural Disturbances in Forested Ecosystems: The Role of GIS and Remote Sensing.* Unpublished manuscript. Joseph W. Jones Ecological Research Center, Newton, GA.

Michener, W. K., and P. F. Houhoulis. 1997. Detection of vegetation changes associated with extensive flooding in a forested ecosystem. *Photogramm. Eng. Rem. Sensing,* 63: 1363–1374.

Mickelson, J. G., Jr., D. L. Civco, and J. A. Silander, Jr. 1998. Delineating forest canopy species in the northeastern United States using multi-temporal TM imagery. *Photogramm. Eng. Rem. Sensing,* 64: 891–904.

Miller, L. D., R. L. Pearson, and C. J. Tucker. 1976. A mobile field spectrometer laboratory. *Photogramm. Eng. Rem. Sensing,* 42: 569–572.

Miller, A. B., E. S. Bryant, and R. W. Birnie. 1998. An analysis of land cover changes in the northern forest of New England using multitemporal Landsat MSS data. *Int. J. Rem. Sensing,* 19: 245–265.

Miller, J., J. Wu, M. G. Boyer, M. Belanger, and E. W. Hare. 1991. Seasonal patterns in leaf reflectance red-edge characteristics. *Int. J. Rem. Sensing,* 12: 1509–1523.

Milton, E. J., K. Lawless, A. Roberts, and S. E. Franklin. 1997. The effect of unresolved scene elements on the spectral response of calibration targets: an example. *Can. J. Rem. Sensing,* 23: 126–130.

Milton, E. J., E. M. Rollin, and D. R. Emery. 1995. Advances in field spectroscopy. Pages 9–32 in F. M. Danson, and S. E. Plummer, Eds. *Advances in Environmental Remote Sensing*. John Wiley & Sons, London.

Mitasova, H., J. Hofierka, M. Zlocha, and L. R. Iverson. 1996. Modelling topographic potential for erosion and deposition using GIS. *Int. J. Geogr. Inf. Syst.,* 10: 629–641.

Molenaar, M. 1998. *An Introduction to the Theory of Spatial Object Modelling for GIS*. Taylor & Francis, London.

Monteith, J. 1972. Solar radiation and productivity in tropical ecosystems. *J. Appl. Ecol.,* 9: 747–766.

Moody, A., and C. E. Woodcock. 1995. The influence of scale and the spatial characteristics of landscapes on land-cover mapping using remote sensing. *Landscape Ecol.,* 10: 363–379.

Moore, I. D., R. B. Grayson, and A. R. Ladson. 1993a. Digital terrain modelling: a review of hydrological, geomorphological and biological applications. *Hydrol. Process.,* 5: 3–30.

Moore, I. D., T. W. Norton, and J. E. Williams. 1993b. Modelling environmental heterogeneity in forested landscapes. *J. Hydrol.,* 150: 717–747.

Moore, W., and T. Polzin. 1990. ER-2 high altitude reconnaissance: a case study. *For. Chron.,* 66: 480–486.

Morain, S. A. 1998. A brief history of remote sensing applications, with emphasis on Landsat. Pages 28–50 in Liverman, D., E. F. Moran, R. R. Rindfuss, and P. C. Stern, Eds. *People and Pixels: Linking Remote Sensing and Social Science*. National Academy Press, Washington, DC.

Morgan, D. J. 1991. Aspen inventory: problems and challenges. Pages 33–38 in Navatril, S., and P. B. Chapman, Eds. *Proc. Aspen Manage. 21st Century.* Poplar Council of Canada and Canadian Forest Service Northern Forestry Centre, Edmonton, Alberta.

Morisette, J. T., S. Khorram, and T. Mace. 1999. Land-cover change detection enhanced with generalized linear models. *Int. J. Rem. Sensing,* 20: 2703–2721.

Muchoney, D. M., and B. M. Haack. 1994. Change detection for monitoring forest defoliation. *Photogramm. Eng. Rem. Sensing,* 60: 1243–1251.

Mukai, Y., and I. Hasegawa. 2000. Extraction of damaged areas of windfall trees by typhoons using Landsat TM data. *Int. J. Rem. Sensing,* 21: 647–654.

Muller, E. 1993. Evaluation and correction of angular anisotropic effects in multidate SPOT and Thematic Mapper data. *Rem. Sensing Environ.,* 45: 295–309.

Muller, J.-P. A. L. 1988. Key issues in image understanding in remote sensing. *Philos. Trans. R. Soc. London,* A324: 381–395.

Mummery, D., M. Battaglia, C. L. Beadle, C. R. A. Turnbull, and R. McLeod. 1999. An application of terrain and environmental modelling in a large-scale forestry experiment. *For. Ecol. Manage.,* 118: 149–159.

Murtha, P. A. 1972. *A Guide to Air Photo Interpretation of Forest Damage in Canada.* Canadian Forest Service Publ. No. 1292, Ottawa.

Murtha, P. A. 1976. Vegetation damage and remote sensing: principal problems and some recommendations. *Photogrammetria,* 32: 147–156.

Murtha, P. A. 1977. Remote sensing in ecological land classification. Pages 157–168 in *Proc. Symp. Ecol. Land Classification Can.,* University of British Columbia, Vancouver, British Columbia.

Murtha, P. A. 1978. Remote sensing and vegetation damage: a theory for detection and assessment. *Photogramm. Eng. Rem. Sensing,* 44: 1147–1158.
Murtha, P. A. 1985. Photo interpretation of spruce beetle-attacked spruce. *Can. J. Rem. Sensing,* 11: 93–102.
Murtha, P. A., and T. M. Ballard. 1983. Foliar nutrients and photo interpretation of Douglas-fir stress. *Can. J. Rem. Sensing,* 9: 99–110.
Murtha, P. A., and R. Cozens. 1985. Color infra-red photo interpretation and ground surveys evaluate spruce beetle attack. *Can. J. Rem. Sensing,* 11: 177–187.
Murtha, P. A., J. Maedel, and J. Morrison. 1996. Ecoregion and biogeoclimatic ecosystem classifications applied to gap analysis in British Columbia using GIS. Pages 239–249 in J. M. Scott, T. H. Tear, and F. W. Davis, Eds. *Gap Analysis: A Landscape Approach to Biodiversity Planning*. American Society for Photogrammetry and Remote Sensing, Bethesda, Maryland.
Murtha, P. A., and R. J. Pollock. 1996. Airborne SAR studies of north Vancouver Island rainforests. *Can. J. Rem. Sensing,* 22: 175–183.
Murtha, P. A., and R. Wiart. 1987. PC-based digital image analysis for mountain pine beetle green attack: preliminary results. *Can. J. Rem. Sensing,* 13: 92–95.
Murtha, P. A., and R. Wiart. 1989a. PC-based digital image analysis of mountain pine beetle current-attacked and non-attacked lodgepole pine. *Can. J. Rem. Sensing,* 15: 70–76.
Murtha, P. A., and R. Wiart. 1989b. Cluster analysis of pine crown foliage patterns aid identification of mountain pine beetle current-attack. *Photogramm. Eng. Rem. Sensing,* 55: 83–86.
Mussio, L., and D.L. Light. 1995. ISPRS Commission I: sensors, platforms, and imagery symposium. *Photogramm. Eng. Rem. Sensing,* 61: 1339–1344.
Myneni, R. B., and J. Ross, Eds. 1991. *Photon-Vegetation Interactions. Applications in Optical Remote Sensing and Plant Ecology.* Springer-Verlag, Berlin.
Nandhakumar, N., and J. K. Aggarwal. 1985. The artificial intelligence approach to pattern recognition — a perspective and overview. *Pattern Recognition,* 18: 383–389.
National Research Council. 1998. *Forested Landscapes in Perspective: Prospects and Opportunities for Sustainable Management of America's Nonfederal Forests.* National Academy Press, Washington, DC.
Natural Resources Canada. 2000. *The State of Canada's Forests 1999–2000.* Canadian Forest Service, Ottawa.
Neale, C. M. U., and B. G. Crowther. 1994. An airborne multispectral video/radiometer remote sensing system: development and calibration. *Rem. Sensing Environ.,* 49: 187–194.
Needham, T. D., and J. A. Smith. 1987. Stem count accuracy and species determination in loblolly pine plantations using 35–mm photography. *Photogramm. Eng. Rem. Sensing,* 53: 1675–1678.
Nelson, R., W. Krabill, and J. Tonelli. 1988. Estimating forest biomass and volume using airborne laser data. *Rem. Sensing Environ.,* 24: 247–267.
Nemani, R., L. L. Pierce, S. W. Running, and L. E. Band. 1993. Forest ecosystem processes at the watershed scale: sensitivity to remotely-sensed leaf area index estimates. *Int. J. Rem. Sensing,* 14: 2519–2534.
Nemani, R., and S. W. Running. 1989. Estimation of regional surface resistance to evapotranspiration from NDVI and thermal-IR AVHRR data. *J. Appl. Meteorol.,* 28: 276–284.
Neville, R., and S. Till. 1991. MEIS-FM, a multispectral imager for forestry and mapping. *IEEE Trans. Geosci. Rem. Sensing,* 29: 184–186.
Nichols, W. F., K. T. Killingbeck, and P. V. August. 1998. The influence of geomorphological heterogeneity on biodiversity. II. A landscape perspective. *Conserv. Biol.,* 12: 371–379.

Nicolin, B., and R. Gabler. 1987. A knowledge-based system for the analysis of aerial photographs. *IEEE Trans. Geosci. Rem. Sensing,* 25: 317–329.

Niemann, K. O. 1995. Remote sensing of forest stand age using airborne spectrometer data. *Photogramm. Eng. Rem. Sensing,* 61: 1119–1127.

Nilson, T. 1992. Radiative transfer in nonhomogeneous plant canopies. Pages 60–88 in G. Stanhill, Ed. *Advances in Bioclimatology 1.* Springer-Verlag, Berlin.

Nilson, T., and U. Peterson. 1991. A forest canopy reflectance model and a test case. *Rem. Sensing Environ.,* 37: 131–142.

Nilson, T., and U. Peterson. 1994. Age dependence of forest reflectance: main driving factors. *Rem. Sensing Environ.,* 48: 319–333.

Nilson, T., and J. Ross. 1997. Modeling radiative transfer through forest canopies: implications for canopy photosynthesis and remote sensing. Pages 23–60 in Gholz, H. L., K. Nakane, and H. Shimoda, Eds. *The Use of Remote Sensing in the Modeling of Forest Productivity.* Kluwer, Dordrecht.

Nilsson, M. 1996. Estimation of tree heights and stand volume using airborne laser data. *Rem. Sensing Environ.,* 56: 1–7.

Nixon, P., D. Escobar, and R. Menges. 1985. Use of a multiband video system for quick assessment of vegetation condition and discrimination of plant species. *Rem. Sensing Environ.,* 17: 203–208.

Noss, R. F. 1990. Indicators of biodiversity: a hierarchical approach. *Conserv. Biol.,* 4: 355–364.

Noss, R. F. 1999. Assessing and monitoring forest biodiversity: a suggested framework and indicators. *For. Ecol. Manage.,* 115: 135–146.

Oderwald, R. G., and R. H. Wynne. 1998. The Virginia Tech resource assessment and inventory consortium. Pages 816–818 in *Proc. ASPRS Resour. Technol. Inst. Annu. Meet.,* American Society of Photogrammetry and Remote Sensing, Bethesda, MD.

Oderwald, R. G., and R. H. Wynne. 2000. Field applications for statistical data and techniques. *J. For.,* 98: 58–60.

Oliver, C. D., M. Boydak, G. Segura, and B. B. Bare. 1999. Forest organization, management, and policy. Pages 556–596 in M. L. Hunter, Jr., Ed. *Maintaining Biodiversity in Forest Ecosystems.* Cambridge University Press, Cambridge, U.K.

Ollinger, S. V., J. D. Aber, and C. A. Federer. 1998. Estimating regional forest productivity and water yield using an ecosystem model linked to a GIS. *Landscape Ecol.,* 13: 323–334.

Olson, C. E., Jr., and F. P. Weber. 2000. Foresters' roles in remote sensing. *J. For.,* 98: 11–12.

Olson, R. J., J. M. Briggs, J. H. Porter, G. R. Mark, and S. G. Stafford. 1999. Managing data from multiple disciplines, scales, and sites to support synthesis and modelling. *Rem. Sensing Environ.,* 70: 99–107.

Olsson, H. 1994. Changes in satellite-measured reflectances caused by thinning cuttings in boreal forest. *Rem. Sensing Environ.,* 50: 221–230.

Olthof, I., and D. J. King. 2000. Development of a forest health index using multispectral airborne digital camera imagery. *Can. J. Rem. Sensing,* 26: 166–176.

O'Neill, N. T., J. R. Miller, and J. R. Freemantle. 1995. Atmospheric correction of airborne BRF to surface BRF: nomenclature, theory, and methods. *Can. J. Rem. Sensing,* 21: 309–327.

O'Neill, R. V., D. L. DeAngelis, J. B. Waide, and T. F. H. Allen. 1986. *A Hierarchical Concept of Ecosystems.* Princeton University Press, Princeton, NJ.

O'Neill, R. V., C. T. Hunsaker, S. P. Timmins, B. L. Jackson, K. B. Jones, K. H. Riitters, and J. D. Wickham. 1996. Scale problems in reporting landscape patterns at the regional scale. *Landscape Ecol.,* 11: 169–180.

O'Neill, R. V., J. R. Krummer, R. H. Gardner, G. Sugihara, B. Jackson, D. L. DeAngelis, B. T. Milne, M. G. Turner, B. Zygmunt, S. W. Christensen, V. H. Dale, and R. L. Graham. 1988. Indices of landscape pattern. *Landscape Ecol.,* 1: 153–162.

O'Neill, R. V., S. J. Turner, V. I. Cullinan, D. P. Coffin, T. Cook, W. Conley, J. Brunt, J. M. Thomas, M. R. Conley, and J. Gosz. 1991. Multiple landscape scales: an intersite comparison. *Landscape Ecol.,* 5: 137–144.

Ong, P. C., and M. Kleine. 1996. DIPSIM: Dipterocarp forest growth simulation model — a tool for forest level management planning. Pages 228–246 in A. Schulte and D. Schone, Eds. *Dipterocarp Forest Ecosystems: Towards Sustainable Forest Management.* World Scientific, Singapore.

Oswald, E. T. 1976. Terrain analysis from Landsat imagery. *For. Chron.,* 56: 274–282.

Ouaidrari, H., and E. F. Vermote. 1999. Operational atmospheric correction of Landsat TM data. *Rem. Sensing Environ.,* 70: 4–15.

Pacala, S. W., C. D. Canham, J. Saponara, J. A. Silander, Jr., R. K. Kobe, and E. Ribbens. 1996. Forest models defined by field measurements: estimation, error analysis and dynamics. *Ecol. Appl.,* 66: 1–43.

Paijmans, K. 1966. Typing of tropical vegetation in northern Papua. *Photogrammetria,* 21: 1–25.

Paijmans, K. 1970. Land evaluation by air photo interpretation and field sampling in Australian New Guinea. *Photogrammetria,* 26: 77–100.

Palo, M., and G. Mery, Eds. 1996. *Sustainable Forestry Challenges for Developing Countries.* Kluwer, Dordrecht.

Papadias, D., and M. Egenhofer. 1996. *Algorithms for Hierarchical Spatial Reasoning.* NCGIA Techn. Rep. 96-2. University of Maine, Orono.

Paris, J. F., and H. H. Kwong. 1988. Characterization of vegetation with combined Thematic Mapper and Shuttle Imaging Radar (SIR-B) image data. *Photogramm. Eng. Rem. Sensing,* 54: 1187–1193.

Parker, W. C., S. J. Colombo, M. L. Cherry, M. D. Flannigan, S. Greifenhagen, R. S. McAlpine, C. Papadopol, and T. Scarr. 2000. Third millennium forestry: what climate change might mean to forests and forest management in Ontario. *For. Chron.,* 76: 445–463.

Payn, T. W., R. B. Hill, B. K. Hock, M. F. Skinner, A. J. Thorn, and W. C. Rijkse. 1999. Potential for the use of GIS and spatial analysis techniques as tools for monitoring changes in forest productivity and nutrition, a New Zealand example. *For. Ecol. Manage.,* 122: 187–196.

Pellikka, P. 1996. Illumination compensation for aerial video images to increase land cover classification in mountains. *Can. J. Rem. Sensing,* 22: 368–381.

Peddle, D.R. 1995a. MERCURY[⊕] An evidential reasoning image classifier. *Comput. Geosci.,* 21: 1163–1176.

Peddle, D. R. 1995b. Knowledge formulation for supervised evidential classification. *Photogramm. Eng. Rem. Sensing,* 61: 409–417.

Peddle, D. R., G. Foody, A. Zhang, S. E. Franklin, and E. F. LeDrew. 1994. Multisource image classification. II. An empirical comparison of the evidential reasoning and neural network approaches. *Can. J. Rem. Sensing,* 20: 396–407.

Peddle, D. R., and S. E. Franklin. 1991. Image texture processing and ancillary data integration for surface pattern discrimination. *Photogramm. Eng. Rem. Sensing,* 57: 413–420.

Peddle, D.R., and S. E. Franklin. 1992. Multi-source evidential classification of surface cover and frozen ground. *Int. J. Rem. Sensing,* 13: 3375–3380.

Peddle, D. R., and S. E. Franklin. 1993. Classification of permafrost active layer depth from remotely sensed and topographic evidence. *Rem. Sensing Environ.,* 44: 67–80.

Peddle, D. R., F. G. Hall, and E. F. LeDrew. 1999. Spectral mixture analysis and geometric-optical reflectance modeling of boreal forest biophysical structure. *Rem. Sensing Environ.,* 67: 288–297.

Peery, M. Z., R. J. Gutierrez, and M. E. Seamans. 1999. Habitat composition and configuration around Mexican spotted owl nest and roost sites in the Tularosa Mountains, New Mexico. *J. Wild. Manage.,* 63: 36–43.

Penner, M., K. Power, C. Muhairwe, R. Tellier, and Y. Wang. 1997. *Canada's Forest Biomass Resources: Deriving Estimates from Canada's Forest Inventory.* Canadian Forest Service Inf. Rep. BC-X-370. Pacific Forestry Centre, Victoria, British Columbia.

Peterson, D. L. 1997. Forest structure and productivity along the Oregon Transect. Pages 173–218 in H. L. Gholz, K. Nakane, and H. Shimoda, Eds. *The Use of Remote Sensing in the Modeling of Forest Productivity.* Kluwer, Dordrecht.

Peterson, D. L., J. D. Aber, P. A. Matson, D. H. Card, N. Swanberg, C. Wessman, and M. A. Spanner. 1988. Remote sensing of forest canopy and leaf biochemical contents. *Rem. Sensing Environ.,* 24: 85–108.

Peterson, D. L. and V. T. Parker, Eds. 1998. *Ecological Scale: Theory and Applications.* Columbia University Press, New York.

Peterson, D. L., M. A. Spanner, S. W. Running, and K. Teuber. 1987. Relationship of Thematic Mapper data to leaf area index of temperate coniferous forests. *Rem. Sensing Environ., 22*: 323–341.

Peterson, D. L. and R. H. Waring. 1994. Overview of the Oregon Transect Ecosystem Research Project. *Ecol. Appl.,* 4: 211–225.

Peterson, E. B. and N. M. Peterson. 1992. *Ecology, Management and Use of Aspen and Balsam Poplar in the Prairie Provinces.* Northwest Region Spec. Rep. 1. Canadian Forest Service, Edmonton, Alberta.

Peterson, U. and T. Nilson. 1993. Successional reflectance trajectories in northern temperate forests. *Int. J. Rem. Sensing,* 14:609–613.

Pettinger, L. R. 1982. *Digital Classification of Landsat Data for Vegetation and Land-cover Mapping in the Blackfoot River Watershed, Southeastern Idaho.* USGS Prof. Pap. 1219. Washington, DC.

Pfaff, A. S. P., S. Kerr, R. Flint Hughes, S. Liu, G. A. Sanchez-Azofeifa, D. Schimel, J. Tosi, and V. Watson. 2000. The Kyoto protocol and payments for tropical forest: an interdisciplinary method for estimating carbon-offset supply and increasing the feasibility of a carbon market under the CDM. *Ecol. Econ.,* 35: 203–221.

Philipson, W., Ed. 1997. *Manual of Aerial Photointerpretation.* 2nd ed. ASPRS, Bethesda, MD.

Phillips, C. G. and J. Randolph. 1998. Has ecosystem management really changed practices on National Forests? *J. For.,* 96: 40–45.

Pickup, G. and V. H. Chewings. 1996. Correlations between DEM-derived topographic indices and remotely sensed vegetation cover in rangelands. *Earth Surf. Process. Landforms,* 21: 517–529.

Pierce, L. L. and S. W. Running. 1988. Rapid estimation of coniferous forest leaf area index using a portable integrating radiometer. *Ecology,* 69: 1762–1767.

Pierce, L. L. and S. W. Running. 1995. The effects of aggregating sub-grid land surface variation on large-scale estimates of net primary production. *Landscape Ecol.,* 10: 239–253.

Pike, R. J. 1988. The geometric signature: quantifying landslide susceptible terrain types from digital elevation models. *Math. Geol.,* 20: 491–511.

Pike, R. J. 1999. Geomorphometry: diversity in quantitative surface analysis. *Prog. Phys. Geogr.,*. 24: 1–20.

Pilon, P. G., P. J. Howarth, R. A. Bullock, and P. O. Adeniyi. 1988. An enhanced classification approach to change detection in semi-arid environments. *Photogramm. Eng. Rem. Sensing,* 54: 1709–1716.

Pilon, P. G. and R. J. Wiart. 1990. Operational forest inventory applications using Landsat TM: the British Columbia experience. *Geocarto Int.,* 1: 25–30.

Pinz, A., M. Prantl, H. Ganster, and H. Kopp-Borotschnig. 1996. Active fusion — a new method applied to remote sensing image interpretation. *Pattern Recognition Lett.,* 17: 1349–1359.

Pitt, D. G., R. G. Wagner, R.J. Hall, D. J. King, D. G. Leckie, and U. Runesson. 1997. Use of remote sensing for forest vegetation management: a problem analysis. *For. Chron.,* 73: 459–477.

Piwowar, J. M. and E. F. LeDrew. 1990. Integrating spatial data: a user's perspective. *Photogramm. Eng. Rem. Sensing,* 56: 1497–1502.

Porter , W. M. and H. T. Enmark. 1987. A system overview of the Airborne Visible/Infrared Imaging Spectrometer (AVIRIS). Pages 22–31 in *Proc. Imaging Spectrosc., II SPIE 834*. SPIE, Bellingham, WA.

Poso, S., T. Häme, and R. Paananen. 1984. A method of estimating the stand characteristics of a forest compartment using satellite imagery. *Silva Fenn.,* 18: 261–292.

Pouch, G. and D. Compagna. 1990. Hyperspherical directional cosines for separation of spectral and illumination information in digital scanner data. *Photogramm. Eng. Rem. Sensing,* 56: 475–479.

Presutti, M. E., L. M. Moskal, E. E. Dickson, and S. E. Franklin. 2000. Forestry and agricultural land cover classification using Radarsat and Landsat TM in SE Buenos Aires Province, Argentina. Pages 715–726 in *Proc. 22nd Can. Symp. Rem. Sensing.* Canadian Aeronautics and Space Institute, Ottawa.

Price, J. C. 1987. Combining panchromatic and multispectral imagery from dual resolution instruments. *Rem. Sensing Environ.,* 21: 119–128.

Price, J. C. 1990. Using spatial context in satellite data to infer regional scale evapotranspiration. *IEEE Trans. Geosci. Rem. Sensing,* 28: 940–948.

Price, K. P. and M. E. Jakubauskas. 1998. Spectral retrogression and insect damage in lodgepole pine successional forests. *Int. J. Rem. Sensing,* 19: 1627–1632.

Price, M. 1986. The analysis of vegetation change by remote sensing. *Prog. Phys. Geogr.,* 10: 473–491.

Prince, S. D. 1991. A model of regional primary production for use with coarse resolution satellite data. *Int. J. Rem. Sensing,* 12: 1313–1330.

Prins, E. and I. S. Kikula. 1996. Deforestation and regrowth phenology in miombo woodland assessed by Landsat Multispectral Scanner System data. *For. Ecol. Manage.,* 84: 263–266.

Proisy, C., E. Mougin, F. Fromard, and M. A. Karam. 2000. Interpretation of polarimetric radar signatures of mangrove forests. *Rem. Sensing Environ.,* 71: 56–66.

Qi, J., A. R. Huete, M. S. Moran, A. Chehbouni, and R. D. Jackson. 1993. Interpretation of vegetation indices derived from multitemporal SPOT images. *Rem. Sensing Environ.,* 44: 80–101.

Qian, J., R. W. Ehrich, and J. B. Campbell. 1990. DNESYS — an expert system for automatic extraction of drainage networks from digital elevation data. *IEEE Trans. Geosci. Rem. Sensing,* 28: 29–45.

Quegan, S. 1995. Recent advances in understanding SAR imagery. Pages 89–104 in F. M. Danson, and S. E. Plummer, Eds. *Advances in Environmental Remote Sensing*. John Wiley & Sons, London.

Raa, D. P., P. S. Desai, D. K. Das, and P. S. Roy. 1997. Remote sensing applications: future research thrust areas. *J. Indian Soc. Rem. Sensing,* 25: 195–224.

Rajan, M. S. 1991. *Remote Sensing and GIS for Natural Resource Management.* Asian Development Bank, Manilla.

Ramirez-Garcia, P., J. Lopez-Blanco, and D. Ocana. 1998. Mangrove vegetation assessment in the Santiago River Mouth, Mexico, by means of supervised classification using Landsat TM imagery. *For. Ecol. Manage.,* 105: 217–229.

Ramsay, E. W., III, D. K. Chappell, D. M. Jacobs, S. K. Sapkota, and D. G. Baldwin. 1998. Resource management of forest wetlands: hurricane impact and recovery mapped by combining Landsat TM and NOAA AVHRR data. *Photogramm. Eng. Rem. Sensing,* 64: 733–738.

Ranson, K. J., J. R. Irons, and C. S. T. Daughtry. 1991. Surface albedo from bidirectional reflectance. *Rem. Sensing Environ.,* 35: 201–211.

Ranson, K. J., J. R. Irons, and D. L. Williams. 1994. Multispectral bidirectional reflectance of northern forest canopies with the Advanced Solid-State Array Spectroradiometer (ASAS). *Rem. Sensing Environ.,* 47: 276–289.

Ranson, K. J., and S. S. Saatchi. 1992. C-band scattering from small balsam fir. *IEEE Trans. Geosci. Rem. Sensing,* 30: 924– 932.

Ranson, K. J., and G. Sun. 1994a. Northern forest classification using temporal multifrequency and multipolarimetric SAR images. *Rem. Sensing Environ.,* 47: 142–153.

Ranson, K. J., and G. Sun. 1994b. Mapping biomass of a northern forest using multifrequency SAR data *IEEE Trans. Geosci. Rem. Sensing,* 32: 388–396.

Ranson, K. J., G. Sun, J. F. Weishampel, R. G. Knox. 1996. Forest biomass from combined ecosystem and radar backscatter modelling. *Rem. Sensing Environ.,* 59: 118–133.

Raske, A. 1986. *Vulnerability Rating of Forests of Newfoundland to Spruce Budworm Attack.* Forest Res. Centre Inf. Rep. N-X-239, Canadian Forest Service, St. John's, Newfoundland.

Rauste, Y. 1990. Incidence angle dependence in forested and non-forested areas in Seasat SAR data. *Int. J. Rem. Sensing,* 11: 1267–1276.

Raven, P. H. and J. A. McNeely. 1998. Biological extinction: its scope and meaning for us. Pages 13–32 in L. D. Guruswamy and J. A. McNeely, Eds. *Protection of Global Biodiversity: Converging Strategies.* Duke University Press, Durham.

Reed, R. A., J. Johnson-Barnard, and W. L. Baker. 1996. Fragmentation of a forested Rocky Mountain landscape, 1950–1993. *Biol. Conserv.,* 75: 267–277.

Rees, G. 1999. *The Remote Sensing Data Book.* Cambridge University Press, Cambridge.

Reid, N. J. 1987. Remote sensing and forest damage. *Environ. Sci. Technol.,* 21: 428–429.

Rempel, R. S., P. C. Elkie, A. R. Rodgers, and M. J. Gluck. 1997. Timber-management and natural-disturbance effects on moose habitat: landscape evaluation. *J. Wild. Manage.,* 61: 517–524.

Rencz, A.N. 1985. Multitemporal analysis of Landsat imagery for monitoring forest cutovers in Nova Scotia. *Can. J. Rem. Sensing,* 11: 188–194.

Ribe, R., R. Morganti, D. Hulse, and R. Shull. 1999. A management driven investigation of landscape patterns of northern spotted owl in the high Cascades of Oregon. *Landscape Ecol.,* 13: 1–13.

Rice, T. 1996. *Changes in Forestry Policies and Practices.* Environmental Research Ltd., London.

Richelson, J. T. 1991. The future of space reconnaissance. *Sci. Am.,* 264: 38–44.

Richards, J. A., and X. Jia. 1999. *Remote Sensing Digital Image Analysis. An Introduction.* 3rd ed. Springer-Verlag, New York.

Richardson, B., M. F. Skinner, and G. West. 1999. The role of forest productivity in defining the sustainability of plantation forests in New Zealand. *For. Ecol. Manage.,* 122: 125–137.

Richter, R. 1990. A fast atmospheric correction algorithm applied to Landsat TM images. *Int. J. Rem. Sensing,* 11: 159–166.

Richter, R. 1997. Correction of atmospheric and topographic effects for high spatial resolution satellite images. *Int. J. Rem. Sensing,* 18: 1099–1111.

Rickers, J. R., L. P. Queen, and G. J. Arthaud. 1995. A proximity-based approach to assessing habitat. *Landscape Ecol.,* 10: 309–321.

Rignot, E. C., C. Williams, J. B. Way, and L. A. Viereck. 1994. Mapping of forest types in Alaskan boreal forests using SAR imagery. *IEEE Trans. Geosci. Rem. Sensing,* 32: 1051–1058.

Riitters, K. H., R. V. O'Neill, C. T. Hunsaker, J. D. Wickham, D. H. Yankee, S. P. Timmins, K. B. Jones, and B. L. Jackson. 1995. A factor analysis of landscape pattern and structure metrics. *Landscape Ecol.,* 10: 23–39.

Riley, J. R. 1989. Remote sensing in entomology. *Annu. Rev. Entomol.,* 34: 247–271.

Ringrose, S., and P. Large. 1983. The comparative value of Landsat print and digitized data and radar imagery for ecological land classification in the humid tropics. *Can. J. Rem. Sensing,* 9: 45–60.

Ringrose, S., and W. Matheson. 1991. Characterization of woody vegetation cover in the south-east Botswana Kalahari. *Global Ecol. Biogeogr. Lett.,* 1: 176–181.

Ringrose, S., and W. Matheson. 1992. The use of Landsat MSS imagery to determine the extent of woody vegetation cover change in the west-central Sahel. *Global Ecol. Biogeogr. Lett.,* 2: 16–25.

Ripple, W. J., D. H. Johnson, K. T. Hershey, and E. C. Meslow. 1991. Old-growth and mature forests near spotted owl nests in western Oregon. *J. Wild. Manage.,* 55: 316–318.

Ripple, W. J., P. D. Lattin, K. T. Hershey, F. F. Wagner, and E. C. Meslow. 1997. Landscape composition and pattern around northern spotted owl nest sites in southwest Oregon. *J. Wild. Manage.,* 61: 151–158.

Ripple, W. J., S. Wang, D. L. Isaacson, and D. P. Paine. 1991. A preliminary comparison of Landsat TM and SPOT-1 HRV multispectral data for estimating coniferous forest volume. *Int. J. Rem. Sensing,* 12: 1971–1977.

Risley, E. M. 1967. Developments in the application of Earth observation satellites to geographic problems. *Prof. Geogr.,* XIX: 130–132.

Roberts, A. 1995. Integrated MSV airborne remote sensing. *Can. J. Rem. Sensing,* 21: 214–224.

Robertson, G. J., and S. R. Cvetkovic. 1991. Remote sensing with small satellites. *Int. J. Rem. Sensing,*. 12: 23–31.

Robinove, C. J. 1979. *Integrated Terrain Mapping with Digital Landsat Images in Queensland, Australia.* USGS Prof. Pap. No. 1102. Washington, DC.

Robinove, C. J. 1981. The logic of multispectral classification and mapping of land. *Rem. Sensing Environ.,* 11: 231–244.

Robinove, C. J. 1982. Computation with physical values from Landsat digital data. *Photogramm. Eng. Rem. Sensing,* 48: 781–784.

Robinson, G. D., H. N. Gross, and J. R. Schott. 2000. Evaluation of two applications of spectral mixing models to image fusion. *Rem. Sensing Environ.,* 71: 272–281.

Rock, B. N., T. Hoshizaki, and J. R. Miller. 1988. Comparison of *in situ* and airborne spectral measurements of the blue shift associated with forest decline. *Rem. Sensing Environ.,* 24: 109–127.

Rock, B. N., and J. E. Vogelmann. 1986. Use of TMS/TM data for mapping forest decline damage in the northeastern United States. Pages 1405–1410 in *Proc. IGARSS'86 Symp.,* ESA SP-254, European Space Agency Publ. Div., Paris, France.

Rock, B. N., J. E. Vogelmann, D. L. Williams, A. F. Vogelmann, and T. Hoshizaki. 1986. Remote detection of forest damage. *BioScience,* 36: 439–445.

Rohde, W. G. and C. E. Olson. 1972. Multispectral sensing of forest tree species. *Photogramm. Eng.,* 38: 1209.

Roller, N. 2000. Intermediate multispectral satellite sensors. *J. For.,* 98: 32–35.

Rosenzweig, M. L., and Z. Abramsky. 1993. How are diversity and productivity related? Pages 52–65 in R. E. Ricklefs, and D. Schluter, Eds. *Species Diversity in Ecological Communities: Historical and Geographical Perspectives.* University of Chicago Press, Chicago, IL.

Rothery, D. A., and P. W. Francis. 1987. Synergistic use of MOMS-01 and Landsat TM data. *Int. J. Rem. Sensing,* 8: 501–508.

Rousseau, D. 1998. Strategies at the national and international levels for sustainable forest management. *For. Chron.,* 74: 220–223.

Rowe, J. P., T. A. Warner, D. R. Dean, Jr., and A. F. Egan. 1999. A remote sensing strategy for measuring logging road length from small-format aerial photography. *Photogramm. Eng. Rem. Sensing,* 65: 697–703.

Rowe, J. S. 1972. *Forest Regions of Canada.* Canadian Forest Service Publ. No. 1300. Environment Canada, Ottawa.

Rowe, J. S. 1996. Land classification and ecosystem classification. *Environ. Monit. Assess.,* 39: 11–20.

Royer, A., P. Vincent, and F. Bonn. 1985. Evaluation and correction of viewing angle effects on satellite measurements of bidirectional reflectance. *Photogramm. Eng. Rem. Sensing,* 51: 1899–1914.

Royle, D. D., and R. Lathrop. 1997. Monitoring hemlock forest health in New Jersey using Landsat Thematic Mapper data and change detection techniques. *For. Sci.,* 43: 327–335.

Rubec, C. D. A., Ed. 1979. *Applications of Ecological (Biophysical) Land Classification in Canada.* Ecological Land Classification Series No. 7. Environment Canada, Ottawa.

Rubec, C. D. A. 1983. Applications of remote sensing in ecological land survey in Canada. *Can. J. Rem. Sensing,* 9: 19–30.

Ruimy, A., B. Saugier, and G. Dedieu. 1994. Methodology for the estimation of net primary production from remotely sensed data. *J. Geophys. Res.,* 99: 5263–5283.

Running, S. W. 1990. Estimating terrestrial primary productivity by combining remote sensing and ecosystem simulation. Pages 65–86 in R. Hobbs, and H. Mooney, Eds. *Remote Sensing of Biosphere Functioning.* Springer-Verlag, New York.

Running, S. W. 1994. Testing FOREST-BGC ecosystem process simulations across a climatic gradient in Oregon. *Ecol. Appl.* 4: 238–247.

Running, S. W., and J. C. Coughlan. 1988. A general model of forest ecosystem processes for regional applications. I. Hydrologic balance, canopy gas exchange and primary production processes. *Ecol. Model.,* 42: 125–154.

Running, S. W., and S. T. Gower. 1991. FOREST-BGC, a general model of forest ecosystem processes for regional applications. II. Dynamic carbon allocation and nitrogen budgets. *Tree Physiol.,* 9: 147–160.

Running, S.W., and E. R. Hunt, Jr. 1993. Generalization of a Forest Ecosystem Process Model for Other Biomes, BIOME–BGC, and an Application for Global–Scale Models. Pages 141–157 in Ehleringer, J., and C. Field, Eds. *Scaling Physiological Processes: Leaf to Globe.* Academic Press, Toronto.

Running, S. W., T. R. Loveland, and L. L. Pierce. 1994. A vegetation classification logic based on remote sensing for use in global biogeochemical models. *Ambio,* 23: 77–81.

Running, S.W., T. R. Loveland, L. L. Pierce, and E. R. Hunt, Jr. 1995. A remote sensing based vegetation classification logic for global land cover analysis. *Rem. Sensing Environ.,* 51: 39–48.

Running, S. W., R. Nemani, and R. D. Hungerford. 1987. Extrapolation of synoptic meteorological data in mountainous terrain and its use for simulating forest evapotranspiration and photosynthesis. *Can. J. For. Res.,* 17: 472–483.

Running, S. W., R. R. Nemani, D. L. Peterson, L. E. Band, D. F. Potts, L. L. Pierce, and M. A. Spanner. 1989. Mapping regional forest evapotranspiration and photosynthesis by coupling satellite data with ecosystem simulation. *Ecology,* 70: 1090–1101.

Running, S. W., L. Queen, and M. Thornton. 2000. The Earth Observing System and forest management. *J. For.,* 98: 29–31.

Runyon, J., R. H. Waring, S. N. Goward, and J. M. Welles. 1994. Environmental limits on net primary production and light-use efficiency across the Oregon Transect. *Ecol. Appl.,* 4: 226–237.

Ryan, M. G., M. B. Lavigne, and S. T. Gower. 1997. Annual carbon cost of autotrophic respiration in boreal forest ecosystems in relation to species and climate. *J. Geophys. Res.,* 102, D24: 28871–28883.

Ryerson, R. 1989. Image interpretation concerns for the 1990s and lessons from the past. *Photogramm. Eng. Rem. Sensing,* 55: 1427–1430.

Ryherd, S. and C. E. Woodcock. 1997. Combining spectral and texture data in the segmentation of remotely sensed images. *Photogramm. Eng. Rem. Sensing,* 62: 181–194.

Sachs, D. L., P. Sollins, and W. B. Cohen. 1998. Detecting landscape changes in the interior of British Columbia from 1975 to 1992 using satellite imagery. *Can. J. For. Res.,* 28: 23–36.

Sader, S. A., D. Ahl, W. S. Liou. 1995. Accuracy of Landsat Thematic Mapper and GIS rule-based methods for forest wetland classification in Maine. *Rem. Sensing Environ.,* 53: 133–144.

Sader, S. A., and S. Vermillion. 2000. Remote sensing education: an updated survey. *J. For.,* 98: 31–37.

Sader, S. A., R. B. Waide, W. T. Lawrence, and A. T. Joyce. 1989. Tropical forest biomass and successional age class relationships to a vegetation index derived from Landsat data. *Rem. Sensing Environ.,* 28: 143–156.

Saint, G. 1980. Multitemporal remote sensing: satellites provide a new tool for Earth resources management. *Acta Astron.,* 7: 373–383.

Sali, E., and H. Wolfson. 1992. Texture classification in aerial photographs and satellite data. *Int. J. Rem. Sensing,* 13: 3395–3408.

Salvador, R., and X. Pons. 1998a. On the applicability of Landsat TM images to Mediterranean forest inventories. *For. Ecol. Manage.,* 104: 193–208.

Salvador, R., and X. Pons. 1998b. On the reliability of Landsat TM for estimating forest variables by regression techniques: a methodological analysis. *IEEE Trans. Geosci. Rem. Sensing,* 36: 1888–1897.

Salvador, R., J. Valeriano, X. Pons, and R. Diaz-Delgado. 2000. A semi-automatic methodology to detect fire scars in shrubs and evergreen forests with Landsat MSS time series. *Int. J. Rem. Sensing,* 21: 655–671.

Sampson, R. N., and D. L. Adams, Eds. 1994. *Assessing Forest Health in the Inland West.* Haworth Press, Binghamton, NY.

Sanchez, J., and M. P. Canton. 1999. *Space Image Processing.* CRC Press, Boca Raton, FL.

Sandmeier, St., and D. W. Deering. 1999. Structure analysis and classification of boreal forests using airborne hyperspectral BRDF data from ASAS. *Rem. Sensing Environ.,* 69: 281–295.

Sandmeier, St., and K. I. Itten. 1997. A physically based model to correct atmospheric and illumination effects in optical satellite data of rugged terrain. *IEEE Trans. Geosci. Rem. Sensing,* 35: 708–717.

Sauer, C. O. 1921. The problem of land classification. *Ann. Assoc. Am. Geographers,* 11: 3–16.
Saunders, D. A., R. J. Hobbs, and C. R. Margules. 1991. Biological consequences of ecosystem fragmentation: a review. *Conserv. Biol.,* 5: 18–32.
Sayn-Wittgenstein, L. 1978. *Recognition of Tree Species on Aerial Photographs.* Forest Management Institute Inf. Rep. FMR-X-118. Canadian Forest Service, Ottawa.
Sayn-Wittgenstein, L., R. de Milde, and C. J. Inglis. 1978. *Identification of Tropical Trees on Aerial Photographs.* Forest Management Institute Inf. Rep. FMR-X-113. Canadian Forest Service, Ottawa.
Scarth, P., and S. Phinn. 2000. Determining forest structural attributes using an inverted geometric-optical model in mixed eucalypt forests, southeast Queensland, Australia. *Rem. Sensing Environ.,* 71: 141–157.
Schama, S. 1995. *Landscape and Memory.* Random House, Toronto.
Scheidegger, A. E. 1986. The catena principle in geomorphology. *Z. Geomorphol.,* 30: 257–273.
Schneider, E. D., and J. J. Kay. 1994. Complexity and thermodynamics: towards a new ecology. *Futures,* 24: 626–647.
Schoonmaker, P. K. 1998. Paleoecological perspectives on ecological scale. Pages 79–103 in D. L. Peterson, and V. T. Parker, Eds. *Ecological Scale: Theory and Applications.* Columbia University Press, New York.
Schriever, J. R., and R. G. Congalton. 1995. Evaluating seasonal variability as an aid to cover type mapping from Landsat Thematic Mapper data in the Northeast. *Photogramm. Eng. Rem. Sensing,* 61: 321–327.
Schroeder, P., S. Brown, J. Mo, R. Birdsey, and C. Cieszewski. 1997. Biomass estimation for temperate broadleaf forests of the United States using inventory data. *For. Sci.,* 43: 424–434.
Schweitzer, G. E. 1982. Airborne remote sensing. *Environ. Sci. Technol.,* 16: 338A-346A.
Scott, H. D., and T. H. Udouj. 1999. Spatial and temporal characterization of land-use in the Buffalo National River Watershed. *Environ. Conserv.,* 26: 94–101.
Scott, J. M., T. H. Tear, and F. W. Davis, Eds. 1996. *Gap Analysis: A Landscape Approach to Biodiversity Planning.* American Society for Photogrammetry and Remote Sensing, Bethesda, MD.
Seely, B., J. P. Kimmins, C. Welham, and K. Scoullar. 1999. Defining stand-level sustainability; exploring stand-level stewardship. *J. For.,* 97: 4–10.
Sellers, P. J. 1985. Canopy reflectance, photosynthesis and transpiration. *Int. J. Rem. Sensing,* 6: 1335–1372.
Shafer, G. 1976. *A Mathematical Theory of Evidence.* Princeton University Press, Princeton, NJ.
Shaffer, L. R. 1996. CEOS holds meeting on integrated global observing strategy. *Comm. Earth Observation Satellites Newsl.,* No. 7, September.
Shaffer, L. R. 1997. CEOS and IGOF — the way forward. *Comm. Earth Observation Satellites Newsl.,* No. 8, February.
Sheen, D. R., and L. P. Johnston. 1992. Statistical and spatial properties on forest clutter measured with polarimetric SAR. *IEEE Trans. Geosci. Rem. Sensing,* 30: 578–588.
Shepard, W. B. 1993. Modern forest management: its about opening up, not locking up. Pages 218–227 in W. W. Covington, and L. F. DeBano, Eds. *Sustainable Ecosystems: Implementing an Ecological Approach to Land Management.* USDA Forest Service Tech. Rep. RM-247. USDA, Fort Collins, CO.
Shettigara, V. K., and G. M. Sumerling. 1998. Height determination of extended objects using shadows in SPOT images. *Photogramm. Eng. Rem. Sensing,* 64: 35–44.
Shimabukuro, Y. E., and J. A. Smith. 1991. The least-squares mixing models to generate fraction images derived from remote sensing multispectral data. *IEEE Trans. Geosci. Rem. Sensing,* 29: 16–20.

Shimabukuro, Y. E., and J. A. Smith. 1995. Fraction images derived from Landsat TM and MSS data for monitoring reforested areas. *Can. J. Rem. Sensing,* 21: 67–74.

Shlien, S. 1979. Geometric correction, registration, and resampling of Landsat imagery. *Can. J. Rem. Sensing,* 5: 74–89.

Shore, T. L., and L. Safranyik. 1992. *Susceptibility and Risk Rating Systems for the Mountain Pine Beetle in Lodgepole Pine Stands.* Pacific Forestry Centre Inf. Rep. BC-C-336, Canadian Forest Service, Victoria, British Columbia.

Shoshany, M. 2000. Satellite remote sensing of natural Mediterranean vegetation: a review within an ecological context. *Prog. Phys. Geogr.,* 24: 153–178.

Shugart, H. H. 1998. *Terrestrial Ecosystems in Changing Environments.* Cambridge University Press, Cambridge, U.K.

Silbaugh, J. M., and D. R. Betters. 1997. Biodiversity values and measures applied to forest management. *J. Sustainable For.,* 5: 235–248.

Silva, L. F. 1978. Radiation and instrumentation in remote sensing. Pages 21–135 in P. H. Swain, and S. M. Davis, Eds. *Remote Sensing: the Quantitative Approach.* McGraw-Hill, New York.

Simberloff, D. 1999. The role of science in the preservation of forest biodiversity. *For. Ecol. Manage.,* 115: 101–111.

Simmons, M. A., V. I. Cullinan, and J. M. Thomas. 1992. Satellite imagery as a tool to evaluate ecological scale. *Landscape Ecol.,* 7: 77–85.

Sims, R. A., I. G. W. Corns, and K. Klinka, Eds. 1996. *Global to Local Ecological Land Classification.* Kluwer, Dordrecht.

Singer, M. J., and D. N. Munns. 1987. *Soils: An Introduction.* Macmillan Publishing Company, New York.

Singh, A. 1989. Digital change detection techniques using remotely sensed data. *Int. J. Rem. Sensing,* 10: 989–1001.

Singh, T., and E. E. Wheaton. 1991. Boreal forest sensitivity to global warming: implications for forest management in western Canada. *For. Chron.,* 67: 342–348.

Sirois, J., and F. J. Ahern. 1988. An investigation of SPOT HRV data for detecting mountain pine beetle mortality. *Can. J. Rem. Sensing,* 14: 104–108.

Skelly, W. C. 1990. *Microwave Backscatter Modelling of Forested Terrain: A Theoretical Approach to Image Interpretation.* Unpublished MSc Thesis. University of Western Ontario, London, Canada.

Skidmore, A. K. 1989. An expert system classifies eucalypt forest types using Thematic Mapper data and a digital terrain model. *Photogramm. Eng. Rem. Sensing,* 55: 1449–1464.

Skidmore, A. K., F. Watford, P. Luckananumy, and P. Ryan. 1996. An operational GIS expert system for mapping forest soils. *Photogramm. Eng. Rem. Sensing,* 62: 501–511.

Skole, D. L., and C. J. Tucker. 1993. Tropical deforestation and habitat fragmentation in the Amazon: satellite data from 1978 to 1988. *Science,* 260: 1905–1910.

Slaymaker, D. M., K. M. L. Jones, C. R. Griffin, and J. T. Finn. 1996. Mapping deciduous forests in southern New England using aerial videography and hyperclustered multi-temporal Landsat TM imagery. Pages 87–101 J. M. Scott, T. H. Tear, and F. W. Davis, Eds. *Gap Analysis: A Landscape Approach to Biodiversity Planning.* American Society for Photogrammetry and Remote Sensing, Bethesda, MD.

Smith, C. T., A. T. Lowe, and M. F. Proe. 1999. Preface: indicators of sustainable forest management. *For. Ecol. Manage.,* 122: 1–5.

Smith, C. T., and R. J. Raison. 1998. Utility of Montreal Process indicators for soil conservation in native forests and plantations in Australia and New Zealand. Pages 121–135 in Soil Science of America Spec. Publ. No. 53. *Contribution of Soil Science to the Development and Implementation of Criteria and Indicators of Sustainable Forest Management.* SSSA, Madison, WI.

Smith, D. M. 1986. *The Practice of Silviculture.* John Wiley & Sons, New York.

Smith, F. W., D. A. Sampson, and J. L. Long. 1991. Comparison of leaf area index estimates from tree allometrics and measured light interception. *For. Sci.,* 37: 1682–1688.

Smith, G. M., and E. J. Milton. 1999. The use of the empirical line method to calibrate remotely sensed data to reflectance. *Int. J. Rem. Sensing,* 20: 2653–2662.

Smith, J. A., T. L. Lin, and K. J. Ranson. 1980. The Lambertian assumption and Landsat data. *Photogramm. Eng. Rem. Sensing,* 46: 1183–1189.

Smith, J. L. 1986. Evaluation of the effects of photo measurement errors on predictions of stand volume from aerial photography. *Photogramm. Eng. Rem. Sensing,* 52: 401–410.

Smith, N., G. Borstad, D. Hill, and R. Kerr. 1991. Using high-resolution airborne spectral data to estimate leaf area and stand structure. *Can. J. For. Res.,* 21: 1127–1132.

Smith, T. R., C. Zhan, and P. Gao. 1990. A knowledge-based, two-step procedure for extracting channel networks from noisy DEM data. *Comput. Geosci.,* 16: 777–786.

Smith, W., T. C. Meredith, and T. Johns. 1999. Exploring methods for rapid assessment of woody vegetation in the Batemi Valley, north-central Tanzania. *Biodiversity Conserv.,* 8: 447–470.

Smits, P. C., and A. Annoni. 1999. Updating land-cover maps by using texture information from very high resolution spaceborne imagery. *IEEE Trans. Geosci. Rem. Sensing,* 37: 1244–1254.

Sohlberg, S., and V. E. Sokolov, Eds. 1986. *Practical Applications of Remote Sensing in Forestry.* Martinus Nijhoff, Dordrecht.

Sohn, Y., E. Moran, and F. Gurri. 1999. Deforestation in north-central Yucatan (1985–1995): mapping secondary succession of forest and agricultural land-use in Sotuta using the cosine of the angle concept. *Photogramm. Eng. Rem. Sensing,* 65: 947–958.

Solberg, A. H. S. 1999. Contextual data fusion applied to forest map revision. *IEEE Trans. Geosci. Rem. Sensing,* 37: 1234–1243.

South, D. B. 1999. How can we feign sustainability with an increasing population? *New For.,* 17: 193–212.

Spanner, M. A., L. Johnson, J. Miller, R. McCreight, J. Freemantle, J. Runyon, and P. Gong. 1994. Remote sensing of leaf area index across the Oregon Transect. *Ecol. Appl.,* 4: 258–271.

Spanner, M. A., L. L. Pierce, S. W. Running, and D. L. Peterson. 1990. Remote sensing of temperate coniferous forest leaf area index: the influence of canopy closure, understory vegetation, and background spectral response. *Int. J. Rem. Sensing,* 11: 95–111.

Speight, M. R., and D. Wainhouse. 1989. *Ecology and Management of Forest Insects.* Clarendon Press, London.

Spencer, R. D., and R. J. Hall. 1988. Canadian large-scale aerial photographic systems (LSP). *Photogramm. Eng. Rem. Sensing,* 54: 475–482.

Spies, T. A. 1997. Forest stand structure, composition, and function. Pages 11–30 in K. A. Kohm, and J. F. Franklin, Eds. *Creating a Forestry for the 21st Century, the Science of Ecosystem Management.* Island Press, Washington, DC.

Spies, T. A., W. J. Ripple, and G. A. Bradshaw. 1994. Dynamics and pattern of a managed coniferous forest landscape in Oregon. *Ecological Appl.* 4: 555–568.

Spies, T. A. and M. G. Turner. 1999. Dynamic forest mosaics. Pages 95–160 in M. L. Hunter, Jr., Ed. *Maintaining Biodiversity in Forest Ecosystems.* Cambridge University Press, Cambridge, U.K.

Spurr, S. A. 1960. *Photogrammetry and Photo-interpretation.* 2nd ed. Ronald Press, New York.

Star, J. L., J. E. Estes, and K. C. McGwire. 1997. Integration of geographic information systems and remote sensing; a background to NCGIA initiative 12. Pages 1–12 in Star, J. L., J. E. Estes, and K. C. McGwire, Eds. *Integration of Geographic Information Systems and Remote Sensing.* Cambridge University Press, Cambridge, U.K.

Steiner, D. 1974. Digital geometric picture correction using a piecewise zero-order transformation. *Rem. Sensing Environ.,* 3: 261–274.

Stellingwerf, D. A. 1966. *Practical Applications of Aerial Photographs in Forestry and Other Vegetation Studies.* ITJ Publ. Ser. B., Delft, The Netherlands.

Stellingwerf, D. A., and Y. A. Hussin. 1997. *Measurements and Estimations of Forest Stand Parameters Using Remote Sensing.* VSPBV, Utrecht, The Netherlands.

Stenback, J. M., and R. G. Congalton. 1990. Using Thematic Mapper imagery to examine forest understory. *Photogramm. Eng. Rem. Sensing,* 56: 1285–1290.

Stoms, D. M. 1992. Effects of habitat map generalization in biodiversity assessment. *Photogramm. Eng. Rem. Sensing,* 58: 1587–1591.

Stoms, D. M., and J. E. Estes. 1993. A remote sensing research agenda for mapping and monitoring biodiversity. *Int. J. Rem. Sensing,* 14: 1839–1860.

Stoms, D. M., and W. W. Hargrove. 2000. Potential NDVI as a baseline for monitoring ecosystem functioning. *Int. J. Rem. Sensing,* 21: 401–407.

St-Onge, B. A., and F. Cavayas. 1995. Estimating forest stand structure from high resolution imagery using the directional variogram. *Int. J. Rem. Sensing,* 16: 1999–2021.

St-Onge, B. A., and F. Cavayas. 1997. Automated forest structure from high resolution imagery based on directional semivariogram estimates. *Rem. Sensing Environ.,* 61: 82–95.

Stoney, W. E., and J. R. Hughes. 1998. A new space race is on. *GIS World,* March 11: 44–46.

Story, R., G. A. Yapp, and A. T. Dunn. 1976. Landsat patterns considered in relation to resource surveys. *Rem. Sensing Environ.,* 4: 281–303.

Stoszek, K. J. 1988. Forests under stress and insect outbreaks. *Northwest Environ. J.,* 4: 247–261.

Stow, D., A. Hope, D. Richardson, D. Chen, C. Garrison, and D. Service. 2000. Potential of color-infrared digital camera imagery for inventory and mapping of alien plant invasions in South African shrublands. *Int. J. Rem. Sensing,* 21: 2965–2970.

Strahler, A. H. 1980. The use of prior probabilities in maximum likelihood classification of remotely sensed data. *Rem. Sensing Environ.,* 10: 135–163.

Strahler, A. H. 1981. Stratification of natural vegetation for forest and rangeland inventory using Landsat digital imagery and collateral data. *Int. J. Rem. Sensing,* 2: 15–41.

Strahler, A. H., T. L. Logan, and N. A. Bryant. 1978. Improving forest classification accuracy from Landsat by incorporating topographic information. Pages 927–942 in *Proc. 12th Int. Symp. Rem. Sensing Environ.,*. Environmental Research Institute of Michigan, Ann Arbor, MI.

Strahler, A. H., C. E. Woodcock, and J. A. Smith. 1986. On the nature of models in remote sensing. *Rem. Sensing Environ.,* 20: 121–139.

Strong, W. L. 1992. *Ecoregions and Ecodistricts of Alberta.* Vol. 1. Alberta Forestry, Lands and Wildlife, Edmonton.

Sun, C., and W. G. Wee. 1983. Neighboring gray level dependence matrix for texture classification. *Comput. Vision, Graphics Image Process.,* 23: 341–352.

Sun, G., and K. J. Ranson. 1998. Radar modelling of forest spatial patterns. *Int. J. Rem. Sensing,* 19: 1769–1791.

Sun, G., D. S. Simonett, and A. H. Strahler. 1991. A radar backscatter model for discontinuous coniferous forests. *IEEE Trans. Geosci. Rem. Sensing,* 29: 639–650.

Sun, X., and J. M. Anderson. 1993. A spatially variable light frequency selective component based airborne pushbroom imaging spectrometer for the water environment. *Photogramm. Eng. Rem. Sensing,* 59: 399–406.

Sutton, W. R. J. 1999. The need for planted forests and the example of radiata pine. *New For.,* 17: 95–109.

Suzuki, D. 1989. *Inventing the Future: Reflections on Science, Technology and Nature.* Stoddart, Toronto.

Swain, P. H. 1978. Fundamentals of pattern recognition in remote sensing. Pages 136–187 in P. H. Swain, and S. M. Davis, Eds. *Remote Sensing: The Quantitative Approach.* McGraw-Hill, New York.

Swain, P. H., and S. M. Davis, Eds. 1978. *Remote Sensing: The Quantitative Approach.* McGraw-Hill, New York.

Swanson, F. J., J. A. Jones, and G. E. Grant. 1997. The physical environment as a basis for managing ecosystems. Pages 229–238 in K. A. Kohm, and J. F. Franklin, Eds. *Creating a Forestry for the 21st Century.* Island Press, Washington, DC.

Swanson, F. J., T. K. Kratz, N. Caine, and R. G. Woodmansee. 1988. Landform effects on ecosystem patterns and processes. *BioScience,* 38: 92–98.

Talbot, S. S. and C. J. Markon. 1988. Intermediate-scale vegetation mapping of Innoko National Wildlife Refuge, Alaska, using Landsat MSS digital data. *Photogramm. Eng. Rem. Sensing,* 54: 377–383.

Taylor, J. E. 1993. Factors causing variation in reflectance measurements from bracken in eastern Australia. *Rem. Sensing Environ.,* 43: 217–229.

Teillet, P. M. 1986. Image corrections for radiometric effects in remote sensing. *Int. J. Rem. Sensing,* 7: 1637–1651.

Teillet, P. M. 1997. A status overview of Earth observation calibration/validation for terrestrial applications. *Can. J. Rem. Sensing,* 23: 291–298.

Teillet, P. M., and G. Fedosejevs. 1995. On the dark target approach to atmospheric correction of remotely sensed data. *Can. J. Rem. Sensing,* 21: 374–387.

Teillet, P. M., B. Guindon, and D. G. Goodenough. 1982. On the slope-aspect correction of multispectral scanner data. *Can. J. Rem. Sensing,* 8: 84–106.

Teillet, P. M., B. Guindon, J. F. Meunier, and D. G. Goodenough. 1985. Slope-aspect effects in synthetic aperture radar imagery. *Can. J. Rem. Sensing,* 11: 39–49.

Teillet, P.M., D. N. H. Horler, and N. T. O'Neill. 1997. Calibration, validation, and quality assurance in remote sensing: a new paradigm. *Can. J. Rem. Sensing,* 23: 401–413.

Thomas, I. L., V. M. Benning, and N. P. Ching. 1987. *Classification of Remotely Sensed Images.* Adam Hilger, Bristol.

Thomasson, J. A., C. W. Bennett, B. D. Jackson, and M. P. Mailander. 1994. Differentiating bottomland tree species with multispectral videography. *Photogramm. Eng. Rem. Sensing,* 60: 55–59.

Thomlinson, J. R., P. V. Bolstad, and W. B. Cohen. 1999. Coordinating methodologies for scaling landcover classifications from site-specific to global: steps toward validating global map products. *Rem. Sensing Environ.,* 70: 16–28.

Thompson, I. D., M. D. Flannigan, B. M. Wotton, and R. Suffling. 1998. The effects of climate change on landscape diversity: an example in Ontario forests. *Environ. Monit. Assess.,* 49: 213–233.

Thompson, J. A., J. C. Bell, and C. A. Butler. 1997. Quantitative soil-landscape modelling for estimating the areal extent of hydromorphic soils. *J. Soil Sci. Soc. Am.,* 61: 971–980.

Thompson, M. D., M. E. Kirby, R. T. Lowry, L. Lalonde, J. B. Mercer, and E. Krakowski. 1994. Comparative effectiveness of airborne and satellite SAR systems for operational mapping programs. Pages 201–216 in Vol. I *Proc. 1st Int. Airborne Rem. Sensing Symp.,* ERIM, Strasbourg, France.

Thompson, M. D., and B. Macdonald. 1995. Operational tropical vegetation mapping programs based on airborne synthetic aperture radar. *Can. J. Rem. Sensing,* 21: 86–95.

Tilman, D., and S. Pacala. 1993. The maintenance of species richness in plant communities. Pages 13–25 in R. E. Ricklefs, and D. Schluter, Eds. *Species Diversity in Ecological Communities: Historical and Geographical Perspectives*. University of Chicago Press, Chicago, IL.

Tiner, R. W. 1990. Use of high-altitude aerial photography for inventorying forested wetlands in the United States. *For. Ecol. Manage.,* 33: 593–604.

Tinker, D. B., C. A. C. Resor, G. P. Beauvais, K. F. Kipfmueller, C. I. Fernandes, and W. L. Baker. 1998. Watershed analysis of forest fragmentation by clearcuts and roads in a Wyoming forest. *Landscape Ecol.,* 13: 149–165.

Titus, S. J., and D. J. Morgan. 1985. Tree height: can large scale photo measurements be more accurate than field measurements? *For. Chron.,* 61: 214–217.

Tiwari, K. P. 1975. Tree species identification on large scale aerial photographs at New Forest. *Indian For.,* 101: 132–136.

Tokola, T., and P. Kilpelainen. 1999. The forest stand margin area in the interpretation of growing stock using Landsat TM imagery. *Can. J. For. Res.,* 29: 303–309.

Tokola, T., J. Pitkanen, S. Partinen, and E. Muinonen. 1996. Point accuracy of a nonparametric method in estimation of forest characteristics with different satellite materials. *Int. J. Rem. Sensing,* 17: 2333–2351

Tom, C. D., and L. D. Miller. 1984. An automated mapping comparison of the Bayesian maximum likelihood and linear discriminant analysis algorithms. *Photogramm. Eng. Rem. Sensing,* 50: 193–207.

Tomlinson, R. F., Ed. 1972. *Geographical Data Handling, Proc. IGU/UNESCO Symp. GIS*. IGU Commission on Geographical Data Sensing and Processing, Ottawa, Canada.

Tømmervik, H., M. E. Johnson, J. P. Pedersen, and T. Gunerisussen. 1998. Integration of remote sensing and in situ data in an analysis of air pollution effects on terrestrial ecosystems in the border area between Norway and Russia. *Environ. Monit. Assess.,* 49: 51–85.

Tomppo, E. 1988. Standwise forest variate estimation by means of satellite images. Pages 103–111 in *Department of Forest Mensuration and Management Research Notes, No. 21*. University of Helsinki, Finland.

Tomppo, E. 1990. Designing a satellite image-aided National Forest Survey in Finland. Pages 43–47 in R. Sylvander, Ed. *SNS/IUFRO Workshop on the Usability of Remote Sensing for Forest Inventory and Planning*. Umeå, Sweden.

Townshend, J. R. G. 1981a. The spatial resolving power of Earth resources satellites. *Prog. Phys. Geogr.,* 5: 32–55.

Townshend, J. R. G. 1981b. Prospect: a comment on the future role of remote sensing in integrated terrain analysis. Pages 219–223 in J. R. G. Townshend, Ed. *Terrain Analysis and Remote Sensing*. George Allen & Unwin, London.

Townshend, J. R. G., Ed. 1981c. *Terrain Analysis and Remote Sensing*. George Allen & Unwin, London.

Townshend, J. R. G., and C. O. Justice. 1981. Information extraction from remotely sensed data. A user view. *Int. J. Rem. Sensing,* 2: 313–329.

Townshend, P. A. 2000. A quantitative fuzzy approach to assess mapped vegetation classifications for ecological applications. *Rem. Sensing Environ.,* 72: 253–267.

Tou, J. T., and R. C. Gonzalez. 1974. *Pattern Recognition Principles*. Addison-Wesley, Reading, MA.

Toutin, Th. 1997. Qualitative aspects of chromo-stereoscopy for depth perception. *Photogramm. Eng. Rem. Sensing,* 63: 193–203.

Toutin, Th., and B. Rivard. 1995. A new tool for depth perception of multi-source data. *Photogramm. Eng. Rem. Sensing,* 61: 1209–1211.

Toutin, Th., and B. Rivard. 1997. Value added Radarsat products for geoscientific applications. *Can. J. Rem. Sensing,* 23: 63–70.

Treitz, P. M., and P. J. Howarth. 1999. Hyperspectral remote sensing for estimating biophysical parameters of forest ecosystems. *Prog. Phys. Geogr.,* 23: 359–390.

Treitz, P. M., and P. J. Howarth. 2000a. High spatial resolution remote sensing data for forest ecosystem classification: an examination of spatial scale. *Rem. Sensing Environ.,* 72: 268–289.

Treitz, P. M., and P. J. Howarth. 2000b. Integrating spectral, spatial, and terrain variables for forest ecosystem classification. *Photogramm. Eng. Rem. Sensing,* 66: 305–317.

Treitz, P. M., P. J. Howarth, O. R. Filho, and E. D. Soulis. 2000. Agricultural crop classification using SAR tone and texture statistics. *Can. J. Rem. Sensing,* 26: 18–29.

Treitz, P. M., P. J. Howarth, and D. G. Leckie. 1985. The capabilities of two airborne multispectral sensors for classifying coniferous forest species. *Proc. 19th Int. Symp. Rem. Sensing Environ. ERIM, Ann Arbor, MI.* Pages 335–350.

Treweek, J. 1999. *Ecological Impact Assessment.* Blackwell Scientific, Oxford, U.K.

Tribe, A. 1992. Problems in automated recognition of valley features from digital elevation models and a new method towards their resolution. *Earth Surf. Process. Landforms,* 17: 437–454.

Trichon, V., D. Ducrot, and J. P. Gastellu-Etchegorry. 1999. SPOT-4 potential for the monitoring of tropical vegetation: a case study in Sumatra. *Int. J. Rem. Sensing,* 20: 2761–2785.

Trotter, C. M., J. R. Dymond, and C. J. Goulding. 1997. Estimation of timber volume in a coniferous forest using Landsat TM. *Int. J. Rem. Sensing,* 18: 2209–2223.

Tucker, C. J. 1979. Red and photographic infrared linear combinations for monitoring vegetation. *Rem. Sensing Environ.,* 8: 127–150.

Turner, D. P., W. B. Cohen, R. E. Kennedy, K. S. Fassnacht, and J. M. Briggs. 1999. Relationships between LAI and Landsat Thematic Mapper spectral indices across three temperate zone sites. *Rem. Sensing Environ.,* 70: 52–68.

Turner, M. G. 1989. Landscape ecology: the effect of pattern on process. *Annu. Rev. Ecol. Syst.,* 20: 171–197.

Turner, M. G. 1990. Landscape change in nine rural counties in Georgia. *Photogramm. Eng. Rem. Sensing,* 56: 379–386.

Ulaby, F. T., F. Kouyate, B. Brisco, and T. H. L. Williams. 1986. Textural information in SAR images. *IEEE Trans. Geosci. Rem. Sensing,* 24: 235–245.

Ulaby, F. T., K. Sarabandi, K. McDonald, M. Whitt, and M. Dobson. 1990. Michigan Microwave Canopy Scattering Model. *Int. J. Rem. Sensing,* 11: 1223–1253.

Um, J. S., and R. Wright. 1999. The analog-to-digital transition and implications for operational use of airborne videography. *Photogramm. Eng. Rem. Sensing,* 65: 269–275.

Unger, D. R., and J. J. Ulliman. 2000. Delineating relative temperature zones in forest ecosystems: an adaptation and evaluation of current methodologies. *Can. J. Rem. Sensing,* 26: 30–37.

United Nations. 1992. *Long Range World Populations: Two Centuries of Population Growth, 1959–2150.* United Nations, New York.

Urban, D. L. 1993. Landscape ecology and ecosystem management. Pages 127–136 in W. W. Covington, and L. F. DeBano, Eds. *Sustainable Ecosystems: Implementing an Ecological Approach to Land Management.* USDA Forest Service Techn. Rep. RM-247. Fort Collins, CO.

Urban, D. L., R. V. O'Neill, and H. H. Shugart, Jr. 1987. Landscape ecology: a hierarchical perspective can help scientists understand spatial patterns. *BioScience,* 37: 119–127.

Ustin, S. L., and A. Trabucco. 2000. Using hyperspectral data to assess forest structure. *J. For.,* 98: 47–49.

Van der Sanden, J. J., P. Budkewitsch, R. Landry, M. J. Manore, H. McNairn, T. J. Pultz, and P. W. Vachon. 2000. Application potential of planned SAR satellites — a preview. Pages 111–118 in *Proc. 22nd Can. Symp. Rem. Sensing.* Canadian Aeronautics and Space Institute, Ottawa.

Van Zyl, J. J. 1993. The effect of topography on radar scattering from vegetated areas. *IEEE Trans. Geosci. Rem. Sensing,* 31: 153–160.

Vanday, J. K. 1996. Lessons from the Queensland Rainforests: steps toward sustainability. *J. Sustainable For.,* 3: 1–25.

Vande Castle, J. 1998. Remote sensing applications in ecosystem analysis. Pages 271–287 in D. L. Peterson, and V. T. Parker, Eds. *Ecological Scale: Theory and Applications.* Columbia University Press, New York.

Vane, G., and A. F. H. Goetz. 1993. Terrestrial imaging spectroscopy: current status and future trends. *Rem. Sensing Environ.,* 44: 117–126.

Varjö, J. 1996. Controlling continuously updated forest data by satellite remote sensing. *Int. J. Rem. Sensing,* 17: 43–67.

Vckovski, A. 1998. *Interoperable and Distributed Processing in GIS.* Taylor & Francis, London.

Vckovski, A. 1999. Interoperability and spatial information theory. Pages 31–37 in M. Goodchild, M. Egenhofer, R. Fegeas, and C. Kottman, Eds. *Interoperating Geographical Information Systems.* Kluwar, Dordrecht.

Venier, L. A., and B. G. Mackey. 1997. A method for rapid, spatially explicit habitat assessment for forest songbirds. *J. Sustainable For.,* 4: 99–118.

Verbyla, D. C., and K. K. Chang. 1997. *Processing Digital Images in Geographical Information Systems.* Onword Press, Sante Fe, NM.

Verstraete, M. M., B. Pinty, and R. B. Myneni. 1996. Potential and limits of information extraction on the terrestrial biosphere from satellite remote sensing. *Rem. Sensing Environ.,* 58: 201–214.

Vincent, R. K. 1997. *Fundamentals of Geological and Environmental Remote Sensing.* Prentice-Hall, Englewood Cliffs, NJ.

Vine, P., and C. Puech. 1999. Cartography of post-fire forest regeneration by coupling a spectral mixture model with a vegetation regrowth model. *Can. J. Rem. Sensing,* 25: 152–159.

Vogelmann, J. E., T. Sohl, and S. M. Howard. 1998. Regional characterization of land cover using multiple sources of data. *Photogramm. Eng. Rem. Sensing,* 64: 45–57.

Vogt, K. A., J. Gordon, J. Wargo, D. Vogt, H. Asbjornsen, P. A. Palmiotto, H. J. Clark, J. L. O'Hara, W. S. Keaton, T. Patel-Weynand, and E. Witten, Eds. 1996. *Ecosystems: Balancing Science and Management.* Springer-Verlag, New York.

Vogt, K. A., B. C. Larson, J. C. Gordon, D. J. Vogt, and A. Fanzeres. 1999. *Forest Certification: Roots, Issues, Challenges, Benefits.* CRC Press, Boca Raton, FL.

Wagner, C. S. 1998. *International Agreements on Cooperation in Remote Sensing and Earth Observation.* Critical Technologies Institute, Washington, DC.

Wald, L., T. Ranchin, and M. Mangolini. 1997. Fusion of satellite images of different spatial resolutions: assessing the quality of resulting images. *Photogramm. Eng. Rem. Sensing,* 63: 691–699.

Walker, R. E., D. E. Stoms, J. E. Estes, and K. D. Cayocca. 1992. Relationships between biological diversity and multi-temporal vegetation index data in California. Pages 562–571 in *Tech. Pap., ASPRS/ACSM Annu. Meet.* ASPRS, Bethesda, MD.

Wallace, J., and N. Campbell. 1998. *Evaluation of the Feasibility for Monitoring National State of the Environment Indicators in Australia.* State of the Environment Tech. Ser. Department of Environment, Canberra.

Walsh, S. J. 1980. Coniferous tree species mapping using Landsat data. *Rem. Sensing Environ.,* 9: 11–26.

Walsh, S. J. 1987. Variability of Landsat MSS spectral responses of forests in relation to stand and site characteristics. *Int. J. Rem. Sensing,* 8: 1289–1299.

Walsworth, N. A., and D. J. King. 1999. Image modelling of forest changes associated with acid mine drainage. *Comput. Geosci.,* 25: 567–580.

Walters, R. 1998. Improving links between ecosystem scientists and managers. Pages 272–286 in M. L. Pace, and P. M. Graffman, Eds. *Success, Limitations, and Frontiers in Ecosystem Science.* Springer-Verlag, New York.

Wang, F. 1990a. Fuzzy supervised classification of remote sensing images. *IEEE Trans. Geosci. Rem. Sensing,* 28: 194–201.

Wang, F. 1990b. Improving remote sensing image analysis through fuzzy information representation. *Photogramm. Eng. Rem. Sensing,* 56: 1163–1169.

Wang, J., and W. Liu. 1994. Road detection from multispectral satellite imagery. *Can. J. Rem. Sensing,* 20: 180–191.

Wang, S., D. B. Elliot, J. B. Campbell, R. W. Erich, and R. M. Haralick. 1983. Spatial reasoning in remotely sensed data. *IEEE Trans. Geosci. Rem. Sensing,* 21: 94–101.

Wang, Y., E. S. Kasischke, J. Melack, F. W. Davis, and N. L. Christensen, Jr. 1994. The effects of changes in loblolly pine biomass and soil moisture on ERS-1 SAR backscatter. *Rem. Sensing Environ.,* 49: 25–31.

Waring, R. H. 1987. Characteristics of trees predisposed to die. *BioScience,* 37: 569–574.

Waring, R. H., J. J. Landsberg, and M. Williams. 1998. Net primary production of forests: a constant fraction of gross primary production? *Tree Physiol.,* 18: 129–134.

Waring, R. H., B . E. Law, M. L. Goulden, S. L. Bassow, R. W. McCreight, S. C. Wofsey, and F. A. Bazzaz. 1995a. Scaling gross ecosystem productivity at Harvard Forest using remote sensing: a comparison of estimates from a constrained quantum-use efficiency model and eddy correlation. *Plant, Cell Environ.,* 18: 1201–1213.

Waring, R. H., and S. W. Running. 1998. *Forest Ecosystems: Analysis at Multiple Scales.* 2nd ed. Academic Press, San Diego, CA.

Waring, R. H., and S. W. Running. 1999. Remote sensing requirements to drive ecosystem models at the landscape and regional scale. Pages 23–38 in J. D. Tenhunen, and P. Kabat, Eds. *Integrating Hydrology, Ecosystem Dynamics, and Biogeochemistry in Complex Landscapes.* John Wiley & Sons, New York.

Waring, R. H., and W. H. Schlesinger. 1985. *Forest Ecosystems: Concepts and Management.* Academic Press, San Diego, CA.

Waring, R. H., W. G. Thies, and D. Muscato. 1980. Stem growth per unit of leaf area: a measure of tree vigor. *For. Sci.,* 26: 112–117.

Waring, R. H., J. B. Way, E. R. Hunt, Jr., L. Morrissey, K. J. Ranson, J. Weishampel, R. Oren, and S. E. Franklin. 1995b. Imaging radar for ecosystem studies. *BioScience,* 45: 715–723.

Warner, T., R. Bell, and V. Singhroy. 1996. Local incidence angle effects on X- and C-band radar backscatter of boreal forest communities. *Can. J. Rem. Sensing,* 22: 269–279.

Warner, T., J. Lee, and J. McGraw. 1999. Delineation and identification of individual trees in the eastern deciduous forest. Pages 81–91 in D. A. Hill, and D. G. Leckie, Eds. *Proc. Int. Forum Automated Interpretation High Spatial Resolution Digital Imagery Forestry.* Canadian Forest Service, Pacific Forestry Centre, Victoria, British Columbia.

Warren, S. D., M. O. Johnson, W. D. Goran, and V. E. Diersing. 1990. An automated, objective procedure for selecting representative field sample sites. *Photogramm. Eng. Rem. Sensing,* 56: 333–335.

Weaver, K. F. 1969. Remote sensing: new eyes see the world. *National Geographic,* 135 (January): 47–73.

Webster, R., and P. H. T. Beckett. 1970. Terrain classification and evaluation using air photography: a review of recent work at Oxford. *Photogrammetria,* 26: 51–75.

Webster, R., P. J. Curran, and J. W. Munden. 1989. Spatial correlation in reflected radiation from the ground and its implications for sampling and mapping by ground-based radiometry. *Rem. Sensing Environ.,* 29: 67–79.

Wehr, A., and U. Lohr. 1999. Airborne laser scanning — an introduction and overview. *ISPRS J. Photogramm. Rem. Sensing,* 54: 68–82.

Weibel, R. 1999. Generalization of spatial data: principles and selected algorithms. Pages 99–144 in M. von Kreveld, J. Nievergelt, T. Roos, and P. Widmayer, Eds. *Algorithmic Foundations of Geographic Information Systems.* Springer-Verlag, Berlin.

Weintreb, A., and A. Cholaky. 1991. A hierarchical approach to forestry planning. *For. Sci.,* 37: 439–460.

Weishampel, J. F., J. B. Blair, R. G. Knox, R. Dubayah, and D. B. Clark. 2000. Volumetric lidar return patterns from an old-growth tropical rainforest canopy. *Int. J. Rem. Sensing,* 21: 409–415.

Weishampel, J. F., R. G. Knox, K. J. Ranson, D. L. Williams, and J. A. Smith. 1997. Integrating remotely sensed spatial heterogeneity with a three-dimensional forest succession model. Pages 109–134 in Gholz, H. L., K. Nakane, and H. Shimoda, Eds. *The Use of Remote Sensing in the Modeling of Forest Productivity.* Kluwer, Dordrecht.

Weishampel, J. F., G. Sun, K. J. Ranson, K. D. LeJeune, and H. H. Shugart. 1994. Forest textural properties from simulated microwave backscatter: the influence of spatial resolution. *Rem. Sensing Environ.,* 47: 120–131.

Wessman, C. A. 1990. Evaluation of canopy biochemistry. Pages 135–156 in R. J. Hobbs, and H. A. Mooney, Eds. *Remote Sensing of Biosphere Functioning.* Springer-Verlag, New York.

Wessman, C. A., and E. M. Nel. 1993. A distant perspective: approaching sustainability in a regional context. Pages 169–177 in W. W. Covington, and L. F. DeBano, Eds. *Sustainable Ecosystems: Implementing an Ecological Approach to Land Management.* USDA Forest Service Tech. Rep. RM-247. Fort Collins, CO.

Weszka, J. S., C. Y. Dyer, and A. Rosenfeld. 1976. A comparative study of texture measures for terrain classification. *IEEE Trans. Syst. Man, Cybern.,* 6: 269–285.

Wewel, F., F. Scholten, and K. Gwinner. 1999. High resolution stereo camera (HRSC) — multispectral 3D data acquisition and photogrammetric data processing. Pages 263–272, Vol. I in *Proc. 4th Int. Symp. Airborne Rem. Sensing 21st Can. Symp. Rem. Sensing.* ERIM/Canadian Aeronautics and Space Institute, Ottawa.

Weyerhaeuser, G. H., Jr. 1998. The challenge of adaptive forest management: aren't people part of the ecosystem too? *For. Chron.,* 74: 865–870.

White, J. D., G. S. Kroh, and J. E. Pinder, III. 1995. Forest mapping at Lassen Volcanic National Park, California, using Landsat TM data and a geographical information system. *Photogramm. Eng. Rem. Sensing,* 61: 299–305.

White, J. D., S. W. Running, R. Nemani, R. E. Keane, and K. C. Ryan. 1997. Measurement and remote sensing of LAI in Rocky Mountain montane ecosystems. *Can. J. For. Res.,* 27: 1714–1727.

Whittaker, R. H. 1975. *Communities and Ecosystems. Second Edition.* Macmillan Publishing Company, New York.

Whyte, A. G. 1996. Multicriteria planning and management of forest sustainability. Pages 189–205 in A. Schulte, and D. Schone, Eds. *Dipterocarp Forest Ecosystems: Towards Sustainable Forest Management.* World Scientific, Singapore.

Wickham, J. D., R. V. O'Neill, K. Riitters, T. Wade, and K. B. Jones. 1997. Sensitivity of selected landscape pattern metrics to land-cover misclassification and differences in land-cover composition. *Photogramm. Eng. Rem. Sensing,* 63: 397–402.

Wickland, D. E. 1991. Mission to Planet Earth: the ecological perspective. *Ecology,* 72: 1923–1933.

Wickware, G. M., and P. J. Howarth. 1981. Change detection in the Peace-Athabasca Delta using digital Landsat data. *Rem. Sensing Environ.,* 11: 9–25.

Wilkinson, G. G. 1996. A review of the current issues in the integration of GIS and remote sensing data. *Int. J. Geogr. Inf. Syst.,* 10: 85–101.

Williamson, R. A. 1997. The Landsat legacy: remote sensing policy and the development of commercial remote sensing. *Photogramm. Eng. Rem. Sensing,* 63: 877–885.

Wilson, B. A. 1996. Estimating coniferous forest structure using SAR texture and tone. *Can. J. Rem. Sensing,* 22: 382–389.

Wilson, B. A., C. F. Ow, M. Heathcott, D. Milne, T. McCaffrey, G. Ghitter, and S. E. Franklin. 1994. Landsat MSS classification of fire fuel types in Wood Buffalo National Park. *Global Ecol. Biogeogr. Lett.,* 4: 33–39.

Wilson, E. O., Ed. 1988. *Biodiversity.* National Academy Press, Washington, DC.

Wolter, P. T., D. J. Mladenoff, G. E. Host, and T. R. Crow. 1995. Improved forest classification in the northern Lake States using multitemporal Landsat imagery. *Photogramm. Eng. Rem. Sensing,* 61: 1129–1143.

Woodcock, C. E., J. B. Collins, V. D. Jakabhazy, X. Li, S. A. Macomber, and Y. Wu. 1997. Inversion of the Li-Strahler canopy reflectance model for mapping forest structure. *IEEE Trans. Geosci. Rem. Sensing,* 35: 405–414.

Woodcock, C. E., and J. Harward. 1992. Nested-hierarchical scene models and image segmentation. *Int. J. Rem. Sensing,* 13: 3167–3187.

Woodcock, C. E., and A. H. Strahler. 1987. The factor of scale in remote sensing. *Rem. Sensing Environ.,* 21: 311–332.

Woodham, R. J. 1989. Determining intrinsic surface reflectance in rugged terrain and changing illumination. Pages 1–5 in *Proc. Int. Geosci. Rem. Sensing Symp. 12th Can. Symp. Rem. Sensing*. Canadian Aeronautics and Space Institute, Ottawa.

Woodham, R. J., and T. K. Lee. 1985. Photometric method for radiometric correction of multispectral scanner data. *Can. J. Rem. Sensing,* 11: 132–142.

Woodley, S. and G. Forbes, Eds. 1997. *Forest Management Guidelines to Protect Native Biodiversity in the Fundy Model Forest.* Fish and Wildlife Research Unit, University of New Brunswick, Fredericton.

Worboys, M. F. 1995. *GIS: A Computing Perspective.* Taylor & Francis, London.

World Bank. 1997. *Russian Forest Policy During Transition: A World Bank Country Study.* IBRD, The World Bank, Washington, DC.

Wright, D. J., M. F. Goodchild, and J. D. Proctor. 1997. GIS: Tool or Science? Demystifying the persistent ambiguity of GIS as 'tool' vs. 'science.' *Ann. Am. Assoc. Geogr.,* 87: 346–362.

Wu, D., and J. Linders. 2000. Comparison of three different methods to select features for discriminating forest cover types using SAR imagery. *Int. J. Rem. Sensing,* 21: 2089–2099.

Wu, S. T. 1990. Assessment of tropical forest stand characteristics with multipolarization SAR data acquired over a mountainous region in Costa Rica. *IEEE Trans. Geosci. Rem. Sensing,* 28: 752–755.

Wu, Y., and A. Strahler. 1994. Remote estimation of crown size, stand density, and biomass on the Oregon transect. *Ecol. Appl.,* 4: 299–312.

Wulder, M. A. 1998a. The prediction of leaf area index from forest polygons decomposed through the integration of remote sensing, GIS, Unix and C. *Comput. Geosci.,* 24: 151–157.

Wulder, M. A. 1998b. Optical remote sensing techniques for the assessment of forest inventory and biophysical parameters. *Prog. Phys. Geogr.,* 22: 449–476.

Wulder, M. A. 1999. Image spectral and spatial information in the assessment of forest structural and biophysical data. Pages 267–281 in *Proc. Int. Forum Automated Interpretation High Spatial Resolution Digital Imagery Forestry.* Canadian Forest Service, Pacific Forestry Centre, Victoria, British Columbia.

Wulder, M. A., S. E. Franklin, and M. B. Lavigne. 1996. High spatial resolution optical image texture for improved estimation of forest stand leaf area index. *Can. J. Rem. Sensing,* 22: 441–449.

Wulder, M. A., M. B. Lavigne, and S. E. Franklin. 1995. Empirical relations between forest stand structure and ERS-1 SAR imagery in northwestern New Brunswick. Pages 663–668 in *Proc. XVII Can. Symp. Rem. Sensing.* Canadian Aeronautics and Space Institute, Ottawa.

Wulder, M. A., E. F. LeDrew, S. E. Franklin, and M. B. Lavigne. 1998. Aerial image texture information in the estimation of deciduous and mixedwood forest leaf area index (LAI). *Rem. Sensing Environ.,* 64: 64–76.

Wulder, M. A., S. Mah, and D. Trudeau. 1996. Mission planning for operational data acquisition campaigns with the casi. Pages 53–62, Vol. III *in Proc. 2nd Airborne Rem. Sensing Conf. Exhibition.* ERIM, San Fransisco, CA.

Wulder, M. A., K. O. Niemann, and D. G. Goodenough. 2000. Local maximum filtering for the extraction of tree locations and basal area from high spatial resolution imagery. *Rem. Sensing Environ.,* 73: 103–114.

Wynne, R. H., and D. B. Carter. 1997. Will remote sensing live up to its promise for forest management? *J. For.,* 95: 23–26.

Wynne, R. H., and R. G. Oderwald. 1998. Forest inventory and assessment in the smallsat era. Pages 810–815 in *Proc. ASPRS Res. Technol. Inst. Annu. Meet.* American Society of Photogrammetry and Remote Sensing, Bethesda, MD.

Wynne, R. H., R. G. Oderwald, G. A. Reams, and J. A. Scrivani. 2000. Optical remote sensing for forest area estimation. *J. For.,* 98: 31–36.

Yafee, S. 1999. Three faces of ecosystem management. *Conserv. Biol.,* 13: 713–725.

Yang, C., and A. Vidal. 1990. Combination of digital elevation models with SPOT-1 HRV multispectral imagery for reflectance factor mapping. *Rem. Sensing Environ.,* 32: 35–45.

Yatabe, S. M., and A. G. Fabbri. 1986. The application of remote sensing to Canadian petroleum exploration: promising and yet unexploited. *Comput. Geosci.,* 12: 597–609.

Yatabe, S. M., and A. G. Fabbri. 1989. Putting AI to work in geoscience. *Episodes,* 12: 10–17.

Yatabe, S. M., and D. G. Leckie. 1995. Clearcut and forest type discrimination in satellite SAR imagery. *Can. J. Rem. Sensing,* 21: 455–467.

Yool, S. R., M. J. Makaio, and J. M. Watts. 1997. Techniques for computer-assisted mapping of rangeland change. *J. Range Manage.,* 50: 307–314.

Young, R. W., and K. L. White. 1994. Satellite imagery analysis of landforms: illustrations from southeastern Australia. *Geocarto Int.,* 2: 33–44.

Yuan, X. 1993. Empirical correction techniques for view-angle effects on airborne multispectral scanner data of forests. Pages 63–68 in D. G. Leckie, and M. D. Gillis, Eds. *Proc. Int. Forum Airborne Multispectral Scanning Forestry Mapping (Emphasis on MEIS).* Inf. Rep. PI-X-113. Canadian Forest Service, Ottawa.

Yuan, X., D. King, and J. Vlcek. 1991. Sugar maple decline assessment based on spectral and textural analysis of multispectral aerial videography. *Rem. Sensing Environ.,* 37: 47–54.

Zagolski, F., V. Pinel, J. Romier, D. Alcayde, J. Fontanari, J. P. Gastellu-Etchegorry, G. Giordano, G. Marty, E. Mougin, and R. Joffre. 1996. Forest canopy chemistry with high spectral resolution remote sensing. *International Journal Remote Sensing* 17: 1107–1128.

Zebker, H. A., and R. M. Goldstein. 1986. Topographic mapping from interferometric SAR observations. *J. Geophys. Res. Solid Earth Planets,* 91: 4993–4999.

Zeff, I. S., and C. J. Merry. 1993. Thematic Mapper data for forest resource allocation. *Photogramm. Eng. Rem. Sensing,* 59: 93–99.

Zevenbergen, L. W. and C. R. Thorne. 1987. Quantitative analysis of land surface topography. *Earth Surf. Processes Landforms,* 12: 47–56.

Zheng, D., E. R. Hunt, Jr., and S. W. Running. 1996. Comparison of available soil water capacity estimated from topography and soil series information. *Landscape Ecol.,* 11: 3–14.

Zheng, D., D. O. Wallin, and Z. Hao. 1997. Rates and patterns of landscape change between 1972 and 1988 in the Changbai Mountain area of China and North Korea. *Landscape Ecol.,* 12: 241–254.

Zhou, Q., M. Robson, and P. Pilesjo. 1998. On the ground estimation of vegetation cover in Australian rangelands. *Int. J. Rem. Sensing,* 19: 1815–1820.

Zhu, A. 1997. Measuring uncertainty in class assignment for natural resource maps under fuzzy logic. *Photogramm. Eng. Rem. Sensing,* 63: 1195–1202.

Zonneveld, I. S. 1989. The land unit — a fundamental concept in landscape ecology, and its applications. *Landscape Ecol.,* 3: 67–86.

Zsilinszky, V. 1964. The practice of photo interpretation for forestry. *Photogrammetria,* 5: 1–17.

Zsilinszky, V. 1970. Supplemental aerial photography with miniature cameras. *Photogrammetria,* 25: 27–38.

Index

A

B

D

E

F

G

H

M

N

O

P

Q

R

S

T

U

V

W

Y